Thermal Sprayed Coatings and their Tribological Performances

Manish Roy
Defence Metallurgical Research Laboratory, India

J. Paulo Davim
University of Aveiro, Portugal

ENGINEERING SCIENCE REFERENCE
An Imprint of IGI Global

Managing Director:	Lindsay Johnston
Managing Editor:	Austin DeMarco
Director of Intellectual Property & Contracts:	Jan Travers
Acquisitions Editor:	Kayla Wolfe
Production Editor:	Christina Henning
Development Editor:	Erin O'Dea
Typesetter:	Cody Page
Cover Design:	Jason Mull

Published in the United States of America by
Engineering Science Reference (an imprint of IGI Global)
701 E. Chocolate Avenue
Hershey PA 17033
Tel: 717-533-8845
Fax: 717-533-8661
E-mail: cust@igi-global.com
Web site: http://www.igi-global.com

Library of Congress Cataloging-in-Publication Data

Thermal sprayed coatings and their tribological performances / Manish Roy and J. Paulo Davim, editors.
pages cm
Includes bibliographical references and index.
ISBN 978-1-4666-7489-9 (hardcover) -- ISBN 978-1-4666-7490-5 (ebook) 1. Flame spraying. 2. Plasma spraying. 3. Surface roughness. 4. Tribology. I. Roy, Manish, 1964- II. Davim, J. Paulo.
TS655.T474 2015
621.8'9--dc23

2014044999

British Cataloguing in Publication Data
A Cataloguing in Publication record for this book is available from the British Library.

Table of Contents

Detailed Table of Contents

The wear resistance of thirty WC-Co coatings, deposited by standard High-Velocity Oxyfuel (HVOF) techniques and a high-temperature variant of HVOF, with standard commercial and experimental nanostructured feedstocks, is examined. It is found that the high-temperature gun produces harder and more wear-resistant coatings than the standard gun. The latter does not generate high enough temperatures to melt the powder and provide good bonding between WC grains and Co binder. All coatings present higher wear resistance than the steel substrate. Coatings deposited with standard feedstock possess generally higher wear resistance than nanostructured coatings. The difference is more pronounced in sliding than in abrasive wear. WC-Co Coatings deposited with nanostructured feedstocks are recommended for use in bearings and other machinery with sliding parts because they inflict much less wear on the material on which they slide than conventional coatings. Coatings with micrometer WC grains are recommended for abrasion resistance applications such as earth moving or slurry processing machinery.

Triboelectrochemical response is concerned with different electrochemical-controlled wear experiments designed to understand tribocorrosion behaviour of materials and coatings. Tribocorrosion is defined as material degradation resulting from simultaneous action of wear and corrosion and it is found in many engineering applications, but the involved mechanisms are still only partially understood. In this chapter, a brief overview of tribocorrosion testing techniques is followed by issues which have helped us in gaining in-depth scientific knowledge of tribocorrosion. The overview is further substantiated by detailed studies and observations on tribocorrosion of thermal sprayed coatings in recent times.

Solution Precursor Plasma Spray (SPPS) technique is an emergent technique which offers significant performance advantages with various functional coatings like YSZ, ZnFe2O4, LSM, metal-doped ZnO, etc. Apart from the above capabilities, efforts towards development of an innovative hybrid approach through simultaneous deposition of powder and solution resulted in coatings with unique microstructures. Some illustrative examples of the large variety of coatings that can be realized through the hybrid route combining the conventional plasma spray and SPPS techniques are discussed. The attractive prospects offered by hybrid technique for spraying nanocomposite coatings are specifically highlighted through a case study. Successful development of hybrid coatings using a Mo-alloy powder and a suitable oxide-forming solution precursor has been shown to exhibit improved sliding wear performance. The relationship between the splats, the ensuing coating microstructure at varied processing conditions, and its tribological behaviour of the coatings is comprehensively discussed.

WC-based thermal-spray and High Velocity Oxy-Fuel (HVOF) coatings are extensively used in a wide range of applications ranging from downhole drilling tools to gas turbine engines. WC-based thermal spray coatings offer improved wear resistance as a result of hard phases dispersed in binder-rich regions. However, the presence of hard and soft phases within the coating can also lead to the formation of micro-galvanic couplings in aqueous environments leading to some reduction in combined wear-corrosion resistance. Furthermore, the coating also responds differently to change in mechanical loading conditions. This chapter examines the wear-corrosion performance of thermal spray coatings in a range of wear, electrochemical, and wear-corrosion tests under varying contact conditions to develop models and establish relationships between wear mechanisms, wear rates, and environmental factors such as pH and applied load.

Functionally Gradient Coatings (FGCs) are emerging materials with an improved service life and have a promising future for the production of (a) tailored components for applications subjected to large thermal gradients, (b) smart coating with improved corrosion and wear resistance, (c) improved fatigue wear, and (d) improved material structures for energy applications like batteries, fuel cells, etc. FGCs may be developed by physical/chemical vapor deposition, electro/electroless deposition, thermal spray deposition technique, etc. Thermal spraying refers to the technique or a group of techniques whereby molten or semi-molten droplets of materials are sprayed onto a solid substrate to develop the coating. In this chapter, detailed overviews of the development of functionally graded coating by thermal spray deposition techniques are presented. In addition, a few research results on the development of functionally graded coating for tribological and thermal barrier applications are presented.

In this chapter, various layers that are formed during sliding wear of thermal sprayed coatings at elevated temperature are discussed. Glazed layers are formed on the worn surfaces during elevated temperature sliding wear of thermal sprayed coatings. These layers have a characteristic physical appearance, mechanical properties, chemical compositions, and failure mechanisms. Wearing conditions, wearing material, and mating material influence formation and characteristics of glazed layers. Among these parameters, wearing material and mating material are most important. These glazed layers are divided into different types of layers, namely Transfer Layer (TL), Mechanically Mixed Layer (MML), Reaction Layer (RL), and Composite Layer (CL). The recent results on friction of thermal sprayed coatings at elevated temperature are rationalised in the light of different types of glazed layer formation.

Solid particle erosion is an important material degradation mechanism. Although various methods of coating are tried and used for protection against erosion, thermal sprayed coating for such purpose is the most widely used method. In this chapter, evolution of thermal sprayed coating, erosion testing methods, and erosive wear of thermal sprayed coatings are discussed extensively with emphasis on recent developments. It is generally found that erosion of thermal sprayed coatings depends on erosion test conditions, microstructural features, and mechanical properties of the coating materials. Most thermal sprayed coatings respond in brittle manner having maximum erosion rate at oblique impact and velocity exponent in excess of 3.0. Erosion rate is also dependent on thermal spraying techniques and post coating treatment. However, little work is done on dependence of erosion rate on coating techniques and coating conditions. Future direction of work is also reported.

With the exception of ZrO_2, the individual oxides and binary compositions in the system Al_2O_3-Cr_2O_3-TiO_2 are the most important oxide materials for thermally sprayed coating solutions. Traditionally, these coatings are prepared by Atmospheric Plasma Spraying (APS), but processes such as Detonation Gun Spraying (DGS) and High Velocity Oxy-Fuel (HVOF) spraying can produce coatings with lower porosity and higher wear resistance. Traditionally, feedstock powders have been used for coating preparation. Recent developments have seen the emergence of suspensions as a new feedstock, but tribological properties of coatings prepared using suspensions have not yet been studied in detail. This chapter summarizes some important issues regarding wear protection applications of coatings in the Al2O3-Cr2O3-TiO2 system, the advantage of alloying the individual oxides, and the influence of different feedstocks and spray processes.

Slurry erosion is a degrading phenomenon usually observed in machineries dealing with particle-laden fluid such as hydro power plants, ship propellers, pump impellers, valves, and connecting pipes. The low erosion resistance of commonly employed structural materials prompts the use of different surface modification techniques. Among several types of surface modification techniques, thermal spraying has achieved a significant recognition worldwide due to its versatile nature. In this chapter, slurry erosion behavior of thermal sprayed coatings has been discussed with special emphasis on the contribution of different coating related parameter. It has been observed that microstructure play an important role in determining the slurry erosion performance of thermal spray coatings. Different microstructural features such as splat boundaries, pores, un-melted particles, and cracks are detrimental for the thermal spray coatings exposed to erosive environment. A parameter useful for identification of primary erosion mechanism for thermal sprayed coatings is also discussed.

The Detonation Spray Coating (DSC) process is a unique variant among the wide choice of thermal spray processes. The typical functionalities of DSC coatings include wear and corrosion resistance, elevated temperature oxidation resistance, thermal barrier, insulative/conductive, abradable, lubricious surface, etc. Among the coatings for wear resistance, the cermet coatings based on WC–Co and Cr3C2–NiCr are the most popular materials of choice and contribute to bulk of the utilization by the industry towards wear resistance. Notwithstanding the above materials, alternative materials involving modifications in both hard and binder phases like TiMo (CN)–NiCo, WC-CrC-Ni, WC-Co-Cr, WC-Ni, Cr_3C_2-Ni, Cr_3C_2-Inconel, etc. exhibit great promise towards tribological applications under diverse wear modes. This chapter on the tribological characteristics of the detonation sprayed coatings provides a comprehensive overview on the characteristics of various cermet coatings generated at varied process conditions and its influence on the tribological properties under abrasive, sliding, and erosive wear modes.

Thermal spray is a generic term used to define a group of coating processes used to apply both metallic and non-metallic coatings. These coatings are usually defined by their hardness, strength, porosity, roughness, and wear resistance. In this chapter, the authors turn their attention to the mechanical and tribological properties of thermal spray coatings. The individual phase plays a very important role in determining the performance of the coating. However, evaluating the mechanical and tribological properties at a nano-level requires new test methods and their validation. In this chapter, elaborate discussion of some techniques to evaluate and analyze the mechanical and tribological properties of different thermal spray coatings is done. This chapter is intended to help the reader to firstly understand the basic principle and methods of characterization of thermal spray coatings using instrumented nanoindentation, nanoscratch, abrasive wear testing techniques, and secondly to get an idea of the recent techniques and review the research and development in the same field.

Foreword

Surface engineering has a major part to play in a world where resources such as materials and energy are becoming increasingly valuable, and thermal spray processes form a very important contribution to the range of surface coating methods available. Thermal spraying provides a uniquely versatile route for the deposition of a very wide range of materials on to almost any substrate to produce a surface with enhanced resistance to corrosion or wear while the key properties of the underlying structure, be they for example stiffness, strength, lightness, creep resistance, or some combination, are preserved.

Although the first thermal spray methods were used more than a century ago, and many of the methods used today were introduced in the 1950s and 1960s, it is only recently that the combination of well-controlled deposition methods and materials has been refined to provide the high performance which can be achieved today. Thermal spraying is now extensively used in industrial applications to provide resistance to corrosion, both in aqueous conditions and at high temperatures, and to wear in a wide range of environments. Sometimes, the coating is exposed to simultaneous wear and corrosion (tribo-corrosion), possibly with synergistic interactions. To achieve optimum performance and a cost-effective coating which will endure for the life of the component requires a good understanding of the methods used to form the coating, the coating materials and their microstructures after deposition, appropriate methods for characterization and quality control of the process and resulting coatings, and methods for testing and understanding the response of the coating under the often complex conditions which give rise to wear or tribo-corrosion. All these topics are covered in this book, which provides a state-of-the-art survey from expert authors. I expect it to provide a valuable resource for users of thermal spray coatings and researchers for many years to come.

Ian Hutchings
University of Cambridge, UK

Ian Hutchings has been GKN Professor of Manufacturing Engineering at the University of Cambridge since 2001 and has a background in materials science, tribology, and surface engineering. Between 1977 and 2001, he was a faculty member in the Department of Materials Science and Metallurgy at Cambridge, becoming Reader in Tribology in 1997. He is the author of more than 330 research papers and a widely used textbook, Tribology: Friction and Wear of Engineering Materials (1992), and was Editor-in-Chief of the journal Wear from 1998 to 2012. He is a Fellow of the Royal Academy of Engineering, a Fellow of the Institute of Physics, and a Fellow of the Institute of Materials. Among other awards, he received the Silver Medal of the UK's Tribology Trust in 1994, the Institute of Materials NPL Award for Materials Metrology in 2000, the Donald Julius Groen Prize from the Institution of Mechanical Engineers Tribology Trust in 2000, and the Staudinger-Durrer Prize and Medal from ETH Zürich in 2007.

Preface

Thermal spray is a generic term (Pawlowski, 1995) for a group of coating processes where the coating is deposited on a substrate by applying a stream of particles, metallic, or ceramic materials, which flatten more or less forming platelets, called splats. Several layers of these splats form the coating. Upon impact, a bond forms with the surface, with subsequent particles causing a build-up of the coating to its final thickness. Since 1910, when Dr. Schoop for the first time conceived thermal spraying, a large development has taken place. A device to spray metal wires was made available in 1912. This technique is known as flame spraying. Later on, in 1930, new technique for thermal spraying known as flame spraying process was introduced. Subsequently, in 1958, Union Carbide developed detonation spraying. Atmospheric plasma spraying was introduced in 1960. This was followed by development of Vacuum Plasma Spraying (VPS) and Low-Pressure Plasma Spraying (LPPS) in late 1970 and 1980. In the year 1980, High Velocity Oxy Fuel (HVOF) technique was invented. James A Browning developed liquid fuel based coating gun "Jet-Kote" in 1982. The most important technology cold spraying was developed in 1990. The first literature on warm spraying was available in 2006. The classification of various thermal spray processes is summarized in Figure 1. Over last several decades, large developments in thermal spraying have made these techniques as an invaluable tool for engineering applications.

There has been an increased dependence of industrial sectors on thermal spraying technologies, not only for key operational purposes but also for gaining strategic advantage. Thermal spraying is a dynamic process and a rapidly changing field, which is used in a wide range of industries to solve increasingly challenging problems. These techniques are used for performance enhancement and extension of life of industrial components, which are subjected to wear, corrosion, etc., and also to meet numerous other practical solutions. As the application for thermal spraying is expanding to performance-critical applications, it has become mandatory to ascertain how the coating affects mechanical properties of the substrate. It is critical to

Figure 1. Classification of various thermal spray processes

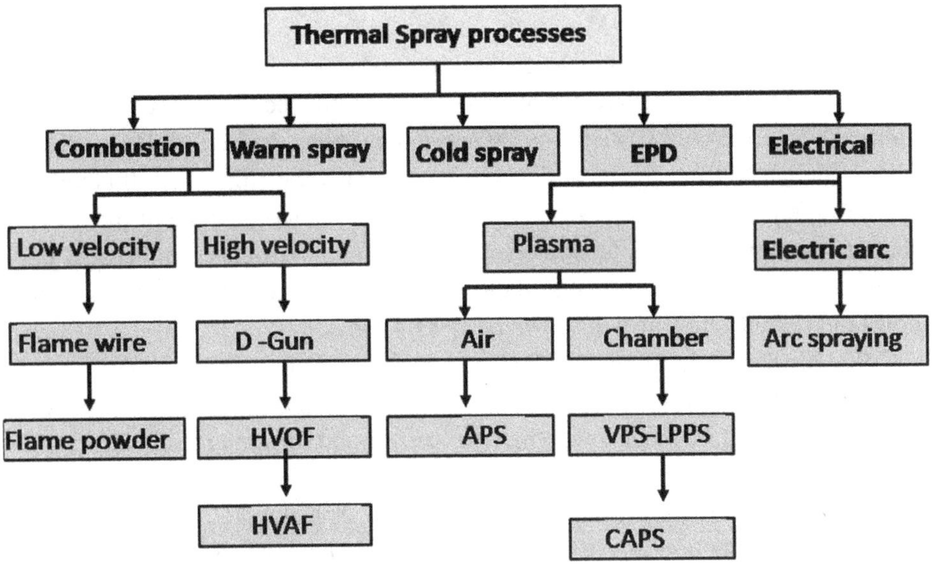

establish standard method and matrix for evaluation of effect of coatings on the substrate. This has resulted in establishing a standard test method for evaluation and comparison.

There are several emerging trends in thermal spraying. Nanostructured coatings by Hypersonic Plasma Particle Deposition (HPPD) have been established (Heberlein, Rao, Neuman, Blum, Tymiak, McMurry, & Girshick, 1997). In this process, vapour phase precursors are injected in to a plasma stream generated by a DC arc. The plasma is quenched by supersonic expansion through a nozzle into a vacuum deposition chamber. Ultrafine particles nucleated in the nozzle are accelerated in the hypersonic free jet downstream of the nozzle and inertially deposited on the substrate. The short transit times between the nozzle and the substrate prevent in-flight agglomeration, while the high particle velocity results in formation of consolidated coatings.

Suspension thermal spraying in which fine powders suspended in suitable liquid phase are sprayed (Bouyer, Gitzhofer, & Boulos, 1997). Suspensions exceeding 50 wt % solid phase content is successfully injected into inductively coupled plasma. Coating is deposited successfully by a novel radio frequency – Suspension Plasma Spraying (RF-SPS). However, this approach still relies on the use of expensive nano/ultra-fine powder feedstock. Difficulties in suspending the particles in a stable manner and maintaining homogeneity also exist. Alternately, a single-step coating process involving *in situ* synthesis of nanoparticles and their subsequent consolidation into an adherent deposit through solution precursor spraying appears promising. Some preliminary reports on High-Velocity Oxy-Fuel (HVOF) spraying of solution precursors is also available (Ma, Roth, Gandy, & Frederick, 2006).

A new electromagnetic powder deposition technique that employs electromagnetic force can accelerate powder particles up to 2 km/sec as opposed to 1km/sec obtained by conventional thermal spray process has been developed (Bacon, Davis, Polizzi, Sledge, Uglum, & Zowarka, 1997). Particles travelling with this velocity have sufficient kinetic energy to melt their own mass and equivalent mass of the substrate on impact. In this technology, argon gas and high-energy electrical pulse provided by a capacitor bank drives the gas to a very high velocity, which ultimately allows the particles to travel at the velocity of 4 km/sec.

Cold Gas Dynamic Spray (CGDS) is a process where coating is deposited by high-velocity jet of solid phase particles, which has been accelerated by a supersonic gas jet at a temperature much lower than the melting and softening temperature of the coating materials (Bhagat, Amateau, Papyrin, Conway, Stutzman, & Jones, 1997). The warm spray process (Kuroda, Watanabe, Kim, & Katanoda, 2011) is based on high velocity impact bonding of powder particles, which is similar to cold spraying, but the temperature of the particles at impact is significantly higher and often very close to the melting point of the material. Therefore, warm spray may be regarded as a process to fill the gap between HVOF and CS.

Considerable effort will be diverted for further development of functionally graded thermal sprayed coatings. A functionally graded coating can be defined as a series of the coating having continuously changing composition from substrate to the surface to achieve combinations of properties (Kawasaki & Watanabe, 1997; Krumova, Klingshirn, Haupert, & Friedrich, 2001). These advanced coatings with gradient in composition, structure/specific properties in the preferred direction/orientation are superior to homogeneous materials composed of similar constituents. Functionally graded coating offers several advantages over conventional monolithic coating system, such as reduction of in-plane and through-the-thickness transverse stresses, improved thermal properties, high toughness, etc. Functionally graded coatings consisting of metallic and ceramic components are established to enhance the properties of thermal-barrier systems by making smooth transition of the properties with depth-leading enhancement of adhesion characteristics of the coating.

Smart thermal sprayed coatings, which respond in a selective way to external factors such as stress, temperature, etc., are another important recent development (Fasching, Prinz, & Weiss, 1995). With the help of thermal spraying, it is possible to deposit novel smart coatings with embedded sprayed sensors by combining coating and shaping methodologies. Various sensor materials can be selectively sprayed and shaped using masking techniques (Weiss, Prinz, Adams, & Siewiorek, 1992). Discreet shapes can be formed with appropriately shaped masks. Sprayed sensors, such as thermocouple, humidity sensor, etc., have been manufactured and performance evaluated successfully. In addition, sensors, like strain gauges, antennas, pressure sensors, etc., have been developed. There are several important properties with sprayed sensors that cannot be easily achieved by other conventional methods. It is possible to have sprayed sensor conforming to complex surface shapes with optimal surface contact by flexible masking. For a given application, the shape of the sensor and its interconnection can be separately designed. Production of sensor array is straightforward, as coating and sensor are deposited simultaneously. Sensors can be protected against the environment by depositing another coating. It is also possible to incorporate new materials for sensor technology by using multi-component powder.

Hybrid thermal spraying will draw a lot of attention in the years to come. Low-pressure plasma spray hybrid system can produce coatings within three distinct regimes. Plasma spray-PVD can produce thick, columnar-structured coatings using high gun enthalpy to vaporize specific types of feedstock materials. Plasma spray-CVD uses modified conventional thermal spray components operated below 0.5 mbar to produce CVD-like coatings at higher deposition rates by using liquid for gaseous precursors as feedstock materials. Plasma spray-thin film can produce thin, dense layers from liquid splats using a classical thermal spray approach but at high velocity and enthalpy.

Another important development for thermal spraying is forming of near-net shape components. Refractory materials have high melting points, high temperature strength, good thermal properties, and high ablation resistance. Parts of near-net shape refractory materials, such as heating element, crucible, and rocket nozzle, etc., have found wide applications in chemical processing, electrical and mechanical engineering, airplane and aerospace industries (French, Hurst, & Marvig, 2001). However, due to their ultra-high melting point and high ductile-to-brittle transition temperature (around room temperature), it is difficult to fabricate large-scale or thin-walled parts with complex shape of these materials by conventional industrial methods such as Powder Metallurgy (PM) and Chemical Vapour Deposition (CVD) (Cockeram, 2006). It has always been intended to develop new and effective fabrication methods to produce refractory metallic parts of desired shapes and density. Capable of making high quality near-net-shape parts, Plasma Spray Forming (PSF) then came into play as one such potential fabrication method of choice for refractory materials. During PSF process, refractory metallic powders are directly fed onto a predesigned mandrel (Park, 1999). Almost any feedstock can be sprayed onto the mandrel in a controllable manner to provide a compact of desired shape and wall thickness with benefits of simplifying fabrication process and cost reduction. Substantial research is expected to be directed in the development of removable mandrel and development of complex shape. A larger variety of coatings for near-net shape tribology-related components and their increased usage is expected to come in the near future.

Over the last decades, various sensors relevant to the thermal spraying process and capable of operating under the harsh environment of spray booths have been developed (Fauchais & Verdelle, 2010). Today, sensors are available to measure trajectories, temperatures, velocities, sizes, and shapes of in-flight particles (Fauchais, 1992; Landes, 2006; Gougeon & Moreau, 1993). Infrared cameras and pyrometers are employed to understand the temperature profile of substrate and coatings during preheating, spray process and cooling down (Matejicek & Sampath, 2003; Clyne & Gill, 1996; Kuroda & Clyne, 1991). Sensors are also developed to measure the stresses within the coatings and evolution of thickness of the coating during spraying (Nadeau, Pouliot, Nadeau, Blain, Berube, Moreau, & Lamontagne, 2006). Further development towards improved precision on measurement is expected. New techniques, such as shadowgraphs and laser, allow precise measurement of particle diameter, which was not hitherto possible. Development of sensors will allow online control of thermal spraying and thus enhance coating quality and reliability.

Over the last several decades, thermal spraying has come a very long way, especially in terms of revolution with regard to automation of process variables and work handling equipment. Process automation has brought about a much higher degree of repeatability and reliability to the thermal sprayed coating process and the coating being produced. As a result, thermal spraying is being considered as a modern and well-respected manufacturing process. The widespread use and growing acceptability of thermal sprayed coating for commercial product is the testimony for that. Although lot of investigation has been carried out and significant information has been generated about the variables involved in the thermal spraying process, much still needs to be learned. Automation has induced more correctness and more adequacies in controlling these variables to a significant extent. Automation has also helped us in mass production of a large number of feasible components.

THE CHALLENGES

It goes without saying that demands for high performance thermal sprayed coatings and coatings with added functionality have been on the increase. And any attempt to deal with the problem demands an adequate understanding of the challenges that exist in the new millennium. Such challenges can be classified into following categories:

- The challenge of establishing newer techniques giving rise to coatings with improved mechanical properties and performances.

- The challenge of establishing new coatings with new composition and new functionality.
- The challenge of establishing correct processing condition for the existing deposition techniques for existing coating materials.
- The challenge of establishing appropriate spraying technology to meet the challenging demand of modern technology for operation in demanding and harsh environments.
- The challenge of establishing automation for the processing conditions and equipment handling to meet the challenging demand of modern technology.
- The challenge of establishing appropriate spraying technology to meet demand for mass production.

Although thermal sprayed coating has evolved as an important tool for large variety of applications, a significant fraction of thermal sprayed coatings are used for protection against tribology in ambient or in corrosive environments. Tribology, which deals with the science of friction, is an important degradation process, protection against which is essential. Commonly occurring tribology-related degradations are erosive wear, abrasive wear, and sliding wear. Erosive wear can be due to impact by solid particles liquid droplets, or it can also be due to liquid metal or cavitations. Abrasive wear can be defined as material loss when a hard particle is made to slide against a relatively soft material. Sliding wear is essentially degradation when two surfaces are made to rub against each other. Sliding wear can be unidirectional or reciprocating. The strain, strain rate, thermodynamic status of deformation of erosive wear, abrasive wear, and sliding wear differ considerably. The unique features of the deformation conditions pertinent to various degradation conditions, such as erosive wear, abrasive wear, and sliding wear are given in Figure 2. The other variety of tribological degradation process is fretting when rubbing takes place in reciprocating motion with very low amplitude. When tribology-related degradation takes place in the presence of a corrosive medium, degradation is an outcome of synergistic effect of tribology and corrosion such as erosion-corrosion, fretting-corrosion, abrasion-corrosion, etc. When such degradation takes place at elevated temperature, material loss is an outcome of interaction between tribology and oxidation, such as erosion-oxidation, etc. Erosion oxidation interaction map for Ni, for feed rate of 0.2 gm/min is illustrated in Figure 3.

Prevention of tribology-related degradation is an age-old phenomenon. Surface hardening by diffusion of carbonaceous materials has been known since Roman times. Vikings were also famous for embedding the leading edges of their ploughs to resist soil erosion. However, scientific investigation to minimize tribology-related problems started in the last 50 years. To be more precise, enhanced understand-

Figure 2. The unique features of the deformation conditions pertinent to various degradation conditions such as erosive wear, abrasive wear and sliding wear (Roy, 2006)

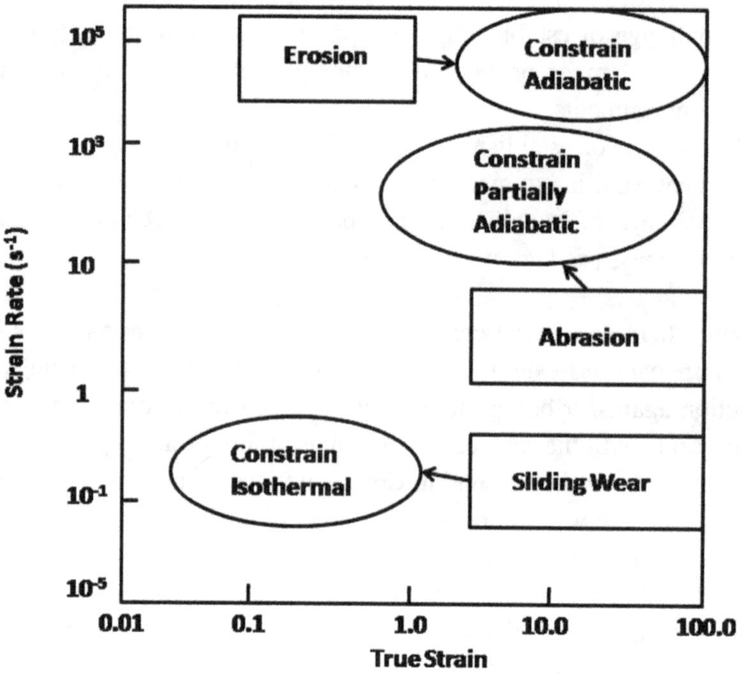

ing for more complicated problems involving tribology and corrosion interaction is merely a decade old. Most manufacturing industries are forced to overcome a loss of 0.5% of their turnover due to tribology and related problem and this figure should be applicable worldwide. Hence, it is an expensive degradation process and there are enough incentives to minimize it by applying a coating by thermal spraying. Since tribology is mostly a surface-related phenomena, thermal sprayed coatings have been, more often than not, successful in improving the performances. Thermal sprayed coating is one of the most widely and commonly used methods for protection against such degradation.

Because of the requirements for protection against the above-mentioned degradation, demand for thermal sprayed coatings is increasing. The book, *Thermal Sprayed Coatings and their Tribological Performances*, has been planned keeping this important aspect in mind for future research. Research on thermal sprayed coatings has gathered huge momentum across the world as evidenced from the number of research publications in this area in recent times. Today, many spheres of life, which includes energy, oil and gas, mining, transportation, and healthcare sectors,

Figure 3. Erosion oxidation interaction map for Ni, for feed rate of 0.2 gm/min (Roy, 2006)

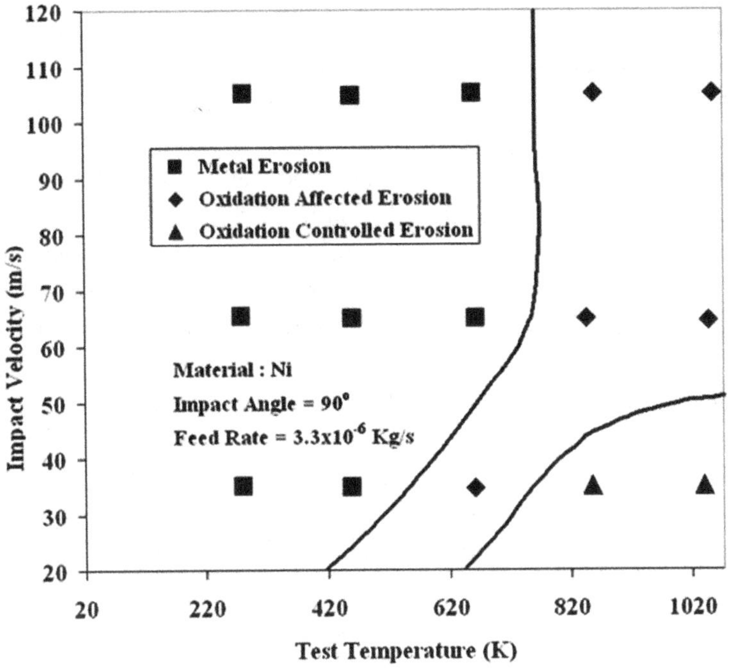

etc., have problems related to tribology. Over the years, numerous studies, reviews, and special issues and books have been published on thermal sprayed coatings. However, no attempt has ever been made to compile a book addressing exclusively the tribological problem. This is the first attempt to compile a book related to the tribology of thermal sprayed coating.

Because of time constraints, we have missed several areas, like elevated temperature erosion, abrasive wear, etc., of thermal sprayed coating. We have also missed tribology of several emerging thermal sprayed coating processes, such as cold sprayed, electromagnetic deposition process, etc., as the literature in these areas are limited. We hope to have a book with all varieties in future. We think that the chapters are a valuable reference and that many researchers will use them as a useful guide to the state of the art in the field of tribology of thermal sprayed coatings.

ORGANIZATION OF THE BOOK

After this introductory preface, it is pertinent to describe the organization of the book. In this book, we have successfully gotten several leading practitioners who are currently involved in research and development of thermal sprayed coatings for tribological application to contribute. The book is organized into 11 chapters. A brief description of each of the chapters follows:

Chapter 1 discusses the effects of feedstock powder and deposition technique on the hardness and tribological performance of thermal-sprayed WC-Co coatings. It was found that the high-temperature gun produces harder and more wear-resistant coatings than the standard gun. WC-Co Coatings deposited with nanostructured feedstocks are useful in bearings and other machinery with sliding parts because they inflict much less wear on the material on which they slide than conventional coatings. Coatings with micrometer WC grains are suitable for abrasion resistance applications such as earth moving or slurry processing machinery

Chapter 2 identifies tribo-corrosion of thermal sprayed coatings. Tribo-corrosion is material degradation resulting from synergistic action of wear and corrosion and it is prevalent in many engineering applications although the involved mechanisms are still little understood. In this chapter, a brief overview of tribo-corrosion testing techniques followed by issues which have helped in gaining in-depth scientific knowledge of tribo-corrosion has been discussed. The overview is further substantiated by detailed studies and observations on tribo-corrosion of thermal sprayed coatings in recent times.

Chapter 3 specifically focuses on the emergent Solution Precursor Plasma Spray (SPPS) technique that has been the subject of considerable research interest in recent times. A variant of the thermal spray family, the SPPS method offers significant advantages by avoiding use of powder-feedstock that has to meet stringent particle size specifications, permitting better control over coating chemistry and yielding interesting features like vertical cracks, nano-sized pore structure, fine splats, etc. Some illustrative examples of the large variety of coatings that can be realized by adopting the above hybrid route that involves combining the conventional plasma spray and SPPS techniques will be discussed. In an attempt to understand coating formation by the proposed hybrid route, splat formation under varied processing conditions has been comprehensively investigated. The relationship between these splats, the resulting coating microstructure and tribological behaviour of the coatings has also been comprehensively discussed.

Chapter 4 reviews research associated with the development of coatings that are designed to withstand combined abrasive wear and corrosion conditions. Understanding how coatings perform under these tribo-corrosion conditions is essential if the service life of equipment is to be predicted. Therefore, the tribo-corrosion performance of coatings deposited by thermal spray techniques is discussed and the main mechanisms associated with their degradation under combined wear and corrosion highlighted. The importance of post-coating deposition treatments such as laser resurfacing and sealing are also discussed. Interactions between abrasion and corrosion mechanisms are identified along with some models and mapping techniques that aim to inform coating selection and predict performance.

Chapter 5 deals with detailed overviews of the development of functionally graded coating by thermal spray deposition techniques. In addition, the present status of research efforts in development of functionally graded coating for tribological and thermal barrier applications is also elaborated.

Chapter 6 is concerned with elevated temperature sliding wear of various thermal sprayed coating actually meant for high temperature application. After introducing the subject in Chapter 1, a brief outline of various thermal spraying techniques is made in subsequently sliding wear of various thermal sprayed coatings is addressed. High temperature application of thermal spayed coatings and future direction of research for high temperature thermal spray coating are eventually furnished.

Chapter 7 provides a comprehensive overview on the characteristics of various cermets coatings intended for erosion resistant applications. In this chapter, evolution of thermal sprayed coating, erosion testing methods and erosive wear of thermal sprayed coatings are discussed extensively with emphasis on recent developments. It is generally found that erosion of thermal sprayed coatings depends on erosion test conditions, microstructural features and mechanical properties of the coating materials. Most thermal sprayed coatings respond in brittle manner having maximum erosion rate at oblique impact and velocity exponent in excess of 3.0. Erosion rate is also dependent on thermal spraying techniques and post coating treatment. However, little work is done on dependence of erosion rate on coating techniques and coating conditions. Future direction of work has also been reported.

Chapter 8 addresses tribology of thermally sprayed coatings in the Al_2O_3-Cr_2O_3-TiO_2 system. This contribution summarizes some important issues, regarding wear protection applications of coatings in the Al_2O_3-Cr_2O_3-TiO_2 system, the advantage of alloying the individual oxides, the influence of different feedstocks and spray processes.

Chapter 9 reviews slurry erosion of thermal sprayed coatings. This chapter presents basics of slurry erosion phenomenon in its first part. Effect of different parameters on slurry erosion process is also discussed. In the second part of chapter, slurry erosion behavior of several thermal sprayed coatings is presented with the help of some case studies. An attempt has been made to understand the role of different coating compositions and deposition processes to control slurry erosion. Slurry erosion mechanisms for these coatings have also been explained.

Chapter 10 deals with on the tribological characteristics of the detonation sprayed coatings. It provides a comprehensive overview on the characteristics of various cermet coatings generated at varied process conditions and its influence on the tribological properties under abrasive, sliding and erosive wear modes. The aim of this chapter is to provide its readers an opportunity to judicially select the right material composition for the perceived tribological application and also to understand the role of processing conditions to achieve the desired coating properties.

Chapter 11 concludes with an elaborate discussion of some techniques to evaluate and analyze the mechanical and tribological properties of different thermal spray. This chapter is intended to help the reader to firstly understand the basic principle and methods of characterization of thermal spray coatings using instrumented nanoindentation, nanoscratch, abrasive wear testing techniques and secondly to get an idea of the recent techniques and review the research and development in the same field.

This research book can be used as a support book for final undergraduate engineering course (for example, materials, mechanical, manufacturing, etc.) or as a subject on thermal spray at the postgraduate level. In addition, this book can serve as a useful reference for academics, materials researchers, materials, physics, mechanical, and manufacturing engineers, as well as professionals in related industries with thermal spray.

Manish Roy
Defence Metallurgical Research Laboratory, India

J. Paulo Davim
University of Aveiro, Portugal

REFERENCES

Bacon, J. L., Davis, D. G., Polizzi, R. J., Sledge, R. L., Uglum, J. R., & Zowarka, R. C. (1997). A new electromagnetic power deposition system. In *Thermal spray: A united forum for scientific and technological advances* (pp. 393–397). Materials Park, OH: ASM International.

Bhagat, R. B., Amateau, M. F., Papyrin, A., Conway, J. C. Jr, Stutzman, B., & Jones, B. (1997). Deposition of nickel-aluminium bronze powder by cold gas-dynami spray method on 2618 Al for developing wear resistant coatings. In *Thermal spray: A united forum for scientific and technological advances* (pp. 361–367). Materials Park, OH: ASM International.

Bouyer, E., Gitzhofer, F., & Boulos, M. I. (1997). Powder processing by suspension thermal spraying. In *Thermal spray: A united forum for scientific and technological advances* (pp. 353–359). Materials Park, OH: ASM International.

Clyne, T. W., & Gill, S. C. (1996). Residual stresses in thermal spray coatings and their effect on interfacial adhesion: A review of recent work. *Journal of Thermal Spray Technology*, *5*(4), 401–418. doi:10.1007/BF02645271

Cockeram, B. V. (2006). The mechanical properties and fracture mechanisms of wrought low carbon arc cast (LCAC), molybdenum–0.5pct titanium–0.1pct zirconium (TZM), and oxide dispersion strengthened (ODS) molybdenum flat products. *Materials Science and Engineering A*, *418*(1-2), 120–136. doi:10.1016/j.msea.2005.11.030

Fasching, M., Prinz, F. B., & Weiss, L. E. (1995). Smart coating: A technical note. *Journal of Thermal Spray Technology*, *4*(2), 133–136. doi:10.1007/BF02646101

Fauchais, P., Coudert, J. F., Vardelle, M., Vardelle, A., & Denoirjean, A. (1992). Diagnostics of thermal spraying plasma jets. *Journal of Thermal Spray Technology*, *1*(2), 117–128. doi:10.1007/BF02659011

Fauchais, P., & Verdelle, M. (2010). Sensors in spray processes. *Journal of Thermal Spray Technology*, *19*(4), 668–694. doi:10.1007/s11666-010-9485-0

French, D., Hurst, H. J., & Marvig, P. (2001). Comments on the use of molybdenum components for slag viscosity measurements. *Fuel Processing Technology*, *72*(3), 215–225. doi:10.1016/S0378-3820(01)00188-6

Gougeon, P., & Moreau, C. (1993). In-flight particle surface temperature measurement: Influence of the plasma light scattered by the particles. *Journal of Thermal Spray Technology*, *2*(3), 229–233. doi:10.1007/BF02650470

Heberlein, J., Rao, N. P., Neuman, A., Blum, J., Tymiak, N., McMurry, P. H., & Girshick, S. L. (1997). Thermal spraying of nanostructured coatings by hypersonic plasma particle deposition. In *Thermal spray: A united forum for scientific and technological advances* (pp. 393–397). Materials Park, OH: ASM International.

Kawasaki, A., & Watanabe, R. (1997). Concept and P/M fabrication of functionally gradient materials. *Ceramics International*, *23*(1), 73–83. doi:10.1016/0272-8842(95)00143-3

Krumova, M., Klingshirn, C., Haupert, F., & Friedrich, K. (2011). Microhardness studies on functionally graded polymer composites. *Composites Science and Technology*, *61*(4), 557–563. doi:10.1016/S0266-3538(00)00228-1

Kuroda, S., & Clyne, T. W. (1991). The quenching stress in thermally sprayed coatings. *Thin Solid Films*, *200*(1), 49–66. doi:10.1016/0040-6090(91)90029-W

Kuroda, S., Watanabe, M., Kim, K., & Katanoda, H. (2011). Current status and future prospect of warm spray technology. *Journal of Thermal Spray Technology, 20*(4), 653–673. doi:10.1007/s11666-011-9648-7

Landes, K. (2006). Diagonistics in plasma spraying techniques. *Surface and Coatings Technology, 201*(5), 1948–1954. doi:10.1016/j.surfcoat.2006.04.036

Ma, X. Q., Roth, J., Gandy, D. W., & Frederick, G. J. (2006). A new high velocity oxy-fuel process for making finely structured and highly bonded alloy layers from liquid precursors. In *Proceedings of International Thermal Spray Conference*. Materials Park, OH: ASM International.

Matejicek, J., & Sampath, S. (2003). In situ measurement of residual stresses and elastic moduli in thermal sprayed coatings: Part 1: Apparatus and analysis. *Acta Materialia, 51*(3), 863–872. doi:10.1016/S1359-6454(02)00478-0

Nadeau, A., Pouliot, L., Nadeau, F., Blain, J., Berube, S. A., Moreau, C., & Lamontagne, M. (2006). A new approach to online thickness measurement of thermal spray coatings. *Journal of Thermal Spray Technology, 15*(4), 744–749. doi:10.1361/105996306X147054

Park, J. J. (1999). Creep strength of a tungsten–rhenium–hafnium carbide alloy from 2200 to 2400 K. *Materials Science and Engineering A, 265*(1-2), 174–178. doi:10.1016/S0921-5093(98)01134-4

Pawlowski, L. (1995). *The science and engineering of thermal spray coatings*. New York: Wiley.

Roy, M. (2006). Mode of deformation during tribological degradation. In *Proceedings of the International Conference on Advances in Materials and Materials Processing* (pp. 202-209). Kharagpur, India: Academic Press.

Roy, M. (2006). Elevated temperature erosive wear of metallic material. *Journal of Physics. D, Applied Physics, 39*(6), R101–R124. doi:10.1088/0022-3727/39/6/R01

Vaidya, A., Streibl, T., Li, L., Sampath, S., Kovarik, O., & Greenlaw, R. (2005). An integrated study of thermal spray process-structure-property correlations: A case study for plasma sprayed molybdenum coatings. *Materials Science Engineering A. Structural Materials: Properties, Microstructure, and Processing, 403*(1-2), 191–204.

Weiss, L. E., Prinz, F. B., Adams, D., & Siewiorek, D. P. (1992). Thermal spray shape deposition. *Journal of Thermal Spray Technology, 1*(3), 231–237. doi:10.1007/BF02646778

Acknowledgment

The editors acknowledge their gratitude to IGI Global and the editorial team for their adequate and professional support throughout the preparation of this book. In addition, we would like to thank all the chapter authors, the reviewers, and the editorial advisory board members for their work on this editorial project.

Manish Roy
Defence Metallurgical Research Laboratory, India

J. Paulo Davim
University of Aveiro, Portugal

Chapter 1
The Effects of Feedstock Powder and Deposition Technique on the Hardness and Tribological Performance of Thermal-Sprayed WC-Co Coatings

Traugott E. Fischer
Stevens Institute of Technology, USA

Yunfei Qiao
Stevens Institute of Technology, USA

YouRong Liu
Stevens Institute of Technology, USA

ABSTRACT

The wear resistance of thirty WC-Co coatings, deposited by standard High-Velocity Oxyfuel (HVOF) techniques and a high-temperature variant of HVOF, with standard commercial and experimental nanostructured feedstocks, is examined. It is found that the high-temperature gun produces harder and more wear-resistant coatings than the standard gun. The latter does not generate high enough temperatures to melt the powder and provide good bonding between WC grains and Co binder. All

DOI: 10.4018/978-1-4666-7489-9.ch001

coatings present higher wear resistance than the steel substrate. Coatings deposited with standard feedstock possess generally higher wear resistance than nanostructured coatings. The difference is more pronounced in sliding than in abrasive wear. WC-Co Coatings deposited with nanostructured feedstocks are recommended for use in bearings and other machinery with sliding parts because they inflict much less wear on the material on which they slide than conventional coatings. Coatings with micrometer WC grains are recommended for abrasion resistance applications such as earth moving or slurry processing machinery.

INTRODUCTION

At the turn of the century, the Office of Naval Research funded a concerted program to investigate the potential of nanostructured coatings for applications in bearings and in abrasive media. The purpose of the research was to find whether coatings made of feedstock with nanometer-sized WC grains possess advantages over the commercially available coatings. It is easy to imagine the potential savings in labor, material and especially down time offered by the possibility of repairing, say, the worn surface of the bearing on a ship's large propeller shaft by depositing material in situ, rather than disassembling, fabricating and installing a new shaft. The research program involved a number or commercial providers of precursor material, companies specializing in various deposition techniques, and universities doing research in material synthesis and in the various deposition technologies; performing structural characterization of the coatings, and evaluating their performance. Our laboratory was charged with measuring the coating's resistance to sliding and abrasive wear and relating this performance to the nature of the feedstock and the deposition technique. The present contribution combines the observations described in previous reports (Qiao, Liu & Fischer, 2002, 2003) and provides a new analysis.

The processing of nanoscale WC had started when Larry McCandlish was investigating the potential of WC as a catalyst at Exxon Research Laboratories. Catalysts are prepared as nanometer grains that are dispersed on a support from the aqueous solution of a salt. His colleague Richard Polizzotti enquired about the possibility of co-dispersing cobalt with the same method. This led to the development of the Spray Conversion Processing of nanometer scale WC/Co composites at Nanodyne, a company founded by Bernard Kear for that purpose (McCandlish, Kear & Kim, 1992). It was known that WC/Co composites with nanometer size WC grains are harder and possess a higher resistance to sliding and abrasive wear than conventional cermets (Jia & Fischer, 1996, 1997, 1998). Since the structure and mechanical properties of coatings sprayed onto a substrate are different from those of bulk materials, the possible advantages of nanostructure in coatings needed to be verified.

The evaluation of the WC/Co coatings proceeded in two phases. In the first phase (Qiao et al., 2002), thirty coatings that several industrial and academic laboratories had prepared by high-velocity oxygen fuel (HVOF) methods, using either commercial or experimental powder feeds with micrometer-sized, nanometer-sized WC grains or powders containing a mixture of micro- and nanometer-sized WC grains, were analyzed for their microstructure and chemical composition and evaluated for hardness and their resistance to abrasive and unlubricated sliding wear. Based on the results obtained, nine nanostructured coatings were deposited under controlled conditions with three designed nanostructured feedstocks in a second phase (Qiao et al., 2003) which examined their microstructure and composition and measured their hardness, fracture toughness, abrasion and sliding wear resistance. The results of both investigations are combined here.

THE COATINGS

The Feedstocks

The powders in this technology consist of a mixture of WC grains and cobalt, which are agglomerated into particles, (20 - 40 μm), large enough to flow in the spray equipment. The structure of the feedstocks is shown in Figure 1.

The nanostructured feedstocks were obtained from Union Minière, New Brunswick, NJ, USA; they are shown in Figure 1a. The nanostructured material is obtained by Spray Conversion in which an aqueous solution of tungsten and cobalt salts is atomized in a fine spray. The grains are then calcined to obtain WC and Co grains of very small size. The Nanocarb powders present a typical spray-drying morphology in which the carbide and cobalt are combined together in 30-50 nm grains due to the chemical reaction; the small grains are porous and present a relatively large surface-to-volume ratio. The Nanocarb® agglomerate particles are hollow spheres; their size is in the range of 5 - 90 μm with an average of 40 μ. Some nanostructured powders were also provided by Osram Sylvania. Coatings prepared with these feedstocks are labeled "Nano";

Figure 1b represents nanostructured powders that have been reprocessed by Inframat Corporation. North Haven, CO, USA ; the WC grains received a proprietary coating aimed at reducing grain growth. These Infralloy® powders are composed of densely packed 200-500 nm carbides that expose flat crystal planes. The size of the agglomerates is 5-60μm with an average of 24 μm. The resulting coatings are labeled "Infra" in the following figures.

Figure 1. Low and high magnification views of the showing the agglomerate and the grain structure of (a) Nano, (b) Infra, (c) Mixed and (d) Micro feedstock. (©Elsevier, 2003, Used with permission)

Feedstocks prepared with a mixture of nanometer sized and micrometer sized WC grains were obtained from Nanopowder Enterprises. These Nanomyte® powders, shown in Figure 1c, are composed of 70% coarse (1-3 μm) and 30% nanocrystalline (30-50 nm) carbides. The powders were agglomerated by sintering; the particle size is 5-60 μm with an average of 26 μm. Coatings deposited with this powder will be labeled "Mixed". (Nanomyte is now a Registered Trademark for a different product).

Coatings were also deposited with commercial Diamalloy® WC-Co powders procured from Sulzer Metco, Westbuty, NY, USA. This feedstock is composed of 2-3 μm carbides with an average particle size of 25 μm as shown in Figure 1d. These coatings are labeled "Micro".

Deposition of the Coatings

Coatings were deposited on a steel substrate with standard High-Velocity Oxyfuel guns or with a modified HVOF gun (DJ2700) that generates higher temperatures (2000 – 2150C) and higher velocities of the particles. During deposition, the particles are partially melted and projected onto the substrate at high velocity. In this process the cobalt is molten and dissolves a fraction of the tungsten carbide (Verdon, Karimi & Martin, 1998 and Stewart, Shipway & McCartney, 2000). Since the particles do not reach equilibrium in flight, they reach a higher temperature at their surface than at their center. Near their surface, the particles loose carbon to oxidation. Upon impact on the substrate, the particles are flattened to form a splat and cool very rapidly. As a consequence, the W- and C- rich cobalt solidifies in nanocrystalline and amorphous form and precipitates W_2C particles around the original WC grains. When decarburization is severe, W particles precipitate near the border of the splats. The result is a classic WC-Co structure in the middle of the splat, increasing decarburization towards the splat boundary and an amorphous phase at the splat boundary.

Verdon and coworkers (1998) have also found a 5-10 nm thick oxygen and carbon-rich layer separating some splats.

Figure 2 shows high-resolution SEM micrographs representing the cross section of six representative coatings. These were obtained by cutting through the thickness of the coatings, polishing the cut and etching in boiling water to reveal the micro-structure. The coatings depicted in Figure 2 were deposited with the feedstocks shown in Figure 1. The relative amounts of WC, W_2C and W in the coatings were measured by X-ray diffraction (not shown).

Figure 2a shows the structure of a "Micro" coating obtained from the commercial powder (Diamalloy 2004) deposited with the high-temperature DJ2700 gun. It has a decarburization rate of 28%. The splat structure is clearly visible. The "cracks" at the splat boundaries may not be real; they may have been caused by dissolution

Figure 2. Micrographs of the cross-sections of selected coatings.(a) Micro powder, DJ2700 gun, (b) Mixed powder, HVOF gun. (c) Nanopowder DJ2700 gun,(d) Nanopowder, HVOF gun, (e): Infra powder, DJ2700 gun, (f) Infra powder, HVOF gun. (©Elsevier, 2003, Used with permission)

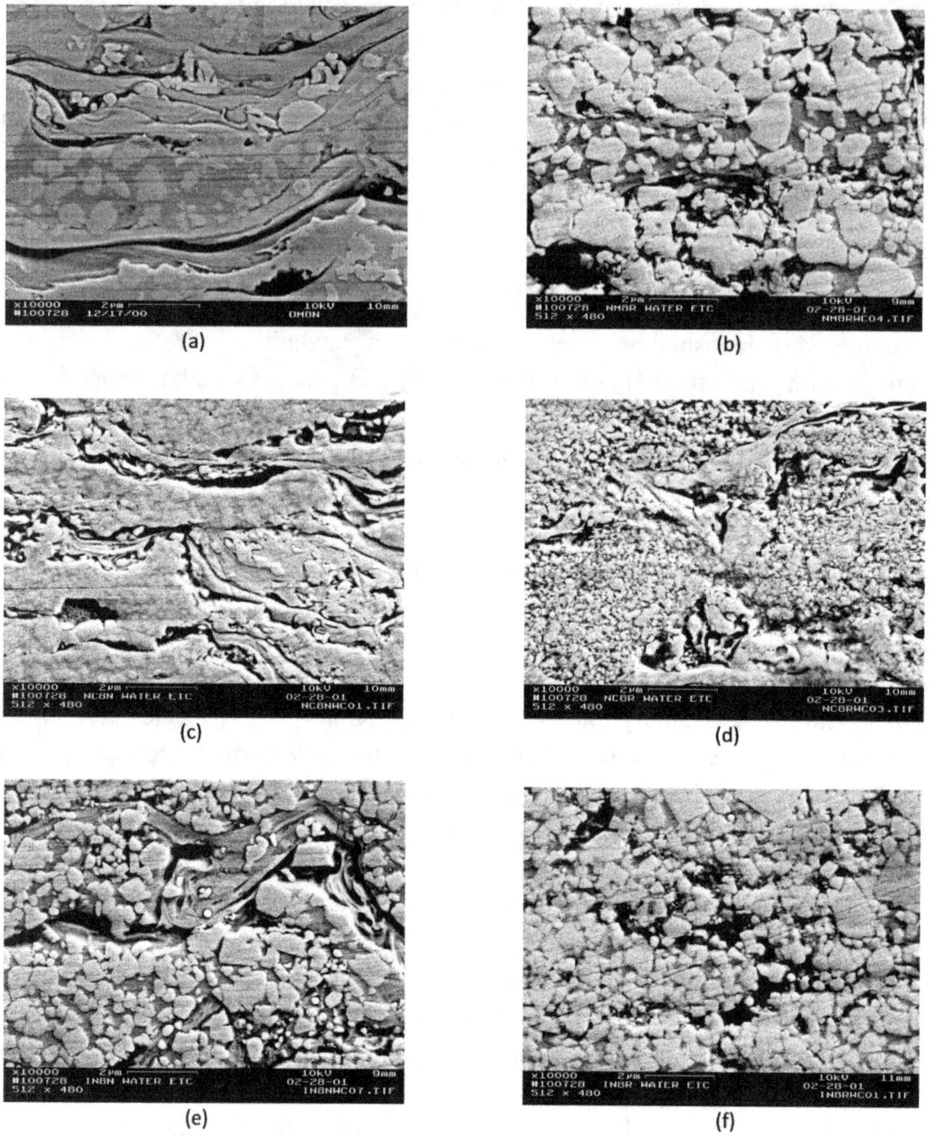

of material (Verdon's layer?) during the etching by hot water. The amorphous phase near the splat boundaries is clearly visible and contains the dendrites described by Verdon et al. (1998). Between these amorphous phases we discern the WC-Co structure and note the excellent cohesion between WC grains and binder.

Figure 2b shows the structure of a coating from "mixed" powder (Figure 1c) deposited with the standard HVOF gun. Its decarburization rate is a low 2.5%. Splat boundaries are not clearly visible and we discern no amorphous phase. There is a 1- 2 μm thick band of well bonded WC-Co in the middle of the picture but we discern a large number of nanopores separating WC grains. Evidently the temperature of the agglomerate in transit was not sufficient to melt all the cobalt and form a well-bonded WC-Co.

Figures 2c and 2d represent the structures of coatings deposited with the Nano powder (Figure 1a). The Nano sample of Figure 2c was deposited with DJ2700 and was 37% decarburized. A low-magnification micrograph (not shown) reveals that the splat boundaries are more numerous than in other materials; this is to be expected from the smaller mass of the very porous particles. Figure 2c shows that these boundaries contain rounded precipitates that we attribute to W as described by Verdon and coworkers (1998). We shall see later that the adhesion between splats in these samples is poor but Figure 2c shows that the adhesion between WC and Co inside the splats is excellent. Deposition of the same powder with the lower-temperature HVOF, Figure 2d, produces a loose structure with weak WC-Co bonds and a large amount of nanoporosity. There is some amorphous phase and evidence of decarburization (10%) but the numerous voids seen in the boundary are evidence for poor adhesion between the splats.

The Infra coating deposited with the DJ2700 gun (Figure 2e), with13% decarburization, shows again a thin amorphous phase at the splat boundaries. Close examination reveals that the WC grains retain their angular structure; we discern very fine pores between them and the binder. Deposition with standard HVOF (Figure 2f) produces no amorphous phase, no evident splat boundaries, poor WC-Co adhesion and a large amount of microporosity. Evidently, the coating designed to limit grain growth also decreases the adhesion with the Co binder.

We shall see later that the properties of the coatings revealed by their microstructure will, to a large extent, explain their behavior in abrasive and sliding wear.

HARDNESS MEASUREMENTS

The surfaces of all samples were polished with 1 μm diamond lapping prior to all measurements. The hardness of the samples was measured by Vickers indentation at loads varying from 50 g to 45 kg resulting in indentation sizes 5 to 200 μm. The hardness used for our comparisons or samples was measured at 1 Kg load, with indents 30 to 60 μm wide depending on the sample. Comparison with Figure 2 shows that these indentations were much larger than individual structural features and thus represent a spatial average. Comparison with Figures 6 to 9 also shows that the indents are much larger than individual wear events.

Six indentations were made on each sample; the values obtained varied by about 20% and the average was recorded on Figures 3 to 5. As usual with ceramics, the measured hardness of WC/Co coatings depends on the applied load. It is highest at low indentation load and decreases as the load increases. As an example, the hardness of a sample was measured at 1500 Kg/mm^2 at 50 g indentation load, as 1300 Kg/mm^2 at 1 Kg and 1150 Kg/mm^2 at 50 Kg load.

With the hardest samples, radial cracks extended from the corners of the indentations, but only when the indentation load exceeded 30 Kg. These cracks allowed a calculation of the fracture toughness with the formula of Anstis et al (1981):

$$K_{1C} = 0.016(E/H)^{1/2}P/C^{3/2} \qquad (1)$$

It turns out that the harder coatings, which were deposited with the DJ2700 gun, also possess a higher toughness. On a coating with H_V = 1250 Kg/mm^2 the toughness was 6 MPam$^{1/2}$; it was 9 MPam$^{1/2}$ on a coating with hardness 1350 Kg/mm^2. This is contrary to the behavior of bulk metals and most materials that possess a measure of ductility. In those materials, the higher toughness is due to the blunting of crack tips by plastic deformation. In the case of the WC/Co coatings, a decrease in hardness is caused by porosity, reduced adhesion between WC and binder and reduced adhesion of the splats; these same defects reduce resistance to crack propagation. This argument is strengthened by the fact that in the softer coatings, deposited with the standard HVOF gun, cracks developed parallel to the edges of the indents, indicating crumbling of the material under the indenter, rather than plastic deformation (Qiao et al., 2003).

WEAR MEASUREMENTS

Sliding wear tests were conducted in laboratory atmosphere with a ball-on-disk tribometer in which a WC/Co-coated steel disk slid, without lubrication, against a commercial Si_3N_4 ball of ¼" diameter with a load of 9.8 N. The sliding speed was 58 mm/s and the diameter of the traveling circle of the pin on the disk was 8.5 mm. The tests ran to a distance of 10,000 m or 250,000 rotations of the disk. The

Figure 3. Wear resistance of thermal-spray deposited WC-co coatings identified by deposition technique. Top: Sliding wear resistance. Bottom: Abrasion resistance.

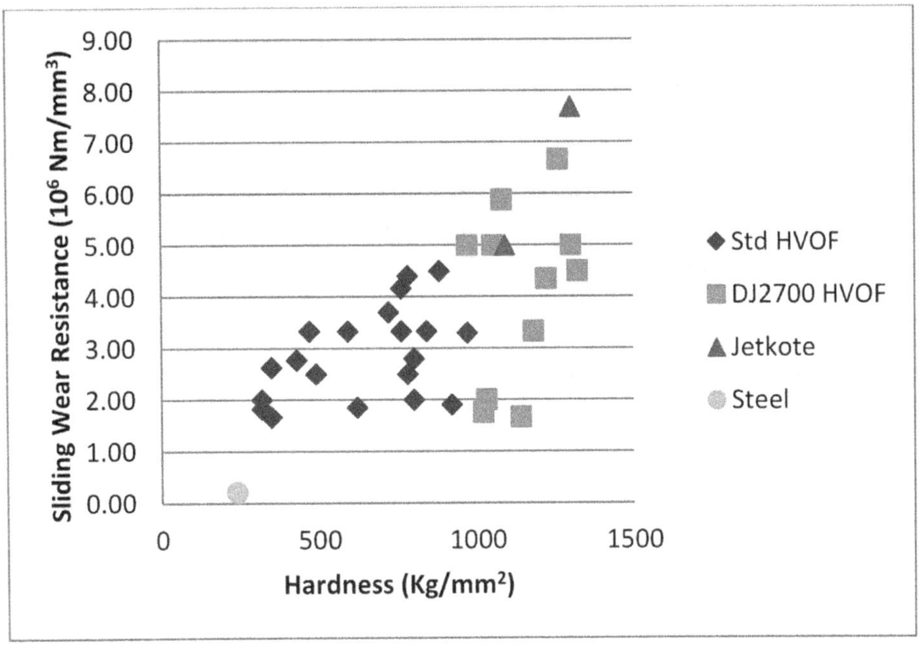

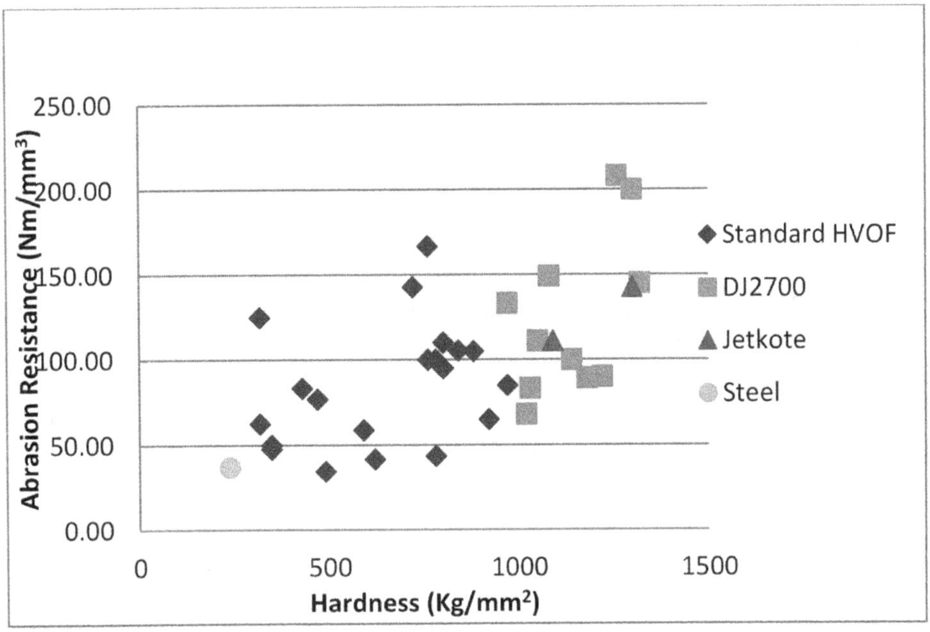

volume of material removed from the coatings was determined by measuring the cross-section of the wear scar with a profilometer at six positions and averaging. The wear of the silicon nitride was calculated from the size of the wear flat on the ball.

Abrasion tests were performed on a ML-100 pin-on-disk tribometer in ambient temperature and humidity. WC/Co-coated steel pins with 8 x 8 mm area were slid under a load of 2 to 17 N against disks of SiC abrasive bonded to paper. The abrasive was LECO 120 GRIT (106 μm) plain abrasive with a hardness of $Hv_{0.1}$2600. The samples moved in a spiral on the abrasive disk in order to encounter fresh abrasive during the whole sliding distance of 38 m. A new abrasive disk was used for each run. The loss of material was determined by weighing the samples before and after the tests after ultrasonic cleaning.

Presentation of the Wear Data

It is usual to express wear in terms of the volume V (in mm^3) of the material removed since that is related to the dimensional modification of machine elements by wear. In most cases the amount of material removed by wear is proportional to the load P (N) and the distance s (m) slid against the counter face.

$$V = W\, P\, s. \tag{2}$$

We have verified experimentally that this applies to the WC-Co coatings as well; this allows us to define a wear rate W in mm^3/Nm by dividing the volume of material lost by the distance and the load.

$$W = V/Ps \tag{3}$$

This wear rate is independent of the sliding distance and load applied, but it depends on other aspects of the sliding conditions such as the nature of the material against which the sample slides, the presence of lubricants, often the sliding velocity and the ambient atmosphere and temperature. The data presented here do not have universal value but serve to compare the various coatings which were all examined under the same conditions.

The wear resistance of many materials is proportional to their hardness; this is expressed by Archard's law

$$V = k\, P.s/H \tag{4}$$

where V is the volume lost to wear, P the normal force on the slider (the load), s the sliding distance, H the hardness and k is the dimensionless wear coefficient.

Therefore we find it more convenient to present the wear resistance R which we define as the inverse of the wear rate:

$$R = 1/W = P.s/V = cH \tag{5}$$

which provides a linear plot of R vs. H.

In order to compare sliding and abrasive wear, we must express the latter in terms of volume removed. Having weighed the mass of material lost to abrasion, we use a density of 14 g/cm^3 to transform the mass into volume so we can compare sliding and abrasive wear. This is the density of compact, sintered, materials. We are aware that the density of the coatings is usually lower because of porosity. It is difficult to measure the density of coatings.

The tribological performance of the coatings is shown in Figures 3 and 4 where we plot the sliding and abrasive wear resistance according to equation (5) vs. the hardness of the coatings. In Figure 3, the data are selected according to the deposition technique employed; Figure 4 identifies the feedstocks.

WEAR OF THE COATINGS

We start by noting that there is a wide spread in the hardness and tribological performances in every group of coatings whether they have the feedstock type or the deposition technique in common. As an example, consider the sliding wear resistance of the Infra samples in Figure 3. Four of the samples have a wear resistance around 4.10^6 Nm/mm^3 and four show a wear resistance slightly below 2.10^6 Nm/mm^3, remarkably independent of their hardness. These samples were all fabricated with feedstocks from the same source; the four samples with higher wear resistance were deposited in the controlled series (Qiao et al., 2003) and the others were deposited in other laboratories (Qiao et al., 2002). Without having sufficient knowledge to state it with authority, we suspect that the cause of the spread lies more in the less easily controlled deposition techniques than in variation of the feedstocks.

It is apparent that the wear resistance of the coatings is limited by their hardness: the highest wear resistance obtained increases linearly with hardness as expressed by equation (5). For a given hardness, however, the wear resistance can be lower than the maximum value by as much as a factor 4 because of defects in the coating, such as brittle phases or porosity. We also note that the sliding wear resistance is at least four orders of magnitude higher than the abrasion resistance. This is related to the contact geometry. Abrasion proceeds by the penetration of sharp grains that generates very large stresses on a small volume; in ductile materials, these stresses cause plastic deformation, pushing the material to the side, and removal of debris

Figure 4. Wear resistance of WC-co coatings identified by the feedstock employed. Top: Sliding wear resistance, Bottom: abrasion resistance.

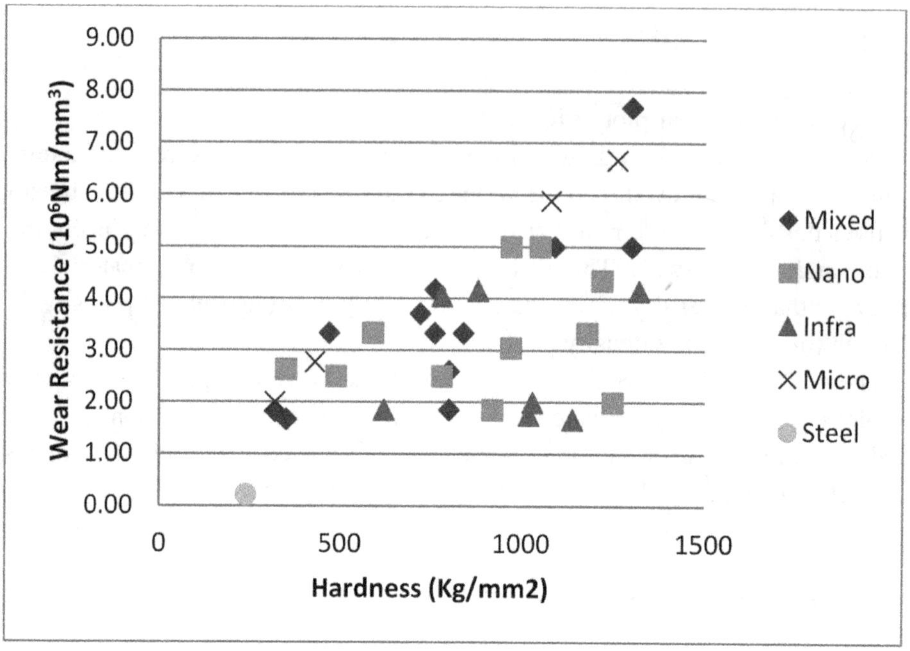

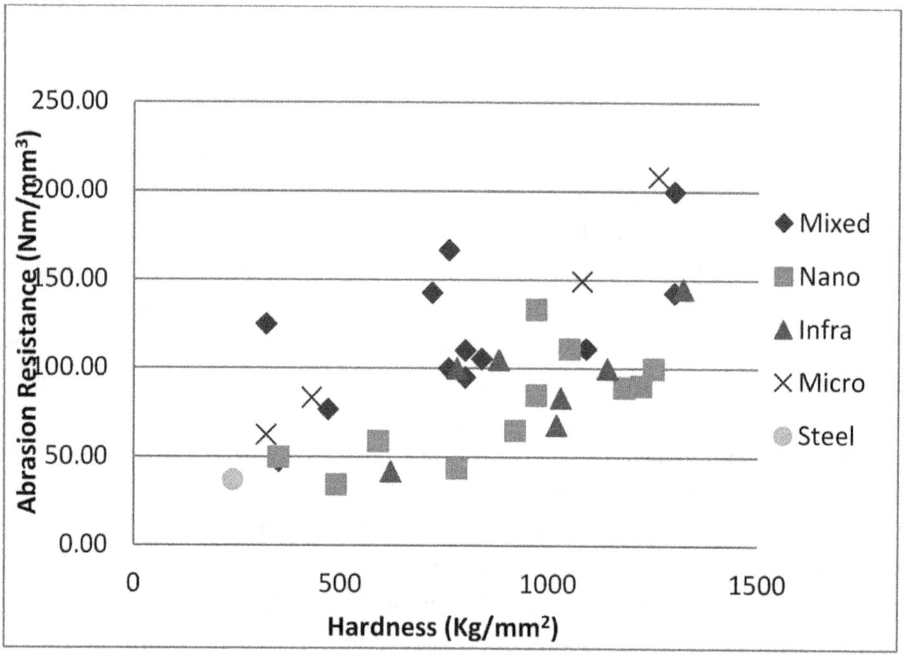

Figure 5. Sliding and abrasive wear resistance of the Nano and Infra coatings with different degrees of decarburization

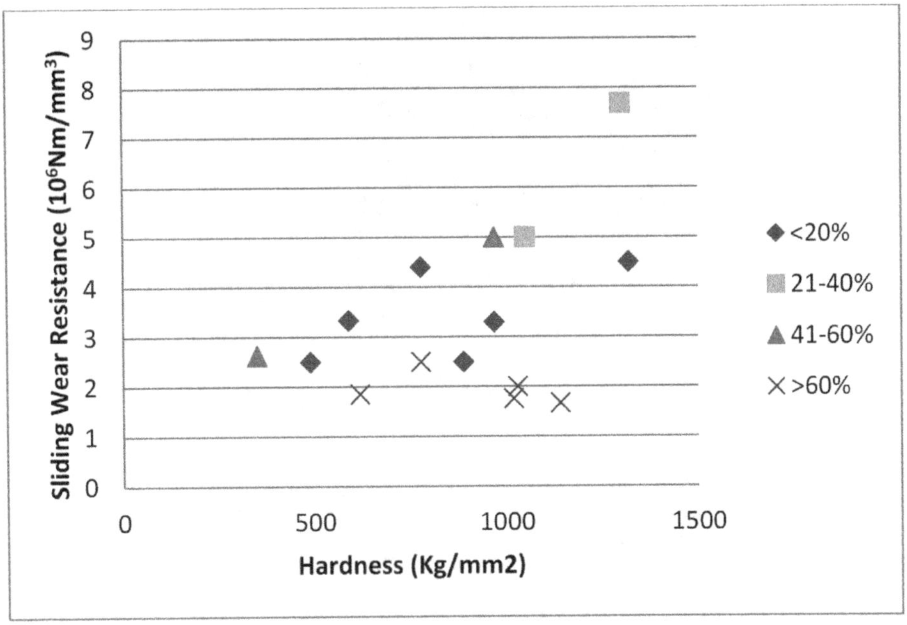

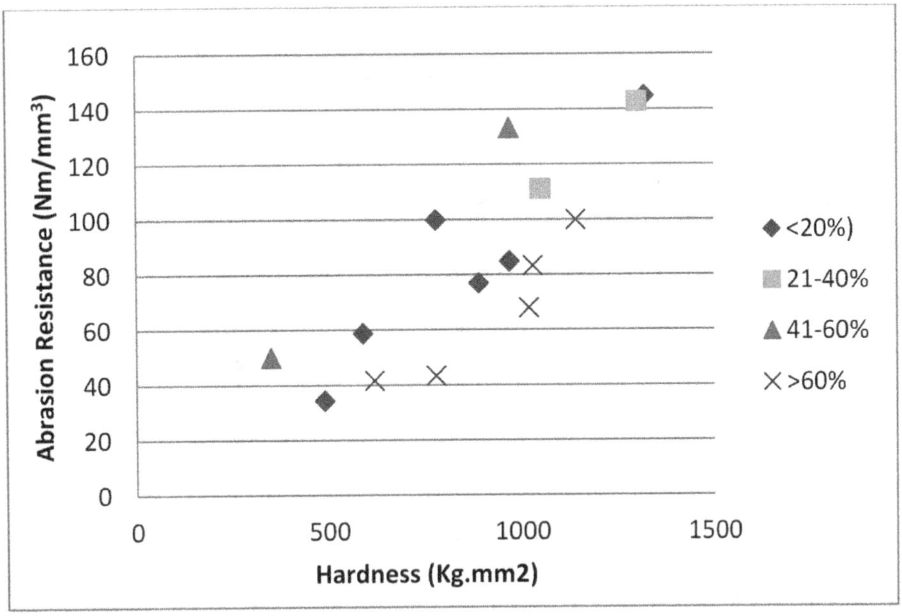

Figure 6. Short abrasion grooves. Samples deposited with DJ2700 gun. (a) Mixed, (b) Infra, (c) Nano feedstock. (©Elsevier, 2003, Used with permission)

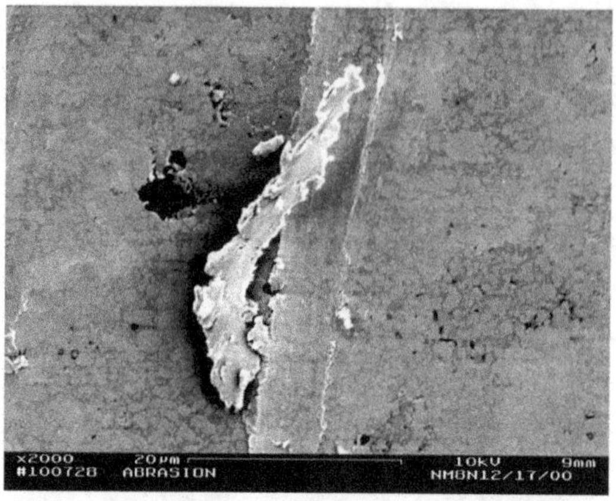

(a)

(b)

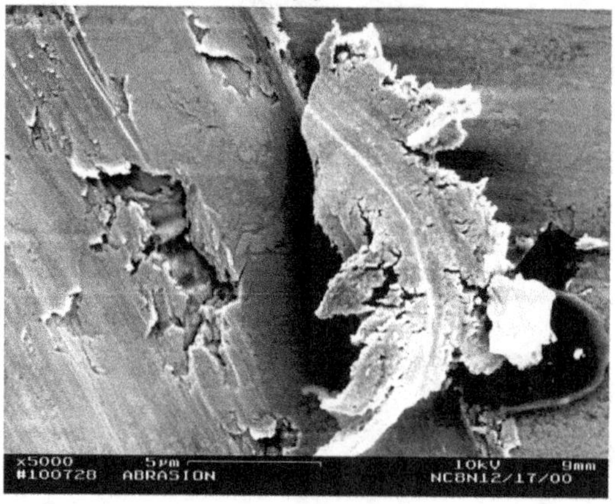

(c)

by cutting. In brittle materials, they remove material by microfracture; it is therefore not surprising that, at equal hardness, brittle materials suffer much more abrasive wear than ductile ones and that the abrasion resistance of the steel substrate is not lower than that of the coatings of similar hardness. In sliding wear, the contact area is larger, the stresses are locally smaller but decrease more slowly with increasing depth. Fatigue plays a larger role in sliding wear than in abrasion.

Effect of Deposition Technique

Figure 3 makes clear that the choice of deposition technique determines the hardness of the coatings. The highest hardness, ranging from 900 to 1270 Kg/mm^2, is obtained with the DJ2700 and the Jetkote guns; the hardness obtained with standard HVOF ranges from 300 to 900 Kg/mm^2. Following equation (5), a higher hardness also produces a higher wear resistance. We have seen in Figure 2 that coatings deposited with standard HVOF have a larger amount of nanometer-sized pores and a poorer adhesion between WC and Co binder than coatings sprayed with DJ2700 and Jetkote guns. This microporosity and poor WC-Co adhesion are responsible for the lower hardness of the HVOF-deposited coatings. In these coatings, the defects that decrease the hardness also degrade the fracture resistance (Qiao et al., 2003). The lower hardness and ductility both contribute to the lower wear resistance of the coatings deposited by HVOF.

Effect of Feedstocks

In Figure 4 we show the same data as in Figure 3 but we identify the feedstocks from which the coatings were deposited. The spread in the measured results makes it unwise to attribute too much importance to the differences between materials. Nevertheless, we note that all WC-Co coatings have a higher wear resistance than the substrate steel.

What strikes the attention first is the excellent abrasion and sliding wear resistance of the Mixed and the Micro samples, both types containing micrometer size WC grains. The coatings deposited with Nano and Infra feedstocks have a somewhat lower, but still acceptable, performance. In sliding wear, the spread of performance is somewhat wider and the conventional Micro coatings exhibit the highest wear resistance. With the large scatter of the data, it is difficult to discern a difference between the coatings deposited with Nano feedstocks; we need to examine the possible influence of decarburization.

Effect of Decarburization

Coatings deposited with conventional micro and mixed feedstocks suffered very little decarburization, therefore we show in Figure 5 the effect of decarburization on the tribological performance only for the Nano and the Infra coatings. The figure shows that this effect is not clear for moderate amounts of WC loss, i.e. below 60%. Above this level, we discern a clear deterioration of the sliding wear resistance, especially for the hard coatings that were deposited with the DJ2700 gun.

Wear Mechanisms

The most convenient way to obtain information about the wear mechanisms is to run very short sliding and abrasion wear tests. This allows us to observe individual wear events and wear particles that have not been further deformed by subsequent passages of the wear agent. Such observations make it possible to relate the wear of the coatings to their microstructure and chemical composition.

Figure 6 shows abrasion grooves on three coatings deposited with the high-temperature DJ2700 gun. These are the same coatings whose feedstock powders and structure are shown in Figures 1 and 2. Abrasion in the Mixed coating (Figure 6a) consists of ductile cutting, similar to what occurs in a metal. We note that the groove is smooth, which indicates that the abrasive cut through the large WC particles as well as through the binder. The shape of the wear particle in the figure, likewise, is representative of ductile cutting. The dark spots in the figure represent pores. In Figure 6b we see a different behavior in the Infra coating: individual grains break away from the material pushed out of the grooves. The difference in the shapes of the wear particles is striking: the wear particle of the Infra coating shows clear evidence of separation of platelets which is absent in the mixed sample in accordance with the microstructures of Figure 2. In the Nano sample, Figure 6c we observe a pit in the ductile grove; its bottom does not show any evidence of cutting. The morphology of this pit indicates that it was caused by the removal of material from a weak interface. If we refer to the microstructure of this sample (Figure 2c) we conclude that this interface is a splat boundary and that the pit was formed by the lifting of a poorly bonded splat fragment.

The sliding wear tracks of the same coatings at short distances (Figure 7) show features that are compatible with those in the abrasion grooves. (In these micrographs, the dark "clouds" represent fine silicon nitride wear particles from the ball). The mixed coating (Figure 7a) is characterized by wear of the binder, leaving the larger WC grains to stand proud. The Infra coating (Figure 7b) exhibits small, shallow, pits that are caused by the breaking out of individual grains, especially at the pit boundary, where local contact stresses are enhanced. The Nano sample (Figure 7b), which suffered 33% decarburization, exhibits a large pit. Its morphology is similar to that on the abrasion groove Figure 6c. We conclude that this pit results from the detachment of a splat as in abrasion. We note that the pit in the Nano sample is much larger and deeper than in the abrasion groove. It reflects the difference in contact geometry: being more spread out in the sliding wear sample, the contact stresses penetrate deeper below the surface. Outside of the pits, this coating also shows evidence of wear of the binder, similar to that of the Mixed coating (Figure 7a).

These features are visible also on surfaces submitted to the long wear tests. The wear tracks after long sliding tests of 5000 m are shown in Figure 8. In the highly decarburized Nano coating, Figure 8a, we discern a high density of pits, 10 to 30 μm in diameter and 3 to 5 μm deep (the latter measured with an optical microscope). The bottoms of these pits are smooth as in Figure 7; they are obviously the amorphous boundaries of splats we have seen in Figure 2. We conclude that wear of these samples consists of the removal of entire splats in addition to the attrition of WC grains. We compare this track with that of the Infra coating, which has experienced very low decarburization, Figure 8b. Low magnification shows a smaller density of pits and high magnification reveals that the bottom of these pits is granular. These pits do not represent lifting of splats but are original pores in the samples that have been enlarged by loss of grains at their border due to high local contact stresses. Such porosity exists in all samples deposited at low temperature and in Infra coatings.

After long abrasion tests the appearance of the surfaces reveals the influence of material properties in a similar way. Figure 9 shows two extreme examples. The hard, ductile coating from mixed feedstock shows smooth and shallow abrasion grooves that reflect the hardness and ductility of this sample. The Nano coating deposited with standard HVOF exhibits deep grooves because of its low hardness and a rough surface caused by the removal of splats.

These figures teach us that sliding wear of the WC-Co coatings consists of three mechanisms: the wear of the binder, the falling out of grains in grooves or around native pores, and the lifting of entire splats in decarburized samples where the splat boundaries are weak. These three mechanisms operate in all samples to different degrees, depending on the material properties that result from the particular powders and processing parameters. Wear of the WC grains undoubtedly exists; in sintered WC/Co composites, we have seen that such wear produces polished WC grains (Jia & Fischer, 1997). In the present coatings, it is impossible to distinguish experimentally the parts played by the falling out and the wear of the WC grains so that we are reduced to lumping them together as grain attrition.

Figure 7. Short sliding tracks. Samples deposited with DJ2700. (a) Mixed, (b) Infra,(c) Nano powder . (©Elsevier, 2003, Used with permission)

(a)

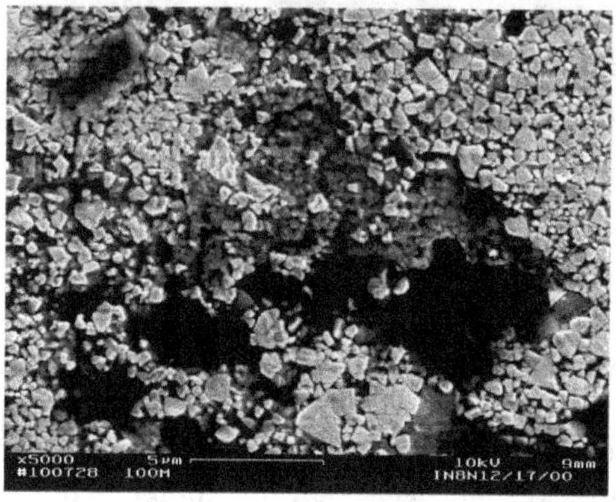

(b)

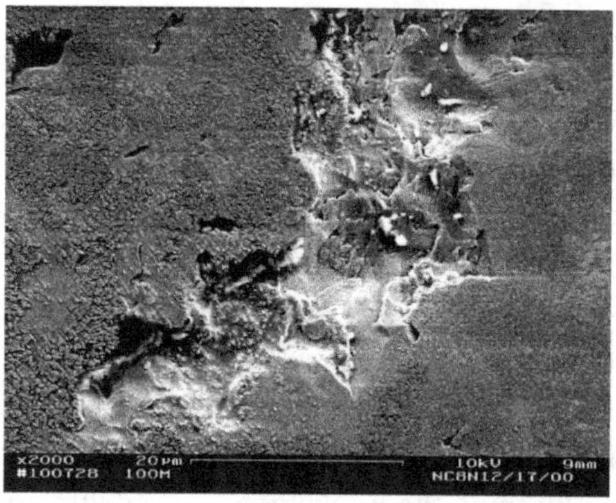

(c)

Figure 8. Wear tracks after long sliding tests. (a) Nano coating deposited with DJ2700 (b) Infra coating deposited with standard HVOF. (©Elsevier, 2003, Used with permission)

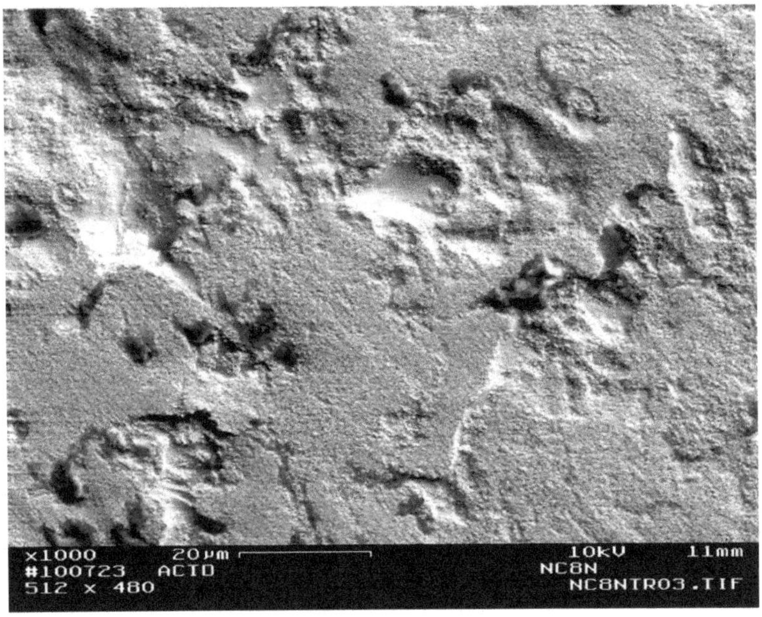

(a)

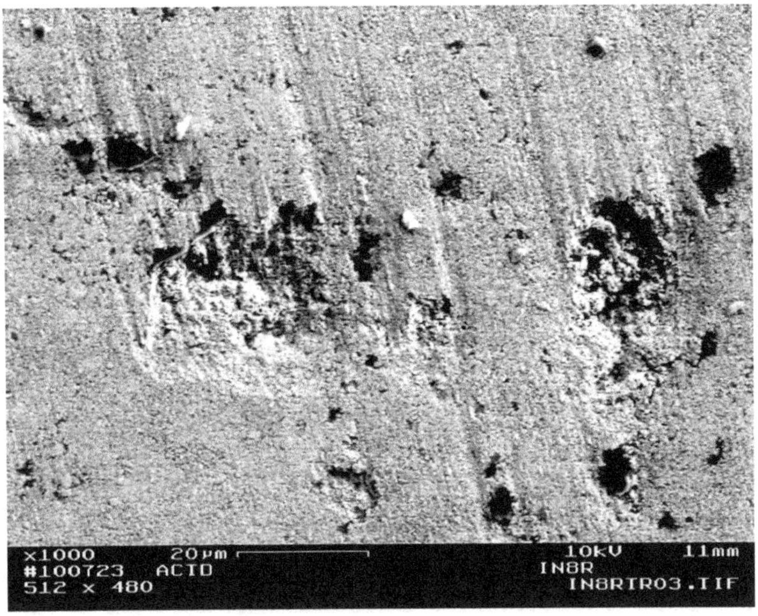

(b)

Figure 9. WC-Co coatings after long abrasion. (a) Mixed coating deposited with DJ2700 (b) Infra coating deposited with standard HVOF. (©Elsevier, 2003, Used with permission)

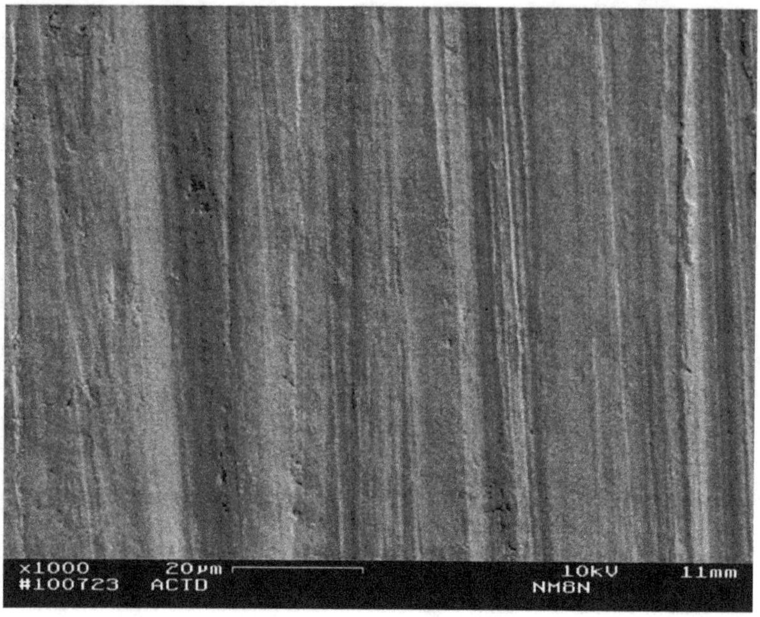

(a)

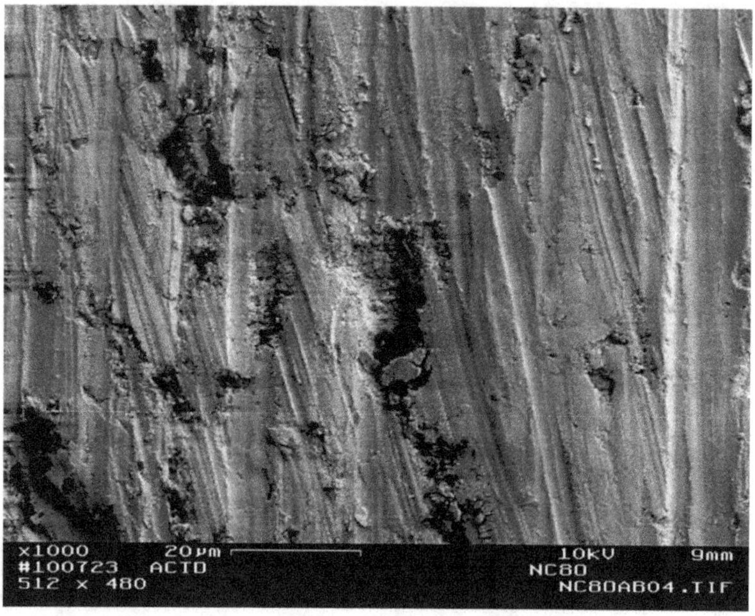

(b)

WEAR OF THE SILICON BALL

In practical applications, sliding wear occurs where machine elements rub against each others, for instance in sliding bearings. It is therefore important to examine not only the wear resistance of the coating itself but the amount of wear it inflicts on the surface against which it slides. We have examined the wear of the silicon balls employed in the sliding tests and the results are shown in Figure 10. It appears that the Mixed and Micro coatings, which contain micrometer size WC grains, inflict severe wear on the silicon. Evidently, the worn coatings, with their WC grains standing proud of the worn-away binder, act as abrasives against the silicon. The Nano coatings cause much less wear of the counter-face and the Infra coatings even less. Clearly coatings with Nano and Infra feedstocks are the material of choice for sliding applications despite their somewhat lower wear resistance which is nevertheless respectable (Figure 4). Conventional (Micro) and Mixed feedstocks are the material of choice when abrasion resistance is desired; in such applications, such as earth moving or slurry processing equipment, there is usually not a machine element rubbing against the coating. If there is, one prefers the Mixed coatings to the Micro.

CONCLUSION

This examination has allowed us to gain some insight into the relative importance of the feedstock properties and deposition techniques of WC-Co coatings. In particular, we have learned that:

The hardness and tribological performance of the coatings cannot be narrowly controlled but vary by more than a factor 2 within a feedstock group or a deposition technique. Nevertheless, the performance of all coatings is quite satisfactory and much better than that of steel.

The hardest and most wear resistant coatings are obtained with a hot, neutral flame. Deposition by standard HVOF generates less decarburization but relatively poor bonding between the WC grains and the Co binder. Evidently, good cohesion of the coatings and adhesion between splats are more important for wear resistance than avoiding decarburization.

Feedstocks containing micrometer-sized WC produce coatings with the best tribological performance, especially in abrasion. These coatings wear mostly by attrition of the binder phase (as do the bulk WC-Co materials (Jia & Fischer, 1997). They act as abrasives themselves and inflict severe wear on a surface against which they slide.

Figure 10. Wear of WC/Co coatings and the silicon balls against which they slide. (©Elsevier, 2003, Used with permission)

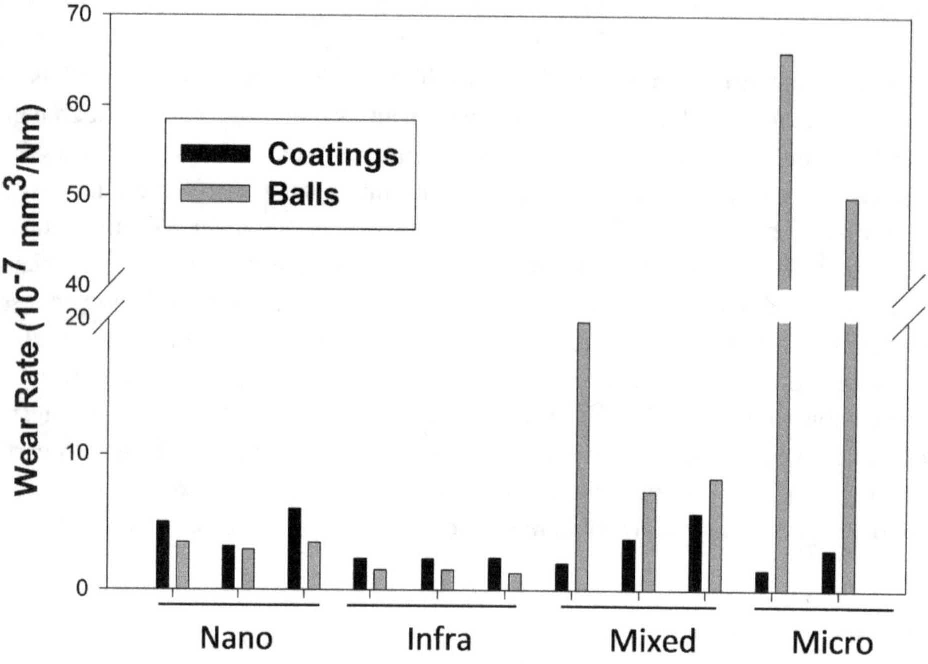

Nano feedstocks, with their large surface-to-volume ratio, are reactive, suffer decarburization and produce coatings with low adhesion between splats. They wear mostly by splat removal.

Infra feedstocks, which are made from Nanocarb powders reprocessed with a coating designed to suppress grain growth, produce coatings with reduced WC-binder adhesion and wear predominately by grain separation.

APPLICATIONS

Coatings deposited with Nano and especially Infra feedstock inflict the least wear on the surface against which they slide and are the coatings of choice for bearings and sliding machine elements.

Coatings deposited with Micro or Mixed feedstock inflict excessive wear on surfaces on which they slide but present excellent abrasion resistance. They are not recommended for bearing applications but they are the coatings of choice for surfaces exposed to abrasion or erosion, such as earth moving or slurry processing machinery.

ACKNOWLEDGMENT

The authors are indebted to the many industrial and academic scientists who provided the feedstocks or deposited the coatings. The whole research program was funded by the Office of Naval Research.

REFERENCES

Anstis, G. R., Chantikul, P., Lawn, B. R., & Marshall, D. B. (1981, September). A critical evaluation of indentation techniques for measuring fracture toughness: I, direct crack measurements. *Journal of the American Ceramic Society*, *64*(9), 533–538. doi:10.1111/j.1151-2916.1981.tb10320.x

Jia, K., & Fischer, T. E. (1996). Abrasion resistance of nanostructured and conventional cemented carbides. *Wear*, *200*(1-2), 206–214. doi:10.1016/S0043-1648(96)07277-8

Jia, K., & Fischer, T. E. (1997). Sliding wear resistance of conventional and nanostructured cemented carbides. *Wear*, *203*, 310–318. doi:10.1016/S0043-1648(96)07423-6

Jia, K., Fischer, T. E., & Gallois, B. (1998). Microstructure, hardness and toughness of nano-structured and conventional WC-Co composites. *Nanostructured Materials*, *10*(5), 875–891K. doi:10.1016/S0965-9773(98)00123-8

McCandlish, L. E., Kear, B. H., & Kim, B. K. (1992). Processing and properties of nanostructured WC-Co. *Nanostructured Materials*, *1*(2), 119–123. doi:10.1016/0965-9773(92)90063-4

Qiao, Y., Fischer, T. E., & Dent, A. (2003). The effects of fuel chemistry and feedstock powder structure on the mechanical and tribological properties of HVOF thermal-sprayed WC-Co coatings with very fines structures. *Journal of Surface Coatings and Technology*, *172*(1), 24–41. doi:10.1016/S0257-8972(03)00242-1

Qiao, Y., Liu, Y. R., & Fischer, T. E. (2001). Sliding and abrasive wear resistance of thermal-sprayed WC/Co coatings. *Journal of Thermal-Sprayed Materials*, *10*(1), 118–125. doi:10.1361/105996301770349583

Stewart, D. A., Shipway, P. H., & McCartney, D. G. (2000). Microstructural evolution in thermally sprayed WC-Co coatings: Comparison between nanocomposite and conventional starting powders. *Acta Materialia*, *48*(7), 1593–1604. doi:10.1016/S1359-6454(99)00440-1

Verdon, C., Karimi, A., & Martin, J.-L. (1998). A study of high velocity oxy-fuel thermally sprayed tungsten carbide based coatings: Part 1: Microstructures. *Materials Science and Engineering A*, *246*(1-2), 11–24. doi:10.1016/S0921-5093(97)00759-4

KEY TERMS AND DEFINITIONS

Abrasion Resistance (Nm/mm^3): The product of distance (m) a body slides on an abrasive surface under a load (N) to cause the removal of 1 mm^3 of its material by wear. It is the inverse of the abrasion rate (mm^3/Nm). It is about 10,000 times smaller than the sliding wear resistance.

Deposition Technique: The choice of gun, flame temperature and gas velocity in the HVOF deposition of coatings.

Hardness: The hardness of the coatings is measured here by the size of the imprint of a Vickers diamond pyramid that is pressed into the surface with a force of 1Kg (9.81N). This choice of force produces a value of the hardness of the composite, allowing for the influence of weaker interfaces between its components.

High-Velocity Oxyfuel (HVOF) Deposition: Technique by which a powder is heated and projected at high velocity against a substrate surface by the flame caused by the burning of a fuel in order to produce a desired coating.

Microstructure: The sizes, shapes and mutual arrangements of the components of a coating as revealed by the microscopic examination of a polished and etched section of that coating.

Nanostructured Feedstock: Material used in the deposition of coatings that is composed of grains smaller than one micrometer.

Sliding Wear Resistance (Nm/mm^3): The product of distance (m) a body slides on a flat surface under a load (N) to cause the removal of 1 mm^3 of its material by wear. It is the inverse of the wear rate (mm^3/Nm).

Chapter 2
Tribocorrosion of Thermal Sprayed Coatings

S. Mischler
Ecole Polytechnique Fédérale de Lausanne (EPFL), Switzerland

Manish Roy
Defence Metallurgical Research Laboratory, India

ABSTRACT

Triboelectrochemical response is concerned with different electrochemical-controlled wear experiments designed to understand tribocorrosion behaviour of materials and coatings. Tribocorrosion is defined as material degradation resulting from simultaneous action of wear and corrosion and it is found in many engineering applications, but the involved mechanisms are still only partially understood. In this chapter, a brief overview of tribocorrosion testing techniques is followed by issues which have helped us in gaining in-depth scientific knowledge of tribocorrosion. The overview is further substantiated by detailed studies and observations on tribocorrosion of thermal sprayed coatings in recent times.

1.0 INTRODUCTION

Tribocorrosion is a form of material degradation or transformation which results from synergistic effect of sliding wear and corrosion. The interaction of wear and corrosion is complex. The extensive knowledge of wear of materials in the absence of corrosive fluid and that of the corrosionof materials in absence of wear does not

DOI: 10.4018/978-1-4666-7489-9.ch002

suffice to predict the nature of degradation under condition of tribocorrosion. It is true that wear influences the sensitivity of material to corrosion.At the same time corrosion does govern the conditions of friction. There is synergy between wear and corrosion. The interaction between corrosion and wear often results in substantial enhancement of material loss.More often than not this material loss is much higher than the sum of the individual contribution of wear and corrosion.A large number of engineering components are subjected to tribocorrosion such as mining equipment, food processing devices, biomedical implants, chemo-mechanical polishing etc.

Thermal spraying has been increasingly used for wear resistance application for the last few decades(Stein *et al.,* 1999; Barbezat & Nicoll, 1993; Spear, 1989; Chen *et al.,* 2005). With this process, the coating material, in powder form or wire, is fed into the combustion chamber of a gun where, a fuel, such as hydrogen, ethylene or kerosene etc, is burned with oxygen, and the heated and softened materials are expelled as a spray with the high speed fluid. Thermal spraying is one of the most versatile techniques for depositing wear resistant materials.It allows rapid and efficient ways of depositing variety of materials on different substrates. It is a fast, easy to apply and efficient way of depositing coating materials. It has the ability to deposit coatings with thicknesses ranging from several micrometers to tens of millimetres.Additionally, it is suitable for a great variety of shapes and sizes and had the advantage of maintaining the substrate temperature relatively low.There are a lot of techniques to melt and propel the coating material. However, to achieve optimal performances, various coatings need to be evaluated separately for individual cases.

Over the years, significant work is done on tribocorrosion as evident from several reviews (Landolt *et al.*, 2001; Wood, 2006). The triboelectrochemical techniques were applied on a variety of sliding contact conditions, varying from unidirectional motion (pin on disk), reciprocating motion contacts, fretting or spinning contacts. The most investigated materials were model alloys for fundamental mechanisms identification(Mischler *et al.,* 1998; More *et al.,* 2011; Stack & Chi, 2003), followed by biomedical alloys(Vieria *et al.,* 2006; Barradja *et al.,* 2006; Diomidis *et al.,* 2011), seals materials (Serre *et al.,* 2002), materials for nuclear reactors (Benea *et al.,* 2004) and chemo-mechanical polishing (Bielmann *et al.,* 2000). Interestingly, all these materials were passive alloys even though passive materials may significantly degrade under the combined action of corrosion and wear when the passive layer is destroyed mechanically. The work on tribocorrosion on coatings is very much limited (Shenhar *et al.,* 2000; Galliano *et al.,* 2001; Vieria *et al.,* 2006) and among these a few work is done for thermal sprayed coatings.

This presentation therefore begins with a brief overview of tribocorrosion testing techniques published in recent literature following introduction. After describing the tribocorrosion test techniques, background and issues which have helped us in gaining in-depth scientific knowledge of tribocorrosion has been discussed. The overview is further substantiated by detailed studies and observations on tribocorrosion of thermal sprayed coatings in recent times. Future study and summary concludes this write up. Although, tribocorrosion involves interaction between corrosion especially in aqueous medium and various forms of wear such as sliding wear, erosive wear, abrasive wear etc. However, only interaction between sliding wear and oxidation will be presented in this communication.

2.0 TRIBOCORROSION TESTING

Wear studies performed in presence of an ionicliquid under controlled electrochemical conditions can be termed as tribocorrosion experiment. The standardisation of the tribocorrosion experiment was reported for thefirst time to ASTM G119 standard guide in 1995. The fact that the electrochemical measurements can be combined with tribological experiments without any compromise of reproducibility and reliability was for the first time in 2001 by publishingthe results of an interlaboratory study (Mischler Ponthiaux, 2001).

An extensive discussion on tribocorrosion testing is carried out by Mischler (Mischler, 2008). Thetechniques used for tribocorrosion experiment can be divided undersevenmain types: (1) corrosion potential technique, (2) galvanic cells, (3) potentiostatic technique, (4) potentiodynamictechnique, 5) electrochemical noise technique (Wu P-Q, Celis 2004; Barradja*et al.* 2006), 6) electrochemical impedance spectroscopy technique (Ponthiaux *et al.*, 2004) and 7) galvanic coupling technique. The most used techniques were the potentiostaticand the corrosion potential ones. Some of these techniques areschematically illustrated in Figure 1.

2.1. Galvanic Cells

In this technique, the sample is subject to rubbing. Another plate of the same material is placed at a certain distance (few cm) from the test material. The plate and the sample are connected to a zero-resistance ammeter to measure the galvanic current. Initially, both the sample and the plate have the same corrosion potential and ammeter indicates zero current. As soon as rubbing starts the corrosion potential of the sample changes and a galvanic cell between sample and plate is formed and an electric current flows between the plate and the rubbed sample. When rubbing stops, the sample surface repassivates and the galvanic current becomes nil. The schematic representation of a galvanic cells is made in Figure 1a. The galvanic current should ideally represent the anodic current between the worn area and the unworn area, i.e. the excess corrosion induced by rubbing. However, the current estimated in this method does not account for the current flowing between the worn and unworn

Figure 1. Schematic representation of various tribocorrosion test arrangements (Mischler, 2008)

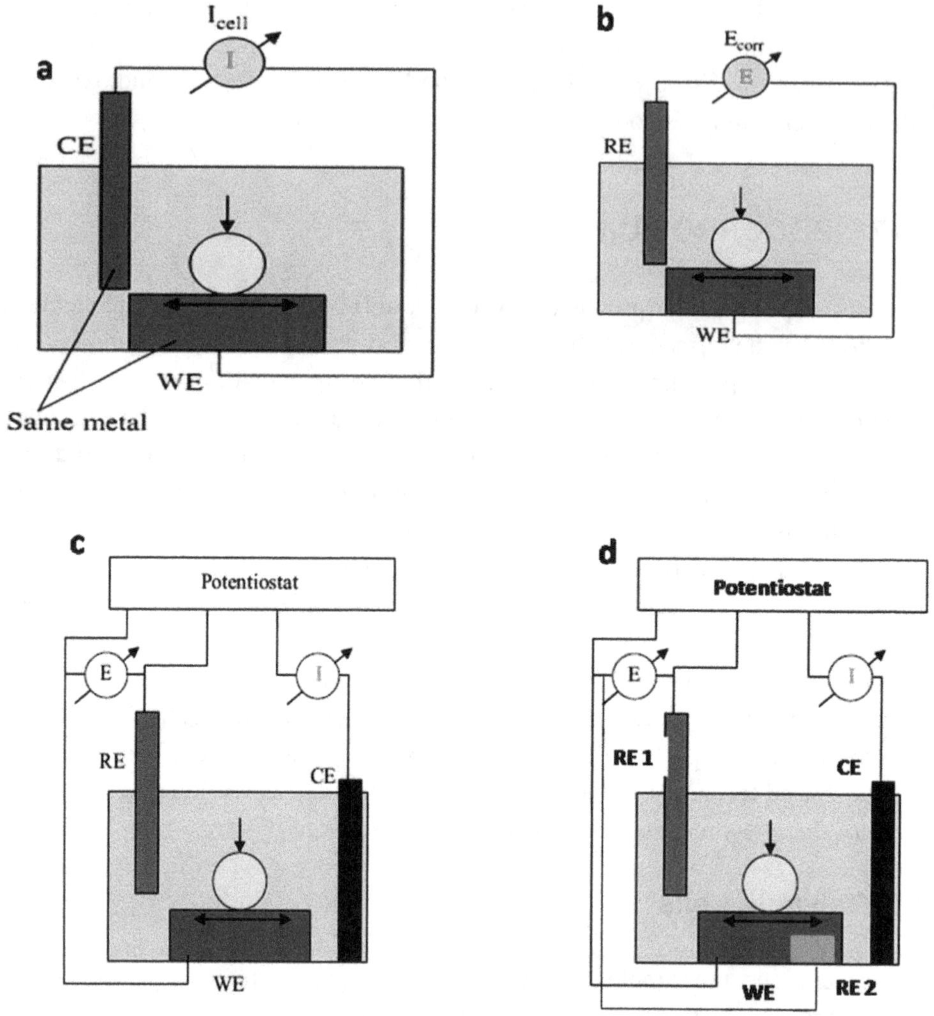

areas of the rubbed sample and therefore it indicates only a fraction of the excess corrosion rate. The current flowing in the galvanic cell depends, in principle on the electrical resistance of the electrolyte and on the potentials established on sample and plate. The electrical resistance of electrolyte, in turns, depends on conductivity of the electrolyte, the distance between sample and plate and their respective size. This method has the advantage of simplicity and to work at the corrosion potential, i.e. under similar conditions as in engineering systems. In addition, due to its semi-quantitative character, it allows comparison of material couples.

2.2. Measurement of Corrosion Potential

This technique is shown schematically in Figure 1b.The corrosion potential, i.e. the potential difference spontaneously established between the working electrode (the metal being investigated) and a reference electrode placed in the solution close to the working electrode is recorded in this technique. The corrosion potential of the passive alloys drops to lower values (cathodic shift) as soon as the rubbing starts. During rubbing, the measured corrosion potential reflects the galvanic coupling of two distinct surface states of the metal: (1) the passive metal (unworn area) and (2) the bare metal (worn area) exposed to the solution by abrasion of the passive film. It is to be stated that Ponthiaux *et al.* (Ponthiaux *et al.,* 2004) found that worn and unworn areas of active metals exhibit similar surface states resulting in insignificant effect of rubbing on corrosion potential. When rubbing is stopped, the corrosion potential comes back to the initial value due to the formation of the passive film on the worn area. Four parameters namely (1) the intrinsic corrosion potentials of the worn and unworn surfaces, (2) the ratio between worn and unworn surface areas, (3) the relative position of worn and unworn areas and (4) the mechanisms and kinetics of the involved reactions which influence the corrosion potential during rubbing: have been identified by Ponthiaux *et al.* (Ponthiaux *et al.,* 2004). Many of these parameters are not defined properly. The situation of the worn area corresponds in principle to an active dissolution characterised by potentials below the passivation potential; however, no precise values can be given. The ratio between worn and unworn areas can be calculated by considering, in general, only a fraction of the rubbed area which becomes effectively depassivated by the asperities of the counter piece. No reliable methods exist to determine this fraction presently. More precise knowledge of the local state of the surface is required to interpret the corrosion potential measured under friction. In a preliminary attempt to determine experimentally the local distribution of the potential around and inside rubbed areas micro-electrodes (Wu & Celis, 2004) or SRET (scanning reference electrode technique) (Serre *et al.*, 2006) were used. Nevertheless, the interpretation of local potential measurements requires complex kinetics models describing the reactions occurring in the different surface areas. However, such models are not yet available. Similar to galvanic cell, the measurement of the corrosion potential during tribocorrosion is a very simple technique to gather information on the surface state of the sliding metal although this technique does not give direct information on the kinetics of the corrosion reactions. Only recently, Vieira *et al.* (Vieira *et al.*, 2012) and Papageorgiu *et al.* (Papageorgiu *et al.*, 2012) proposed kinetic models based on the buil up of a galvanic cell between the depassiavted areas in the wear track and the surrounding passive areas. This models allow determining the wear accelerated corrosion rates by known cathodic kinetics and ratio between worn and unworn areas.

2.3. Potentiostatic Tests

A selectedpotential E is imposed to the metal samples by using a three electrodeset-up in potentiostatic tribocorrosion tests. The set up is shown in Figure 1c. These three electrodes are the working electrode (the metalbeing investigated), the reference electrode and the counter electrode. The counter electrode is generally made from inert materials such as platinum or graphite. These three electrodes are connected to an electronic device known as potentiostat, that maintains the selected potential between working and reference electrodes by passing an appropriate current between working and counter electrodes. The current is measured at fixed potential as a function of the time in order to record and monitor the evolution of the electrochemical kinetics of the involved reactions. At the onset of rubbing a sudden increase in current due to the oxidation of the bare specimen exposed to the solution after abrasion of the passive film is noted. After rubbing, the current decreases again to the initial value as all the metal passivates again. It is to be noted that the current measured during rubbing flows mainly through the wear track area, an area in general, much smaller than the overall metal area exposed to the solution. Representative current densities are expected to be of the order of 0.008 and 5 mA/cm^2 before and during rubbing, respectively. This is a demonstration of the fact that under tribo-corrosion conditions the corrosion rate of passive metals can increase by several orders of magnitude compared to static situations. The current measured, $I_{measured}$, during potentiostatictests is equal to the sum of the cathodic, $I_{cathodic}$, currents and anodic, $I_{anodic,i}$ due to all the electrochemical reactions taking place on the working electrode, as described by Equation (1):

$$I_{measured} = \sum I_{anodic,i} + I_{cathodic,i} \tag{1}$$

Generally cathodic currents are considered to be negative while anodic currents are considered to be positive. The prevailing electrochemical reactions are determined by the potential. At the corrosion potential, the measured current is zero while the anodic and cathodic reactions occur at the same rate. At cathodic potentials, which is well below the corrosion potential, the dissolution rate of the metal is negligible and the kinetics of cathodic reactions determines the measured current. In contrast when the potentials are well above the corrosion potential, the rate of the cathodic reactions becomes negligible and the kinetics of metal oxidation determines the current, for potential below a threshold potential (the passivation potential) or above it.

The relation between loss of mass of metal and the current in the anodic area can be determined by using Faraday's law as given below:

$$m = \frac{I \times M \times t}{n \times F} \qquad (2)$$

where m is metal mass oxidized during time t, I is anodic current, F is Faraday's constant (approximately 96,500 C/mol),n is oxidation valence and M is the atomic mass of the metal. According to Equations (1) and (2), the conversion of current into mass of oxidized metal is possible only by known oxidation valence and if the oxidation of the metal is the prevailing contribution to the measured current. A continuous regime of depassivation (establishment of active dissolution in the worn areas) and repassivation of the activated areas determines the overall current during tribo-corrosion,. The transient coexistence of a active and passive state makes it difficult to estimate the valence of dissolution for metals such as Cr, Fe or Ti which present different valences for active dissolution (2 for Cr, Fe and Ti) and for the passive state (3 for Cr and Fe, 4 for Ti). This yields uncertainties on the relative contribution of the corrosion reaction to the overall material loss (Spear, 1989). The electrochemical conditions and reactions may be modified by the special conditions prevailing within contacts.

The electric resistance R of the solution separating working and reference electrode causes the actual electrode potential E to differ from the applied one $E_{applied}$ accordingto Equation (3):

$$E = E_{applied} - RI \qquad (3)$$

The ohmic resistance R is proportional to the resistivity of the solution, the distance between reference and working electrode and depends in case of tribocorrosion experiments on the masking effect exerted by the inert counterbody. The latter factor is controlled by the size and geometry of the counter body and by the dimension of the wear track.

2.4. Potentiodynamic Tests

This experiment is similar to potentiostatic tests and instead of maintaining a well-defined potential, the potential is swept at constant rate using a function generator to drive the potentiostat. It is possible to observe the effect of friction on the different electrochemical reactions taking place depending on potential employing this method. This method is very useful. This technique also suffers of some major limitations. The first limitation is related to the non-stationary conditions involved in the experiment. Secondly, the effect of rubbing on polarization curves depends on the ratio between worn and unworn area of the electrodes. The wear transitions (running in, transformation of the counter body) are dependent on time. This may interfere with the potential scan, so that features in the current response cannot always be attributed to a pure electrochemical effect. Indeed, Landolt *et al.*(Landolt *et al.*, 2001) observed (stainless steel under constant applied potential and rubbing against alumina in sulphuric acid) a run-in period.

2.5 Electrochemical Noise Technique

Electrochemical noise (EN) was introduced as a complement to open-circuit potential and galavnic cell measurements to investigate corrosion–wear phenomena in aqueous solutions (Waterhouse, 1972). The main advantage of the electrochemical noise technique is that, it provides simultaneously in situ data on potential and current fluctuations, whereas with the open-circuit potential technique only potential data are obtained. There are three major modes for measuring potential and current noise in a corrosion system, namely the two identical working electrode (WE) mode (Eden, 1986), one WE coupled to a microelectrode (e.g. Pt) (Chen & Bogaerts, 1996), and two identical WEs with a bias potential (Benish *et al.*, 1998). The two samples were connected through a zero-resistance ammeter (ZRA) to monitor galvanic current during corrosion–wear. A reference electrode was used to measure the open-circuit potential of the worn sample. Apparently, the two originally identical samples may become significantly different once a wear process is started on one of them. A galvanic corrosion between the two samples may result from the wear on one sample, and this may significantly accelerate the corrosion–wear process on the worn sample. Similar problems may be encountered in applying the EN technique consisting of two identical WEs with a bias potential, to corrosion–wear. In addition, the application of a bias potential imposes the use of a more complex electrochemical setup. This problem may be overcome by using an appropriate microelectrode instead of a second identical electrode coupled to the working electrode through a ZRA. This method is known as electrochemical emission spectroscopy (EES) (Chen & Bogaerts, 1996). The integration of this EES technique into sliding corrosion–wear tests is not straightforward. Indeed, the microelectrode needs to be optimized, and counter-measures must be taken to combat external influences, especially those from the fretting wear test equipment, overwhelming the real electrochemical signal. Moreover, there are numerous differences between the EN in a corrosion system and in a corrosion–wear system. One of the major differences is that the mechanical interaction in a corrosion–wear test may induce EN that is not appearing in a corrosion test.

The experimental setup used for electrochemical noise measurements is schematically shown in Figure 1d. A potentiostat which allows voltage and current measurements at a resolution of 1μV and 1 pA, respectively is used. The microelectrode coupled to the working electrode is used to sense the current flowing between them. Coaxial cables are used for electrical connections wherever possible in order to reduce the noise coming from the surrounding environment. Data acquisition on potential and current was done for 3000 s at a sampling rate of 1 Hz, initiated 10 min after immersion of the samples into the test solution. The counter body is lifted away immediately from the WE at the end of fretting tests. The electrochemical noise data are reported according to ASTM conventions (ASTM Standard). For the potential of the working electrode, the positive direction is denoted as the noble direction and the negative direction as the active direction. Anodic currents at the working electrode are considered positive and cathodic currents are negative.

2.6 Impedance Spectroscopic Technique

EIS is a most performing method for a detailed analysis of electrochemical reactions mechanisms and kinetics. Impedance diagrams give data on the elementary steps occurring in an electrochemical reaction and on their kinetics. They allow a thorough study of the role of intermediate species adsorbed on the surface and of reaction mechanisms, as well as a study of the properties of passive films. Consequently, EIS is an interesting method to analyze the wiping by friction of adsorbed films or passive films from materials. Electrochemical impedance diagrams are represented in a Nyquist plane. A classical interpretation of impedance diagrams consists in relating the corrosion current, I_c, on a given area A, to the polarization resistance, R_{pol}, which is a parameter of the electrochemical impedance that can be deduced from the experimental diagrams:

$$I_C = \frac{B}{R_{pol}} \tag{4}$$

B is a constant factor related to the corrosion or passivation mechanism. The polarization resistance, R_{pol}, is defined as the limit of the impedance Z_i of metal–electrolyte system when the frequency tends to zero:

$$Rpol = \lim_{\omega \to 0} Z_i \tag{5}$$

The polarization resistance corresponds to the value of the impedance at very low frequencies:

The impedance measurements carried out during tribo-corrosion tests are useful to study the mechanism of electrochemical reactions involved in tribo-corrosion processes, and the interaction between corrosion and friction. More detailed information on tribo-corrosion processes are expected from a systematic study of impedance diagrams recorded at varying tribological test conditions (variation of the normal force, sliding speed, rotation frequency, etc.). The interpretation of impedance measurements recorded during sliding tests is difficult due to the heterogeneous state of the surface, as in the case of polarization measurements. In fact, a non-uniform distribution of the electrochemical impedance over the disksurface must be considered. The action of friction can be thoroughly analyzed, only if this distribution is known. In tribo-corrosion experiments, a local analysis of the electrochemical state is thus necessary to interpret impedance measurements. Research work on electrochemical systemswith a non-uniform distribution of the electrochemical impedance is available now (Wenger & Galland, 1990). Such measurements and models can be adapted to tribo-corrosion conditions. In that field, microelectrodes could help to map the electrochemical impedance on disk surfaces. Electrical equivalent circuit models or finite element models could be used to get distributions of impedance, and to calculate the overall impedance.

3.0 THERMAL SPRAYING PROCESS

Numerous investigations have been carried out over the last decades for wear resistance of thermally sprayed coatings in sliding conditions (Savarimuthu *et al.*, 2001; Jacobs *et al.,* 1999). Different coating materials, spraying systems, and a number of tribological tests have been used in previous investigations. Among the most commonly applied are flame spraying and plasma spraying. The flame sprayed coatings exhibit a high porosity and a bad bonding strength in comparison to other thermal spraying process. Plasma spray is suitable for hard and thick coatings especially ceramic coatings and it is used in applications requiring wear and corrosion resistant surfaces.

The vacuum plasma spraying (VPS) eliminates many problems by reducing the interaction between the plasma jet and the environment by spraying in a reduced-pressure or in an inert gas. Typically, VPS involves plasma spraying at 2-25 kPa (20-250 mbar), which results in plasma flames as long as 20-30 cm, compared to APS lengths of 7-14 cm. The longer plume results in a lower energy density of the flame, which requires higher power and longer spray distances for the varied materials to ensure complete melting of particles. VPS plasma jet and particle velocities are typically two to three times those in APS. The advantages of the VPS process are rapid deposition rates, high substrate temperatures that result in interfacial diffusion between the coating and the substrate during spraying, thus favoring the formation of metallurgical bonds (Sampath, 1988) low residual stress levels in the coatings, and reduced amounts of impurities in the coatings (Gruner, 1986). The limitations of the VPS technology include only powders with a stable liquid phase can be used as feedstock materials; it is a line-of-sight coating process and maximum substrate dimensions are dictated by the chamber size. The VPS process has been used for corrosion, wear, and oxidation resistance applications in the automotive, transportation, aerospace, and aircraft industries because of the dense, well-bonded coatings that can be produced.

Controlled atmosphere plasma spraying (CAPS) is an powerful process for highly demanding applications enabling a large range of plasma spraying variants by varying the selected spray conditions (pressure and composition of the spraying chamber atmosphere, temperature, surrounding gas, power, etc.). The major advantage of CAPS is the ability to prevent oxidation and contamination of coatings. This reduces porosity and improve density, uniformity and engineering properties of the coating (Abdel-Samad, 2000). Reactive plasma spraying is an alternative process that promises to eliminate many intermediate processes of powder manufacturing and thus producing wear resistance coatings by incorporating fine and uniformly dispersed carbide phases.

The latest HVOF deposition systems shows that the resulting coatings are more durable in wear applications due to their high hardness, density, low tensile or compressive residual stress when compared to the older HVOF or atmospheric plasma systems (APS). Latest solutions in HVOF thermal spraying technology, e.g. liquid fuel HVOF spraying, have been shown to produce dense coatings with minimum decarburisation and this exhibist superior wear resistance. However, in thermally spray coatings, even those sprayed by the latest liquid fuel JP-5000 systems, partial melting of the particles causes poor bonding at the interfaces between the unmelted and semi-melted particles, and phase composition changes due to in-flight chemical reactions. Very recently, the attention has also been devoted to modifying the high-velocity oxygen-fuel (HVOF) flame spraying technique in order to spray suspension feedstock, with very promising results (Killinger *et al.*, 2006, Ma *et al.* 2006, Gadow *et al.* 2008). In the new high velocity suspension flame spraying (HVSFS) process (Killinger *et al.*, 2006; Gadow *et al.* 2008), where a gas-fuelled HVOF torch has been adapted for liquid feedstock, the problem of feeding the suspension to the jet core is largely solved by the axial injection system of the torch. It has been shown that other problems may arise, like deposition of the suspension inside the combustion chamber resulting in process instabilities and defects in the coatings (Bolelli *et al.* 2008), but these troubles can be overcome, resulting in layers with excellent quality. Moreover, compared to the plasma-spraying technique, low coating defective is also ensured by the very high particle velocity in the gas jet.

Cold spraying is a thermal spray process that fabricates thick, dense, adherent coatings by accelerating small metallicor metal alloy particles (1– 50 μm) to high velocities (300–1200m/s) at low temperatures (Papyrin, 2001). As the temperature of the working gas is low, particle temperatures is much lower than their melting point. The high velocities are attained by using a special nozzle. The high velocities of the sprayed particles give rise to severe plastic deformation of the particles and this results in bonding through adiabatic shear instabilities at the particle-substrate or particle–particle interface during deposition,along with interlocking between the substrate and particles (Grujicic *et al.*, 2004; Assadi *et al.*, 2003; Schmidt *et al.*, 2006; Zou *et al.*, 2010; Hussain *et al.* 2009).

Thermal spray metallic coatings have a lamellar microstructure consisting of splats, which are generated by flattening of molten metal droplets as they hit the surface. There exist oxide layers and inclusions between the splats. The sliding wear of thermal sprayed steel coatings has been attributed to splat delamination (Hartfield-Wunsch & Tung, 1994) due to the 'weak links' caused by the oxide veins (Rabiei *et al.*, 1999). Formation and propagation of subsurface cracks within the oxide veins, resulting in removal of whole splats during the sliding process is generally responsible for the high wear rates. Cermets are intensively used in thermal spraying industry for wear resistance applications. Tungsten carbide based cermets coatings are, the most attractive among these coatings. This is because of a certain degree of ductility of tungsten carbide, in addition to the high hardness, compared to the other carbides. Further, the toughness provided by Co, CoCr, Ni or NiCrBSi matrix, forms not only very hard but also tough cermet system, making it suitable for numerous industrial applications. This resulted in numerous efforts to prepare the cermet coatings containing WC by various techniques such as high-velocity oxyfuel (HVOF), plasma spraying (PS), and detonation gun spraying (DGS) (Aw & Tan, 2006; Yang Q*et al.*, 2006; Du *et al.*, 2007; Ji *et al.*, 2006; Sahraoui *et al.*, 2003). Several researchers have investigated the relationship among process conditions, feedstock powder characteristics, microstructural parameters and wear resistance of thermally sprayed WC–Co coatings (Berget *et al.,* 1998; Wayne *et al.,* 1992; Naerheim *et al.,* 1995; Voyer and Marple, 1999; Usmani *et al.* 1997; Stewart *et al.* 1999; Karimi and Verdon, 1993; Vinayo *et al.* 1985). It has been reported that the starting powder chemistry, powder size, processing condition, initial phase content and carbide grain size determine the phase distribution in the coatings. However, studies relating carbide grain size to the wear performance of the coatings are more limited (Usmani *et al.* 1997; Stewart *et al.* 1999). Thermal-spraying processes are also effective methods to produce nanoscale coatings (Wang *et al.*, 2000; Zeng *et al.* 2002; Zhu *et al.*, 1999). Nanostructured coatings exhibit better tribological properties, than conventional coatings of the same composition. Improved wear resistance, bond strength, and microhardness for nanostructured YPSZ coatings, was reported by Zeng *et al.*(Zeng *et al.* 2002), Chen and coworkers (Chen *et al.,* 2000, 2003a, 2003b, 2002)and Lima and co-workers (Lima *et al.* 2001a, 2001b, 2002).

There are extensive literature on the wear behaviour of plasma sprayed oxide coatings and the corresponding wear mechanisms. Both Cr_2O_3 and Al_2O_3–TiO_2 coatings are reported to form a tribofilm under dry sliding conditions. This film is formed by plastically deformed and compacted wear debris, responsible of the low measured friction coefficients (Bolelli *et al.*, 2006) with the Cr_2O_3 coating. XPS analysis reveals that the film formed at room temperature is constituted by CrO_3 and Cr_2O_3(Ahn & Kwon, 1999). On the contrary, on Al_2O_3–TiO_2 coatings the tribofilm has a rather loose structure which does not adequately protect the underlying material from wear (Bolelli *et al.*, 2006). Roy et al. (Roy *et al.*, 2008) classified such films under five different categories. On these ceramic coatings different wear mechanisms are detected, involving abrasive wear (Vargas *et al.*, 2010; Liu *et al.*, 2009), delamination of weakly adherent successive lamellae (Psyllaki *et al.*, 2001), crack nucleation and delamination (Normand *et al.*, 2000) and adhesive wear (Psyllaki *et al.*, 2001).

The wear resistance of potential cermet coatings in corrosive media strongly depends on the corrosion resistance of the metallic binder. WC–Co cermets undergo a significant corrosion attack dominated by Co phase dissolution, except in alkaline electrolytes where dissolution of WC is also important (Hochstrasser-Kurz *et al.,* 2007). Several efforts have been made to reduce the corrosion susceptibility of cermets by modifying the cermet composition. It was found that using Ni instead of Co as a binder material or alloying Ni into the Co binder phase results in improvement of corrosion resistance (Koon *et al.,* 2008; Bozzini *et al.*, 2002). But this alloying degrades the mechanical properties (*Kny et al.,* 1986). Further, addition of Ni/Cr_3C_2 to the Co binder improves the corrosion resistance (Bozzini *et al.*, 2002). Cr alloying to Co as in WC–CoCr or WC/CrC–CoCr coating does not alter passivity and extensive carbide/matrix interface attack is detected in neutral chloride solution (Souza & Neville, 2003; Perry *et al.*, 2001). On the contrary, both WC–CrNi (Souza and Neville, 2003) and Cr_3C_2–CrNi (Suarez *et al.* 2007; Fedrizzi *et al.* 2007) coatings exhibit passive conditions and low corrosion rates.

The corrosion behaviour of thermally sprayed oxide coatings on steel has been reporteded to a limited extent (Zhang & Kobayashi, 2009; Liu *et al.*, 2009; Zhijian *et al.*, 2009). The clear dependence of the corrosion resistance of the coated materials on the coating porosity has been demonstrated (Celik *et al.*, 1999, 1997).Coating porosity, however can be reduced by a proper choice of the spraying parameters (Sarikaya, 2005; Nusair & Lu, 2009). In some cases, interconnected porosity decreases at increasing coating thickness (Zhang & Kobayashi, 2009). However, the coating porosity is often found to increase with the deposition time and coating thickness (Sarikaya, 2005). In general, corrosion resistance is higher with thin coatings (Celik *et al.*, 1999, 1997). The composition of the sprayed powders also influences porosity of the coatings. For example, ZrO_2 addition to the Al_2O_3 powder tends to increase the coating porosity (Zhang & Kobayashi, 2009), while reverse trend is obtained by TiO_2 addition (Bolelli *et al.*, 2006, Yılmaz *et al.*, 2007). However, literature on tribocorrosion of thermal sprayed coatings is very much limited.

4.0 ISSUES CONTROVERSIES AND PROBLEM

The interaction between mechanical and chemical factors governing tribocorrosion is still little understood and it is still in developmental stage. There is a need to enhance the understanding of the mechanisms and to have greater insight in to critical mechanical, material and chemical factors. It is therefore imperative to control both the mechanical and the chemical test conditions during tribo-corrosion studies. It is possible to control in situ and in real time the surface reactivity of electronic conductive materials employing electrochemical techniques in aqueous ionic electrolytes. Electrochemical techniques are extensively used in corrosion studies. These techniques are increasingly used in tribocorrosion experiments to understand how wear can affect the kinetics of corrosion reactions (Landolt *et al.*, 2001). It is also possible to assess the influence of these reactions on the mechanical behaviour of the contact (Landolt, 2006). However, the interpretation of results obtained in tribo-corrosion tests is much more delicate and complicated than that in the case of pure corrosion. Over the years sufficient experience has been generated on electrochemically controlled tribo-corrosion experiments and it is now possible to compare the different techniques and interpret the data generated.

It is a challenging task to interpret the data from tribo-corrosion experiments. In order to improve the material performance it is important to identify the contribution of corrosion and wear to material removal by tribo-corrosion. The tribo-electrochemical tests have permitted to establish quantitative approaches to describe wear–corrosion interactions. Various approached adopted to interpret tribo-corrosion data can broadly be classified under three heads namely a) Synergistic approach, b) Mechanistic approach and c) Third-body approach. They are briefly discussed below

4.1 Mechanistic Approach

Uhlig (Uhlig, 1954) proposed a simple mechanistic model for fretting corrosion and later others authors applied it to sliding conditions (Mischler *et al.*, 1998, 1999, 2001; Barril *et al.*, 2001,). (1) mechanical wear and (2) wear-accelerated corrosion are two distinct mechanisms which contribute to material deterioration in this approach. The mechanical wear takes place when the metal particles are detached by a hard indenter digging below the metal surface. The surface of clean metal generated by the indenter is more prone to corrosion than the passive metal. This leads to accelerated corrosion on the clean metal until the passive film forms again. Thus, the total volume V_{tot} removed by tribocorrosion according to these mechanisms is given by Equation (7):

$$V_{tot} = V_{mech} + V_{chem} \tag{7}$$

where the term V_{mech} represents the metal volume removed by mechanical wear while V_{chem} corresponds to the equivalent metal volume removed by wear-accelerated corrosion. V_{chem} further consists of two contributions: the corrosion in passive areas and wear accelerated corrosion in depassivated areas. Since the wear accelerated corrosion is usually much more significant, the corrosion in passive areas is neglected. Although V_{mech} and V_{chem} are mechanical and chemical mechanisms, respectively, they are governed by the interplay of mechanical, chemical and material factors. For example, the extent of wear-accelerated corrosion is closely influenced to depassivated surface area that in turn depends on the loading conditions, mechanical and chemical properties of the metal and the abrasivity of the counter piece. On the other hand, corrosion products deposited onto the surface can alter the friction and thus the stress field responsible for mechanical wear.

This model does not take into account the fact that not only the entire rubbed area becomes chemically active (Mischler *et al.*, 1998b; Ponthiaux *et al.*, 2004) but corrosion enhancement due to rubbing can also occur outside the rubbed area depending on the mechanical and electrochemical conditions. Further, surface chemistry and thus reactivity can change due to contact transformations during rubbing (Ponthiaux *et al.*; 2004, Barril *et al.*, 2005). Consequently, many phenomena that are observed such as the build-up of third bodies cannot be anticipated using this simple model.

4.2 The Synergistic Approach

Total materials loss T due to tribocorrosion is described as the sum of three components in this approach according to Equation (6), where C is material loss due to corrosion in absence of wear, W is material loss due to pure wear in absence of corrosion and S is the incremental factor of degradation due to the combined effect of corrosion and wear (synergy).

$$T = C + W + S \tag{6}$$

The term S can be further devided into two more terms: the effect of wear on corrosion and the effect of corrosion on wear. This approach was used for ranking of materials (Watson *et al.*, 1995; Assi & Bo, 1999) establishment of tribo-corrosion maps (Stack & Chi, 2003; Jiang & Stack, 2006) and for the evaluation of the synergistic term S (Assi & Bo, 1999; Yan *et al.*, 2006).

Despite its merits, this approach does not allow a direct mechanistic interpretation of the critical phenomena. Further, current experimental techniques do not allow measuring separately the three contributions and thus arbitrarily chosen external references for mechanical wear and corrosion are needed. As a consequence, the determination of the synergistic term depends critically on the reference taken. Finally, the prevailing electrochemical conditions affect the mechanical behaviour of the metal and thus wear. This renders conceptually difficult to isolate pure mechanical or pure electrochemical phenomena in tribocorrosion such as W or C and to find external references for it.

4.3 Third-Body Approach

During tribocorrosion experiments, third bodies are formed by the accumulation between the contacting bodies of debris particles detached by wear from one or both contacting surfaces. These third bodies play animportant role in the wear process and phenomenological models describing their build-up and elimination were described previously(Godet *et al.*, 1991). The formation of third bodies is very important in tribocorrosion. A small change in surface chemistry can influence the surface mechanical behaviour of passive metals considerably. Different wear mechanisms yield the different amount of mechanical wear. Mischler *et al*.(Mischler *et al.* 2001) postulated that the system behaviour was controlled by the formation of a third body composed by detached metal particles trapped within the contact to address this problem. While maintained in the contact the particles experience breakdown, corrosion, and transfer and, eventually, they are ejected from the contact once they have reached a critical size. There are two different paths through which materials are removed. Sometimes the particles are directly ejected and ions are detached from the metal as postulated in mechanistic approach model. The second one involves the transfer of detached metal particles to the third body and, after transformations occurring during their permanence within the contact, their ejection from the contact as particles (or ion). These mechanisms are controlled by different parameters. The ejection of particles depends on the openness (geometry, stiffness) and kinematics (vibrations) of the contact. It depends thus on the specific tribometer and it is not necessarily linked to the mechanical behaviour of the metal or of the chemical properties of the solution. Landolt *et al.*(Landolt *et al.*, 2004) attempted to include electrochemical aspects in the third-body phenomenological model proposed by Godet *et al.* (Godet *et al.*, 1991) and this in order to describe tribo-corrosion phenomena in term of material flows. By analysing published tribo-corrosion results, they concluded that the third-body approach constitutes a promising way to systematically appraise the effect of different mechanical, material, chemical, electrochemical, physical and system-dependant factors on tribo-corrosion. In the light of the complexity of tribological and tribo-corrosion phenomena, such a benefit is very welcomed. However, the approach still does not offer the possibility of quantitative interpretation or predictions.

5.0 PRESENT UNDERSTANDING OF TRIBOCORROSION OF THERMAL SPRAY COATING

In this section an attempt will be made to examine the state of art of tribo-corrosion of thermal sprayed coating. Under tribo-corrosion conditions, WC-based coatings have mainly shown abrasive–adhesive wear mechanisms. Moreover, due to their relatively low fracture toughness, cracks can develop during both initial surface grinding and subsequent tribo-corrosion tests and this can induce substrate corrosion (Fedrizzi *et al.*, 2007). Fedrizzi *et al.*(Fedrizzi *et al.*, 2004) studied the tribo-corrosion mechanisms of the cermet coatings in a sodium chloride solution under sliding wear. They combined electrochemical measurement with weight loss measurement to estimate specific wear rate K_a as given below.

$$K_a = \frac{\Delta V}{S \times L} \tag{8}$$

Where ΔV is the volume loss in m^3, S is the total sliding distance in m and L is the applied load in N. The values of K_a as determined by them under current control anodic polarisation and to potential control anodic polarisation are presented in Figure 2. When chromium is alloyed into the cobalt matrix, tribo-corrosion behaviour of WC-based systems is improved and lower tribo-corrosion rates are obtained. Cr_3C_2-based coatings containing up to 40% Ni–Cr matrix exhibit higher fracture toughness than WC–Co systems. Coating cracking during surface grinding and tribo-corrosion tests is reduced. Thus, these coatings offer enhanced barrier protection towards substrate corrosion. However, the wear corrosion response of Cr_3C_2-based coatings is inferior to that of WC–Co coatings, owing to their lower hardness. It is to be stated that nickel–chromium matrix allows easier and faster repassivation than cobalt matrix, even when coating is subjected to wear.

Basak *et al.* (Basak *et al.*, 2006) investigated the corrosion–wear behaviour of a thermal sprayed nanostructured FeCu/WC–Co coating and compared the tribo-corrosion behaviour with stainless steel AISI 304 and nanostructured WC–Co coatings. Electrochemical noise and potentiodynamic polarization measurements were conducted. Corundum balls with a diameter of 10 mm, a hardness of 2000 HVN, and a surface roughness $R_a < 0.02$ μm, were used as counter body. Under corrosion–wear conditions, the nanostructured FeCu/WC–Co coating exhibits a depassivation/repassivation behaviour comparable to the behaviour of stainless steel AISI 304 and nanostructured WC–Co coatings. A potential drop and a current rise take place at the start of fretting followed by a steady state condition and finally a potential rise and current drop at the end of fretting tests. The extent current evolutions before and during fretting is illustrated in Figure 3.

Monticelli *et al.*(Monticelli *et al.*, 2011) investigated the tribo-corrosion behaviour of thermally sprayed ceramic coatings deposited on steel specimens. The coating was exposed to a 3.5% NaCl solution. The coatings have been obtained by plasma spraying Cr_2O_3 and Al_2O_3+13% TiO_2 powders on a Ni/20% Cr bond coating. Combined wear corrosion conditions have been monitored by sliding an alumina counter body on the lateral surface of coated steel cylinders, in 3.5% NaCl solution. Sliding increases corrosion rates of the coating marginally as a consequence of an increase in the defect population of the ceramic coatings. On Cr_2O_3 coated specimens, tribo-corrosion is more severe than on Al_2O_3+13% TiO_2 coated specimens Profilometer analysis and wear track observations by optical and scanning electron microscopes supports that abrasion of the surface asperities produce surface polishing effect on both coatings. At high loads, the formation of a tribofilm is noted. This film is more continuous on Al_2O_3+13% TiO_2 film than Cr_2O_3 film. On this coating the tribofilm reduces the amount of surface defects and the corrosion attack to a certain extent. This observation is just opposite to what is noted for these coatings during sliding wear in the absence of corroding fluid (Bolelli *et al.*, 2006).

The polarization curves of the coatings were recorded by Monticelli *et al.* (Monticelli *et al.*, 2011) at the end of the immersion period in the neutral chloride solution. Figure 4a and Figure 4b report the corresponding E_{COR} and i_{COR} values of the coatings, as estimated by the Tafel method by extrapolation from the cathodic polarization curve. Al_2O_3+13% TiO_2 coating exhibits higher anodic current densities than Cr_2O_3 coating, while the cathodic polarization curves and the i_{COR} values are quite close to each other. The differences in the anodic curve suggest that the slightly higher porosity of Al_2O_3+13% TiO_2 coatings facilitates the migration of iron corrosion products through the coatings driven by the anodic polarization. The presence of quite similar cathodic characteristics and i_{COR} values suggests that on these specimens the corrosion process is mainly under cathodic control. Wear application on both coatings slightly stimulates both the anodic and the cathodic reactions of

Figure 2. Specific wear rate under current control anodic polarisation and to potential control anodic polarisation (Fedrizzi et al., 2007)

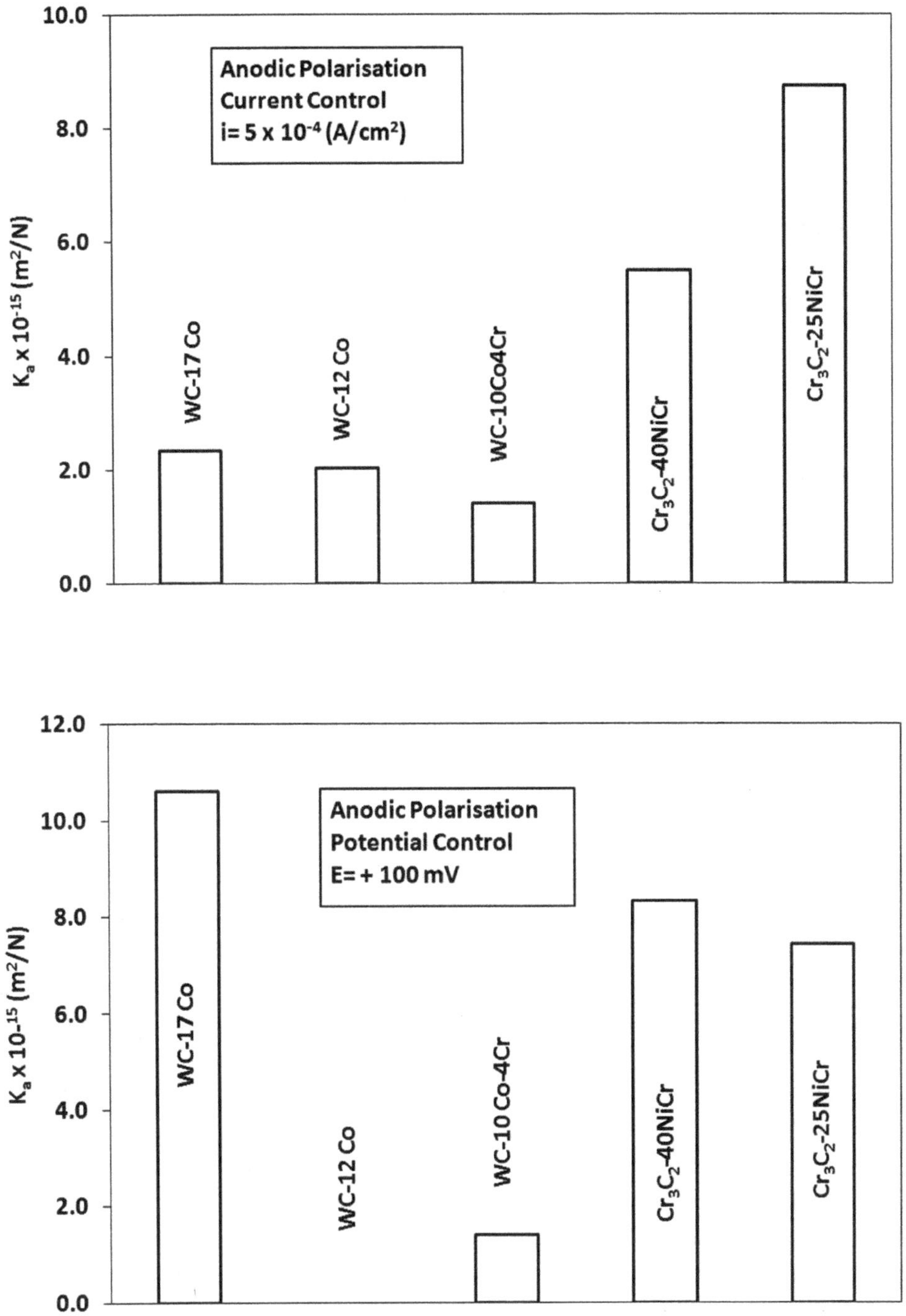

Figure 3. The extent current evolutions before and during fretting(Basak et al., 2006)

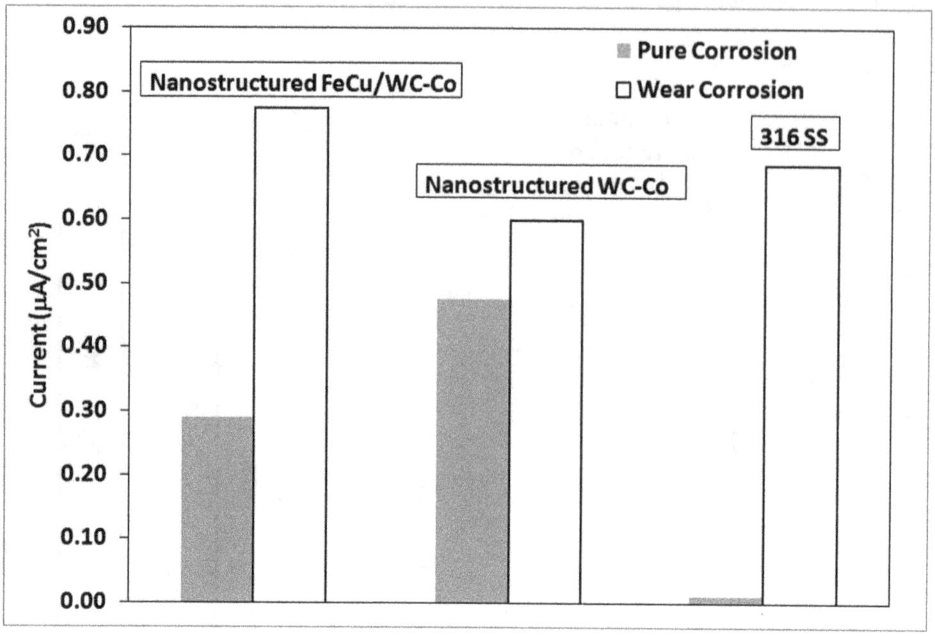

steel corrosion. In the case of Cr_2O_3 coating, i_{COR} is twice the value measured under pure corrosion conditions. On the contrary, in the case of Al_2O_3+13% TiO_2 coating, tribo-corrosion under 2 N and 20 rpm is more severe than that experienced at 10 N load and 100 rpm.

Polarization curve analysis suggests that the corrosion of Cr_2O_3 and Al_2O_3+13% TiO_2coated specimens is mainly under cathodic control. This means that the substrate corrosion rate depends on the rate of oxygen diffusion towards the metallic surfaces through the coating pores, which in turn is affected by the coating thickness, the defect population and the interconnection degree of the defects themselves. On the contrary, it is noted that corrosion rate is independent of rotational speed. This can be attributed to the fact that the oxygen diffusion rate inside the pores cannot be altered by higher electrode rotation speeds. Sliding stimulates both the anodic and the cathodic process on both Cr_2O_3 and Al_2O_3+13% TiO_2 coatings and responsible for limited increase of the corrosion rates. The dependence of corrosion and wear rate of the coating on the wear condition can be explained by taking the modification of the coating surfaces in the track in to account as a result of wear. It has been noticed that on both coatings abrasion of surface asperities occurs, with a consequent surface polishing. This smoothing effect is more evident at high load and rotation speed and particularly on Al_2O_3+13% TiO_2 coating, because it is characterized by the lowest microhardness value. This justifies the lower i_{COR} value

Figure 4. ECOR and iCOR values of the Cr2O3 and Al2O3+13% TiO2coatings, as estimated by the Tafel method by extrapolation from the cathodic polarization curve in 3.5% NaCl solutions under pure corrosion and wear corrosion condition (Monticelli et al., 2011)

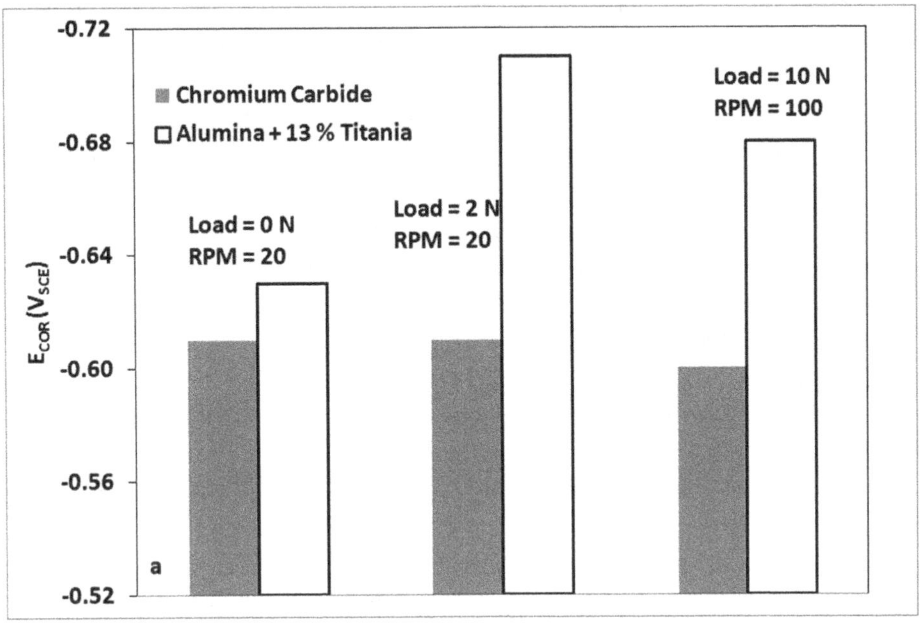

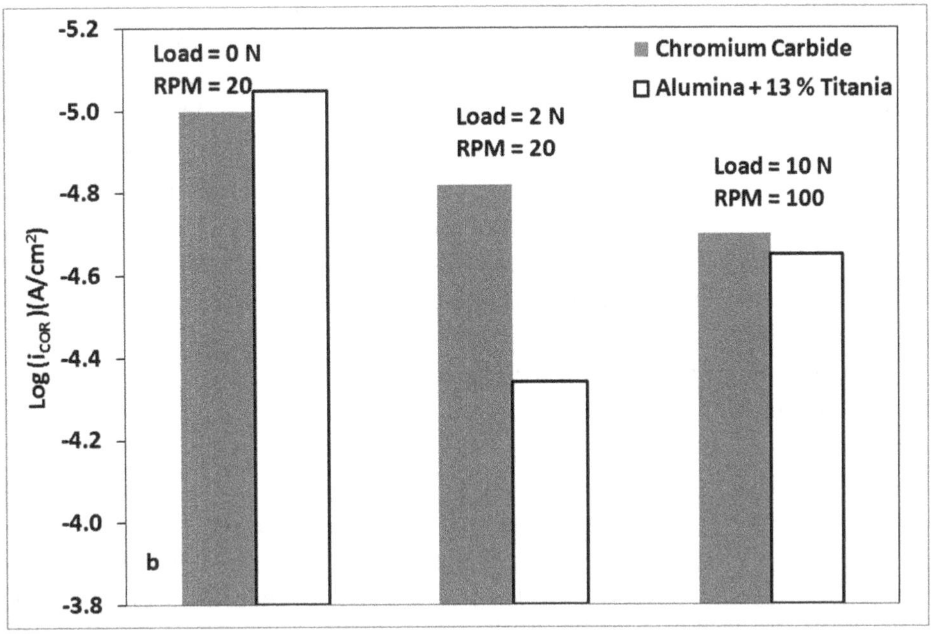

measured on Al_2O_3+13% TiO_2 coatings under these experimental conditions, with respect to those evaluated at 2 N and 20 rpm. Microscale plastic deformation of Cr_2O_3 and Al_2O_3/13% TiO_2 coatings under dry sliding wear has been often reported in the literature (Bolelli *et al.*, 2006; Ahn & Kwon, 1999; Fervel *et al.* 1999) and it has been attributed to plastic slipping at splat boundaries. The progressive polishing effect observed under sliding wear may justify the decrease of the wear rates, going from low to high rotation speeds, as the variation of this parameter implies an increase in the sliding distances which in turn involves smoother and smoother wear paths. A decrease in the friction coefficient going from 20 to 100 rpm could also occur and play a role in reducing the measured wear rates (Woydt & Wäsche, 2010; Monticelli *et al.*, 2004). The observed formation of a tribofilm is the cause of the significant decrease of the wear rates at loads higher than 2 N. In fact when the load applied is sufficiently high, the smearing of the wear debris on thecoating surfaces limits the material removal.

Monticelli *et al.*(Monticelli *et al.*, 2010) studied the tribo-corrosion behaviour of thermally sprayed coatings with nominal composition WC–12Co and Cr_3C_2–37WC– 18Me in 3.5% NaCl solutions. It is to be stated that Me stands for is a metallic Cr–Ni–W–Co alloy containing 3.6% Co, 11.5% Ni, balance W and Cr. Both kinds of coatings exhibit high tribo-corrosion rates which increase with time owing to the partial removal of the cobalt oxide layer in the case of the former coating and the abrasion of the passive film in the case of the latter coating. Both carbide decomposition and carbide dissolution in the matrix are known to hinder the material ductility of Cr_3C_2-37WC–18Me coating. This favours the development of pre-existing cracks, nucleated during the surface grinding process, and corrosion. It is to be stated that the choice of a slightly coarser powder and selection of suitable spraying parameters, such as spraying distance and O_2/H_2 ratio etc. could reduce the diffusion processes and carbide decomposition and this would have monitored the development of cracks. After 1 day, tribo-corrosion rates reach a common limit value due to the accumulation of cobalt corrosion products within grooves in the wear track of WC–12Co coating. In the case of Cr_3C_2–37WC–18Me coating, the limit corrosion rate is measured when a constant crack population is reached. Under these conditions none of the studied thermal spray coatings can avoid the substrate corrosion.

The observation of Monticelli *et al.* (Monticelli *et al.*, 2010) is illustrated in Figure 5 in the form of bar diagram showing the corresponding E_{COR} and i_{COR} values of the coatings. WC–12Co coated specimens show a i_{COR} value slightly lower than that of bare steel. This current value suggests that the coating oxidation forms a porous corrosion product layer. As WC reinforcement is an electro-conducting material which undergoes oxidation only at potentials more positive than +0.190 V [109] and at the E_{COR} value of the WC–12Co coated steel electrodes is about –0.320 V, the reinforcement is immune from corrosion. Thus the anodic reaction process of the coated specimens is essentially sustained by the oxidation of the cobalt matrix (Monticelli *et al.*, 2004). Cr_3C_2–37WC–18Me coating exhibits a passive corrosion behaviour with much lower i_{COR} values and nobler corrosion potentials. Like WC, Cr_3C_2 is also immune from corrosion (Fedrizzi *et al.*, 2004; 2007; Guilemany *et al.*, 2002) and the coating polarization curves reflect the passive behaviour of the Ni–Cr matrix. Wear stimulates both the anodic and the cathodic process of all studied materials. On WC–12Co coatings, i_{COR} increases only slightly and in the case of the passive Cr_3C_2–37WC–18Me coating the i_{COR} increases by a factor of 3–4; From Figure 5, it can be inferred that corrosion of WC–12Co coating occurs by cobalt oxidation and formation of a scarcely protective corrosion product layer. On the other hand, abrasion of the passive film on Cr_3C_2–37WC–18Me cermet coating, effectively hinders the material corrosion, and markedly stimulates i_{COR}.

6.0 CONCLUDING REMARK

Electrochemical techniques can be used to govern the corrosion processes of thermal sprayed coatings and to interpret the main degradation mechanisms. Electrochemical techniques are well suited for the investigation of tribo-corrosion response of thermal sprayed coatings, because they permit one to monitor and control the corrosion conditions during wear and to quantify in situ and in real time the corrosion kinetics. The strong interaction of wear and chemical effects in tribo-corrosion makes difficult to isolate pure mechanical effects. This technique is useful for studying tribo-corrosion of thermal sprayed coatings irrespective of whether the coating is conductive or not. Depending on the conductivity of the coating, the counter body will be conductive or non conductive.

The use of electrochemical methods for characterisation of tribo-corrosion behaviour of thermal sprayed coatings has few limitations, related to the particular conditions of tribo-corrosion tests. These limitations are mainly due to the heterogeneous state of the surface subjected to sliding, and the time evolution of the rubbed surfaces resulting from the wear process.

Greater characterisation of the electrochemical conditions prevailing in contacts of thermal sprayed coating and the counter body is needed to improve the reliability and the general applicability of tribo-electrochemical techniques. In particular, the relevance of ohmic drop effects, secondary electrochemical reactions and oxidation kinetics during active/passive transitions needs to be addressed. Theoretical models need therefore to consider only measurable phenomena and to avoid the need for reference experiments carried out under different chemical conditions. Theoretical models were developed to describe wear-accelerated corrosion as a function

Figure 5. The bar diagram showing the corresponding ECOR and iCOR values of WC-12 and Cr3C2-37WC-19Me coatings in 3.5% NaCl solutions under pure corrosion and wear corrosion condition (Monticelli et al., 2010)

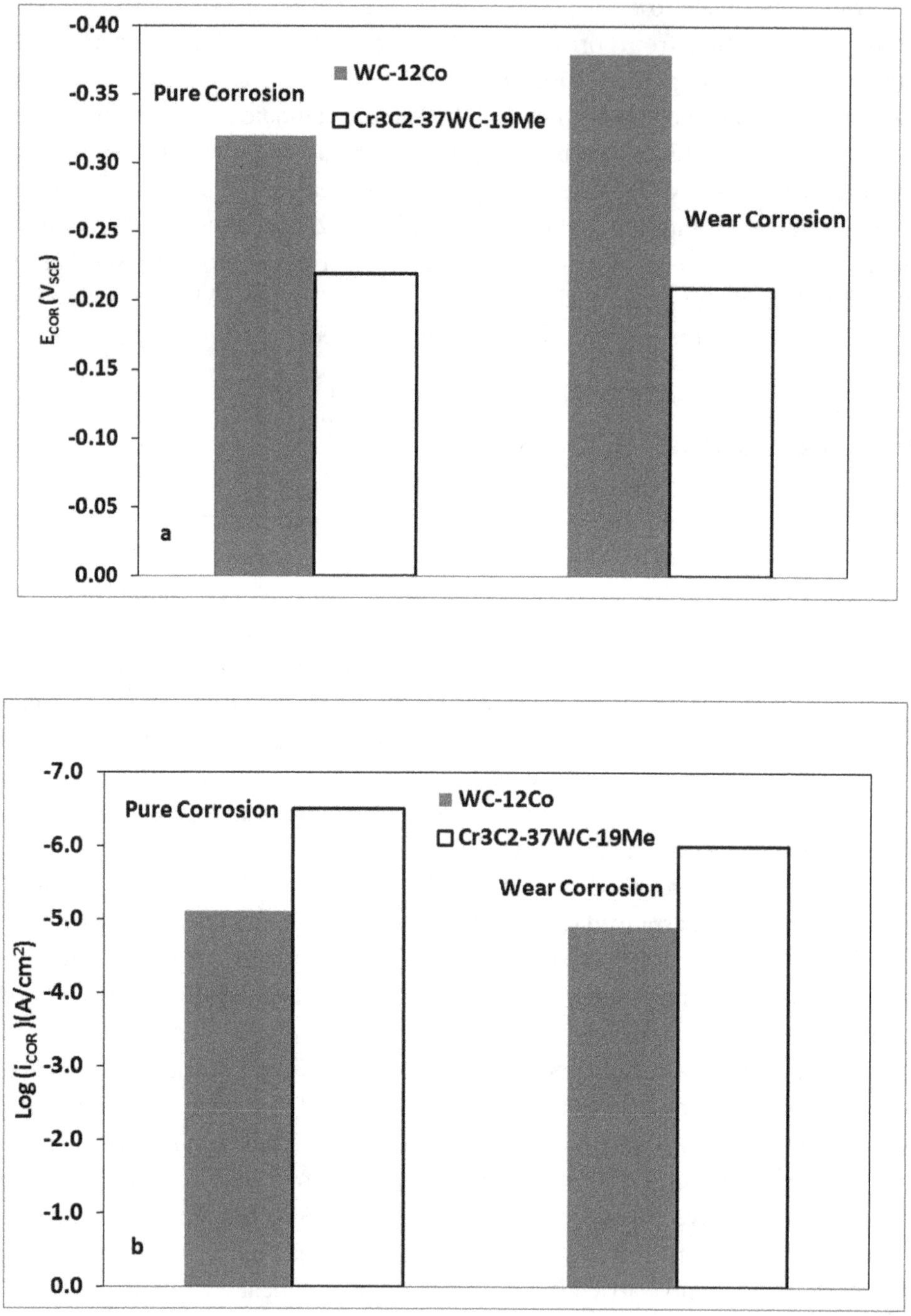

of materials, mechanical and electrochemical parameters. However, hitherto no effort is made to develop such models for thermal sprayed coatings. Such models need to be improved especially for thermal sprayed coatings to describe interaction between corrosion and wear characterised by the formation of third bodies that can greatly influence the corrosion and wear response of tribo-corrosion contacts. Existing phenomenological third body models for thermal sprayed coatings should be completed by a quantitative description of the different material flows occurring within the contact. This will require a better integration in tribo-corrosion modelling of system-dependant mechanical factors such as contact stiffness and vibrational behaviour. Last but not least, more comprehensive data base on tribo-corrosion behaviour of thermal sprayed coating with respect to various coating materials, spraying techniques and spraying condition of each techniques are to be generated.

REFERENCES

Abdel-Samad, A. A., El-Bahloul, A. M. M., Lugscheider, E., & Rassoul, S. A. (2000). A comparative study on thermally sprayed alumina based ceramic coatings. *Journal of Materials Science*, *35*(12), 3127–3130. doi:10.1023/A:1004824104162

Ahn, H. S., & Kwon, O. K. (1999). Tribological behaviour of plasma-sprayed chromium oxide coating. *Wear*, *225–229*, 814–824. doi:10.1016/S0043-1648(98)00390-1

Assadi, H., Gärtner, F., Stoltenhoff, T., & Kreye, H. (2003). Bonding mechanism in cold gas spraying. *Acta Materialia*, *51*(15), 4379–4394. doi:10.1016/S1359-6454(03)00274-X

Assi, F., & Bohni, H. (1999). Study of wear–corrosion synergy with a new microelectrochemical technique. *Wear*, *233–235*, 505–514. doi:10.1016/S0043-1648(99)00237-9

Aw, P. K., & Tan, B. H. (2006). Study of microstructure, phase and microhardness distribution of HVOF sprayed multi-modal structured and conventional WC–17Co coatings. *Journal of Materials Processing Technology*, *174*(1-3), 305–311. doi:10.1016/j.jmatprotec.2006.02.006

Barbezat, G., & Nicoll, A. R. (1993). Abrasion, erosion and scuffing resistance of carbide and oxide ceramic thermal sprayed coatings for different applications. *Wear*, *162*, 529–537. doi:10.1016/0043-1648(93)90538-W

Barradja, A., Bratu, F., Benea, L., Willems, G., & Celis, J.-P. (2006). Effect of sliding wear on tribocorrosion behaviour of stainless steels in a ringer's solution. *Wear*, *261*(9), 987–993. doi:10.1016/j.wear.2006.03.003

Barradja, A., De'forge, D., Nogueira, R. P., Ponthiaux, P., & Celis, J.-P. (2006). An electrochemical noise study of tribocorrosion processes of AISI 304 L in Cl^-□and media. *Journal of Physics. D, Applied Physics*, *39*(15), 3184–3192. doi:10.1088/0022-3727/39/15/S08

Barril, S., Mischler, S., & Landolt, D. (2001). Triboelectrochemical investigation of the friction and wear behaviour of TiN coatings in a neutral solution. *Tribology International*, *34*(9), 599–608. doi:10.1016/S0301-679X(01)00052-4

Barril, S., Mischler, S., & Landolt, D. (2005). Electrochemical effects on the fretting corrosion behaviour of Ti6Al4V in 0.9% sodium chloride solution. *Wear*, *259*(1-6), 282–291. doi:10.1016/j.wear.2004.12.012

Basak, A. K., Matteazzi, P., Vardavoulias, M., & Celis, J.-P. (2006). Corrosion–wear behaviour of thermal sprayed nanostructured FeCu/WC–Co coatings. *Wear*, *261*(9), 1042–1050. doi:10.1016/j.wear.2006.03.026

Benea, L., Ponthiaux, P., Wenger, F., Galland, J., Hertz, D., & Malo, J. Y. (2004). Tribocorrosion of stellite 6 in sulphuric acid medium: Electrochemical behaviour and wear. *Wear*, *256*(9-10), 948–953. doi:10.1016/j.wear.2003.06.003

Benish, M. L., Sikora, J., Shaw, B., Sikora, E., Yaffe, M., Krebs, A., & Martinchek, G. (1998), A new electrochemical noise technique for monitoring the localised corrosion of 304 stainless steel in chloride containing solutions. In *Proceedings of the CORROSION '98* (Paper No. 370). National Association of Corrosion Congress.

Berget, J., Bardal, E., & Ronge, T. (1998). Article. *Surface Modification Technology*, *11*, 501.

Bielmann, M., Mahajan, U., Singh, R. K., Agarwal, P., Mischler, S., Rosset, E., & Landolt, D. (1999). Tribological experiments applied to tungstein chemical mechanical polishing. In *Material Research Symposium Proceedings*. Academic Press. doi:10.1557/PROC-566-97

Bolelli, G., Cannillo, V., Lusvarghi, L., & Manfredini, T. (2006). Wear behaviour of thermally sprayed ceramic oxide coatings. *Wear*, *261*(11-12), 1298–1315. doi:10.1016/j.wear.2006.03.023

Bolelli, G., Rauch, J., Cannillo, V., Killinger, A., Lusvarghi, L., & Gadow, R. (2008). Investigation of high-velocity suspension flame sprayed (HVSFS) glass coatings. *Materials Letters*, *62*(17-18), 2772–2775. doi:10.1016/j.matlet.2008.01.049

Bozzini, B., De Gaudenzi, G. P., Serra, M., Fanigliulo, A., & Bogani, F. (2002). Corrosion behaviour of WC-Co based hardmetal in neutral chloride and acid sulphate media. *Werkstoffe und Korrosion*, *53*(5), 328–334. doi:10.1002/1521-4176(200205)53:5<328::AID-MACO328>3.0.CO;2-G

Celik, E., Demirkıran, A. S., & Avc, E. (1999). Effect of grit blasting of substrate on the corrosion behaviour of plasma-sprayed Al_2O_3 coatings. *Surface and Coatings Technology*, *116–119*, 1061–1064. doi:10.1016/S0257-8972(99)00238-8

Celik, E., Sengil, I. A., & Avc, E. (1997). Effects of some parameters on corrosion behaviour of plasma-sprayed coatings. *Surface and Coatings Technology*, *97*(1-3), 355–360. doi:10.1016/S0257-8972(97)00208-9

Chen, H., & Ding, C. X. (2002). Nanostructured zirconia coating prepared by atmospheric plasma spraying. *Surface and Coatings Technology*, *150*(1), 31–36. doi:10.1016/S0257-8972(01)01525-0

Chen, H., Xu, C., Qu, J., Hutchings, I. M., Shipway, P. H., & Lie, J. (2005). Sliding wear behaviour of laser clad coatings based upon a nickel-based self-fluxing alloy co-deposited with conventional and nanostructured tungsten carbide–cobalt hardmetals. *Wear*, *259*(7-12), 801–806. doi:10.1016/j.wear.2005.02.066

Chen, H., Zeng, Y., & Ding, C. (2003). Microstructural characterization of plasma-sprayed nanostructured zirconia powders and coatings. *Journal of the European Ceramic Society*, *23*(3), 491–497. doi:10.1016/S0955-2219(02)00096-1

Chen, H., Zhang, Y., & Ding, C. (2002). Tribological properties of nanostructured zirconia coatings deposited by plasma spraying. *Wear*, *253*(7-8), 885–893. doi:10.1016/S0043-1648(02)00221-1

Chen, H., Zhou, X., & Ding, C. (2003). Investigation of the thermomechanical properties of a plasma-sprayed nanostructured zirconia coating. *Journal of the European Ceramic Society*, *23*(9), 1449–1455. doi:10.1016/S0955-2219(02)00345-X

Chen, J. F., & Bogaerts, W. F. (1996). Electrochemical emission spectroscopy for monitoring uniform and localized corrosion. *Corrosion*, *52*(10), 753–759. doi:10.5006/1.3292068

Diomidis, N., Mischler, S., More, N. S., & Roy, M. (2012). Tribo-Eelectrochemical characterisation of metallic biomaterial for total joint replacement. *Acta Biomaterialia*, *8*(2), 852–859. doi:10.1016/j.actbio.2011.09.034 PMID:22005332

Du, H., Sun, C., Hua, W. G., Wang, T. G., Gong, J., Jiang, X., & Lee, S. W. (2007). Structure, mechanical and sliding wear properties of WC–Co/MoS_2–Ni coatings by detonation gun spray. *Materials Science and Engineering A*, *445-446*, 122–134. doi:10.1016/j.msea.2006.09.011

Eden, D. A., Hladky, K., John, D. G., & Dawson, J. L. (1986). Simultaneous monitor-ingof potential and current noise signals from corroding electrodes. In *Proceedings of the CORROSION'86* (Paper 274). National Association of Corrosion Engineers.

Fedrizzi, L., Rossi, S., Cristel, R., & Bonora, P. L. (2004). Corrosion and wear behaviour of HVOF cermet coatings used to replace hard chromium. *Electrochimica Acta*, *49*(17-18), 2803–2814. doi:10.1016/j.electacta.2004.01.043

Fedrizzi, L., Valentinelli, L., Rossi, S., & Segna, S. (2007). Tribocorrosion behaviour of HVOF cermet coatings. *Corrosion Science*, *49*(7), 2781–2799. doi:10.1016/j.corsci.2007.02.003

Fervel, V., Normand, B., & Coddet, C. (1999). Tribological behavior of plasma sprayed Al_2O_3-based cermet coatings. *Wear*, *230*(1), 70–77. doi:10.1016/S0043-1648(99)00096-4

Gadow, R., Killinger, A., & Rauch, J. (2008). New results in high velocity suspension flame spraying (HVSFS). *Surface and Coatings Technology*, *202*(18), 4329–4336. doi:10.1016/j.surfcoat.2008.04.005

Galliano, F., Galvanetto, E., Mischler, S., & Landolt, D. (2001). Tribocorrosion behavior of plasma nitrided Ti–6Al–4V alloy in neutral NaCl solution. *Surface and Coatings Technology*, *145*(1-3), 121–131. doi:10.1016/S0257-8972(01)01309-3

Godet, M., Berthier, Y., Lancaster, J., & Vincent, L. (1991). Wear modelling: Using fundamental understanding or practical experience. *Wear*, *149*(1-2), 325–345. doi:10.1016/0043-1648(91)90383-6

Grujicic, M., Zhao, C. L., DeRosset, W. S., & Helfritch, D. (2004). Adiabatic shear instability based mechanism for particles/substrate bonding in the cold-gas dynamic-spray process. *Materials & Design*, *25*(8), 681–688. doi:10.1016/j.matdes.2004.03.008

Gruner, R. (1986), Vacuum plasma sprayed composite coatings: Advances in thermal spraying. In *Proceedings of the Eleventh International Thermal Spraying Conference (ITSC86)*. Montreal, Canada: ASM International.

Guilemany, J. M., Fernandez, J., Delgado, J., Benedetti, A. V., & Climent, F. (2002). Effects of thickness coating on the electrochemical behaviour of thermal spray Cr_3C_2–NiCr coatings. *Surface and Coatings Technology*, *153*(2-3), 107–113. doi:10.1016/S0257-8972(01)01679-6

Hartfield-Wunsch, S. E., & Tung, S. C. (1994). The effect of microstructure on the wear behaviour of thermal spray coatings. In C. C. Berndt & S. Sampath (Eds.), *Thermal spray industrial applications* (pp. 19–29). Materials Park, OH: ASM International.

Hochstrasser-Kurz, S., Mueller, Y., Latkoczy, C., Virtanen, S., & Schmutz, P. (2007). Analytical characterization of the corrosion mechanisms of WC–Co by electrochemical methods and inductively coupled plasma mass spectroscopy. *Corrosion Science*, *49*, 2002–2020. doi:10.1016/j.corsci.2006.08.022

Hussain, T., McCartney, D. G., Shipway, P. H., & Zhang, D. (2009). Bonding mechanisms in cold spraying: The contributions of metallurgical and mechanical components. *Journal of Thermal Spray Technology*, *18*(3), 364–379. doi:10.1007/s11666-009-9298-1

Jacobs, L., Hyland, M. M., & De Bonte, M. (1999). Study of the influence of microstructural properties on the sliding-wear behavior of HVOF and HVAF sprayed WC-cermet coatings. *Journal of Thermal Spray Technology*, *8*(1), 125–132. doi:10.1361/105996399770350656

Ji, G. C., Li, C. J., Wang, Y. Y., & Li, W. Y. (2006). Microstructural characterization and abrasive wear performance of HVOF sprayed Cr_3C_2–NiCr coating. *Surface and Coatings Technology*, *200*(24), 6749–6757. doi:10.1016/j.surfcoat.2005.10.005

Jiang, J., & Stack, M. M. (2006). Modelling sliding wear: From dry to wet environments. *Wear*, *261*(9), 954–965. doi:10.1016/j.wear.2006.03.028

Karimi, A., & Verdon, C. (1993). Hydroabrasive wear behaviour of high velocity oxyfuel thermally sprayed WC-M coatings. *Surface and Coatings Technology*, *62*(1-3), 493–498. doi:10.1016/0257-8972(93)90289-Z

Killinger, A., Kuhn, M., & Gadow, R. (2006). High-velocity suspension flame spraying (HVSFS), a new approach for spraying nanoparticles with hypersonic speed. *Surface and Coatings Technology*, *201*(5), 1922–1929. doi:10.1016/j.surfcoat.2006.04.034

Kny, E., Bader, T., Hohenrainer, Ch., & Schmid, L. (1986). Korrosionsresistente, hochverschleißfeste Hartmetalle. *Werkstoffe und Korrosion*, *37*(5), 230–235. doi:10.1002/maco.19860370505

Koon Aw, P., & Kuan Tan, A. (2008). Corrosion resistance of tungsten carbide based cermet coatings deposited by high velocity oxy-fuel spray process. *Thin Solid Films*, *516*(16), 5710–5715. doi:10.1016/j.tsf.2007.07.065

Landolt, D. (2006). Electrochemical and materials aspects of tribo-corrosion systems. *Journal of Physics. D, Applied Physics*, *39*(15), 3121–3127. doi:10.1088/0022-3727/39/15/S01

Landolt, D., Mischler, S., Stemp, M., & Barril, S. (2004). Third body effects and material fluxes in tribocorrosion systems involving a sliding contact. *Wear*, *256*(5), 517–524. doi:10.1016/S0043-1648(03)00561-1

Landolt, D., Stemp, M., & Mischler, S. (2001). Electrochemical methods in tribo-corrosion: A critical appraisal. *Electrochimica Acta*, *46*(24-25), 3913–3929. doi:10.1016/S0013-4686(01)00679-X

Lima, R. S., Kucuk, A., & Berndt, C. C. (2001). Integrity of nanostructured partially stabilized zirconia after plasma spray processing. *Materials Science and Engineering A*, *313*(1), 75–82. doi:10.1016/S0921-5093(01)01146-7

Lima, R. S., Kucuk, A., & Berndt, C. C. (2001). Evaluation of microhardness and elastic modulus of thermally sprayed nanostructured zirconia coatings. *Surface and Coatings Technology*, *135*(2-3), 166–172. doi:10.1016/S0257-8972(00)00997-X

Lima, R. S., Kucuk, A., & Berndt, C. C. (2002). Bimodal distribution of mechanical properties on plasma sprayed nanostructured partially stabilized zirconia. *Materials Science and Engineering A*, *327*(2), 224–232. doi:10.1016/S0921-5093(01)01530-1

Liu, H., Tao, J., Xu, J., Chen, Z., & Gao, Q. (2009). Corrosion and tribological behaviors of chromium oxide coatings prepared by the glow-discharge plasma technique. *Surface and Coatings Technology*, *204*(1-2), 28–36. doi:10.1016/j.surfcoat.2009.06.020

Ma, X. Q., Roth, J., Gandy, D. W., & Frederick, G. J. (2006). A new high-velocity oxygen fuel process for making finely structured and highly bonded inconel alloy layers from liquid feedstock. *Journal of Thermal Spray Technology*, *15*(4), 670–675. doi:10.1361/105996306X147090

Mischler, S. (2008). Triboelectrochemical techniques and interpretation methods in tribocorrosion: A comparative evaluation. *Tribology International*, *41*(7), 573–583. doi:10.1016/j.triboint.2007.11.003

Mischler, S., Debaud, S., & Landolt, D. (1998). Wear-accelerated corrosion of passive metals in tribocorrosion systems. *Journal of the Electrochemical Society*, *145*(3), 750–758. doi:10.1149/1.1838341

Mischler, S., & Ponthiaux, P. (2001). A round robin on combined electrochemical and friction tests on alumina/stainless steel contacts in sulphuric acid. *Wear*, *248*(1-2), 211–225. doi:10.1016/S0043-1648(00)00559-7

Mischler, S., Spiegel, A., & Landolt, D. (1999). The role of passive oxide films on the degradation of steel in tribocorrosion systems. *Wear*, *225–229*, 1078–1087. doi:10.1016/S0043-1648(99)00056-3

Mischler, S., Spiegel, A., Stemp, M., & Landolt, D. (2001). Influence of passivity on the tribocorrosion of carbon steel in aqueous solutions. *Wear*, *251*(1-12), 1295–1307. doi:10.1016/S0043-1648(01)00754-2

Monticelli, C., Balbo, A., & Zucchi, F. (2010). Corrosion and tribocorrosion behaviour of cermet and cermet/nanoscale multilayer CrN/NbN coatings. *Surface and Coatings Technology*, *204*(9-10), 1452–1460. doi:10.1016/j.surfcoat.2009.09.046

Monticelli, C., Balbo, A., & Zucchi, F. (2011). Corrosion and tribocorrosion behaviour of thermally sprayed ceramic coatings on steel. *Surface and Coatings Technology*, *205*(12), 3683–3691. doi:10.1016/j.surfcoat.2011.01.023

Monticelli, C., Frignani, A., & Zucchi, F. (2004). Investigation of corrosion process of carbon steel coated by HVOFWC-Co cermet in neutral solution. *Corrosion Science*, *46*(5), 1225–1237. doi:10.1016/j.corsci.2003.09.013

More, N., Diomidis, N., Paul, S. N., Roy, M., & Mischler, S. (2011). TriboCorrosion behaviour of β titanium alloys in physiological solution containing joint lubricants. *Materials Science and Engineering C*, *31*(2), 400–408. doi:10.1016/j.msec.2010.10.021

Naerheim, Y., Coddet, C., & Droit, P. (1995). Effect of thermal spray process selection on tribological performances WC–Co and Al_2O_3–TiO_2 coatings. *Surface Engineering*, *11*(1), 66–70. doi:10.1179/sur.1995.11.1.66

Normand, B., Fervel, V., Coddet, C., & Nikitine, V. (2000). Tribological properties of plasma sprayed alumina–titania coatings: Role and control of the microstructure. *Surface and Coatings Technology*, *123*(2-3), 278–287. doi:10.1016/S0257-8972(99)00532-0

Nusair Khan, A., & Lu, J. (2009). Manipulation of air plasma spraying parameters for the production of ceramic coatings. *Journal of Materials Processing Technology*, *209*(5), 2508–2514. doi:10.1016/j.jmatprotec.2008.05.045

Papageorgiou, N., & Mischler, S. (2012). Electrochemical Simulation of the Current and Potential Response in Sliding Tribocorrosion. *Tribology Letters*, *48*(3), 271–283. doi:10.1007/s11249-012-0022-9

Papyrin, A. (2001). Cold spray technology. *Advanced Materials and Processes*, *159*(9), 49–51.

Perry, J. M., Neville, A., Wilson, V. A., & Hodgkiess, T. (2001). Assessment of the corrosion rates and mechanisms of a WC–Co–Cr HVOF coating in static and liquid–solid impingement saline environments. *Surface and Coatings Technology*, *137*(1), 43–51. doi:10.1016/S0257-8972(00)01062-8

Ponthiaux, P., Wenger, F., Drees, D., & Celis, J.-P. (2004). Electrochemical techniques for studying tribocorrosion processes. *Wear*, *256*(5), 459–468. doi:10.1016/S0043-1648(03)00556-8

Rabiei, A., Mumm, D. R., Hutchinson, J. W., Schweinfest, R., Ruhle, M., & Evans, A. G. (1999). Microstructure, deformation and cracking characteristics of thermal spray ferrous coatings. *Materials Science and Engineering A*, *269*(1-2), 152–165. doi:10.1016/S0921-5093(99)00132-X

Roy, M., Pauschitz, A., Polak, R., & Franek, F. (2008). High temperature wear of thermal sprayed Cr3C2-25(Ni-20Cr) coating against Fe based and Ni based mating surfaces. In *Proc. 16th Intl. Colloquium Tribology*. Stuttgart, Germany: Academic Press.

Sahraoui, T., Fenineche, N. E., Montavon, G., & Coddet, C. (2003). Structure and wear behaviour of HVOF sprayed Cr_3C_2–NiCr and WC–Co coatings. *Materials & Design*, *24*(5), 309–313. doi:10.1016/S0261-3069(03)00059-1

Sampath, S. (Ed.). (1988). *Proceedings of the national thermal spray conference*. Cincinnati, OH: ASM International.

Sarikaya, O. (2005). Effect of some parameters on microstructure and hardness of alumina coatings prepared by the air plasma spraying process. *Surface and Coatings Technology*, *190*(2-3), 388–393. doi:10.1016/j.surfcoat.2004.02.007

Savarimuthu, A. C., Taber, H. F., Megat, I., Shadley, J. R., Rybicki, E. F., Cornell, W. C., & Nuse, J. D. et al. (2001). Sliding wear behaviour of tungstein carbide thermal spray coatings for replacement of chromium elctroplate in aircraft applications. *Journal of Thermal Spray Technology*, *10*(3), 502–510. doi:10.1361/105996301770349286

Schmidt, T., Gärtner, F., Assadi, H., & Kreye, H. (2006). Development of a generalized parameter window for cold spray deposition. *Acta Materialia*, *54*(3), 729–742. doi:10.1016/j.actamat.2005.10.005

Serre, I., Celati, N., & Pradeilles-Duval, R. M. (2002). Tribological and corrosion wear of graphite ring against Ti_6Al_4V disk in artificial sea water. *Wear*, *252*(9-10), 711–718. doi:10.1016/S0043-1648(02)00030-3

Serre, I., Pradeilles-Duval, R. M., & Celati, N. (2006). Tribological and corrosion experiments of graphite ring against Ti–6Al–4V disk: Influence of electrochemical and mechanical parameters. *Wear*, *260*(9-10), 1129–1135. doi:10.1016/j.wear.2005.07.007

Shenhar, A., Gotmana, I., Radinb, S., Ducheyne, P., & Gutmanasa, E. Y. (2000). Titanium nitride coatings on surgical titanium alloys produced by a powder immersion reaction assisted coating method: Residual stresses and fretting behavior. *Surface and Coatings Technology*, *126*(2-3), 210–218. doi:10.1016/S0257-8972(00)00524-7

Souza, V. A. D., & Neville, A. (2003). Linking electrochemical corrosion behaviour and corrosion mechanisms of thermal spray cermet coatings (WC–CrNi and WC/CrC–CoCr). *Materials Science and Engineering A*, *352*(1-2), 202–211. doi:10.1016/S0921-5093(02)00888-2

Spear, K. E. (1989). Diamond—ceramic coating of the future. *Journal of the American Ceramic Society*, *72*(2), 171–191. doi:10.1111/j.1151-2916.1989.tb06099.x

Stack, M. M., & Chi, K. (2003). Mapping sliding wear of steels in aqueous conditions. *Wear*, *255*(1-6), 456–465. doi:10.1016/S0043-1648(03)00203-5

ASTM Standard: G3. (n.d.). *Annual book of ASTM standards* (vol. 03.02). Author.

Stein, K. J., Schorr, B. S., & Marder, A. (1999). Erosion of thermal spray $MCr–Cr_3C_2$ cermet coatings. *Wear*, *224*(1), 153–159. doi:10.1016/S0043-1648(98)00298-1

Stewart, D. A., Shipway, P. H., & McCartney, D. G. (1999). Abrasive wear behaviour of conventional and nanocomposite HVOF-sprayed WC–Co coatings. *Wear*, *225–229*, 789–798. doi:10.1016/S0043-1648(99)00032-0

Suarez, M., Bellayer, S., Traisnel, M., Gonzalez, W., Chicot, D., Lesage, J., & Staia, M. H. et al. (2008). Corrosion behavior of Cr_3C_2–NiCr vacuum plasma sprayed coatings. *Surface and Coatings Technology*, *202*(18), 4566–4571. doi:10.1016/j.surfcoat.2008.04.043

Uhlig, H. (1954). Mechanism of Fretting corrosion. *Journal of Applied Mechanics*, *21*(4), 401–407.

Usmani, S., Sampath, S., Houck, D. L., & Lee, D. (1997). Effect of carbide grain size on the sliding and abrasive wear behavior of thermally sprayed WC-Co coatings. *Tribology Transactions*, *40*(3), 470–478. doi:10.1080/10402009708983682

Vargas, F., Ageorges, H., Fournier, P., Fauchais, P., & López, M. E. (2010). Mechanical and tribological performance of Al_2O_3-TiO_2 coatings elaborated by flame and plasma spraying. *Surface and Coatings Technology*, *205*(4), 1132–1136. doi:10.1016/j.surfcoat.2010.07.061

Vieira, A. C., Rocha, L. A., Papageorgiou, N., & Mischler, S. (2012). Mechanical and electrochemical deterioration mechanisms in the tribocorrosion of Al alloys in NaCl and in NaNO3 solutions. *Corrosion Science*, *54*, 26–35. doi:10.1016/j.corsci.2011.08.041

Vieria, A. C., Ribeiro, A. R., Rocha, L. A., & Celis, J. P. (2006). Influence of pH and corrosion inhibitors on the tribocorrosion of titanium in artificial saliva. *Wear*, *261*(9), 994–1004. doi:10.1016/j.wear.2006.03.031

Vinayo, M. E., Kassabji, F., Guyonnet, J., & Fauchais, P. (1985). Plasma sprayed WC-Co coating: Influence of spray condition (atmospheric and low pressure plasma spraying) on the crystal structure, porosity and hardness. *Journal of Vacuum Science & Technology. A, Vacuum, Surfaces, and Films*, *3*(6), 2483–2489. doi:10.1116/1.572863

Voyer, J., & Marple, B. R. (1999). Sliding wear behavior of high velocity oxy-fuel and high power plasma spray-processed tungsten carbide-based cermet coatings. *Wear*, *225–229*, 135–145. doi:10.1016/S0043-1648(99)00007-1

Wang, Y., Jiang, S., Wang, M., Wang, S., Xiao, T. D., & Strutt, P. R. (2000). Abrasive wear characteristics of plasma sprayed nanostructured alumina/titania coatings. *Wear*, *237*(2), 176–185. doi:10.1016/S0043-1648(99)00323-3

Waterhouse, R. B. (1972). *Fretting corrosion*. Oxford, UK: Pergamon Press.

Watson, S. W., Friedersdorf, F. J., Madsen, B. W., & Cramer, S. D. (1995). Methods of measuring wear-corrosion synergism. *Wear*, *181–183*, 476–484. doi:10.1016/0043-1648(94)07108-X

Wayne, S. F., & Sampath, S. (1992). Structure/property relationships in sintered and thermally sprayed WC-Co. *Journal of Thermal Spray Technology*, *1*(4), 307–315. doi:10.1007/BF02647158

Wenger, F., & Galland, J. (1990). Analysis of local corrosion of large metallic structures or reinforced concrete structures by electrochemical impedance spectroscopy (EIS). *Electrochimica Acta*, *35*(10), 1573–1578. doi:10.1016/0013-4686(90)80012-D

Wood, R. K. (2006). Erosion–corrosion interactions and their effect on marine and offshore materials. *Wear*, *261*(9), 1012–1023. doi:10.1016/j.wear.2006.03.033

Woydt, M., & Wäsche, R. (2010). The history of the Stribeck curve and ball bearing steels: The role of Adolf Martens. *Wear*, *268*(11-12), 1542–1546. doi:10.1016/j.wear.2010.02.015

Wu, P.-Q., & Celis, J.-P. (2004). Electrochemical noise measurements on stainless steel during corrosion–wear in sliding contacts. *Wear*, *256*(5), 480–490. doi:10.1016/S0043-1648(03)00558-1

Yan, Y., Neville, A., Dowson, D., & Williams, S. (2006). Tribocorrosion in implants—Assessing high carbon and low carbon Co–Cr–Mo alloys by in situ electrochemical measurements. *Tribology International*, *39*(12), 1509–1517. doi:10.1016/j.triboint.2006.01.016

Yang, Q., Senda, T., & Hirose, A. (2006). Sliding wear behavior of WC–12% Co coatings at elevated temperatures. *Surface and Coatings Technology*, *200*(14-15), 4208–4212. doi:10.1016/j.surfcoat.2004.12.032

Yılmaz, R., Kurt, A. O., Demir, A., & Tatlı, Z. (2007). Effects of TiO_2 on the mechanical properties of the Al_2O_3–TiO_2 plasma sprayed coating. *Journal of the European Ceramic Society*, *27*(2-3), 1319–1323. doi:10.1016/j.jeurceramsoc.2006.04.099

Zeng, Y., Lee, S. W., Gao, L., & Ding, C. X. (2002). Atmospheric plasma sprayed coatings of nanostructured zirconia. *Journal of the European Ceramic Society*, *22*(3), 347–351. doi:10.1016/S0955-2219(01)00291-6

Zhang, J., & Kobayashi, A. (2008). Corrosion resistance of the Al_2O_3+ZrO_2 thermal barrier coatingson stainless steel substrate. *Vacuum*, *83*(1), 92–97. doi:10.1016/j.vacuum.2008.03.090

Zhijian, Y., Shunyan, T., & Xiaming, Z. (2010). Effect of the thickness on properties of Al_2O_3 coatings deposited by plasma spraying. *Materials Characterization*, *62*, 90–93.

Zhu, Y., Huang, M., Huang, J., & Ding, C. (1999). Vacuum-plasma sprayed nanostructured titanium oxide films. *Journal of Thermal Spray Technology*, *8*(2), 219–222. doi:10.1361/105996399770350430

Zou, Y., Goldbaum, D., Szpunar, J., & Yue, S. (2010). Microstructure and nanohardness of cold-sprayed coatings: Electron backscattered diffraction and nanoindentation studies. *Scripta Materialia*, *62*(6), 395–398. doi:10.1016/j.scriptamat.2009.11.034

KEY TERMS AND DEFINITIONS

Corrosion: Degradation due to reaction with the chemical environment.

Electrochemical Techniques: Techniques for measurement of kinetics and degradation rates employing principles of electrochemistry.

Sliding Wear: Degradation due relative motion between two contacting surfaces.

Thermal Spraying Coatings: Coating obtained by a group of processes that apply a consumable powders or wires in the form of finely divided molten and semi molten droplets.

Tribo-Corrosion: Interaction between tribology and corrosion.

Tribology: Science of friction.

Wear Corrosion: Synergy between wear and corrosion.

Chapter 3
Composite Coatings Employing a Novel Hybrid Powder and Solution-Based Plasma Spray Technique for Tribological Applications

G. Sivakumar
International Advanced Research Center for Powder Metallurgy and New Materials, India

S. V. Joshi
International Advanced Research Center for Powder Metallurgy and New Materials, India

ABSTRACT

Solution Precursor Plasma Spray (SPPS) technique is an emergent technique which offers significant performance advantages with various functional coatings like YSZ, $ZnFe_2O_4$, LSM, metal-doped ZnO, etc. Apart from the above capabilities, efforts towards development of an innovative hybrid approach through simultaneous deposition of powder and solution resulted in coatings with unique microstructures. Some illustrative examples of the large variety of coatings that can be realized through the hybrid route combining the conventional plasma spray and SPPS techniques are discussed. The attractive prospects offered by hybrid technique for

DOI: 10.4018/978-1-4666-7489-9.ch003

spraying nanocomposite coatings are specifically highlighted through a case study. Successful development of hybrid coatings using a Mo-alloy powder and a suitable oxide-forming solution precursor has been shown to exhibit improved sliding wear performance. The relationship between the splats, the ensuing coating microstructure at varied processing conditions, and its tribological behaviour of the coatings is comprehensively discussed.

INTRODUCTION

Among various coating methods, the thermal spray based techniques have gained the most widespread industrial acceptance till date. Ever since the first application of thermal spraying was demonstrated in 1909 by Dr. M.U. Schoop (Schoop, 1909), the technique has been widely used for applying coatings on various components to protect them particularly from corrosive attack or wear. The availability of numerous variants of the thermal spray technique, the diversity of spray materials and continuing advances in spray-control systems have together created new opportunities for the thermal spray industry. Above all, the versatility of the thermal spray process is unmatched by any other surface modification technology (Holmberg and Matthews, 1994, Pawlowski, 1995, Schneider *et al.*, 2006, Fauchais *et al.* 2001).

Thermal spraying is a generic term which relates to processes in which the desired coating material is deposited in semi-molten or molten or solid state on a substrate to form a layer. The feedstock material used conventionally has been in the form of powder, rod or wire with use of powder based feedstock being most popular. Numerous spray variants, such as flame spray, arc spray, plasma spray, HVOF, detonation spray, etc. (Holmberg and Matthews, 1994, Pawlowski, 1995, Schneider *et al.*, 2006) are available today, with each process employing a unique method of generating a high enthalpy stream to heat and accelerate the particles. It may be noted that the spray variants are distinct, particularly as far as the gas temperature and gas velocity profiles in the high enthalpy zone are concerned. These most crucially determine the physical state of particles at the time of deposition (molten, semi-molten or solid) and the velocity to which the injected particles are accelerated prior to impact with the substrate to be coated.Typical deposition efficiencies of most thermal spray variants range between 30-60%. This is a major concern from a commercial standpoint, due to the use of typically expensive powder feedstock, which alone accounts for nearly 60% of the overall coating costs, and the low deposition efficiencies make thermal spray a relatively expensive deposition technique.

In recent times, there is also a growing interest in spraying nanostructured or nano-sized powders, mainly due to the characteristic advantages that are expected to accrue compared to micron-sized powders. Thermal spraying is a rapid consolidation process with a potential to realize nanostructured features in a single step (Fauchais *et al.* 2008). Accordingly, it has been presumed that use of nano-sized feedstock can improve the characteristics and tribological performance of various coatings (Stewart *et al.*, 1999, Lima and Marple, 2007, Qiao *et al.* 2003). However, the use of nanosized particles for spraying presents numerous challenges, with some prominent ones being,(i) high production costs associated with manufacture of nanosized feedstock, (ii) difficulty in injecting nanosized particles into the core of the spray plume/flame, and (iii) lack of inertia of fine-sized particles after melting to impact the substrate.

One method that has been attempted to overcome the problem of feeding is to agglomerate the nano/sub-micron sized particles prior to feeding (Lima and Marple, 2007). However, this usually results in an inhomogeneous melting behavior, yielding a combination of nano-sized and micron-sized (Qiao *et al.* 2003). An alternative approach that has been adopted by various researchers involves use of a liquid instead of gas as a carrier medium and therefore, introduce the nanoparticles in the form of a suspension to eliminate feeding related problems (Fauchais *et al.* 2008, Bouyer *et al.*, 1996, Fauchais *et al.*, 2011, Bolelli *et al.*, 2010).

LIQUID BASED SPRAYING

The liquid based spraying process can be classified as suspension based spraying and solution precursor based spraying. The former technique uses suspensions of fine starting powders directly without the need for an additional agglomeration step (Fauchais *et al.* 2008, Fauchais *et al.*, 2011, Fauchais *et al.*, 2010). It is pertinent to note that there are several issues to be addressed even while using suspensions as feedstock. Foremost, this approach still relies on use of expensive nano/ultra-fine powder feedstock. Difficulties in suspending the particles in a stable manner and maintaining homogeneity also exist. In view of the above, a single-step coating process involving *in situ* synthesis of nanoparticles and their subsequent consolidation into an adherent deposit through solution precursor spraying appears promising. Although some preliminary reports on high-velocity oxy-fuel (HVOF) spraying of solution precursors have been reported (Ma *et al.*, 2006) usually availability of significant thermal energy is conducive for coating formation using such precursors. Consequently, a majority of prior efforts exploring this approach have relied on plasma spraying (Fauchais *et al.* 2008, Fauchais *et al.*, 2011, Karthikeyan *et al.* 1997, Karthikeyan *et al.* 1997, Padture *et al.*, 2001, Jordan *et al.*, 2004, Gell *et al.*, 2004, Dom *et al.*, 2012, Garcia *et al.*, 2007, Ma *et al.*, 2004, Coyle and Wang, 2007, Tummala *et al.*, 2012, Gell *et al.*, 2008).

SOLUTION PRECURSOR PLASMA SPRAY (SPPS) PROCESS

SPPS is an exciting method to produce a wide variety of functional oxide ceramic coatings, starting with appropriate solution precursors in contrast to powder feed-stock. The technique utilizes aqueous/organic chemical precursor solutions fed into the high temperature plasma plume employing a dedicated delivery device as shown in Figure 1. As may be noted from the schematic, the deposition setup is essentially identical to the conventional atmospheric plasma spraying (APS) process, except that the powder feeder assembly is replaced by a solution precursor delivery device comprising a pressurized precursor tank and an atomizer. The SPPS process also opens up new avenues for developing compositionally complex functional oxide coatings. Some of the main potential benefits are, (i) ability to create nanosized microstructures without any feeding problems normally associated with powder-based systems, (ii) flexible, rapid exploration of novel precursor compositions and their combinations, (iii) circumvention of expensive powder feedstock, (iv) better control over chemistry of the deposit.

Features of SPPS Process

Prior research efforts by few research groups on SPPS coatings have demonstrated that the process is versatile and allows convenient deposition of many diverse oxides such as ZrO_2-Y_2O_3, TiO_2, CeO_2, $BaTiO_3$, $ZnFeO_4$, NiO-YSZ, $LaSrMnO_3$, hydroxyapatite (Karthikeyan *et al.* 1997, Karthikeyan *et al.* 1997, Padture *et al.*, 2001, Jordan *et al.*, 2004, Gell *et al.*, 2004, Dom *et al.*, 2012, Garcia *et al.*, 2007, Ma *et al.*, 2004, Coyle and Wang, 2007, Tummala *et al.*, 2012, Gell *et al.*, 2008), etc. The various reported studies on SPPS coatings have included presentation of a plausible deposition mechanism by Bhatia *et al* (Bhatia *et al.*, 2002), successively involving break-up of atomized precursor droplets, precipitation, gelation, pyrolysis and sintering, followed by melting of the particles and subsequent splat formation upon impact with the substrate to form the coating.

Thermal barrier coatings (TBCs) developed employing solution precursor plasma spray (SPPS) technique have gained considerable research interest in recent times due to the inherent advantages that this route offers over conventional powder based atmospheric plasma sprayed coatings (Fauchais *et al.* 2008, Padture *et al.*, 2001, Jordan *et al.*, 2004, Gell *et al.*, 2004, Gell *et al.*, 2008). In terms of coating characteristics, YSZ coatings shown to possess interesting intrinsic features like fine grains, vertical cracks, distributed porosity, reduced inter-splat boundary sizes between the lamellae, etc make it suitable for TBC applications. The better durability

Figure 1. Schematic of solution precursor plasma spraying

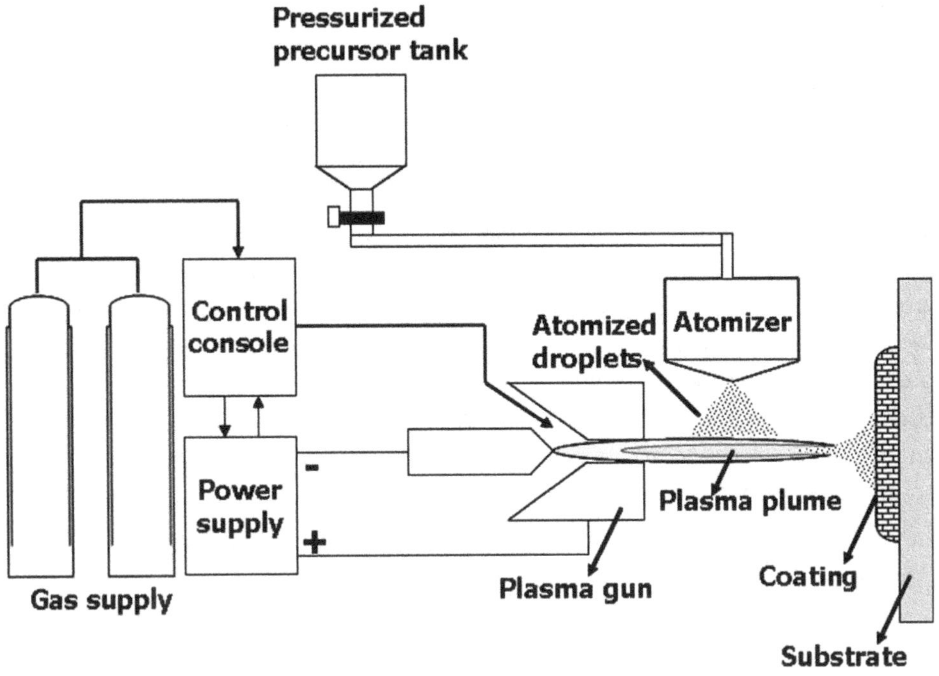

of SPPS TBCs over that of other coatings is attributed to the microstructure which provides better strain tolerance and toughness for the SPPS YSZ coating (Padture *et al.*, 2001, Jordan *et al.*, 2004, Gell *et al.*, 2004). The SPPS technique, apart from affording many other previously listed benefits (Padture *et al.*, 2001, Jordan *et al.*, 2004, Gell *et al.*, 2004, Gell *et al.*, 2008), provides a convenient pathway to deposit nanostructured coatings.

SPPS process is also limited by the low deposition rates, which significantly affects the industrial acceptance. Therefore, SPPS process was required to be combined with an industrially proven process like powder based APS to complement the coating characteristics as well as productivity. The above concept of hybrid processing can harness the mutual benefits of the two processes. The above novel approach involves sequential/simultaneous feeding of powder feedstock and a solution precursor with independent control of their feed rates, appears more feasible and cost-effective (Sivakumar and Joshi, 2013). An exploratory study in understanding the coating formation and subsequent performance studies implied that such hybrid based concept can also be implemented for possible wear resistant applications involving erosive and sliding wear modes.

HYBRID PROCESSING USING POWDER AND SOLUTION PRECURSOR

The hybrid processing involves a method of producing either a layered or composite coating architecture using sequential or simultaneous feeding of powder and liquid feedstock, respectively through a modified attachment as shown in Figure 2. Independently controlled micron sized powder feedstock and liquid precursor solution are fed through separate injection systems, into the plasma spray plume to form the desired coatings. As shown in Figure 2, the plasma spray gun is fitted with an atomizer to spray solution precursor feedstock and powder injector to spray powder feedstock into the plasma plume. Although radial injection of powder and solution precursor feedstock perpendicular to the central line of the plasma spray plume axis is depicted, injection of both feedstock at varying and independently controllable angles, is possible to yield the best coating characteristics for a specific powder or solution precursor feedstock. Accordingly, the feedstock delivery attachment for the plasma spray gun was fabricated to accommodate the atomizer as well as a powder feeding attachment (Sivakumar and Joshi, 2013).

Various architecture of coating configuration is achievable as schematically shown in Figure 3 including layered, composite and graded structures, with varied composition of APS and SPPS. In case of the layered YSZ architecture, a double layer of desired thickness can be deposited sequentially, whereas, the composite coating involves simultaneousspray of powder and solution precursor based feedstock. The composition can be tailored specific to the end applications. Similarly, the graded structure involves simultaneous feeding of solution precursor and powder feedstock, while continuously varying the feed rates of one or both feedstocks.

IN SITU PARTICLE FORMATION AND SPLAT CHARACTERISTICS

In Situ Particle Characteristics

During SPPS coating deposition, the in flight formation of particles and their subsequent heat-up and acceleration to ensure good splat formation upon impact with the substrate are key factors determining the microstructure of the coating as well as its characteristics (Kulkarni *et al.* 2003). Subsequent to the injection of precursor droplets just downstream of the plasma gun nozzle, energy transfer between the plasma and droplets takes place and several events involving solvent vaporization, fragmentation, pyrolysis, particle formation and melting could occur (depending

Figure 2. Schematic of hybrid SPPS process

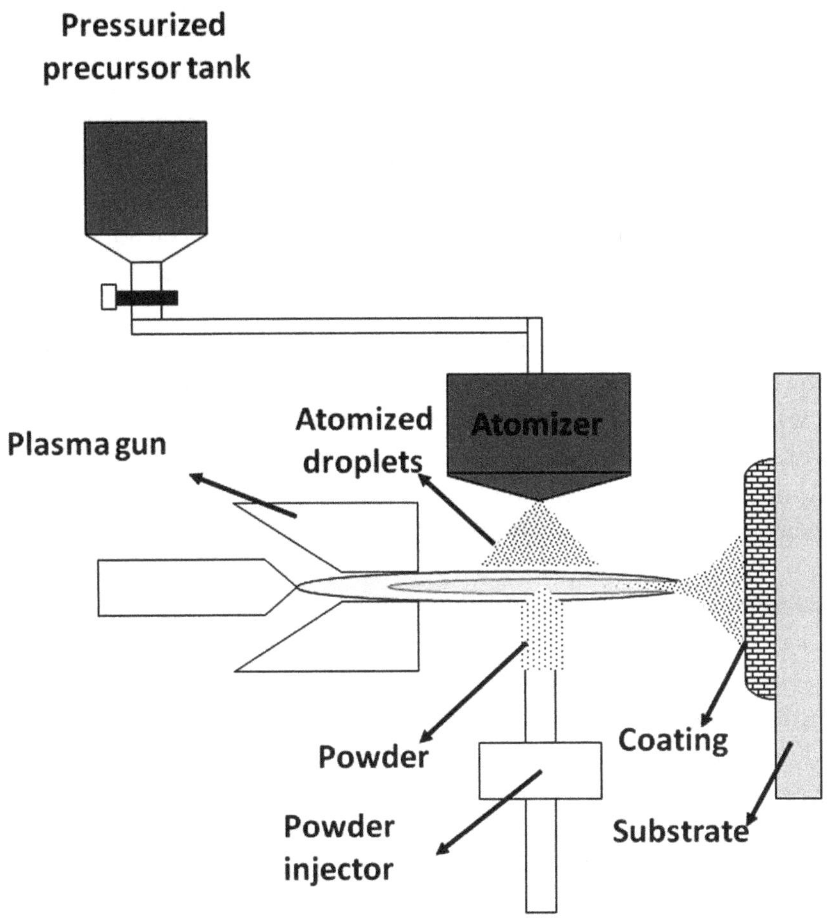

Figure 3. Configuration of coating architecture achievable through hybrid APS-SPPS processing

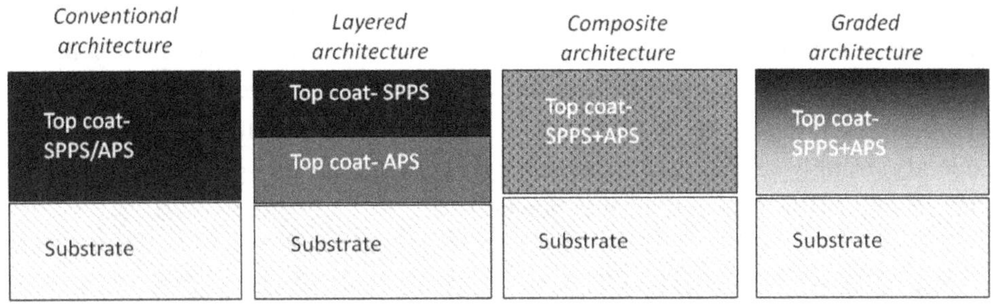

upon the process conditions) before the droplets/particles impact the substrate to form a coating. In conventional plasma spraying, in-flight particle diagnostic tools can provide intrinsic details of the particle status at the time of impact, in terms of its velocity, diameter and surface temperature (Gaona *et al.*, 2008). However, the inability of such tools to detect particles less than 5μm in size, the extremely large particle counts and unreliable pyrometric measurements within first 50 mm of the plasma plume preclude the use of these particle diagnostic systems in case of SPPS (Fauchais *et al.* 2005). Hence, investigation of particles collected in-flight before impact and the splats formed after these particles impact the substrate is extremely educative to develop an understanding of the coating formation mechanism.

Deposition of composite coatings by simultaneous feeding of powder and solution precursor in to the plasma plume opens up exciting prospects for generating layers with unusual properties. However, it also presents a significant challenge in terms of choosing appropriate spraying conditions. The process parameters need to be chosen such that conversion of the solution precursor as well as melting of simultaneously fed powder are both accomplished during the short residence time available. To serve as a proof of concept as well as in view of its relevance for possible wear resistant applications, coatings with a "composite" structure arising using a powder feedstock as well as a precursor solution was attempted with YSZ+YSZ based system. The spray distance was maintained at 50 mm as in case of solution processing instead of 62 mm normally set for powders. Based on a detailed study carried out with the YSZ precursor solution (Sivakumar et al., 2011), a plasma power of 46 kW (preferred for complete pyrolysis of the precursor) as against 39 kW employed for conventional YSZ powder spraying. The precursor droplets and powder particles were injected at the same axial location. It was deemed meaningful to assess the characteristics of *in situ* formed particles and the splat morphologies for academic interest since such co-deposition from powder and solution feedstock has not been previously attempted.

Typical morphologies of the conventional powder feedstock, carefully collected *in situ* formed YSZ particles during SPPS and hybrid processing are shown in Figure 4. It can be noted that the powder feedstock used for conventional APS comprises particles in the range of 10-100 μm as against the size range of 10 nm-2 μm in case of *in situ* formed SPPS particles as shown in Figure 4. The as-collected mid-stream particle specimens in case of SPPS were observed to contain a mix of features corresponding to unpyrolyzed precursor and well-formed spherical particles as illustrated in Figure 4b. It may be noted that the relative content of these features depends strongly on the process variables employed during spraying, particularly the precursor flow rate and plasma power.

Figure 4. Morphologies of a) original YSZ powder feedstock, b) in situ formed YSZ particles when only solution precursor is sprayed, c)-e) typical spectrum of particles collected during hybrid spraying of above YSZ powder feedstock particle along with solution precursor

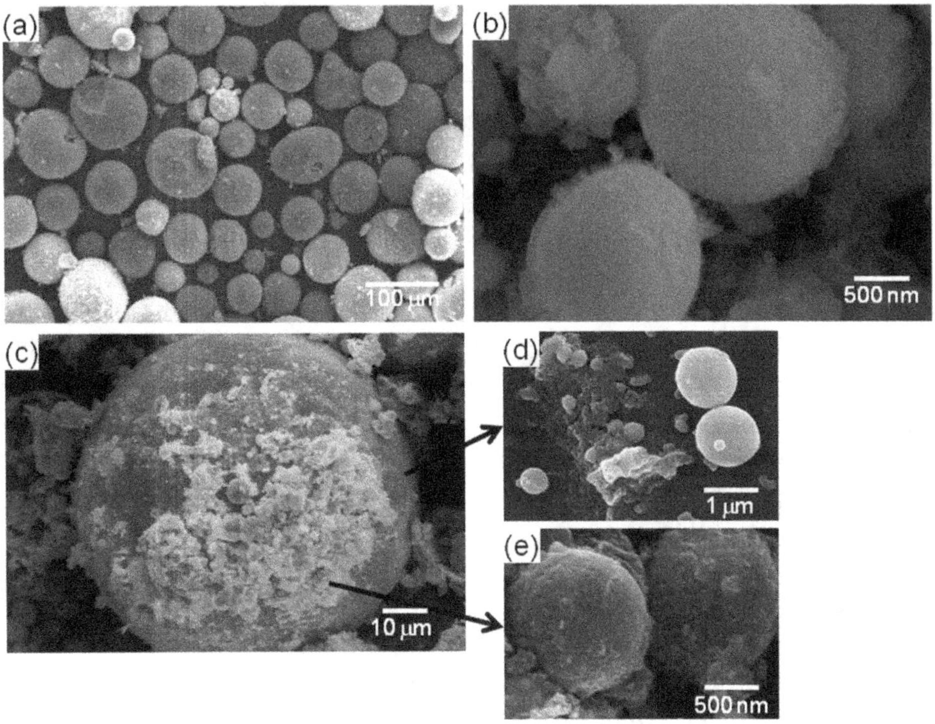

In case of the "composite" YSZ particles collected mid-stream during simultaneous spraying of powder and solution precursor, coarse spherical YSZ particles originating from the powder feedstock as well as numerous fine-sized particles formed *in situ* from the solution precursor are clearly noticed in Figure 4c. High magnification observation of each powder particle clearly reveals the above as illustrated in Figure 4c-d. As in case of particles collected from SPPS alone, the as-collected 'composite' particle specimens were also observed to contain a mix of features corresponding to unpyrolyzed precursor and well-formed spherical particles (Figure 4d). Phase analysis of these "composite" particles revealed that all the major peaks identified corresponded to tetragonal ZrO_2, with only minor presence of monoclinic ZrO_2 peaks being evident. However, the *in situ* formed SPPS YSZ particles exhibited single phase of non-transformable tetragonal ZrO_2.

Splat Formation Studies

Case Study I: Composite YSZ

As a logical extension of the above investigation on particle samples collected in-flight, splat formation upon impact of these particles on the substrate was studied. The splats were collected for three distinct cases, namely (a) APS with powder feedstock alone, (b) SPPS with solution precursor alone and (c) APS-SPPS hybrid deposition with simultaneous feeding of powder and solution precursors. All splats were collected under identical spray conditions on pre-heated SS substrates to 500°C. Typical splats using powder feedstock collected on a SS substrate are found to be almost 100 µm in diameter as shown in Figure 5a. Despite pre-heating the substrates, the YSZ powder splats show extensive cracking and fragmentation than the solution precursor based splats shown in Figure 5b. The SPPS YSZ splats were substantially smaller in size and never larger than 2 µm in diameter as observed in Figure 5b. The SPPS YSZ splats were also clearly seen to have undergone melting, with some spherical particles from secondary pyrolysis on the substrate or a pre-existing splat, as well as few unpyrolyzed precipitates typical of SPPS process (Sivakumar et al., 2011), also being observed.

A typical APS-SPPS composite splat shown in Figure 5c-eis seen to consist of a relatively large single particle splat resulting from powder feedstock along with virtually superimposed numerous fine-sized splats that can be attributed to the molten *in situ* formed particles from the solution precursor impacting the above large splat. The high magnification micrographs (Figure 5d-e) clearly serve to illustrate the marked difference in the relative splat sizes resulting from the powder and solution precursor feedstock. This is eventually responsible for the bimodal architecture noted subsequently in the coating microstructure. It is pertinent to observe that the shapes of molten splats from the powder and the solution precursor were near disk-shaped, which are preferred for better adhesion-cohesion within the microstructure (Bianchi *et al.*, 1997). Similar to the SPPS YSZ splats, the composite YSZ splats also revealed some unpyrolyzed features and extremely fine particles that have undergone secondary pyrolysis. It is clear that the splats originating from the solution precursor were found to deform and spread well over the larger conventional powder-derived splats, indicating low thermal contact resistance.

Case Study II: Composite Mo alloy+YSZ

Encouraged by the above results, a new kind of coating was attempted specifically to address the wear resistance characteristics of hybrid coatings. Therefore, coatings formed using a conventional molybdenum based alloy powder (Mo +NiAlMo

Figure 5. Morphology of YSZ splats collected under identical process conditions for a) APS, b) SPPS, c) APS-SPPS hybrid (low magnification) and d)-e) APS-SPPS hybrid (high magnification)

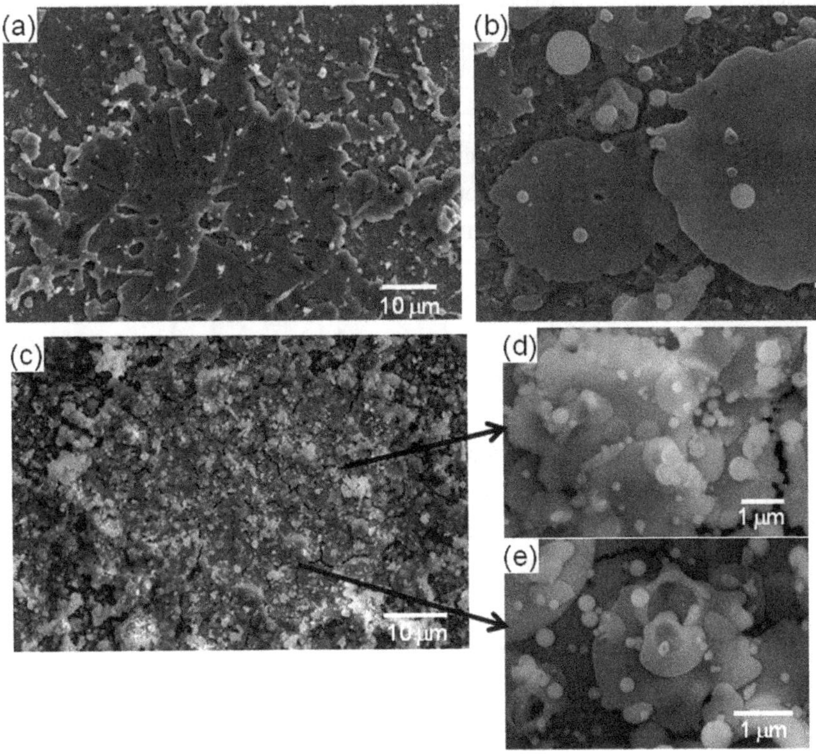

+NiCrBSiFe) along with the solution precursor to form nanosized YSZ *in situ* was investigated for possible tribological applications. Pure molybdenum coatings, although possessing good scuffing resistance and widely employed for piston ring applications, are usually blended with Ni based alloys like NiCrBSiFe, NiAlMo, etc. to reduce high temperature embrittlement as well as to further enhance wear resistance (Hwang *et al.*, 2004, Mrdak *et al.*, 2009). Incorporation of nano-sized features of oxides such as Al_2O_3, Cr_2O_3, and ZrO_2 is expected to further enhance tribological behaviour by virtue of strengthening through hard ceramic particles within the metal matrix (Dejang *et al.*, 2011).

From the preliminary studies, it was observed that the lower power level that are typically employed for powder based spraying was inadequate to form a nanocomposite microstructure, because (i) it did not permit defragmentation of the injected solution precursor droplet, (ii) enable pyrolysis to form YSZ particles *in situ* and their subsequent melting to ensure good splat formation. Secondly, the possibilities of developing suitable splats at optimum spray distance. At the longer standoff distance, well-molten splats of NiCrBSiFe/NIAlMo/Mo could be observed (Figure 6a), however, the solution precursor feedstock manifested in the form of fine spherical YSZ particles in the size range 100 - 1500 nm, were found on the larger powder particle splats. Negligible splats of the above fine YSZ particles were observed. At long spray distances, the spherical particles tend to re-solidify before impacting the substrate. Hence, mostly fine-sized spherical particles with no splatting are observed, as seen from Figure 6b. However, at short spray distance and high plasma power, fine sized YSZ splats could be clearly seen over the fully molten NiCrBSiFe splats and the partially molten NiAlMo and/or molybdenum splats as shown in Figure 6d. Here as well, the large difference between the size of YSZ based splats and powder based splats can be observed in Figure 6, with the former being nearly two orders of magnitude smaller in size. These microstructural features are responsible for the nanocomposite microstructure in the hybrid APS-SPPS coatings. The above splat formation studies are thus indicative that simultaneous injection of YSZ forming solution precursor and the desired powder feedstock into a plasma plume employing appropriate processing conditions can ensure good splat formation and bear promise to deposit nanocomposite coatings with novel microstructure.

COATING CHARACTERISTICS AND PERFORMANCE

Composite YSZ Coatings for Erosive Wear Resistance

The feasibility of hybrid processing in depositing composite coatings was carried out through simultaneous feeding of precursor solution as well as powder at different ratios. The disc rotational speed of the powder feeder was varied to inject different amounts of powder into the plasma plume, while keeping the solution precursor feed rate constant so as to generate composite coatings with varying constitution of coarse and fine features. Accordingly, typical cross-sectional microstructures of composite YSZ coatings with varied proportions of powder are shown in Figure 7. Since the variation in powder feed rate results in significant changes in deposition rate, the number of passes was chosen accordingly to achieve a constant coating thickness of about 250 μm. The composite YSZ coatings were observed to possess unique microstructures, comprising distributed nano-pores along with micron-sized features typical of APS. The porosity of the composite YSZ coatings was found to vary between 12-15%.

One of the motivation for developing composite YSZ coating by the hybrid APS + SPPS process was to tailor the YSZ microstructure for increased resistance to sintering while retaining pores, cracks, etc., to allow for strain tolerance. Accord-

Figure 6. FESEM images of splat formation during hybrid processing with powder and solution precursor feedstockcollected at, (a): 46 kW, 130 mm, (b) 46 kW, 130 mm showing spherical YSZ particles, (c): 46 kW, 50 mm, (d) 46 kW, 50 mm showing YSZ splats over NiCrBSiFe splat

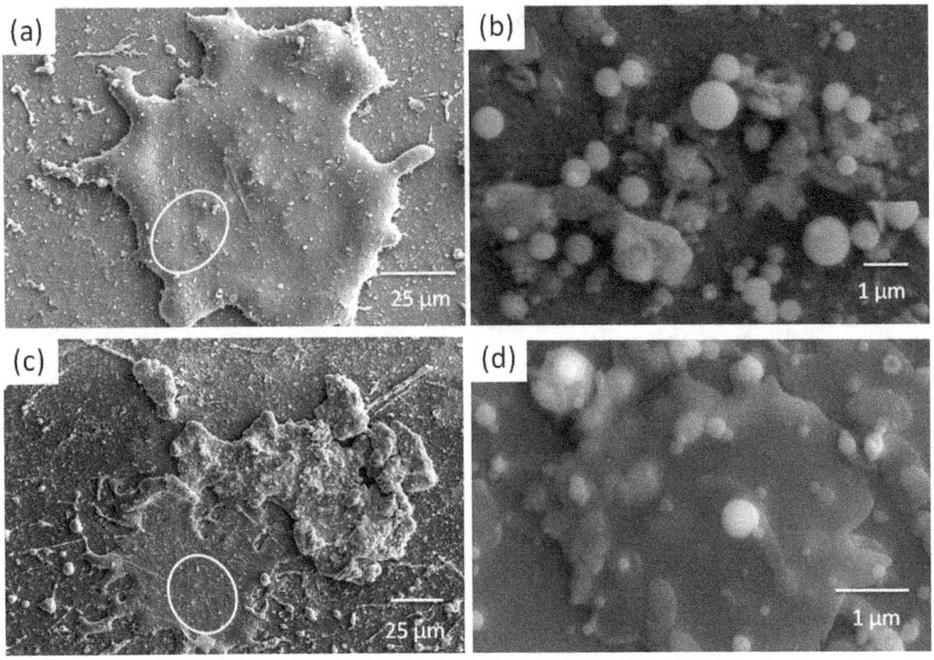

ingly, the concept of hybrid processing through simultaneous feeding of powder and solution precursor was found to yield bimodal microstructures, as illustrated in Figure 7. The nano-or sub-micron sized features of solution based YSZ in combination with large micron sized splats from YSZ powder enable creation of such a novel microstructure. It is pertinent to note that the distributed presence of the nano-zones has been reported to reduce the effective stress within the ceramic top coat and also hinder densification upon high temperature exposure (Lima and Marple, 2008, Lima *et al.*, 2002, Yu *et al.*, 2011, Shukla *et al.*, 2003). Additionally, the sintering resistance is also reported to be higher in case of nanostructured YSZ coatings (Lima and Marple, 2008, Lima *et al.*, 2002) due to the differential sintering characteristics of nano-zones and the micron-sized molten YSZ particles ((Lima and Marple, 2008, Lima *et al.*, 2002).

The microstructures of composite YSZ consisted of splats from powder based YSZ as well as fine YSZ grains/unmelted particles distributed throughout the coating as observed in Figure 7. Interestingly, the microhardness measurements on the composite coatings exhibited improvement in hardness over APS YSZ for

Figure 7. Cross-sectional micrographs of composite YSZ coatings with varied powder content a) 67 wt%, b) 75 wt%, and c) 80 wt%

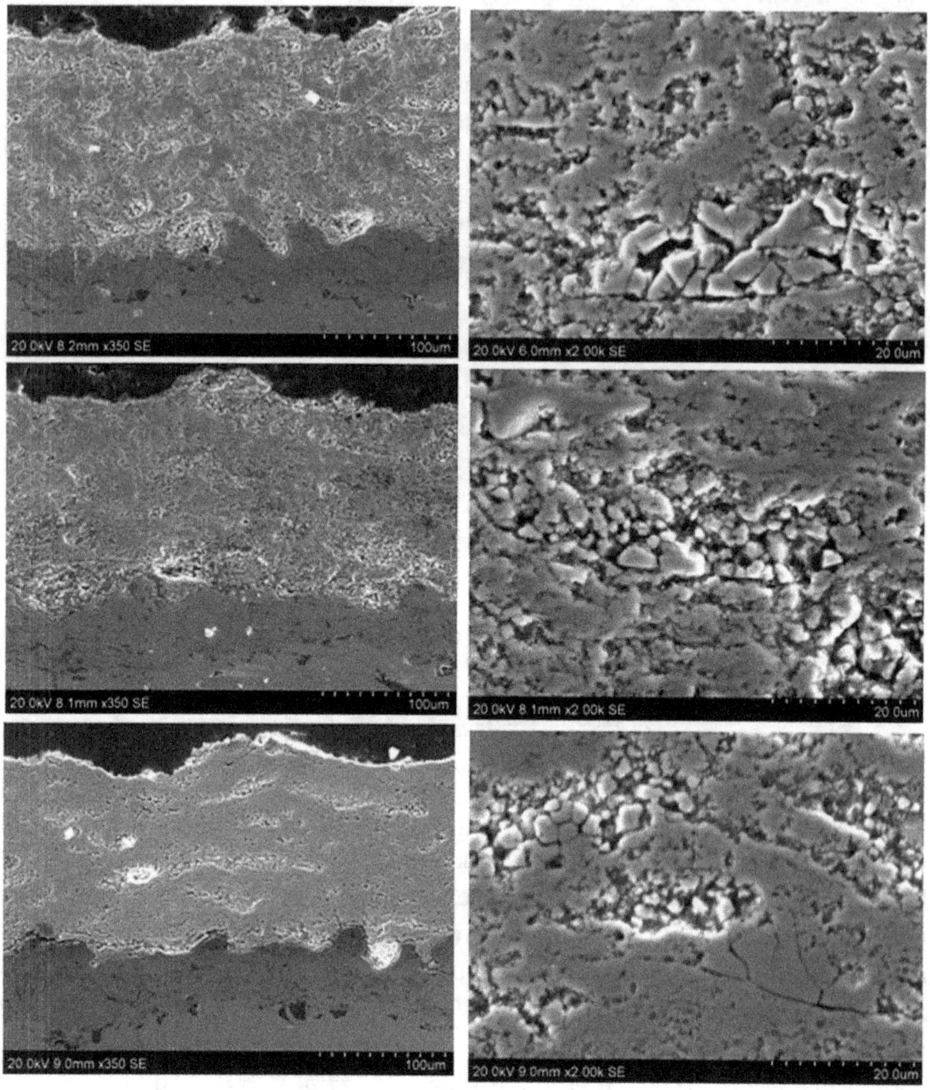

powder YSZ content of 75% or more as shown in Figure 8. This is attributable to two possible factors namely, (a) discrete distribution of nanostructured YSZ grains along with the larger YSZ powder splats and (b) a marginal reduction in porosity. Similar observations have been made in case of hybrid YSZ coatings deposited by laser assisted plasma spraying, which comprise a composite structure involving fine grained YSZ within a molten YSZ matrix (Ouyang and Sasaki, 2002). Phase analysis on composite YSZ coatings revealed presence of *m*-ZrO_2 in minor quantities along with the preferred *t'*-ZrO_2 phase.

TBCs in real-time applications are required to resist various modes degradation like oxidation, sintering, spallation under cyclic loading, erosion, foreign object damage (FOD), CMAS infiltration, etc. The most critical being the spallation resistance under thermal cycling and the damage caused by the impact of ingested particles, ash etc. Therefore, evaluating performance of TBC under erosive wear conditions can be most educative.Erosion resistance was measured through room temperature dry sand erosion test rig designed as per ASTM G76 standards. The tests were conducted at 90° impact angle using alumina particles of size 80-120 µm. The test conditions employed are shown in Table 1. The impact velocity was calibrated for 60 m/s through the double-disk method with the erodent flow maintained at 1 g/min. The weight loss experienced by the specimen after every minute is recorded through a 0.01 mg accuracy weighing balance. The erosion rate is reported in terms of weight loss occurred for a kg of erodent impacting the specimen.

Figure 8. Microhardness values of composite YSZ coatings generated with three different feedstock combinations in comparison with APS and SPPS YSZ coatings

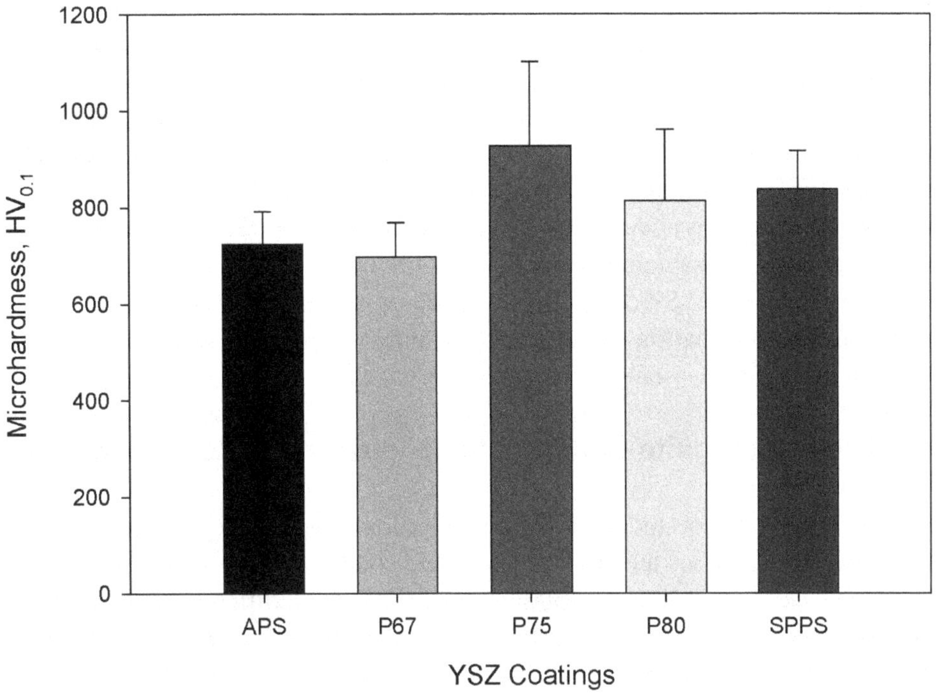

Table 1. Room temperature erosive wear test conditions

Nomenclature	Value
Erodent type and size (μm)	Alumina, 80-120
Particle velocity (ms^{-1})	60
Erodent feed rate (g min^{-1})	1.3
Duration of each run (min)	1
Erodent impact angle	90°
Nozzle to sample distance (mm)	10

It is pertinent to note that alumina is significantly harder than silica, which is a major constituent in typical ash or desert/beach sand that serve as ingested particles impacting TBC coated components in land-based gas turbines as well as aero-engines operating in desert/marine environments. The erosive wear rates of various YSZ coatings are shown in Figure 9.The relative hardness of the erodent compared to the target material plays a significant role in governing the erosion rate (Wada and Ritter, 1992). In the present case, the erodent alumina has an approximate hardness of 21 GPa (Riedel, 2000) which is significantly higher than YSZ coating hardness values of around 6-8 GPa, for which erosion rate is reportedly governed by fracture (Wada and Ritter, 1992). SPPS YSZ coatings exhibited the highest erosion rate amongst the various YSZ coatings studied. The erosion rate of TBCs is influenced by many factors namely porosity, the lamellar interface bonding ratio (Li *et al.*, 2006), splat boundaries, relative hardness, etc. The SPPS YSZ coating is inherently defective with abundant splat boundaries, vertical cracks and presence of unmelted particles which results in higher erosion rates. In case of composite YSZ coatings, the sub-micron/nano sized splats from SPPS based YSZ serve to toughen the APS YSZ coatings under suitable deposition conditions. Hence, the erosion rate was relatively lower for composite coatings containing lesser fine-sized YSZ grains. Thus, the composite YSZ coatings can indeed be practically relevant when the application demands that the TBC should not only yield good thermal cycling life, but also be erosion resistant.

Mo Alloy Composite Coatings for Sliding Wear Resistance

Subsequent to the splat studies, the hybrid Mo alloy composite coatings were deposited using simultaneous feeding of powder and solution precursor feedstock under processing conditions of 46 kW plasma power and at two different spray distances of 130 mm and 50 mm. The hybrid composite coating microstructures shown in

Figure 9. Erosive wear performance of SPPS + APS composite YSZ coatings in comparison with APS and SPPS YSZ coating

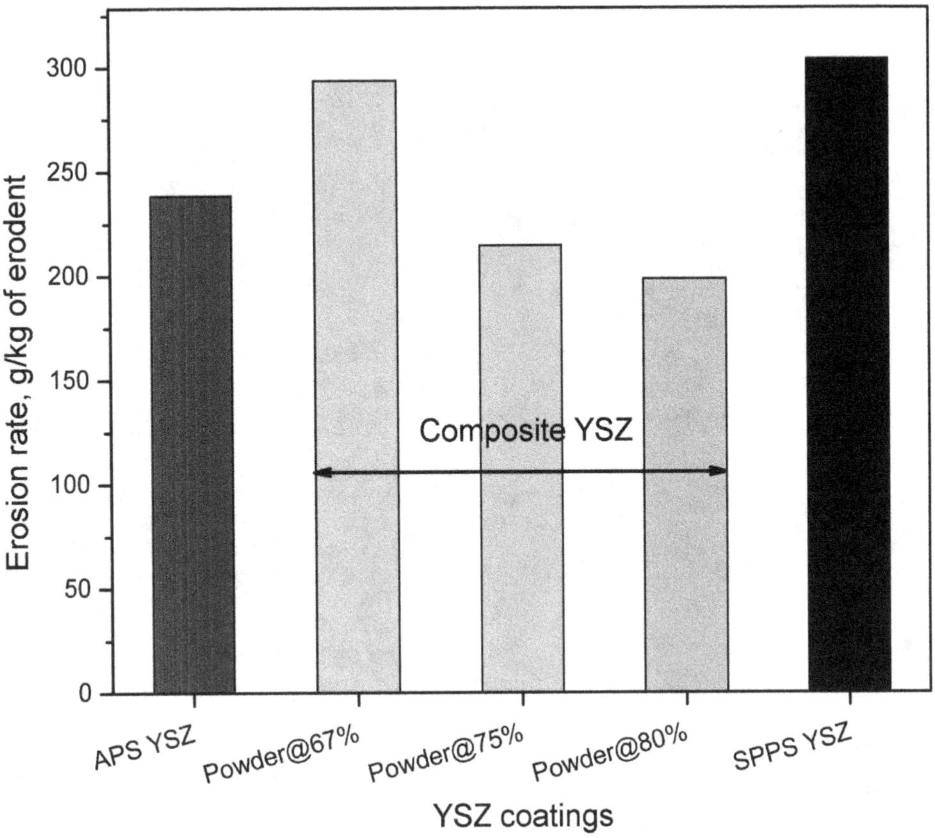

Figure 10 appear dense, although some porosity is evident compared to the pure Mo alloy coating. As observed earlier from the splat studies, at the shorter spray distance of 50 mm, some of the powder feedstock particles do not completely melt; on the other hand, at the higher spray distance of 130 mm, the finer particles formed *in situ* from the solution precursor plausibly re-solidify. At higher spray distances, better density was observed than at lower spray distances. Therefore, there is a trade-off between these extreme processing conditions and there is scope to further tune the spray parameters to achieve a range of microstructures. EDS analysis carried out on the APS-SPPS hybrid coatings clearly indicates the incorporation of YSZ in the hybrid coated specimens.

Fractographic analysis was carried out to investigate the distribution of YSZ-based features in the hybrid composite coatings and also to examine the nature of

Figure 10. SEM images of microstructure of powder feedstock (Mo-alloy powder) and solution precursor (YSZ) coating deposited at, (a): 30 kW, 130 mm, (b): 46 kW, 130 mm, (c): 46 kW, 50 mm at low magnification, (d): 46 kW, 50 mm at high magnification

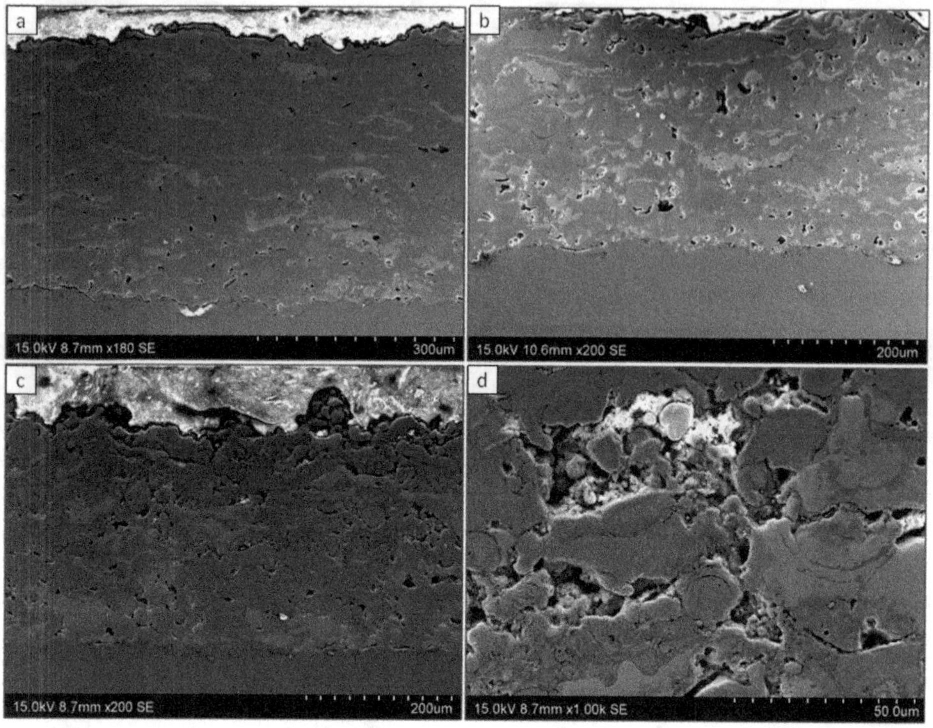

inter-splat bonding between successive layers, including that between the large sized powder-based splats and the fine solution precursor based splats. As shown in Figure 11a-b, the hybrid coating microstructure consists of predominantly two phases of molten, lamellar features corresponding to the Ni-based blend of NiCrBSiFe/NiAlMo phases and the columnar Mo based phases. The powder based lamellae are significantly larger in length and in thickness. The inter-lamellar gap between the adjacent splats appears nearly similar in both cases. Typical inter-splat voids are seen in Figure 11c and are also characterized by the presence of a large number of very fine (100-1,500 nm size) spherical particles of YSZ, rather than flattened splats, arising from the solution precursor as clearly evident in the high magnification micrograph depicted in Figure 11d & g. In case of high plasma power and low standoff distance conditions, there were fewer YSZ particles but there was clear evidence of fine YSZ splats from the solution precursor as shown in Figure 11f. These observations correlates well with the molten splats derived from the solution

Figure 11. Fracture surface analysis of hybrid nanocomposite coatings deposited at a-b) 30 kW, 130 mm, c-d) 46 kW, 130 mm (inset figure (g) showing fine sized YSZ particles) and e-f) 46 kW, 50 mm conditions

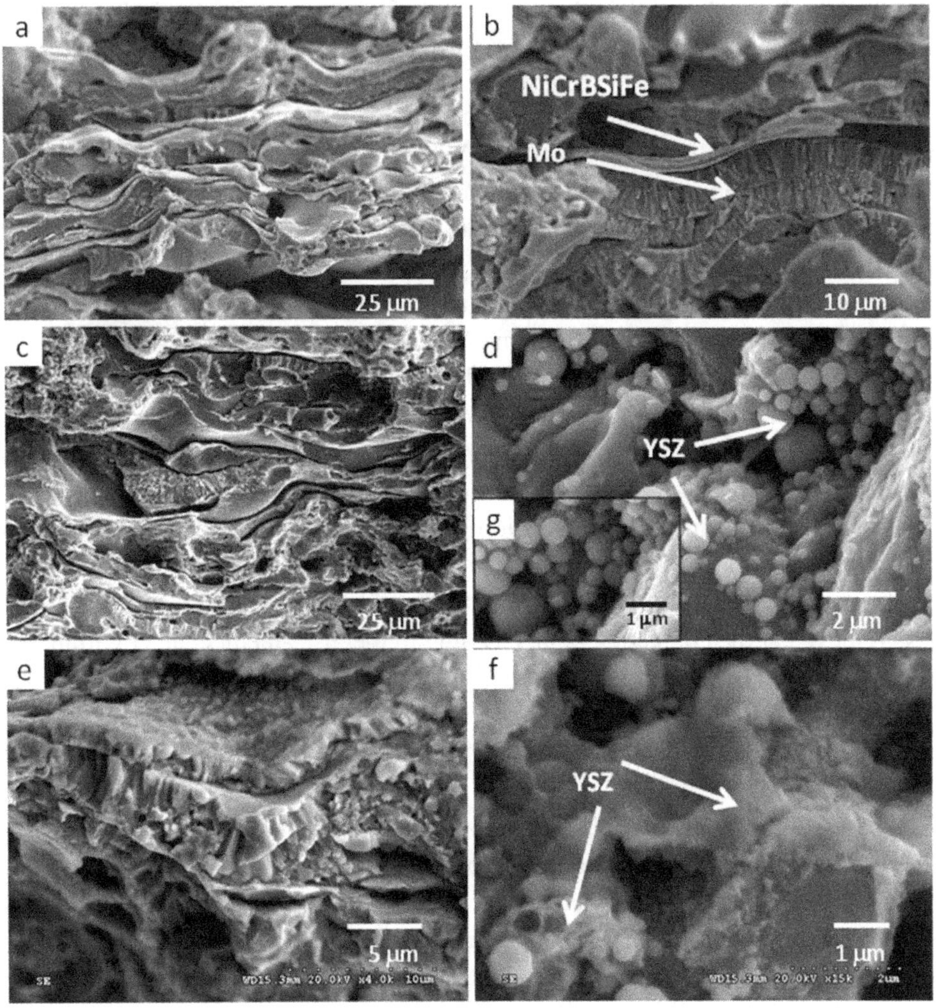

precursor noted at a plasma power of 46 kW and varied stand-off distances. The presence of YSZ based features was also confirmed through the EDS analysis at the respective locations.

TEM investigation was carried out to observe the presence of fine-grained YSZ aggregates within the solidified feature corresponding to the metallic matrix. The fine-grained YSZ aggregates consisted of numerous polycrystalline YSZ grains in the range of 30 – 50 nm along with few unmelted spherical particles. Phase analysis also corroborated the incorporation of YSZ grains which revealed at higher plasma power of 46 kW and spray distance of 130 mm, the t`-$ZrO_2$was of very low intensity but, increased remarkably at the lower spray distance of 50 mm. The above results demonstrates that the desired hybrid nanocomposite microstructure can be realized through the proposed novel APS-SPPS hybrid processing route under suitable processing conditions.

Motivated by the above characteristics, sliding wear test for the molybdenum alloy based APS coatings and the hybrid APS-SPPS nanocomposite coatings under various test environmentswas carried. The sliding wear behaviour of the coatings was assessed using a pin-on-disc apparatus conforming to ASTM G-99 standard at a load of 5 kgf. For these tests, the coatings were deposited on stainless steel pins of 6 mm diameter and 30 mm length, and tested against a grey cast iron disc at a sliding speed of 2.5 m/s. All pin specimens were first identically polished using fine SiC paper to ensure similar coating roughness and the pin-on-disc wear tests then performed under lubricated, starved and dry conditions. During the lubricated tests, the flow rate of the SAE 20W40 oil was maintained at 5 ml/min and the wear tests were performed for about 20 km of sliding distance. The more aggressive starved and dry tests were performed for a 4 km sliding distance. During the starved lubrication conditions, the lubricating oil was dispensed before the test and as soon as the wear test started, the flow was cut off. During the course of wear tests, the oil completely gets removed at the contacting points leading to metal-to-metal contact within few seconds itself. The wear rate is reported in terms of cumulative weight loss (mg) as a function of the total sliding distance, with the weight loss being measured at sliding distance intervals of 5 km for lubricated tests and at every 1 km for starved and dry conditions.

The hardness levels in both the conventional powder based APS coatings and the hybrid APS-SPPS nanocomposite coatings were similar, however the role of the additional ceramic phase on the wear rate of the hybrid coatings was the focus of this investigation. Figure 12 demonstrates that the wear rate of hybrid coatings was significantly lower compared to that of the molybdenum alloy blend powder based coatings under dry as well as lubricated sliding conditions. However, under starved lubricating conditions, the hybrid was comparatively inferior. This is essential to assess under diverse conditions akin to the operational conditions that exist in a cylinder bore-piston ring assembly. An improvement in wear rate to the extent of about 40% under lubricated conditions and about 20% under dry test conditions was noted for the hybrid nanocomposite coatings. The improvements in the wear performance can be attributed to the role of fine-sized, hard and tough t`-ZrO_2 splats within the metallic matrix.

Figure 12. Wear test results for powder-based and hybrid powder + solution precursor based coatings tested under lubricated, starved and dry conditions

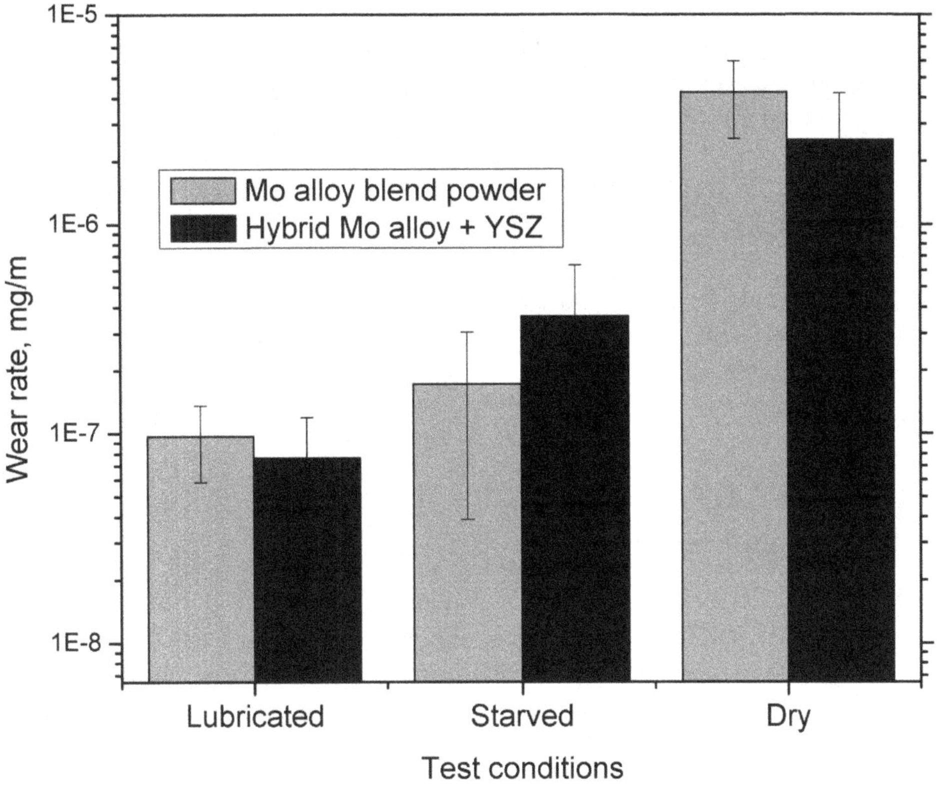

The friction coefficient values in the case of both the APS powder coatings and the hybrid coatings were, however, found to be similar. The typical friction coefficient values were less than 0.1 under lubricated conditions and in the range of 0.2-0.4 under starved and dry sliding conditions. The increased frictional force was plausibly responsible for the higher wear rate of the coated pins under starved and dry sliding conditions as shown in Figure 12.

CONCLUSION

A novel method employing a hybrid approach combining the conventional APS and SPPS processes was used to deposit wear resistant nanocomposite coatings. Case studies involving YSZ based and Molybdenum alloy blend powder-YSZ was presented to demonstrate the prospects for the formation of a nanocomposite coating by the above route. The coating formation mechanism was understood through detailed splat studies under varied processing conditions. There was a good agreement between understanding developed from the splat studies and microstructural characteristics exhibited by the nanocomposite coatings. The limited tribological studies indicated the hybrid approach can be adopted to realize nanocomposite coatings for varied wear resistant applications.

REFERENCES

Bhatia, T., Ozturk, A., Xie, L., Jordan, E. H., Cetegen, B. M., Gell, M., & Padture, N. P. et al. (2002). Mechanisms of ceramic coating deposition in solution-precursor plasma spray. *Journal of Materials Research*, *17*(9), 2363–2372. doi:10.1557/JMR.2002.0346

Bianchi, L., Leger, A. C., Vardelle, M., Vardelle, A., & Fauchais, P. (1997). Splat formation and cooling of plasma-sprayed zirconia. *Thin Solid Films*, *305*(1-2), 35–47. doi:10.1016/S0040-6090(97)80005-3

Bolelli, G., Bonferroni, B., Cannillo, V., Gadow, R., Killinger, A., Lusvarghi, L., & Stiegler, N. et al. (2010). Wear behaviour of high velocity suspension flame sprayed (HVSFS) Al_2O_3 coatings produced using micron- and nano-sized powder suspensions. *Surface and Coatings Technology*, *204*(16-17), 2657–2668. doi:10.1016/j.surfcoat.2010.02.018

Bouyer, E., Gitzhofer, F., & Boulos, M. (1996). Parametric study of suspension plasma sprayed hydroxyapatite. In BerndtC.C. (Ed.), *Proceedings of International Thermal Spray Conference 1996* (pp. 683-691). Materials Park, OH: ASM International.

Coyle, T. W., & Wang, Y. (2007). Solution precursor plasma spray (SPPS) of Ni-YSZ SOFC anode coatings. In B.R. Marple, M.M. Hyland, Y.-C. Lau, C.-J. Li, R.S. Lima, & G. Montavon (Eds.), Thermal spray 2007: Global coating solutions (pp. 699-704). Materials Park, OH: ASM International.

Dejang, N., Limpichaipanit, A., Watcharapasorn, A., Wirojanupatump, S., Niranatlumpong, P., & Jiansirisomboon, S. (2011). Fabrication and properties of plasma-sprayed Al_2O_3/ZrO_2composite coatings. *Journal of Thermal Spray Technology*, *20*(6), 1259–1268. doi:10.1007/s11666-011-9672-7

Dom, R., Sivakumar, G., Hebalkar, N. Y., Joshi, S. V., & Borse, P. H. (2012). Deposition of nanostructured photocatalytic zinc ferrite films using solution precursor plasma spraying. *Materials Research Bulletin*, *47*(3), 562–570. doi:10.1016/j.materresbull.2011.12.044

Fauchais, P., Etchart-Salas, R., Rat, V., Coudert, J. F., Caron, N., & Wittmann-Teneze, K. (2008). Parameters controlling liquid plasma spraying: Solutions, sols, or suspensions. *Journal of Thermal Spray Technology*, *17*(1), 31–59. doi:10.1007/s11666-007-9152-2

Fauchais, P., & Montavon, G. (2010). Latest developments in suspension and liquid precursor thermal spraying. *Journal of Thermal Spray Technology*, *19*(1-2), 226–239. doi:10.1007/s11666-009-9446-7

Fauchais, P., Montavon, G., Lima, R. S., & Marple, B. (2011). Engineering a new class of thermal spray nano-based microstructures from agglomerated nanostructured particles, suspensions and solutions: An invited review. *Journal of Physics. D, Applied Physics*, *44*(9), 1–53. doi:10.1088/0022-3727/44/9/093001

Fauchais, P., Rat, V., Delbos, C., Coudert, J. F., Chartier, T., & Bianchi, L. (2005). Understanding of suspension DC plasma spraying of finely structured coatings for SOFC. *IEEE Transactions on Plasma Science*, *33*(2), 920–930. doi:10.1109/TPS.2005.845094

Fauchais, P., Vardelle, A., & Dussoubs, B. (2001). Quo vadis thermal spraying? *Journal of Thermal Spray Technology*, *10*(1), 44–66. doi:10.1361/105996301770349510

Gaona, M., Lima, R. S., & Marple, B. (2008). Influence of particle temperature and velocity on the microstructure and mechanical behaviour of high velocity oxy-fuel (HVOF)-sprayed nanostructured titania coatings. *Journal of Materials Processing Technology*, *198*(1-3), 426–435. doi:10.1016/j.jmatprotec.2007.07.024

Garcia, E., Zhang, Z. B., Coyle, T. W., Hao, S. E., & Mu, S. L. (2007). Liquid precursors plasma spraying of TiO2 and Ce- Doped Ba(Zr0.2Ti0.8)O3 coatings. In B.R. Marple, M.M. Hyland, Y.-C. Lau, C.-J. Li, R.S. Lima, & G. Montavon (Eds.), Thermal spray 2007: Global coating solutions (pp. 650-654). Materials Park, OH: ASM International.

Gell, M., Jordan, E. H., Teicholz, M., Cetegen, B. M., Padture, N. P., Xie, L., & Roth, J. et al. (2008). Thermal barrier coatings made by the solution precursor plasma spray process. *Journal of Thermal Spray Technology*, *17*(1), 124–135. doi:10.1007/s11666-007-9141-5

Gell, M., Xie, L., Ma, X., Jordan, E.H., & Padture, N.P. (2004). Highly durable thermal barrier coatings made by the solution precursor plasma spray process. *Surface and Coating Technology*, *177 –178*, 97–102.

Holmberg, K., & Matthews, A. (1994). *Coatings tribology: Contact mechanisms, deposition techniques and application*. Amsterdam: Elsevier.

Hwang, B., Lee, S., & Ahn, J. (2004). Correlation of microstructure and wear resistance of molybdenum blend coatings fabricated by atmospheric plasma spraying. *Materials Science and Engineering A*, *366*(1), 152–163. doi:10.1016/j.msea.2003.09.062

Jordan, E. H., Xie, L., Ma, X., Gell, M., Padture, N. P., Cetegen, B., & Bryant, P. E. C. et al. (2004). Superior thermal barrier coatings using solution precursor plasma spray. *Journal of Thermal Spray Technology*, *13*(1), 57–65. doi:10.1007/s11666-004-0050-6

Karthikeyan, J., Berndt, C. C., Tikkanen, J., Reddy, S., & Herman, H. (1997). Plasma spray synthesis of nanomaterial powders and deposits. *Materials Science and Engineering A*, *238*(2), 275–286. doi:10.1016/S0921-5093(96)10568-2

Karthikeyan, J., Berndt, C. C., Tikkanen, J., Wang, J. Y., King, A. H., & Herman, H. (1997). Nanomaterial powders and deposits prepared by flame spray processing of liquid precursors. *Nanostructured Materials*, *8*(1), 61–74. doi:10.1016/S0965-9773(97)00066-4

Kulkarni, A., Vaidya, A., Goland, A., Sampath, S., & Herman, H. (2003). Processing effects on porosity-property correlations in plasma sprayed yttria-stabilized zirconia coatings. *Materials Science and Engineering A*, *359*(1-2), 100–111. doi:10.1016/S0921-5093(03)00342-3

Li, C.-J., Yang, G.-J., & Ohmori, A. (2006). Relationship between particle erosion and lamellar microstructure for plasma-sprayed alumina coatings. *Wear*, *260*(2), 1166–1172. doi:10.1016/j.wear.2005.07.006

Lima, R. S., Kucuk, A., & Berndt, C. C. (2002). Bimodal distribution of mechanical properties on plasma sprayed nanostructured partially stabilized zirconia. *Materials Science and Engineering A*, *327*(2), 224–232. doi:10.1016/S0921-5093(01)01530-1

Lima, R. S., & Marple, B. (2007). Thermal spray coatings engineered from nanostructured ceramic agglomerated powders for structural, thermal barrier and biomedical applications. *Journal of Thermal Spray Technology*, *16*(1), 40–63. doi:10.1007/s11666-006-9010-7

Lima, R. S., & Marple, B. (2008). Nanostructured YSZ thermal barrier coatings engineered to counteract sintering effects. *Materials Science and Engineering A*, *485*(1-2), 182–193. doi:10.1016/j.msea.2007.07.082

Ma, X. Q., Ge, S. H., Zhang, Y. D., Roth, J., & Xiao, T. D. (2004). Solution plasma spray synthesis of niznfe2o4 magnetic nanocomposite thick films. In *Proceedings of International Thermal Spray Conference 2004*. Materials Park, OH: ASM International.

Ma, X. Q., Roth, J., Gandy, D. W., & Frederick, G. J. (2006). A new high velocity oxy-fuel process for making finely structured and highly bonded alloy layers from liquid precursors. In *Proceedings of International Thermal Spray Conference*. Materials Park, OH: ASM International.

Mrdak, M., Vencl, A., & Ćosić, M. (2009). Microstructure and mechanical properties of the Mo-NiCrBSi coating deposited by atmospheric plasma spraying. *FME Transactions*, *37*, 27–32.

Ouyang, J. H., & Sasaki, S. (2002). Microstructure and tribological characteristics of ZrO_2–Y_2O_3 ceramic coatings deposited by laser-assisted plasma hybrid spraying. *Tribology International*, *35*(4), 255–264. doi:10.1016/S0301-679X(02)00005-1

Padture, N. P., Schlichting, K. W., Bhatia, T., Ozturk, A., Cetegen, B., Jordan, E. H., & Osendi, M. I. et al. (2001). Towards durable thermal barrier coatings with novel microstructures deposited by solution precursor plasma spray. *Acta Materialia*, *49*(12), 2251–2257. doi:10.1016/S1359-6454(01)00130-6

Pawlowski, L. (1995). *The science and engineering of thermal spray coatings*. London: Wiley.

Qiao, Y., Fischer, T. E., & Dent, A. (2003). The effects of fuel chemistry and feedstock powder structure on the mechanical and tribological properties of HVOF thermal-sprayed WC–Co coatings with very fine structures. *Surface and Coatings Technology*, *172*(1), 24–41. doi:10.1016/S0257-8972(03)00242-1

Riedel, R. (Ed.). (2000). *Handbook of ceramic hard materials 1*. Weinheim: WILEY-VCH Verlag GmbH. doi:10.1002/9783527618217

Schneider, K. E., Belashchenko, V., Dratwinski, M., Siegmann, S., & Zagorski, A. (2006). *Thermal spraying for power generation components*. Weinheim: WILEY-VCH. doi:10.1002/3527609342

Schoop, M. U. (1909). *Verfahren zur Herstellung von dichten metallischen Schichten*. CH-Patent No. 49278.

Shukla, S., Seal, S., Vij, R., & Bandyopadhyay, S. (2003). Reduced activation energy for grain growth in nanocrystalline Yttria-Stabilized Zirconia. *Nano Letters*, *3*(3), 397–401. doi:10.1021/nl0259380

Sivakumar, G., Dusane, R. O., & Joshi, S. V. (2011). *In situ* particle generation and splat formation during solution precursor plasma spraying of yttria-stabilized zirconia coatings. *Journal of the American Ceramic Society*, *94*(12), 4191–4199. doi:10.1111/j.1551-2916.2011.04773.x

Sivakumar, G., & Joshi, S. V. (2013). *An improved hybrid methodology for producing composite, multi-layered and graded coatings by plasma spraying utilizing powder and solution precursor feedstock*. Indian Patent No. 2965/DEL/2011 dt. 17th Oct, 2011 and United Patent Application No. US2013/0095340A1 dt. 18th Apr, 2013.

Stewart, D. A., Shipway, P. H., & McCartney, D. G. (1999). Abrasive wear behaviour of conventional and nanocomposite HVOF-sprayed WC-Co coatings. *Wear, 225-229*, 789–798. doi:10.1016/S0043-1648(99)00032-0

Tummala, R., Guduru, R. K., & Mohanty, P. S. (2012). Solution precursor plasma deposition of nanostructured CdS thin films. *Materials Research Bulletin, 47*(3), 700–707. doi:10.1016/j.materresbull.2011.12.018

Wada, S., & Ritter, J. E. (Eds.). (1992). *Erosion of ceramic materials*. Zurich: Trans Tech Publications.

Yu, Q., Rauf, A., Wang, N., & Zhou, C. (2011). Thermal properties of plasma-sprayed thermal barrier coating with bimodal structure. *Ceramics International, 37*(3), 1093–1099. doi:10.1016/j.ceramint.2010.11.033

KEY TERMS AND DEFINITIONS

Hybrid APS+SPPS Spray: The process utilizes dual injection system to feed conventional powder as well as solution precursor feedstock either sequentially or simultaneously to form unique coating architecture including multilayered, composite and gradient.

In Situ Formed Particles: The precursor droplet upon interaction with the plasma plume undergoes various stages of transformation to form a particle. Primarily, the droplet gets fragmented, wherein the solvent gets evaporated, followed by pyrolysis and sintering to form a fine sized particle.

Solution Precursor Plasma Spray (SPPS): A thermal spray variant specifically uses metal organic/inorganic precursor salts as the feedstock to produce various oxide based coatings. Coatings are formed through rapid conversion of precursor droplets into fine sized splats which coalesce to form thick coatings on the substrate. Examples include yttria stabilized zirconia, alumina, zinc ferrite on metallic or alloy substrates.

Splat: The basic building block of a coating which essentially occurs through the impact of *in situ* formed particle on the substrate.

Spray Grade Powder: The quality requirement of the powder feedstock used for thermal spraying in terms of its morphology, size, and flow behavior.

Thermal Barrier Coating: A thermal insulating coating, preferably of ceramic material with low thermal conductivity applied over the alloy substrates to reduce the surface temperature during its operation.

Thermal Spray: A generic term used for surface engineering variants which essentially employs thermal energy as a source for heating and accelerating the feedstock to get deposited over a substrate. Thermal spray variants include flame, wire-arc, detonation, plasma, high velocity oxy-fuel (HVOF), cold spray, etc.

Chapter 4
Abrasion–Corrosion of Thermal Spray Coatings

Robert J. K. Wood
University of Southampton – Highfield, UK

Mandar R. Thakare
Element Six Global Innovation Centre, UK

ABSTRACT

WC-based thermal-spray and High Velocity Oxy-Fuel (HVOF) coatings are extensively used in a wide range of applications ranging from downhole drilling tools to gas turbine engines. WC-based thermal spray coatings offer improved wear resistance as a result of hard phases dispersed in binder-rich regions. However, the presence of hard and soft phases within the coating can also lead to the formation of micro-galvanic couplings in aqueous environments leading to some reduction in combined wear-corrosion resistance. Furthermore, the coating also responds differently to change in mechanical loading conditions. This chapter examines the wear-corrosion performance of thermal spray coatings in a range of wear, electrochemical, and wear-corrosion tests under varying contact conditions to develop models and establish relationships between wear mechanisms, wear rates, and environmental factors such as pH and applied load.

INTRODUCTION

Using ceramic–metallic (cermet) or hard oxide coatings to protect metallic components is an effective method to reduce wear and corrosion. Modern surface engineering research is looking into depositing a wide range of hard phases such as carbides along with corrosion resistant metal binder elements with the aim of achieving good

DOI: 10.4018/978-1-4666-7489-9.ch004

adhesion between carbides and binders and therefore successfully combining corrosion and wear resistant elements together. Generally cermet coatings consist of WC or Cr_3C_2 particles embedded in a metal binder, which can be a pure metal or a mixture consisting of Ni, Cr or Co. Hard oxide coatings are typically based on Cr_2O_3.

The High Velocity Oxy-fuel (HVOF) spray technique is commonly used for depositing wear resistant WC-based coatings and has the advantage of generating higher particle velocities and the relatively low temperatures involved which minimise degradation of both the coating and substrate (Sudaprasert, Shipway, & McCartney, 2003).

The Detonation gun (D-gun) process shown schematically in Figure 1 is a modification of the conventional HVOF process in which the coating is deposited by means of a detonation caused in the detonation-gun barrel. The advantage of the D-gun process is that it produces a dense coating with minimum porosity (less than 1%) and high adhesion with the substrate (Wood, Mellor, & Binfield, 1997). A mixture of oxygen and acetylene, along with a pulse of pulverised WC, Co and Cr (in correct proportion) are introduced into a barrel and detonated using a spark. The resulting high temperature, high-pressure detonation wave heats the powder particles to around 3000 °C and accelerates them at a velocity of about 750 m s^{-1} towards the substrate while maintaining relatively low substrate temperatures between 95-150 °C (Tucker, 1999). Although the mechanism of bonding of the particles to the substrate is not fully understood, it is thought to be largely due to mechanical interlocking of the solidifying and shrinking "splats" (lamellar structure) with the asperities on the surface being coated (Tucker, 1999). These splats are approximately 50 µm wide and 10 µm thick and can vary depending on factors like velocity of deposition and the rate of cooling.

Coating Microstructure

The HVOF WC coating microstructure is extremely complex and consists of WC grains in an amorphous matrix consisting of Co with W and C in solution, see Figure 2. Compared to the uniform distribution of carbides and binder observed in sintered hardmetals, the coating shows a random distribution of carbides and binder rich areas referred to as cobalt-lakes.

During the spraying process, the WC particles partially melt and react with the binder to form metallic W and complex WC-M (where, M is the binder) compounds (Verdon, Karimi, & Martin, 1998; Wood et al., 1997).

The formation of W or WC-M compounds is more likely to occur at the periphery of the carbide particles where the temperature is expected to be highest. Alongside the main WC hard phase, W_2C and more complex carbides are often observed in the carbide phase. The formation of amorphous matrix, W_2C and complex carbides

Figure 1. Schematic of the detonation gun process (Praxair Surface Technologies Ltd.)

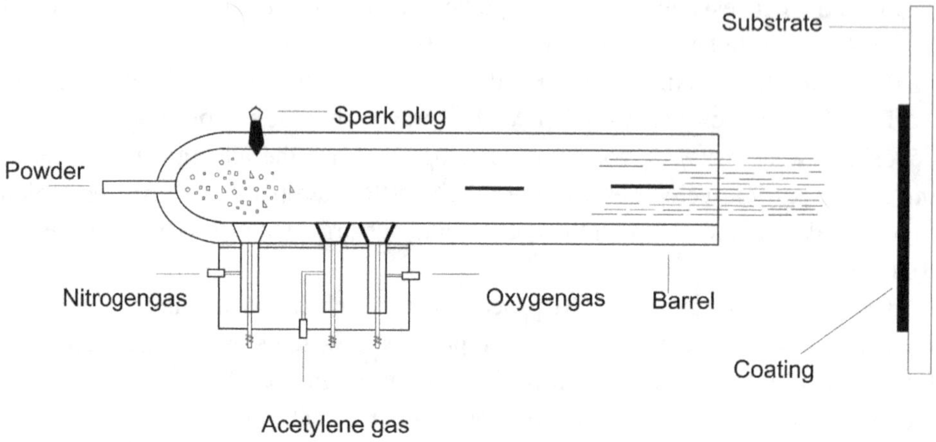

Figure 2. SEM-BSE image of a D-gun sprayed WC-10Co-4Cr coating (a) polished surface showing the distribution of carbide particles and soft binder-phase and (b) cross-section showing splat boundary and voids in the coating

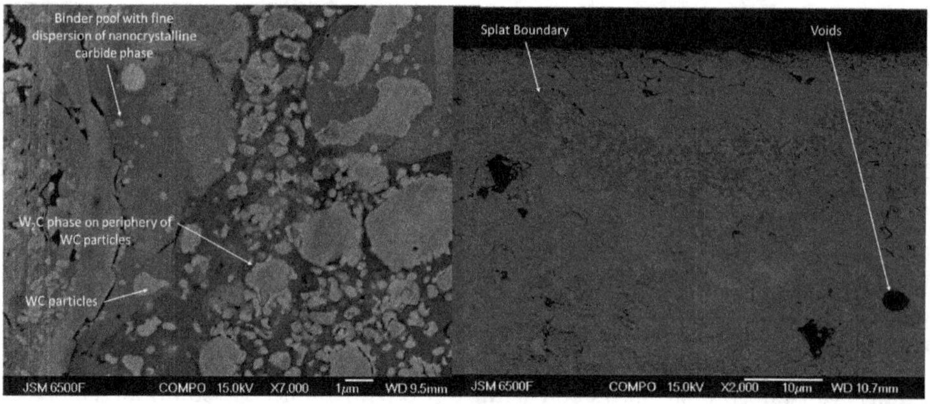

result from dissolution of WC into the molten binder phase during spraying and subsequent oxidation and precipitation reactions(Stewart, Shipway, & McCartney, 2000). The decomposition of WC is thought to proceed in three stages (Guilemany, Paco, Nutting, & Miguel, 1999; Vinayo, Kassabji, Guyonnet, & Fauchais, 1985).

$$2WC \rightarrow W_2C + C \tag{1}$$

$$W_2C + ½ O2 \rightarrow W_2 (C, O) \qquad (2)$$

$$W_2 (C, O) \rightarrow 2W + CO \qquad (3)$$

The decarburised W_2C phase is precipitated along the WC grain boundary along with a ring of metallic W between the WC grain and W_2C phase(Verdon et al., 1998). Higher degree of oxidation is observed along the coating splat boundary which results in the formation of nano-scale W precipitates resulting in a nano-crystalline structure (Stewart et al., 2000)as shown schematically in Figure 3. This is due the higher temperatures and higher cooling rates observed along the splat boundaries during the spray process. Hence unlike the sintered WC hardmetals, the composition of the sprayed WC-based coating does not correspond to the nominal powder composition used for the spraying process and the response to wear and corrosion can be very different from that observed in sintered WC-based materials. Although the parameters governing the wear performance of WC-based sprayed coatings should ideally be similar to sintered hardmetals, this is not the case since the ductile binder in the sintered material has been replaced by a relatively brittle, inhomogeneous binder phase in the HVOF-sprayed coating (Stewart et al., 2000).

Stewart *et al.*(Stewart et al., 2000)compared the X-ray diffraction (XRD) patterns of nanostructured and conventional powders with the coatings and observed that the nanostructured coatings showed a higher degree of decomposition of WC grains during spraying. A higher degree of decomposition of WC grains leads to lowering of carbon content in the WC grain (decarburisation) and formation of higher amounts of amorphous W-C-M compounds in the binder. This occurred due to the larger surface area exposed to react in the nanostructured WC particles.

Residual stresses are formed in the coating and the substrate in contact with the coating due to the high temperatures reached during the spraying process and subsequent cooling of the coating(Ibrahim & Berndt, 2007; Pina, Diasb, & Lebrun, 2003). The process of stress formation in the coating is complex and may result in the formation of either tensile or compressive residual stresses in the coating and substrate (Miguel, Guilemany, Mellor, & Xu, 2003). Tensile stresses can be formed in the coating due to temperature effects caused by quenching of the splats (Stokes & Looney, 2004). The presence of tensile residual stresses in the coating can be detrimental as they aid in propagation of cracks in the coating and also limit the thickness of the coating(Miguel et al., 2003). Compressive stresses in WC–Co coatings/ substrate are produced by a 'peening' action of WC particles during the spraying process which can potentially lead to an increase in the fatigue strength of the component (Ibrahim & Berndt, 2007; Miguel et al., 2003). Another property which strongly influenced by the microstructure of the WC-based coatings is its

Figure 3. Schematic of a molten droplet and a solidified splat structure of the WC-Co coating [Adapted from (Stewart et al., 2000)]

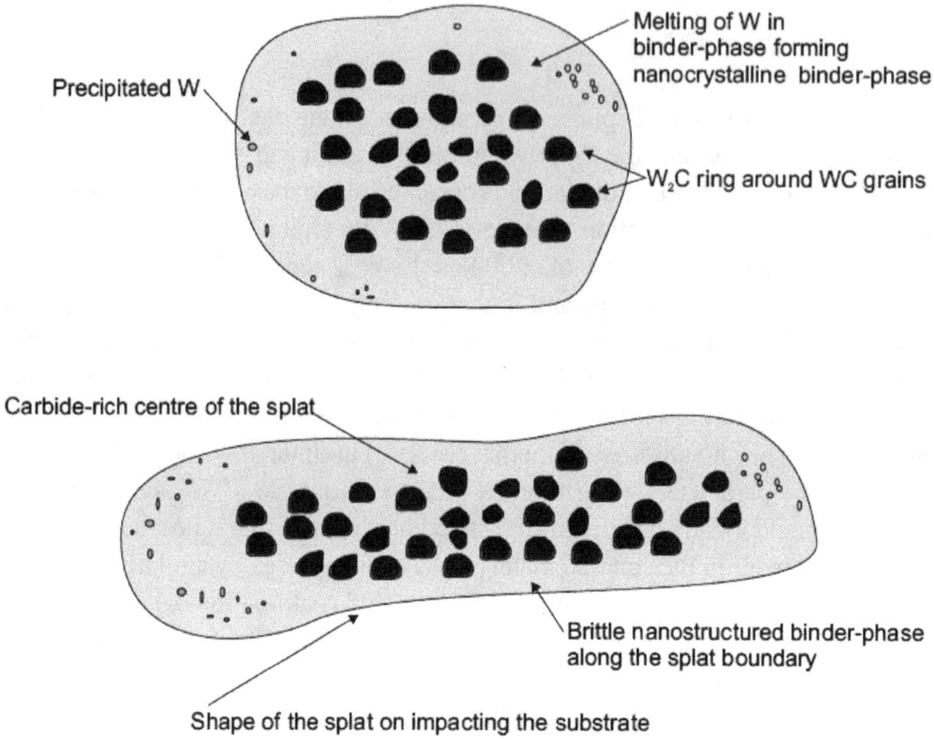

fracture toughness. Lopez-Cantera and Mellor (Cantera & Mellor, 1998) investigated the indentation fracture toughness of WC-10Co-4Cr coatings and found that crack propagation parallel to the substrate is more than the crack propagation transverse to the substrate. They observed that crack propagation preferentially occurred in the nano-crystalline zone between the splat boundary and the carbide rich centre of the coating splat. This was due to the fact that W_2C and other W-C-M phases created during the spraying process are more brittle than WC (Tu, Chang, Chao, & Lin, 1985).

Wear Mechanisms

The wear of thermal spray coatings is a complex process with hard and soft phases within the structure responding differently depending on factors such as the nature of the wear process and the size of the abrasive particles involved. A model published for explaining wear in sintered carbide structure can be used as a basis of understanding some of the likely interactions in the wear of thermal spray coatings.

The abrasive wear of sintered WC occurs by a combination of plastic grooving, binder phase extrusion and fracture of WC grains (Gant & Gee, 2006). Axén and Jacobson (Axen & Jacobson, 1994) introduced a model for the upper and lower limits of the wear resistance of composites. They introduced the concept of equal wear (EW) of the phases and equal pressure (EP) on the phases. EW mode of wear implies that the hard and soft phases in the composite are worn in the same rate and hence in this mode, the hard phase offer maximum protection to the softer phase and result in minimum wear. In EP mode of wear, equal pressure is exerted on the hard as well as the soft phases and as a result, preferential removal of the softer phase occurs since it has a lower wear rate. In the EP mode, the reinforcement phase particles provide minimum protection and are only able to take load in proportion to the area they cover. In this mode of equal pressure of the phases, the wear rate is substantially higher than in the EW mode of wear. The two modes of wear are shown schematically in Figure 4. In reality however, wear mechanisms observed often tend to be a combination of these two simple mechanisms.

Although the parameters governing the wear performance of WC-based sprayed coatings should ideally be similar to sintered hardmetals, this is not the case since the ductile binder in the sintered material has been replaced by a relatively brittle, inhomogeneous binder phase in the HVOF-sprayed coating (Stewart et al., 2000). As observed for sintered hardmetals, the abrasive wear of WC-based coatings using abrasives similar in size to the carbide grains occurs by the preferential removal of the binder phase by rolling and grooving mode of abrasion (Chen et al., 2005; Shipway & Howell, 2005a; Ramnathan & Jayaraman, 1989). Removal of the binder-phase around the carbides leads to the pull-out of carbides as they are undermined. The preferential removal of the binder-phase is easier in WC-based coatings due to the presence of distinct binder-rich regions and the uneven distribution of the carbide-phase. No evidence of sub-surface damage is reported in the

Figure 4. Schematic showing the Axén and Jacobson (Axen & Jacobson, 1994) model of wear in mult-phase materials involving hard and soft phases – originally developed for WC-based hardmetals

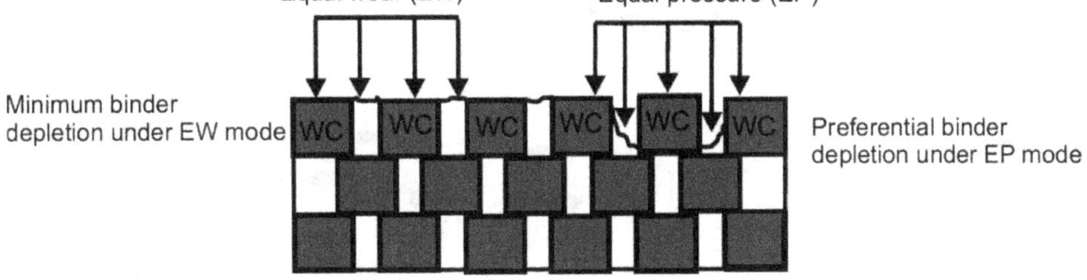

literature during the micro-scale abrasion of WC-Co based coating (abrasives similar in size as the carbides) (Shipway & Howell, 2005b). Abrasive wear using abrasives larger than the carbide grains result in the damage of the coating by the combination of plastic grooving and sub-surface cracking of the coating. As shown in the Figure 5, abrasive wear by an abrasive particle larger than the carbide grains results in ploughing of the binder-phase and fragmentation of carbides present within the grooves (Liao, Normand, & Coddet, 2000). The indentation-induced sub-surface cracking develops in the coating by the formation of lateral cracks and median cracks underneath the plastic zone (Shetty, Wright, Mincer, & Clauer, 1985; Niihara, 1983; Lima, Godoy, Avelar-Batista, & Modenesi, 2003; Yoomin Ahn, Cho, Lee, & Lee, 2003). The subsurface cracks in the coating propagate through the tungsten-rich binder phase present along the splat boundary and lead to the delamination of discrete sections of the coating as shown in Figure 5.

Similar wear mechanism has been observed during the erosion of WC-based sprayed coatings (Barber, Mellor, & Wood, 2005). Wood *et al.*(Wood et al., 1997) investigated the erosion performance of WC-10Co-4Cr coating and proposed two mechanisms depending on the angle of impact and energy of the erodent. At low impact angle and particle energy, micro-cutting and ploughing of the binder-phase

Figure 5. Schematic view of brittle fracture of the HVOF coating during abrasive wear (adapted from (Yoomin Ahn et al., 2003))

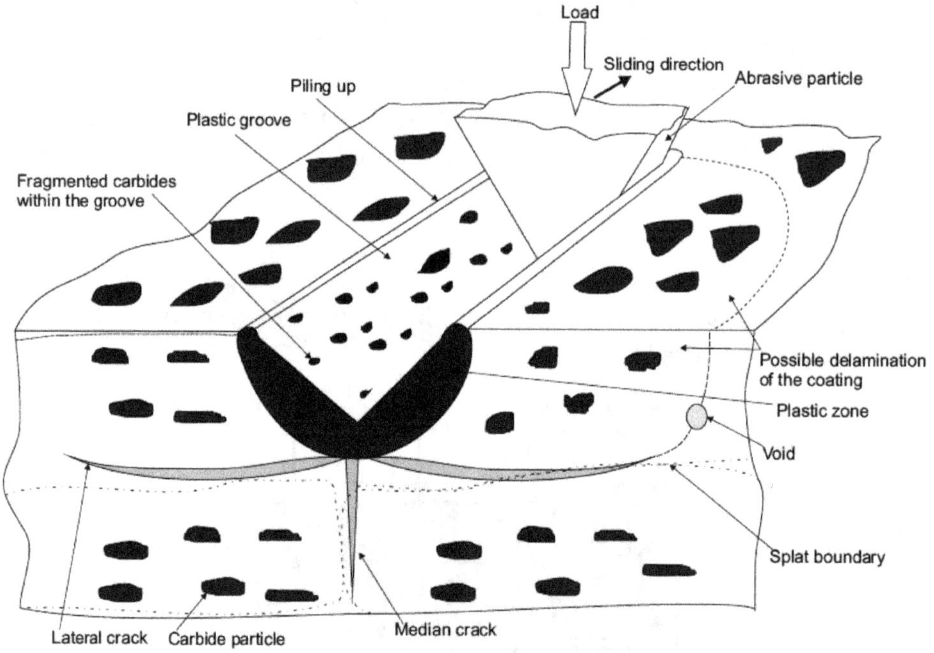

resulted in the embedded carbide particles being exposed. Subsequent erosion led to the gouging of the carbides as they were undermined. At high impact angle and higher kinetic energies, development and propagation of sub-surface cracks lead to the removal of large sections of the coating. Clearly, wear of WC-based coatings due sub-surface cracking and delamination is strongly dependent on its fracture toughness (Barber et al., 2005; Wheeler & Wood, 2005). As discussed in the previous section, lowering of fracture toughness occurs by the dissolution of metallic W in the binder-phase. Hence, unlike the sintered hardmetals, use of nanostructured carbide particles does not result in the improvement in wear resistance for WC-based coatings(Shipway, McCartney, & Sudaprasert, 2005; Stewart, Shipway, & McCartney, 1999).

Corrosion Mechanisms in WC-Based Thermal Spray Coatings

Presence of inhomogeneity within the thermal spray coating structure does make it susceptible to corrosion, particularly for the binder-phase. Corrosion of the binder-phase or the binder-carbide interface results in the weakening of the skeletal carbide structure and results in accelerated wear of the hardmetal. Kalish (Kalish, 1987) investigated the corrosion performance of WC-bases sintered hardmetals and observed that the corrosion rates for WC-based hardmetals are much higher in acidic conditions than in alkaline conditions. This was due to the fact that the binders Co and Ni have a lower corrosion resistance at lower pH values i.e. in acidic environments as compared to the hard WC particles, see Pourbaix diagrams for Co and Ni (Pourbaix, 1974). The removal of the binder phase is likely to be enhanced if the environment is corrosive which could also lead to accelerated loss of material. Tomlinson *et al.*(Tomlinson & Linzell, 1988)studied sintered tungsten carbides with Co and Ni binders in acidic conditions and reported that the electrochemical performance of WC based hardmetals reflected the performance of the binder and was representative of the high dissolution rate of pure Co as opposed to a lower dissolution rate of Ni. In the presence of corrosive environment, micro-galvanic cells are likely to be formed between the hard carbide phase and the soft binder. Carbides are likely to be cathodic to the binder in such micro-galvanic cells resulting in an increase in the binder loss during corrosion as proposed by Hochstrasser(-Kurz) (Hochstrasser (-Kurz), 2006). The galvanic coupling between Co and WC results in the anodic dissolution of Co, given by the following reaction.

The corrosion performance of the sprayed WC-based coatings is very different from their sintered counterparts due to the complex microstructure of the coating. Unlike the skeletal carbide structure observed in sintered hardmetals, WC grains in the coatings are randomly distributed in the metallic binder. Hence corrosion of the binder-phase has a stronger influence on the wear performance of the coating.

However, the use of a corrosion resistant binder improves the corrosion as well as wear performance of the coating (Karimi, Verdon, Martin, & Schmid, 1995; Toma, Brandl, & Marginean, 2001; Barbezat, Nicoll, & Sickinger, 1993; Rogne, Bjordal, Solem, & Bardal, 1996). The influence of binder composition on the corrosion performance of WC-based coatings and have been studied previously (Cho, Hwang, & Kim, 2006; Souza & Neville, 2003b). Cho et al.(Cho et al., 2006) studied the corrosion behaviour of WC-based coatings with different binder-compositions in strong H_2SO_4 solutions. Prolonged exposure to acidic conditions revealed micro-galvanic coupling between WC particles and the binder-phase leading to the preferential dissolution of the binder-phase. It was also found that the presence of Cr in the binder-phase led to a slight decrease in the corrosion rate of the binder-phase. They also noted that the presence of micro-cracks and defects on the surface led to accelerated corrosion attack penetrating the surface. Bolelli et al. (Bolelli, Giovanardi, Lusvarghi, & Manfredini, 2006) also observed that the HVOF WC-based coating with CoCr binder possesses superior corrosion resistance than ordinary thick chromium plating. However, it was also observed that presence of microstructural defects such as oxide inclusions and splat boundaries on the coating surface results in crevice corrosion.

Perry *et al.*(Perry, Neville, Wilson, & Hodgkiess, 2001) studied the effects of prolonged exposure of WC-Co-Cr coating in artificial seawater (3.5% NaCl solutions). SEM micrographs of exposed coating revealed that corrosion occurred at the binder-carbide interface. They concluded that the interface between binder and carbide provided a site for micro-galvanic and or crevice corrosion. It was also concluded that as the matrix corrodes at the binder-carbide interface, the carbide particles then fall out and leave behind small pits. This was also observed during the anodic polarisation of the WC-Co-Cr samples in artificial seawater. These corrosion features observed at the binder-carbide interface were confirmed by Souza and Neville (Souza & Neville, 2003b) by comparing the corrosion performance of WC-Co-Cr coating and WC-CrNi coating. Souza and Neville (Souza & Neville, 2003a, 2005)further studied the corrosion characteristics of HVOF WC-10Co-4Cr in order to understand the effects of corrosion on the erosion-corrosion behaviour of sprayed WC-based coatings. They compared the electrochemical behaviour of UNS S31603 stainless steel with a HVOF WC-10Co-4Cr coating and found that unlike the continuous passive film of Cr_2O_3 present on stainless steels, only the CoCr matrix in the HVOF coating passivates and develops a Cr_2O_3film in a 3.5% NaCl solution. They observed corrosion of small WC grains (< 1 μm) on coating samples subjected to an accelerated corrosion test and worn coating on a sleeve operating in a solid-free environment. It was concluded that the corrosion feature observed at the binder-carbide interface results from the corrosion of the binder-phase as well as WC grains. Similarly, Takeda *et al.*(Takeda, Morihiro, Ebara, & Harada,

2002) observed a Cr_2O_3 passive film on a WC-10Co-4Cr coating after immersion in Na_2SO_4 solution and concluded that this film does not cover the entire surface and was present in the form of islands. This passive film is likely to be weakest at the binder-carbide interface and hence in presence of a corrosive electrolyte, corrosion in HVOF WC-10Co-4Cr is likely to be initiated at this interface (Souza & Neville, 2003a). Further study by Thakare et al.(Thakare, Wharton, Wood, & Menger, 2007) revealed that the static corrosion of thermal spray WC-Co coatings in alkaline and neutral conditions is a two-step process. Using SEM, TEM analysis to examine corroded surfaces of WC-Co coating, they confirmed that corrosion progressed towards the carbide phase which is likely to occur in presence of metallic W formed along the carbide grain boundary due to decarburising during the spray process. Figure 6 shows the SEM micrographs of HVOF WC-10Co-4Cr coating exposed to alkaline conditions and the subsequent formation of a trench-like corrosion feature around the carbide grain. Focussed ion beam sectioning of this sample across the carbide grain revealed the carbide grain to be loosely held in the corrosion trench formed. XPS analysis also revealed that the binder-rich areas were protected by stable passive films, thus leaving the carbide exposed to corrosion. Presence of re-precipitated metallic W and decarburised W_2C phases along the periphery of WC grains has been identified by Verdon *et al.*(Verdon, Karimi, & Martin, 1997).

Tribo-Corrosion of Thermal-Spray Coatings

The need to select or design new surfaces for future equipment as well as minimize the operating costs and extend the life of existing machinery has led to demands

Table 1. XPS results identifying the phases present on the surface of a HVOF WC-10Co-4Cr coating after exposure to alkaline and neutral solutions

Sample Condition	Element	Detected Species	Binding Energies (eV)*
Fresh	Co(2p)	Co	778.3
	Cr(2p)	Cr	574.2
		Cr_2O_3	577.1
	C(1s)	WC	283.0
Exposed (pH 7)	Co(2p)	$Co(OH)_2$.	781.0
	Cr(2p)	Cr_2O_3	577.1
Exposed (pH 11)	Co(2p)	$Co(OH)_2$.	781.0
	Cr(2p)	Cr_2O_3	577.1

*Binding energies compared to NIST Online Database (Wagner et al., 2006).

for a much better understanding of surface degradation processes particularly when tribological components are operating in corrosive environments. This has given rise to the active research area of tribo-corrosion which seeks to address the concerns above and understand the surface degradation mechanisms when mechanical wear and chemical/electrochemical processes interact with each other.

The subject started to be researched in the late 1980s and has now emerged as an active research area as advanced experimental techniques have been developed to yield substantial insight into the complex processes present in tribo-corrosion contacts. The development of in situ electrochemical techniques and post-test analysis techniques for surface film examination for example are powerful tools that can be deployed in tribo-corrosion experimental programmes. The subject, therefore, includes the interaction of corrosion and erosion (solids, droplets or cavitation bubbles), abrasion, adhesion, fretting and fatigue wear. Wear-corrosion interaction can lead to either increase or a decrease in the overall mass loss. The change in the mass loss due to the synergistic effects of coupling wear and corrosion is often referred to as synergy (*S*). Positive synergy results in accelerated material loss due to the combined action of wear and corrosion and is an undesirable material property. On the other hand, negative synergy results in a decrease in the overall loss of material due to improvement in either wear or corrosion resistance and is a desirable material property. According to the ASTM standard guide for determining synergism between wear and corrosion(ASTM G119-93, 1998), the total wear during the process of abrasive wear-corrosion is defined by the following equations

$$\text{Total Wear } (AC) = \text{Pure Abrasion } (PA) + \text{Pure Corrosion } (PC) + \text{Synergy } (S) \quad (4)$$

where,

- $S = \Delta PC_A + \Delta PA_C$ (5)
- ΔPC_A = Change in corrosion due to abrasion.
- ΔPA_C = Change in abrasion due to corrosion

Understanding the tribo-corrosion of coatings is far harder compared with many monolithic surfaces as the coating processes used can unintentionally degrade the coating microstructure by adding significant porosity and pockets of oxides into the coating and the coating process can also affect the substrate properties. Also remnants from the pre-coating substrate treatments (such as shot or grit blasting)

Figure 6. Back scattered electron micrographs of WC-10Co-4Cr HVOF coating showing (a) polished section identifying the different phases formed during spraying, (b) corrosion of metallic W after exposure to alkaline environments, (c) focussed ion beam trench made across one of the WC particles exposed to corrosion and (d) high magnification micrograph of the carbide grain loosely held in the corrosion 'trench'

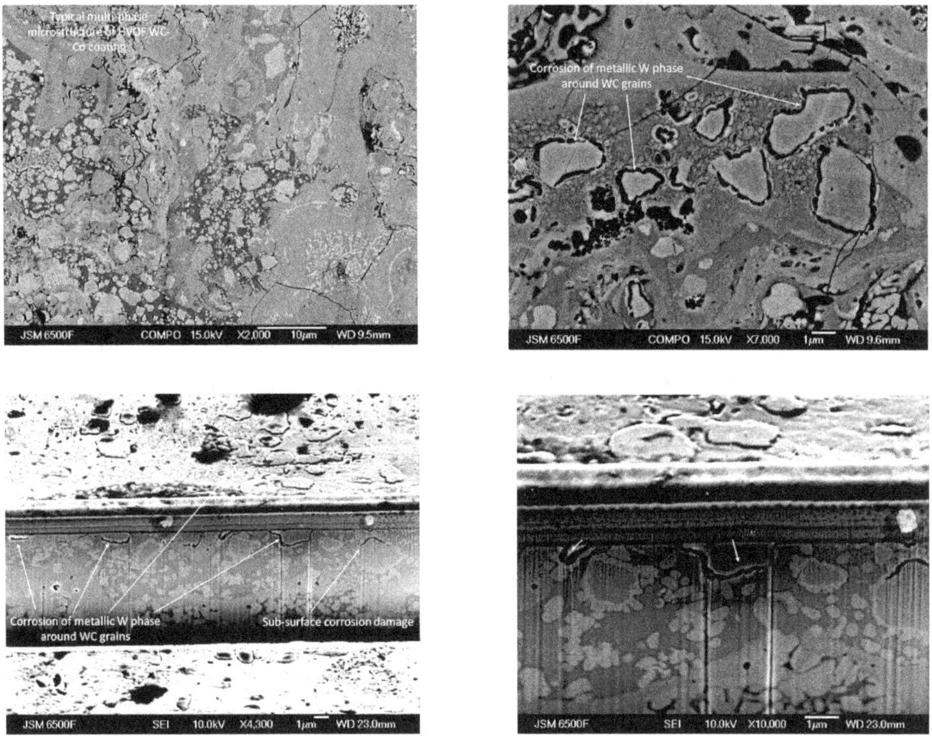

Figure 7. A model proposed by Thakare et al for the corrosion mechanism in WC-based thermal spray coatings in neutral and alkaline conditions (Wood, 2007)

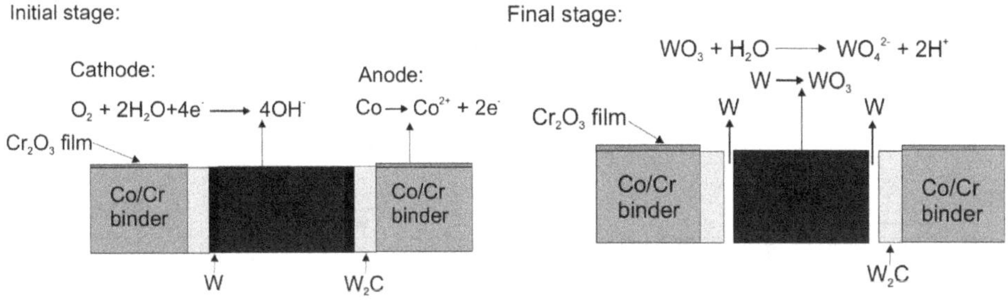

can be embedded into the surface and then subsequently coated, thereby degrading the adhesive properties of the coating/substrate interface or the coating cohesion by weakening the bonding between any inclusions and the matrix. Tribo-corrosion includes the interaction of corrosion with:

- Solid particle erosion,
- Abrasion,
- Cavitation erosion,
- Fretting,
- Biological solutions,
- Sliding wear and tribo-oxidation.

These interactions are represented schematically in Figure 8.

Batchelor *et al.* (Batchelor & Stachowiak, 1988) studied the effects of passive film on the abrasive wear of mild steel zinc and magnesium and found that the regular removal of passive film by abrasion and its regrowth results in a greater material loss than undisturbed corrosion. However, they also noted that the synergistic effect largely depended on the physical nature of the passive film formed which could offer additional protection against mechanical wear, see Figure 9. Table 2 tabulates the mechanisms resulting from a conjoint action of wear and corrosion which could result in either positive or negative wear-corrosion interactions.

The degradation of tungsten carbide (WC) based hardmetals and coatings due to the combined effects of wear and corrosion is complex and it is often difficult to differentiate between the effects of wear and corrosion.

The wear-corrosion synergy performance of WC-based coatings have been studied previously (Toma et al., 2001; Perry et al., 2001; Valentinelli, Valente, Casadei, & Fedrizzi, 2004; Shipway & Howell, 2005a; Shipway & Wirojanupatump, 2002; Rogne et al., 1996; Souza & Neville, 2003a, 2005)mainly focussing on the performance of WC-based coatings in acidic environments. As discussed in the previous

Figure 8. Possible interactions between corrosion and the various wear mechanisms (Wood, 2007)

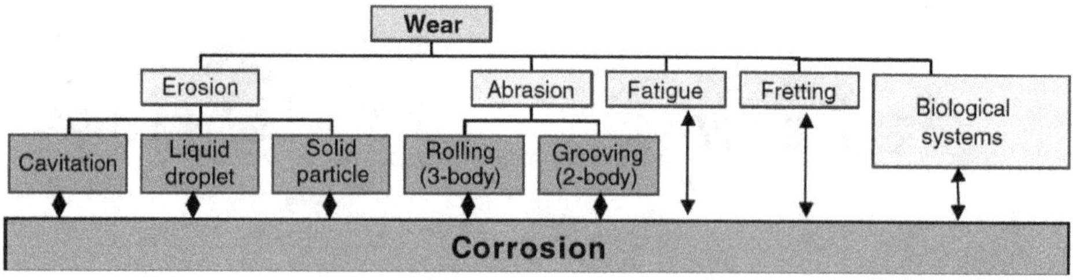

Figure 9. Illustration of the different mechanisms interacting during abrasion-corrosion on a metallic surface in corrosive environments(Wood, 2007)

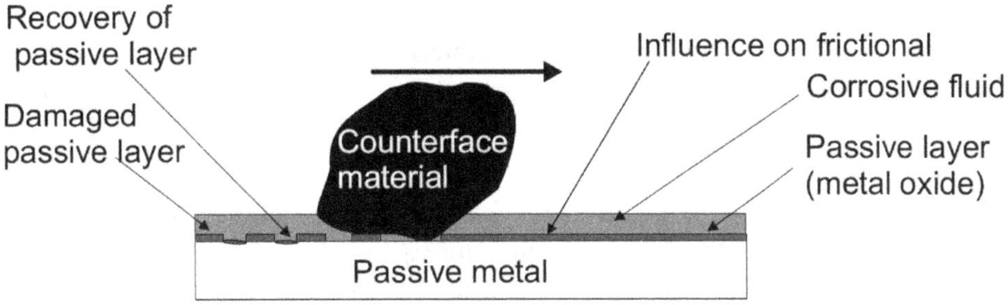

Table 2. Tribo-corrosion mechanisms with possible negative and positive synergistic mechanisms

Mechanism	Positive Interaction	Negative Interaction
Passive film state	Depassivation (removal) of the passive film exposes fresh surface for further corrosion attack and accelerates material removal	Oxide layer could reduce the friction between abrasives and bulk material. Presence of denser oxide film can reduce corrosion as well as abrasion
Dissolution of active species in multiphase surfaces	Could lead to loss of bond integrity between the hard and binder phase. Preferential corrosion of binder undermines the hard phase and facilitates its easy removal by abrasive wear.	Corrosion of the interface can increase the compliance of the surface to abrasive particles.
Plastic deformation/strain	Plastically deformed and stressed surfaces enhance corrosion. Corrosion can lead to premature detachment of plastically deformed or strain hardened splats.	Strain hardness would result in increased hardness and lead to a lower wear. Change in microstructure, grain orientation could lead to lower corrosion rate.
Surface roughness	Could lead to easier removal of oxide film by asperity peaks.	Surface roughness influences the contact mechanics of angular particles and the surface. Increase in R_a could reduce contact stresses and stress distribution

section, the uneven distribution of the binder-phase and the hard-phase within the coating making the coatings vulnerable to corrosion attack. However, increasing the content of the more corrosion resistant binder improves the wear and corrosion performance of the coating (Karimi et al., 1995; Toma et al., 2001; Barbezat et al., 1993; Rogne et al., 1996). As found by Souza and Neville(Souza & Neville, 2003a, 2005), corrosion of the WC-Co-Cr coating initiates at the binder-carbide interface and results in the accelerated removal of the carbide particles by subsequent erosive wear. A similar observation was made by Stack and El Badia (Stack & Badia, 2006) during the erosion-corrosion study of WC/Co-Cr coating. They found that preferential corrosion at the binder-carbide interface led to an increase in the mass loss during erosion-corrosion test. However, they also observed that under conditions when the Co-Cr binder was likely to passivate, there was a mild decrease in the mass loss during erosion-corrosion.

It was proposed that this was probably due to the additional protection offered by the passivation of the binder-phase. Valentinelli *et al.* (Valentinelli et al., 2004) noted that the wear-corrosion performance was strongly influenced by the passivation of the binder-phase and suggested the selection of the binder-composition in consultation with the Pourbaix diagrams (potential vs. pH) to ensure better performance. As discussed in the previous section, Kalish (Kalish, 1987) observed that the corrosion rates for WC-based hardmetals are much higher in acidic conditions than in alkaline conditions due to the lower corrosion resistance of Co and Ni binders as compared to the hard WC particles in acidic environments. This has led to recent studies of wear-corrosion performance of WC-based hardmetals and coatings focussing on evaluating the performance of WC-based hardmetals in acidic conditions. Toma *et al* (Toma et al., 2001) examined the erosion and corrosion resistance of coatings produced by HVOF spraying at ambient temperatures in 0.1M NaOH and 0.1M H2SO4 solutions containing sand. Wear–corrosion tests showed that the corrosion properties of the sprayed coatings strongly affect the material loss rate under wear–corrosion conditions. This work emphasises the importance of the corrosion resistance and passivity of the binder as coatings with less corrosion resistant matrix enhanced erosion. Also the erosion mechanism of the carbide coatings seems to be controlled by the skeletal network of the carbides. For comparison, flame and plasma sprayed hard Cr2O3 coatings were examined. Due to the high impedance of the coating, the corrosion rate was very low. Table 3 shows corrosion and erosion–corrosion rates for the coatings tested by Toma (Toma et al., 2001) which were exposed to solutions of NaOH or H2SO4 as well as seawater and show that the environment has a major influence on corrosion rates and that the addition of sand impingement increases material loss rates except for the Cr_3C_2-NiCr coating in NaOH.

The role of binder corrosion in the erosion–corrosion degradation of HVOF thermally sprayed WC–Co–Cr coatings has been studied by Souza and Neville (Souza & Neville, 2005) using electrochemical polarization techniques in a 3.5% NaCl solution at varying temperatures. They report that initially corrosion proceeds primarily by dissolution of the Co phase, depending on the integrity and thickness of the passive film (thought to be mainly Cr_2O_3) on the binder phase. Similar observations are reported for other compositions of thermally sprayed cermet coatings

Table 3. Some observed results – influence of pH on wear-corrosion rates: results from the electrochemical measurements and erosion–corrosion tests at 2.35ms^{-1}, 100 g l^{-1} silica sand (size between 100–500µm), 20 oC (Toma et al., 2001)

Coating	Electrochemical Measurement Corrosion Rate (mm yr $^{-1}$)			Erosion - Corrosion Rate (mm yr $^{-1}$)	
	NaOH	H_2SO_4	Seawater	NaOH	Seawater
WC-Cr_3C_2-Ni	0.38	0.15		0.4	
Cr_3C_2-NiCr	0.17	0.077		0.17	
WC-Co			0.76		1.6
WC-Co-Cr			0.32		0.55
Cr_3C_2-Al_2O_3-TiO_2	3.2 x 10^{-5}	3.6 x 10^{-5}		0.27	
Cr_3C_2	7.6 x 10^{-3}	1.5 x 10^{-3}		0.35	

(WC–CrNi and WC/CrC–CoCr) (Souza & Neville, 2003b). The dissolution of the binder matrix phase at the hard phase/matrix interface is a known mechanism for synergy contributing to ΔEc. The action of erosion is also likely to weaken the hard phase/matrix interface and after cyclic loading from repeated solid particle impacts could lead to crack initiation at the interface. Binder dissolution issues are also seen to control the erosion–corrosion rates for 50(WC +12Co) balance NiCrSiFeC and WC–17Co coatings studied by Bjordal *et al* (Bjordal, Bardal, Rogne, & G, 1995).

Permeation of electrolyte into these cracks could induce localized environments which are dramatically different (i.e. low pH) from the bulk conditions and induce crevice corrosion, which in turn contributes to increased ΔCe levels. The influence of the spraying parameters (spraying distance, barrel length and kerosene flow rate) on the tribo-corrosion performance of HVOF sprayed WC–17Co coatings has been studied by Valentinelli*et al*(Valentinelli et al., 2004). They also found that binder dissolution controlled wear–corrosion levels and therefore coating compositions should be designed with reference to the relevant Pourbaix diagrams (potential versus pH regime maps based on thermodynamic considerations only) to select compositions that are passive in the service environment and in addition have strong repassivation tendencies. Cr_3C_2 coatings typically have lower wear resistance but greater fracture toughness than WC–Co based coatings as seen by Fedrizzi *et al* (Fedrizzi, et al., 2007). They used a sliding contact (block-on-ring) with a silicon carbide counter face in 3.5% NaCl solutions at 0.3ms^{-1} sliding speeds. The increased fracture toughness was seen to limit crack initiation during surface grinding compared with WC–Co systems. They also found the composition of binders important, with the NiCr alloy exhibiting faster repassivation characteristics compared with the Co-based binders.

HVOF coatings with compositions of 86WC-10Co-4Cr or corrosion resistant alloys such as nickel aluminium bronze have been considered for wear–corrosion applications, Souza and Neville(Souza & Neville, 2005), Wood and Speyer (K & J, 2004) and Barik *et al* (Barik, Wharton, Wood, & Strokes). Of particular concern is the level of porosity (>2%) within the coatings which can accelerate crack propagation and coating removal under erosion and can also be interconnected such that electrolyte can permeate into the coating/substrate interface accelerating through thickness corrosion of the coating and corrosion-driven coating–substrate disbonding, see Figure 10.

The erosion–corrosion performance of these coatings has been studied using free or submerged solid-liquid slurry jets along with a mixture of gravimetric and electrochemical techniques. An example of monitoring the electrochemical potential of a coated steel specimen during erosion–corrosion jet impingement is shown in Figure 11. The potential signal detects electrolyte ingress and corrosion activity at the coating/substrate interface, see Figure 12, as well as coating penetration by erosion. An additional concern with multiphase coatings (carbide/metal matrix) is the potential for debonding between the hard phase and the softer matrix which can accelerate surface degradation.

Coatings deposited with powders containing narrow grain size distributions are preferred as this provokes more controllable and homogeneous particle melting and avoids the particle size influence at a given flame temperature causing over or under heating of powder particles, Berget *et al* (Berget J, E, & T, 1998; Berget J, T, & E, 2007). The influence of the corrosion characteristics of the substrate is also important and can play a vital role in the coated system performance. Al-Fadhli *et al.* (Al-Fadhli H Y, Stokes J, J, & S, 2006) studied the erosion–corrosion characteristics of HVOF thermally sprayed Inconel 625 coatings on different metallic sur-

Figure 10. Disbonded HVOF aluminium bronze coating under erosion–corrosion conditions due to electrolyte ingress to the coating/substrate interface followed by corrosion of the carbon steel substrate and the generation of voluminous corrosion product accelerating crack propagation at the coating/substrate interface(Wood, 2007)

Figure 11. Potential–time curve of a HVOF aluminium bronze coating on AISI 1020 steel substrate under slurry erosion conditions in 3.5% NaCl (Wood, Wharton, Speyer, & Tan, 2002)

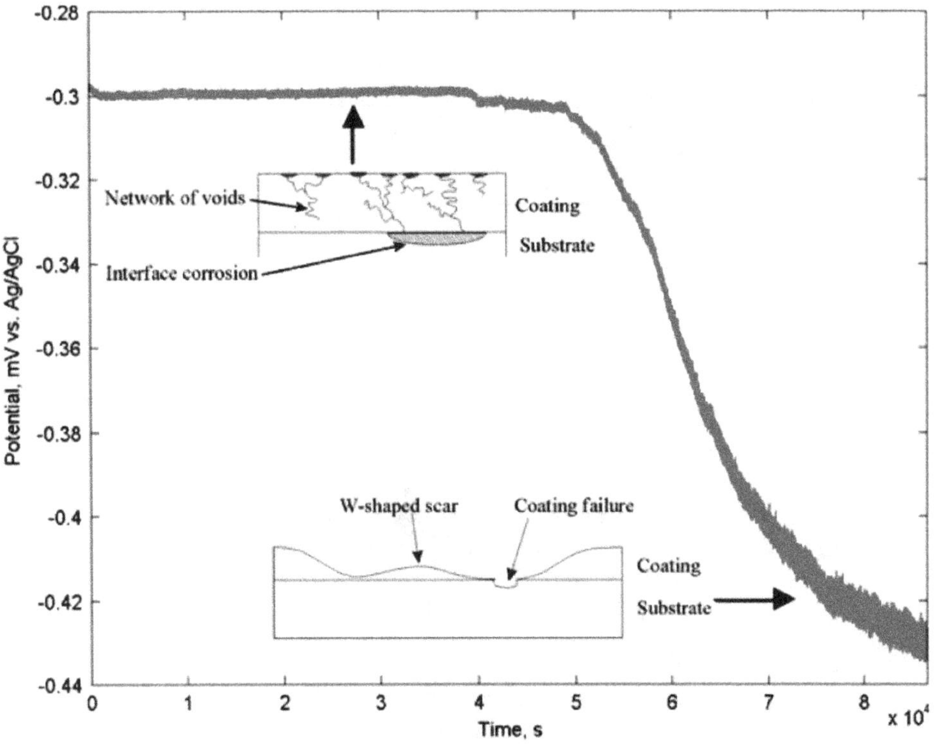

faces that simulated coating over weld repaired or refurbished surfaces found in the offshore industry, namely, bulk stainless steel and a spot welded stainless steel and a composite surface of welded stainless steel and carbon steel. Their results indicate that the coating over both spot-welded and plain stainless steel surfaces exhibited a similar degree of weight loss. However, the coating on the composite surface experienced a greater degree of weight loss and spallation due to electrolyte ingress through the porous coating and subsequent corrosion of the carbon steel substrate. Microscopic observations of the fracture surfaces showed that the metal removal from the surface was concentrated around the unmelted and the semi-melted particles of the weld deposit. Post-coating treatments are a potential way to remove some of the detrimental microstructural features and offer a way to improve the tribo-corrosion performance of coatings. Liu *et al* (Liu Z, Cabrero J, S, & T, 2007) show the potential benefits of laser treated HVOF Inconel 625- and WC-based

Figure 12. (a) Schematic of crack systems in coating generated during erosion–corrosion and the subsequent substrate corrosion and ejection of corrosion products from through-coating corrosion onto the coating surface (Wood, 2007)

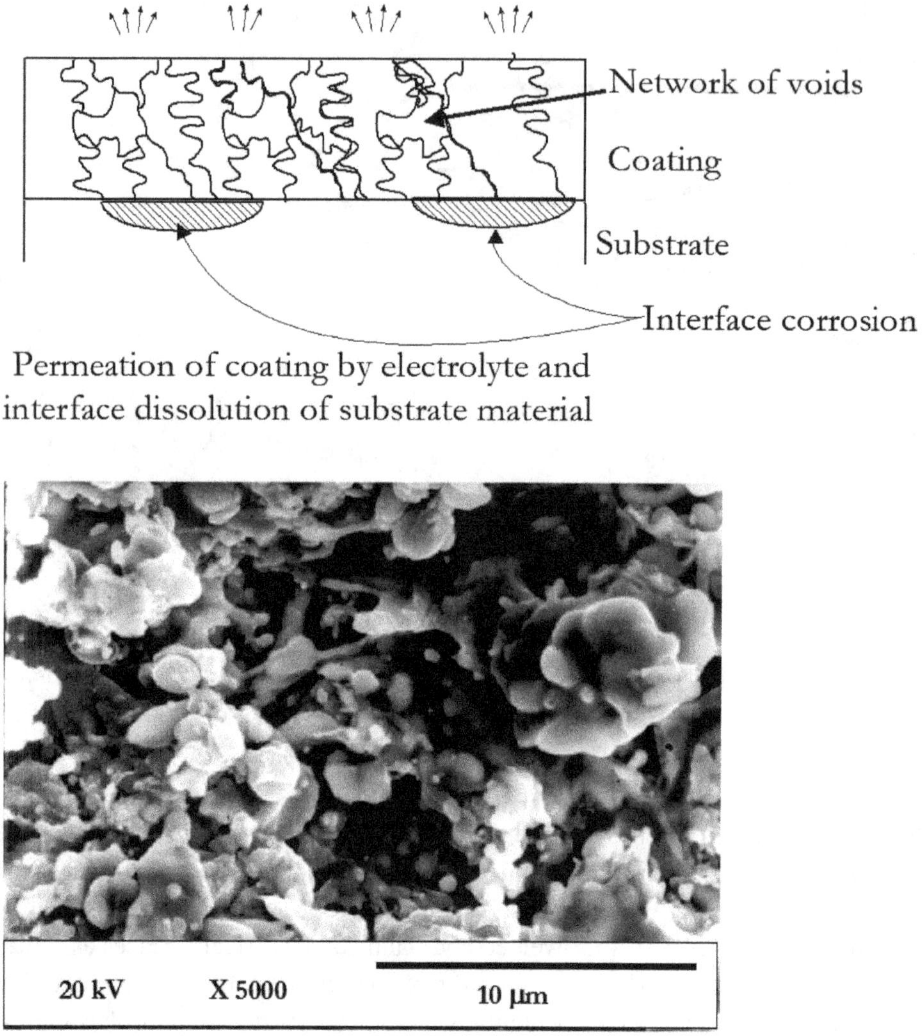

coatings. Improved corrosion and wear resistance was related to the removal of the discrete splat-structures, porosity and micro-crevices as well as the reduction of micro-galvanic activity by compositional gradients between the WC and the matrix due to the formation of interfacial phases. In addition the improved wear resistance was also associated with the formation of rapid solidification microstructure with faceted dendritic WC-phases. A significant reduction in wear was observed when the laser treatment completely dissolved the WC. Inconel 625–42WC coatings appeared the best for wear but Inconel 625–22WC was better against static corrosion in 3.5% NaCl solution.

As many coatings have complex compositions the possibility of micro-galvanic corrosion activity between the different microstructure and the composition is likely to undermine surface integrity. This has been seen in the corrosion and corrosion–wear behaviour of HVOF sprayed WC cermet coatings with metallic binders of Co, Co–Cr, CrC–Ni and Ni, in a strong acidic environment by Cho *et al* (Cho et al., 2006). Considerable micro-galvanic corrosion occurred between the WC particles and the binder, and uniform corrosion occurred in the binder materials of WC–Co, WC–Ni. In spite of the presence of Cr, the overall corrosion resistance of the WC–CrC–Ni coating was inferior to that of the WC–Co–Cr coating due to electrolyte permeation to the coating/substrate interface through interconnecting micro-cracks. The friction coefficient of the coatings without Cr was lower than that of the coatings containing Cr under the corrosion–wear conditions, probably due to lubricious corrosion products within the contact.

The tribo-corrosion performance of nanostructured coatings has also been recently studied for biomedical applications. Basak *et al.* (Basak, 2006) have studied the corrosion and wear–corrosion behaviour of a thermal sprayed nanostructured FeCu/WC–Co coating in a simulated human body fluid (Hank's solution) and compared this with stainless steel AISI 304 and nanostructured WC–Co coatings. The multiphase structure of the FeCu/WC–Co coating induced complex corrosion processes but exhibited depassivation/repassivation behaviour comparable to that of stainless steel AISI 304 and the nanostructured WC–Co coatings. The ability of the coating surface to repassify quickly after the passive film is mechanically damaged is clearly key to being able to minimize the dissolution losses during wear–corrosion. To understand the repassivation kinetics in such complex mechanical and electrochemical environments is not straightforward but some electrochemical techniques may be applicable. For example, electrochemical noise measurements have been used to investigate the interactions between erosion and corrosion of HVOF aluminium, aluminium bronze and aluminium-silicon coatings, Tan (Tan, Wharton, & Wood, 2005) and Wood (Thakare, Wharton, Wood, & Menger, 2012).

Although perturbations (noise) on the corrosion current measurements represented multiple sand particle impacts, valuable insights into the mass transport and repassivation kinetics occurring under erosion–corrosion were obtained. Examples of electrochemical current noise are given in Figures 13, 14 and 15 for flow corrosion, erosion–corrosion and abrasion-corrosion respectively. Ratios of standard deviation of noise have been used to infer repassivation properties of coatings under erosion–corrosion conditions, see Figure 16.

Figure 13. Flow corrosion currents for HVOF NAB coating held at OCP under 90o jet impingement conditions with jet velocities of 3.1 and 6.7 ms^{-1}(Wood, 2007)

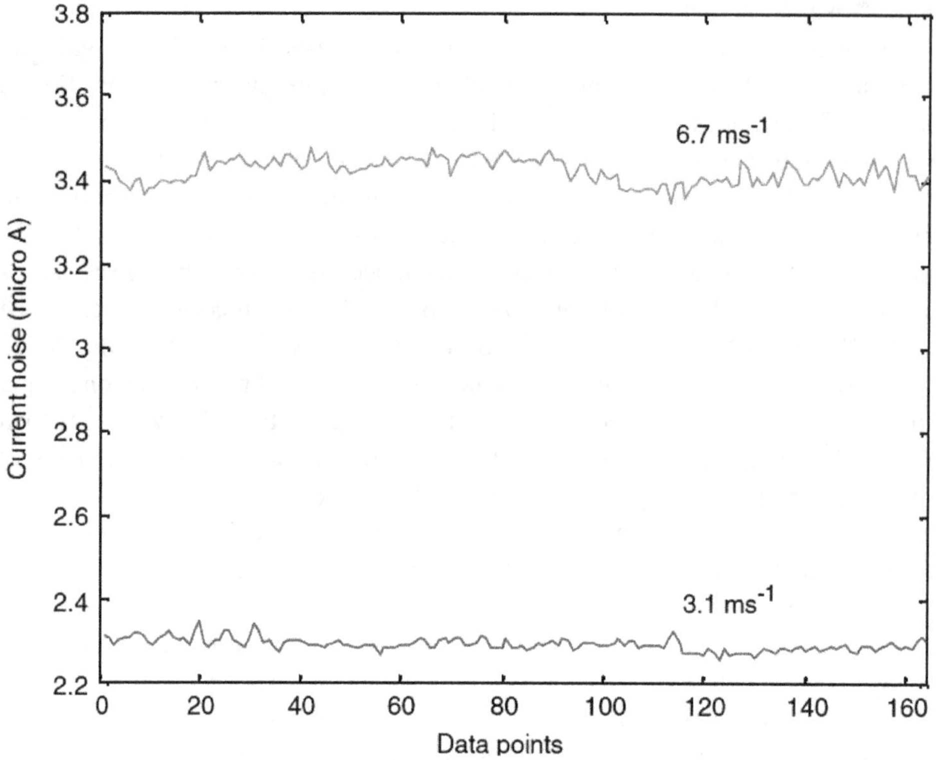

The effect of pH on the overall wear rates was examined by Thakare *et al.* (Thakare et al., 2009). The wear rates observed for HVOF WC-10Co-4Cr coating using a micro-abrasion tester clearly show a strong link between wear-corrosion mechanism and pH, see Figure 17. Tests were carried out using an in situ electrochemical cell capable of holding the sample surface at its open circuit potential. These results along with the static corrosion mechanism of WC-Co coating discussed in the previous section (Figures 6 and 7), can be used to explain the influence of local pH on the overall wear-corrosion mechanism. Selective passivation of the binder-phase, the formation of corrosion trenches around the carbide grains, alkaline dissolution of species and formation of micro-galvanic coupling between active and passive phases within the coating microstructure can easily alter the local conditions resulting in a significant change in the overall wear rates with change in pH. The overall mechanism has been summarised schematically in Figure 18.

Figure 14. Erosion–corrosion currents for HVOF NAB coating held at OCP under 90o slurry jet impingement conditions with jet velocities of 3.1 and 6.7ms^{-1} and 2% w/w sand in seawater slurry (Wood, 2007)

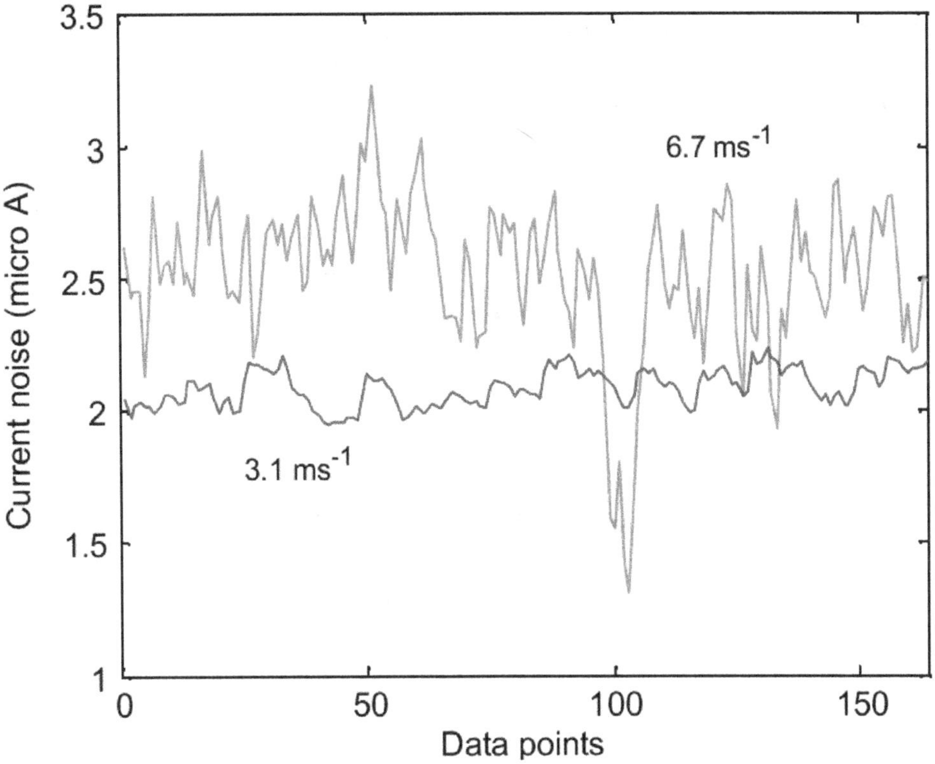

Tribo-Corrosion Models

The use of *in situ* electrochemical measurement techniques has offered considerable insights into the local interactions during tribo-corrosion as discussed earlier. These techniques have also made it possible to examine the various components of the synergy equation (Equations 4 and 5) and map the electrochemical and mechanical components of tribo-corrosion under different operating conditions, resulting in the creation of tribo-corrosion "maps".

Figure 19 shows an example for a HVOF nickel aluminium bronze coating on carbon steel, uncoated carbon steel and cast nickel aluminium bronze. These maps are useful for considering compositions for improved resistance by highlighting

Figure 15. Abrasion–corrosion currents for HVOF WC-10Co-4Cr coating held at OCP during micro-abrasion testing under different pH conditions (Thakare, Wharton, Wood, & Menger, 2009).Figure 16. Standard deviation ratios of electrochemical current noise from erosion–corrosion divided by flow corrosion levels plotted against erosion mass loss by gravimetric analysis for carbon steel AISI 1020, HVOF NAB coating and cast NAB under jet impingement conditions with and without 2% w/w sand in seawater (Barik et al.)

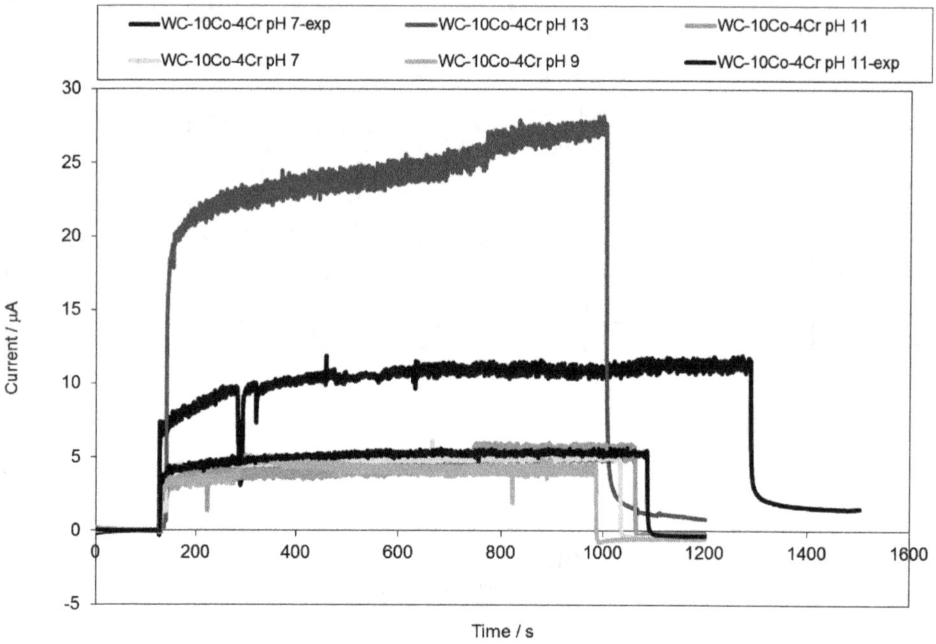

whether mechanical or electrochemical aspects need to be inhibited. Figure 20 illustrates a similar map for abrasion–corrosion of various surfaces including sintered cermets, stainless and duplex stainless steels as well as thermally sprayed tungsten carbide coatings. The trend at present in tribo-corrosion is away from quantifying synergy as it usually involves strong debates over how it was quantified. Sometimes gravimetric and electrochemically derived mass losses are needed and scant regard is given to the cumulative errors that arise when calculating any interactive term such as ΔPC_A and ΔPA_C The current author prefers the approach of isolating mechanical and electrochemical or chemical contributions as the balance between these can inform surface selection or suggest improved coating microstructure/compositions. The focus should be on the resultant total wear rate (*AC*) as this needs to be minimized to yield optimum component life.

Figure 16. Standard deviation ratios of electrochemical current noise from erosion–corrosion divided by flow corrosion levels plotted against erosion mass loss by gravimetric analysis for carbon steel AISI 1020, HVOF NAB coating and cast NAB under jet impingement conditions with and without 2% w/w sand in seawater (Barik et al.)

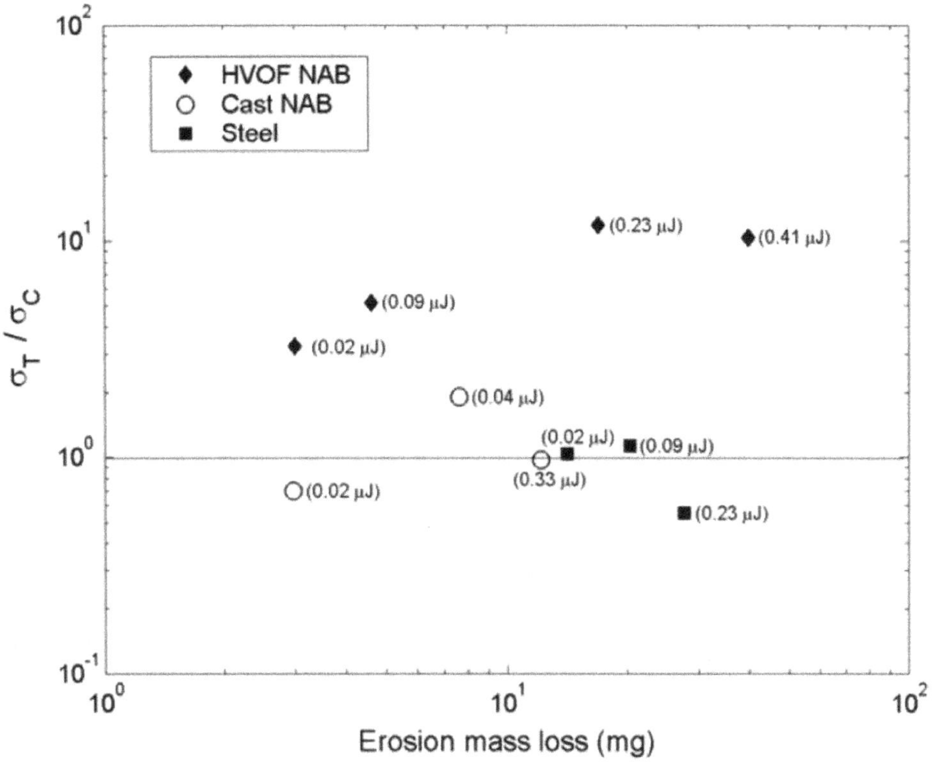

SUMMARY

Overall there are relatively few papers reporting research into the tribo-corrosion of coatings but, of those reviewed, thermally sprayed HVOF coatings are the most common being the subject of erosion–corrosion and sliding wear–corrosion investigations while PVD coatings have also received some attention for sliding wear–corrosion applications. It is, therefore, extremely difficult to gain generic insight into the fundamental behaviour from the published work. Most experiments test existing coatings that have been developed for either wear resistance or corrosion resistance. Coupling this to the fact that most corrosion resistant elements are not wear resistant and vice versa, these coatings, therefore, perform very badly when tested under combined wear–corrosion conditions. There is still considerable scope

Figure 17. Wear rates observed for WC-10Co-4Cr coating using ma micro-abrasion tester under different pH conditions(Thakare et al., 2009)

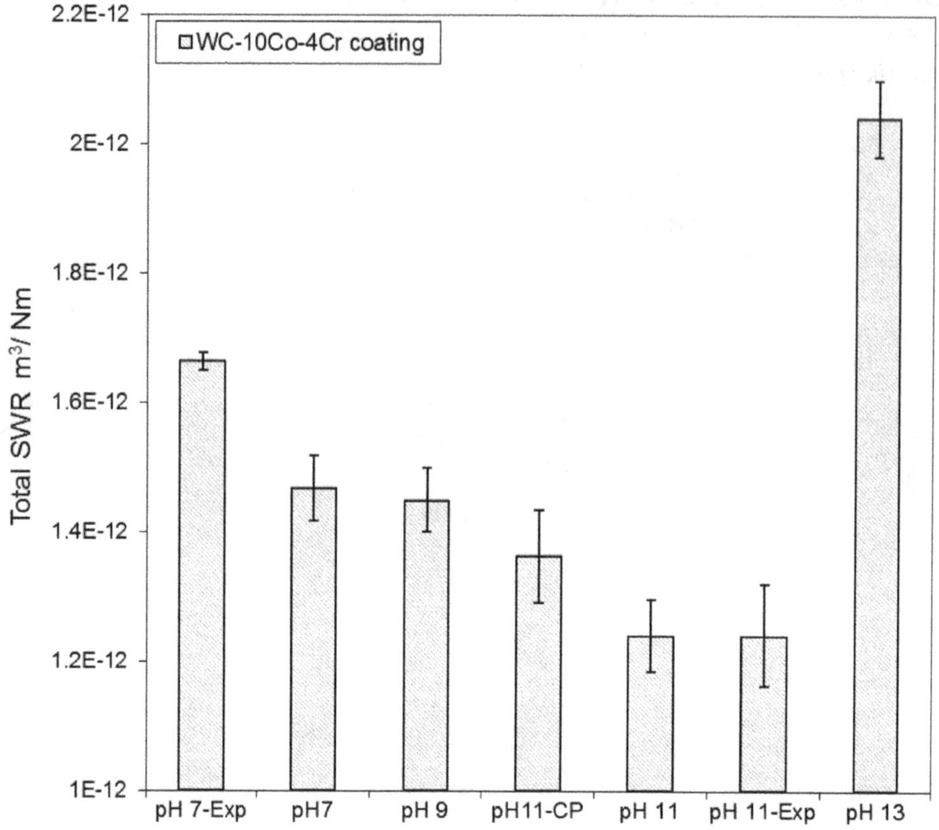

to design coatings specifically for wear–corrosion resistance with coating composition, microstructure, defect level, adhesion, cohesion and substrate properties being seen as some of the critical elements in coating performance in tribo-corrosion contacts. The influence of the corrosive environment is also shown to be important in controlling wear and friction as well as the evolution of near surface tribo-induced layers which are critical in both erosion–corrosion and wear–corrosion situations. This chapter has identified some of the key interactions between wear and corrosion mechanisms that occur in thermal spray coating systems along with some models and mapping techniques that aim to inform microstructure selection and predict performance. However, these need major development before robust predictions can be made as our understanding of the mechano-electrochemical interactions that are present in tribo-corrosion is incomplete.

Figure 18. Schematic of wear-corrosion process for thermal-spray coating showing the individual mechanisms such as depassivation of the binder-rich areas, micro-galvanic coupling between phases, and re-embedment of wear debris, all of which would influence the overall material loss in the coatings (Thakare et al., 2009)

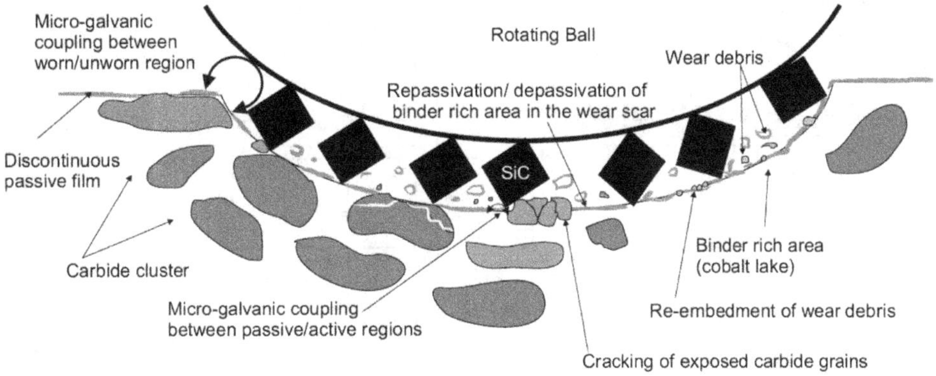

Figure 19. Erosion–corrosion regime map with electrochemical material loss versus mechanical material loss for erosion–corrosion of carbon steel AISI 1020, HVOF nickel–aluminium bronze coating and cast nickel–aluminium bronze under jet impingement conditions with 2% w/w sand in seawater

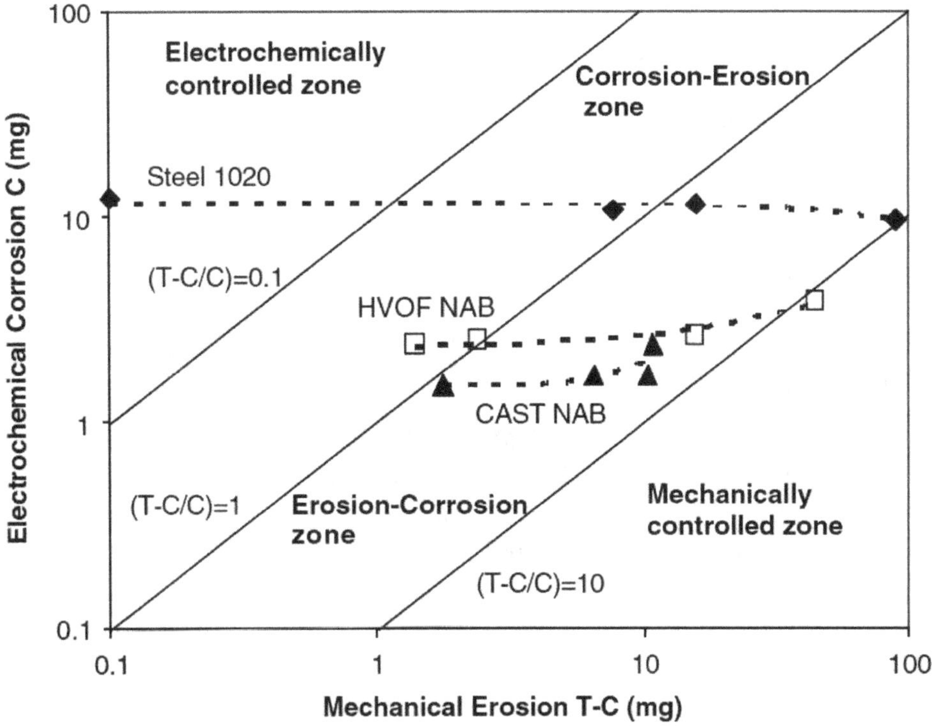

Figure 20. Tribo-corrosion mechanism regime map (Wood, 2007)

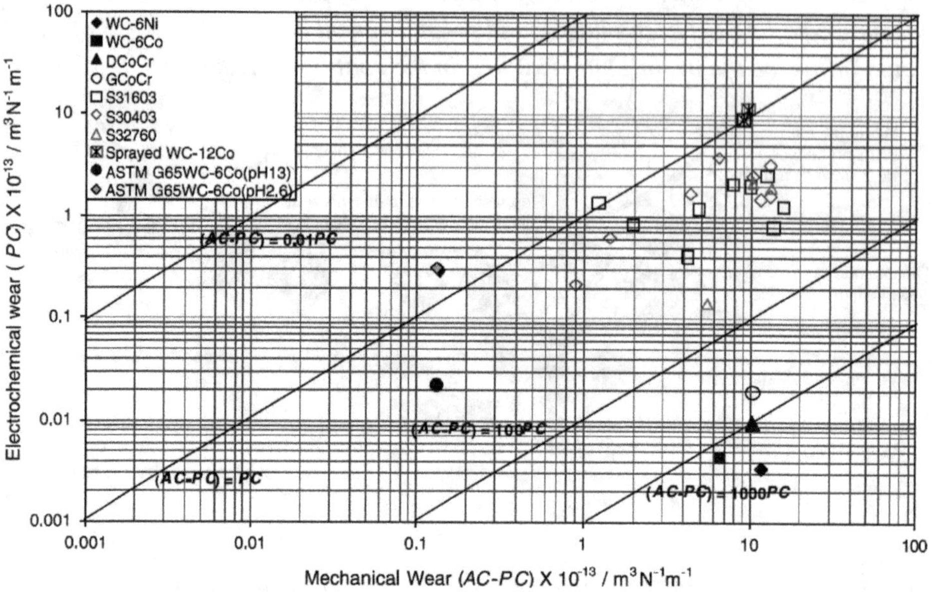

There is also very limited research on the performance of some of the newer thermal spray technologies such as the nano-structured coatings and cold spray systems. These newer systems aim to improve the homogeneity of the microstructure, reduce porosity and minimise formation of secondary phases – all of which should lead to an improved wear performance. However, the effects of these on the synergistic wear-corrosion interactions remain less understood to an extent and should be the focus of further investigation.

REFERENCES

Ahn, Y., Cho, N.-G., Lee, S.-H., & Lee, D. (2003). Lateral crack in abrasive wear of brittle materials. *JSME International Journal, Series A, 46*(2), 140-144.

Al-Fadhli, H. Y., Stokes, J., Hashmi, M. S. J., & Yilbas, B. S. (2006). The erosion–corrosion behaviour of high velocity oxy-fuel (HVOF) thermally sprayed inconel-625 coatings on different metallic surfaces. *Surface and Coatings Technology, 200*(20-21), 5782–5788. doi:10.1016/j.surfcoat.2005.08.143

ASTM G119-93. (1998). Standard guide for determining synergism between wear and corrosion. In *ASTM handbook* (pp. 529-534). West Conshohocken: ASTM.

Axen, N., & Jacobson, S. (1994). A model for abrasive wear resistance for multiphase materials. *Wear, 174*(1-2), 187–199. doi:10.1016/0043-1648(94)90101-5

Barber, J., Mellor, B. G., & Wood, R. J. K. (2005). The development of sub-surface damage during high energy solid particle erosion of a thermally sprayed WC-Co-Cr coating. *Wear, 259*(1-6), 125–134. doi:10.1016/j.wear.2005.02.008

Barbezat, G., Nicoll, A., & Sickinger, A. (1993). Abrasion, erpsopn and scuffing resistance of carbide and oxide ceramic thermal sprayed coatings for different applications. *Wear, 162-164*, 529–537. doi:10.1016/0043-1648(93)90538-W

Barik, R.C., Wharton, J., Wood, R. J. K., & Strokes, K. (n.d.). *Erosion and erosion-corrosion performance of cast and thermally sprayed nickel-aluminium bronze (NAB).* Surface Engineering and Tribology Group, University of Southampton.

Basak, A. K., Matteazzi, P., Vardavoulias, M., & Celis, J.-P. (2006). Corrosion–wear behaviour of thermal sprayed nanostructured FeCu/WC–Co coatings. *Wear, 261*(9), 1042–1050. doi:10.1016/j.wear.2006.03.026

Batchelor, A. W., & Stachowiak, G. W. (1988). Predicting synergism between corrosion and abrasive wear. *Wear, 123*(3), 281–291. doi:10.1016/0043-1648(88)90144-5

Berget, J. E. B. (1998). *Effects of powder composition on the erosion, corrosion and erosion–Corrosion properties of HVOF sprayed WC based coatings.* Paper presented at the 15th Int. Thermal Spray Conference - Meeting the Challenges of 21st Century, Nice, France.

Berget, J., Rogne, T., & Bardal, E. (2007). Erosion–corrosion properties of different WC–Co–Cr coatings deposited by the HVOF process—Influence of metallic matrix composition and spray powder size distribution. *Surface and Coatings Technology, 201*(18), 7619–7625. doi:10.1016/j.surfcoat.2007.02.032

Bjordal, M., Bardal, E., Rogne, T., & Eggen, T. G. (1995). Erosion and corrosion properties of WC coatings and duplex stainless steel in sand-containing synthetic sea-water. *Wear, 186–187*, 508–514. doi:10.1016/0043-1648(95)07148-2

Bolelli, G., Giovanardi, R., Lusvarghi, L., & Manfredini, T. (2006). Corrosion resistance of HVOF sprayed coatings for hard chrome replacment. *Corrosion Science, 48*(11), 3375–3397. doi:10.1016/j.corsci.2006.03.001

Cantera, E. L., & Mellor, B. G. (1998). Fracture toughness and crack morphologies in eroded WC-Co-Cr thermally sprayed coatings. *Materials Letters*, *37*(4-5), 201–210. doi:10.1016/S0167-577X(98)00092-5

Chen, H., Xu, C., Zhou, Q., Hutchings, I. M., Shipway, P. H., & Liud, J. (2005). Micro-scale abrasive wear behaviour of HVOF sprayed and laser-remelted conventional and nanostructured WC-Co coatings. *Wear*, *258*(1-4), 333–338. doi:10.1016/j.wear.2004.09.044

Cho, J. E., Hwang, S. Y., & Kim, K. Y. (2006). Corrosion behaviour of thermal sprayed WC cermet coatings having various metallic binders in strong alkaline environment. *Surface and Coatings Technology*, *200*(8), 2653–2662. doi:10.1016/j.surfcoat.2004.10.142

Fedrizzi, L., Valentinelli, L., Rossi, S., & Segna, S. (2007). Tribocorrosion behaviour of HVOF cermet coatings. *Corrosion Science*, *49*(7), 2781–2799. doi:10.1016/j.corsci.2007.02.003

Gant, A. J., & Gee, M. G. (2006). Abrasion of WC hardmetals using hard countefaces. *International Journal of Refractory Metals & Hard Materials*, *24*(1-2), 189–198. doi:10.1016/j.ijrmhm.2005.05.007

Guilemany, J. M., Paco, J. M., Nutting, J., & Miguel, J. R. (1999). Characterisation of W_2C phase formed during HVOF spraying of a WC-12Co powder. *Metallurgical and Materials Transactions. A, Physical Metallurgy and Materials Science*, 1913–1921. doi:10.1007/s11661-999-0002-3

Hochstrasser, S. (2006). Mechanistic study of corrosion reactions on WC-Co hardmetal in aqueous solution- an investigation by electrochemical methods and elemental solution analysis. (Doctor of Science Thesis). ETH Zurich, Zurich, Switzerland.

Ibrahim, A., & Berndt, C. C. (2007). Fatigue and deformation of HVOF sprayed WC–Co coatings and hard chrome plating. *Materials Science and Engineering A*, *456*(1-2), 114–119. doi:10.1016/j.msea.2006.12.030

K, W. R. J., & J, S. A. (2004). Erosion–corrosion of candidate HVOF aluminium-based marine coatings *Wear, 256*, 545–556.

Kalish, H. S. (1987). *Corrosion of cemented carbides* (9th ed.; Vol. 13). ASM International.

Karimi, A., Verdon, C., Martin, J., & Schmid, R. (1995). Slurry erosion behaviour of thermally sprayed WC-M coatings. *Wear*, *186-187*, 480–486. doi:10.1016/0043-1648(95)07154-7

Liao, H., Normand, B., & Coddet, C. (2000). Influence of coating microstructure on the abrasive wear resistance of WC/Co cermet coatings. *Surface and Coatings Technology, 124*(2-3), 235–242. doi:10.1016/S0257-8972(99)00653-2

Lima, M. M., Godoy, C., Avelar-Batista, J. C., & Modenesi, P. J. (2003). Toughness evaluation of HVOF WC-Co coatings using non-linear regression analysis. *Materials Science and Engineering A, 357*(1-2), 337–345. doi:10.1016/S0921-5093(03)00204-1

Liu, Z., Cabrero, J., Niang, S., & Al-Taha, Z. Y. (2007). Improving corrosion and wear performance of HVOF-sprayed Inconel 625 and WC-Inconel 625 coatings by high power diode laser treatments. *Surface and Coatings Technology, 201*(16-17), 7149–7158. doi:10.1016/j.surfcoat.2007.01.032

Miguel, J. M., Guilemany, J. M., Mellor, B. G., & Xu, Y. M. (2003). Acoustic emission study on WC-Co thermal sprayed coatings. *Materials Science and Engineering A, 352*(1-2), 55–63. doi:10.1016/S0921-5093(02)00546-4

Niihara, K. (1983). A fracture mechanics analysis of indentation induced Palmqvist cracks in ceramics. *Journal of Materials Science Letters, 2*(5), 221–223. doi:10.1007/BF00725625

Perry, J. M., Neville, A., Wilson, V., & Hodgkiess, T. (2001). Assessment of the corrosion rates and mechanisms of a WC-Co-Cr HVOF coatings in static and liquid-solid impingement saline environments. *Surface and Coatings Technology, 137*(1), 43–51. doi:10.1016/S0257-8972(00)01062-8

Pina, J., Diasb, A., & Lebrun, J. (2003). Study by x-ray diffraction and mechanical analysis of the residual stress generation during thermal spraying. *Materials Science and Engineering A, 347*(1-2), 21–31. doi:10.1016/S0921-5093(02)00580-4

Pourbaix, M. (1974). *Atlas of electrochemical equilibria in aqueous solutions*. Houston, TX: NACE-Cebelcor.

Ramnathan, V., & Jayaraman, N. (1989). Characterisation and wear performance of plasma sprayed WC-Co coatings. *Materials Science and Technology, 5*(4), 382–388. doi:10.1179/mst.1989.5.4.382

Rogne, T., Bjordal, M., Solem, T., & Bardal, E. (1996). *The importance of corrosion on the erosion-corrosion performance of thermal spray ceramic-metallic coatings*. ASM International.

Shetty, D. K., Wright, I. G., Mincer, P. N., & Clauer, A. H. (1985). Indentation fracture of WC-Co cermets. *Journal of Materials Science, 20*(5), 1873–1882. doi:10.1007/BF00555296

Shipway, P. H., & Howell, L. (2005a). Microscale abrasion-corrosion behaviour of WC-Co hardmetals and HVOF sprayed coatings. *Wear*, *258*(1-4), 303–312. doi:10.1016/j.wear.2004.04.003

Shipway, P. H., & Howell, L. (2005b). Microscale abrasion-corrosion behaviour of WC-Co hardmetals and HVOF sprayed coatings. *Wear*, *258*(1-4), 303–312. doi:10.1016/j.wear.2004.04.003

Shipway, P. H., McCartney, D. G., & Sudaprasert, T. (2005). Sliding wear behaviour of conventional and nanostructured HVOF sprayed WC-Co coatings. *Wear*, *259*(7-12), 820–827. doi:10.1016/j.wear.2005.02.059

Shipway, P. H., & Wirojanupatump, S. (2002). The role of lubrication and corrosion in abrasion of materials in aqueous environments. *Tribology International*, *35*(10), 661–667. doi:10.1016/S0301-679X(02)00057-9

Souza, V. A. D., & Neville, A. (2003a). Corrosion and erosion damage mechanisms during erosion-corrosion of WC-Co-Cr cermet coatings. *Wear*, *255*(1-6), 146–156. doi:10.1016/S0043-1648(03)00210-2

Souza, V. A. D., & Neville, A. (2003b). Linking electrochemical corrosion behaviour and corrosion mechanisms of thermal spray cermet coatings (WC-CrNi and WC/CrC-CoCr). *Materials Science and Engineering A*, *352*(1-2), 202–211. doi:10.1016/S0921-5093(02)00888-2

Souza, V. A. D., & Neville, A. (2005). Corrosion and synergy in a WC-Co-Cr HVOF thermal spray coating-understanding their role in erosion-corrosion degradation. *Wear*, *259*(1-6), 171–180. doi:10.1016/j.wear.2004.12.003

Stack, M. M., & Badia, T. M. A. E. (2006). On the construction of erosion-corrosion maps for WC/Co-Cr based coatings in aqueous conditions. *Wear*, *261*(11-12), 1181–1190. doi:10.1016/j.wear.2006.03.038

Stewart, D., Shipway, P., & McCartney, D. G. (1999). Abrasive wear behaviour of conventional and nanocomposite HVOF-sprayed WC-Co coatings. *Wear*, *225-229*, 789–798. doi:10.1016/S0043-1648(99)00032-0

Stewart, D., Shipway, P., & McCartney, D. G. (2000). Microstructural evolution in thermally sprayed WC-Co coatings: Comparison between nanocomposite and conventional starting powder. *Acta Materialia*, *40*, 1596–1604.

Stokes, J., & Looney, L. (2004). Residual stress in HVOF thermally sprayed thick deposits. *Surface and Coatings Technology*, *177-178*, 18–23. doi:10.1016/j.surfcoat.2003.06.003

Sudaprasert, T., Shipway, P. H., & McCartney, D. G. (2003). Sliding wear behaviour of HVOF sprayed WC-Co coatings deposited with both gas-fuelled and liquid fuelled systems. *Wear*, *255*(7-12), 943–949. doi:10.1016/S0043-1648(03)00293-X

Takeda, M., Morihiro, N., Ebara, E., Harada, Y., Wang, R., & Kido, M. (2002). Corrosion behavior of thermally sprayed WC coating in Na_2SO_4 aqueous solution. *Materials Transactions*, *33*(11), 2860–2865. doi:10.2320/matertrans.43.2860

Tan, K. S., Wharton, J. A., & Wood, R. J. K. (2005). Solid particle erosion-corrosion behaviour of a novel HVOF nickel aluminium bronze coating for marine applications-correlation between mass loss and electrochemical measurements. *Wear*, *258*(1-4), 629–640. doi:10.1016/j.wear.2004.02.019

Thakare, M. R., Wharton, J. A., Wood, R. J. K., & Menger, C. (2007). Exposure effects of alkaline drilling fluid on the microscale abrasion-corrosion of WC-based hardmetals. *Wear*, *263*(1-6), 125–136. doi:10.1016/j.wear.2006.12.047

Thakare, M. R., Wharton, J. A., Wood, R. J. K., & Menger, C. (2009). Investigation of micro-scale abrasion-corrosion of WC-based sintered hardmetals and sprayed coatings using *in situ* electrochemical current-noise measurements. *Wear*, *267*(11), 1967–1977. doi:10.1016/j.wear.2009.06.006

Thakare, M. R., Wharton, J. A., Wood, R. J. K., & Menger, C. (2012). Effect of abrasive particle size and the influence of microstructure on the wear mechanisms in wear-resistant materials. *Wear*, *276-277*, 16–28. doi:10.1016/j.wear.2011.11.008

Toma, D., Brandl, W., & Marginean, G. (2001). Wear and corrosion behaviour of thermally sprayed cermet coatings. *Surface and Coatings Technology*, *138*(2-3), 149–158. doi:10.1016/S0257-8972(00)01141-5

Tomlinson, W. J., & Linzell, C. R. (1988). Anodic polarisation and corrosion of cemented carbides with cobalt and nickel binders. *Journal of Materials Science*, *23*(3), 914–918. doi:10.1007/BF01153988

Tu, D., Chang, S., Chao, C., & Lin, C. (1985). Tungsten carbide phase transformation during the plasma spray process. *Journal of Vacuum Science & Technology. A, Vacuum, Surfaces, and Films*, *3*(6), 2479–2482. doi:10.1116/1.572862

Tucker, R. C. (1999). Thermal spray coatings. In *ASM handbook for surface engineering* (Vol. 5). ASM International.

Valentinelli, L., Valente, T., Casadei, F., & Fedrizzi, L. (2004). Mechanical and tribocorrosion properties of HVOF sprayed WC-Co coatings. *Corrosion Engineering, Science, and Technology*, *39*, 301–307.

Verdon, C., Karimi, A., & Martin, J. (1998). A study of high velocity oxy-fuel thermally sprayed tungsten carbide based coatings: Part I: Microstructures. *Materials Science and Engineering A, 246*(1-2), 11–24. doi:10.1016/S0921-5093(97)00759-4

Verdon, C., Karimi, A., & Martin, J.-L. (1997). Microstructural and analytical study of thermally sprayed WC-Co coatings in connection with their wear resistance. *Materials Science and Engineering A, 234-236*, 731–734. doi:10.1016/S0921-5093(97)00377-8

Vinayo, M., Kassabji, F., Guyonnet, J., & Fauchais, P. (1985). Plasma sprayed WC-Co coatings: Influence of spray conditions (atmospheric and low pressure plasma spraying) on the crystal structure, porosity and hardness. *Journal of Vacuum Science & Technology. A, Vacuum, Surfaces, and Films, 3*(6), 2483–2489. doi:10.1116/1.572863

Wagner, C. D., Naumkin, A. V., Kraut-Vass, A., Allison, J. W., & Powell, C. J., Jr. (2006). NIST x-ray photoelectron spectroscopy database. Measurement Services Division of the National Institute of Standards and Technology (NIST) Technology Services.

Wheeler, D. W., & Wood, R. J. K. (2005). Erosion of hard surface coating for use in offshore gate valves. *Wear, 258*(1-4), 526–536. doi:10.1016/j.wear.2004.03.035

Wood, R. J. K. (2007). Tribo-corrosion of coatings: A review. *Journal of Physics. D, Applied Physics, 40*(18), 5502–5521. doi:10.1088/0022-3727/40/18/S10

Wood, R. J. K., Mellor, B. G., & Binfield, M. L. (1997). Sand erosion performance of detonation gun applied tungsten carbide/cobalt-chromium coatings. *Wear, 211*(1), 70–83. doi:10.1016/S0043-1648(97)00071-9

Wood, R. J. K., Wharton, J. A., Speyer, A. J., & Tan, K. S. (2002). Investigation of erosion–corrosion processes using electrochemical noise measurements. *Tribology International, 35*(10), 631–641. doi:10.1016/S0301-679X(02)00054-3

KEY TERMS AND DEFINITIONS

Corrosion: Material attrition due to electrochemical processes.

Wear: Material attrition by mechanical processes involving a combination of load and energy in presence of external media, such as fluid flow and / or abrasive particles.

Wear-Corrosion Synergy: Combined effects of mechanical wear and electrochemical corrosion which could result in a mass loss or gain depending on the nature of the interaction.

Chapter 5
Development of Functionally Graded Coating by Thermal Spray Deposition

Jyotsna Dutta Majumdar
Indian Institute of Technology, India

Indranil Manna
Indian Institute of Technology, India & Indian Institute of Technology Kanpur, India

ABSTRACT

Functionally Gradient Coatings (FGCs) are emerging materials with an improved service life and have a promising future for the production of (a) tailored components for applications subjected to large thermal gradients, (b) smart coating with improved corrosion and wear resistance, (c) improved fatigue wear, and (d) improved material structures for energy applications like batteries, fuel cells, etc. FGCs may be developed by physical/chemical vapor deposition, electro/electroless deposition, thermal spray deposition technique, etc. Thermal spraying refers to the technique or a group of techniques whereby molten or semi-molten droplets of materials are sprayed onto a solid substrate to develop the coating. In this chapter, detailed overviews of the development of functionally graded coating by thermal spray deposition techniques are presented. In addition, a few research results on the development of functionally graded coating for tribological and thermal barrier applications are presented.

DOI: 10.4018/978-1-4666-7489-9.ch005

INTRODUCTION

Surface composition and microstructures need to be optimized for an improved performance of engineering components (Budinski, 1988). Surface coating aims at tailoring the surface dependent engineering properties by applying another layer on the surface of substrate using physical, chemical/electrochemical, thermal and high energy surface modification routes. A functionally graded coating refers to the coating consisting of continuously changing composition from substrate to the surface to achieve combinations of toughness and bond strength (Kawasaki & Watanabe, 1997). The gradients may be continuous change in microstructure, microstructure and composition both or porosity distribution. These graded coatings are purposefully developed with an objective to improve the performance and are superior to homogeneous materials composed of similar constituents. The advantages of functionally graded coating over conventional monolithic coating system include a significant reduction in residual stress level, improved toughness, improved wear resistance and thermal properties. Functionally graded coatings are also well-known to enhance the thermal barrier properties due to smooth transition of the properties with depth leading to minimum probability of delaminating of coating from the surface.

Functionally graded coatings are used in many diverse areas and some examples include hydroxyapatite/titanium oxide graded coatings for bio-implant application (Kumar Roop, & Wang, 2002) graded polymer composites reinforced with ceramic particles (Krumova *et al.*, 2001) Ti-Al_2O_3 artificial tooth roots (Iwasaki *et al.*, 1997), reusable high-performance engines (Moro *et al.*, 2002), coatings for tribological application, thermal barrier coating, coatings for cutting tool application and coating for thermal shock resistance application (Schulz *et al.*, 2003; Koizumi & Niino, 1995).

Advantages associated with functionally graded coatings include (i) Reduction of the stress gradient between the coating and the coated component. (b)improvement of adhesion between the substrate and a protective coating, and thus reducing the probability of delamination during thermal or mechanical loading of the coating-substrate system, (iii)improved multi-functionality of composites and FGMs - stable abrasion response combined with a high temperature protection, iv) Improved damage tolerance for brittle ceramic coatings, during abrasive or impact loading and (v) tailored wear-frictional response over the coating lifetime, while the coating is partially removed by wear process (in joints of heavy machinery).

In the present review chapter, detailed overviews of the development of functionally graded coating by thermal spray deposition techniques are presented. In addition, the case studies on development of functionally graded coating for tribological and thermal barrier applications are presented. Finally, the future scope of research in the direction of functionally graded coatings is elaborately stated.

HISTORY OF FUNCTIONALLY GRADED COATINGS

The concept of compositional and microstructural gradient in material microstructure was presented by Bever (Bever & Duwez, 1972) who studied various gradient composites, and reviewed their potential applications. The induction of gradation of polymeric materials by the variation of the chemical nature of the monomers, the molecular constitution of the polymers and the supra-molecular structure or morphology of the polymers was introduced by Shen (Shen & Bever, 1972). The effective chemical, mechanical, biomedical and transport properties were studied and possible applications, including gasoline tank and damping materials, were considered. However, the design, fabrication and evaluation of this gradient structure were not studied.

Fundamental studies on the relaxation of thermal stress by tailoring graded structures, was pioneered in the area of thermal barrier coating for a space plane in Japan (Kieback *et al.*, 2003). The capabilities of withstanding a surface temperature of 1700 °C and a temperature gradient of 1000 °C across a thickness of 10 mm section were achieved by development of functionally graded thermal barrier coating (Shen & Bever, 1972; Kieback *et al.*, 2003). Due to their unique properties, the application of functionally graded coatings are enormous and hence, summarized in Figure 1. In addition to its application in thermal barrier coatings and joints in aerospace, functionally graded coatings had also been developed for other novel applications in the field of biomedical implants, such as artificial bones and dental implants. Hydroxyapatite (HAP) ceramics with a graded porosity was produced by Tampieri et al. (Tampieri *et al.*, 2001) which not only provided good and fast bone in-growth but also improved the capability of withstanding early physiological stress, as an implant to replace natural bone. For dental implants, titanium/hydroxyapatite (Ti/HAP) FGMs exhibited good biocompatibility and mechanical toughness (Watari *et al.*, 1995). No inflammation was observed in both traditional pure Ti and improved Ti/HAP FGM dental implants after eight weeks. But, Ti/HAP FGM implants showed better biocompatibility for newly formed bone (Watari *et al.*, 1997).

FABRICATION OF FUNCTIONALLY GRADED

The processing techniques employed for the development of functionally graded component include powder metallurgy, plasma spraying, physical and chemical vapor deposition, self propagating high temperature synthesis (SHS) and galvanoforming. Since 1991, various new methods have been invented and developed. The processing of FGMs has been categorized in different ways (Mortensen & Suresh, 1995; Miyamoto *et al.*, 1999). Miyamoto et al. (Miyamoto *et al.*, 1999) classified

Figure 1. Field of application of functionally graded coating

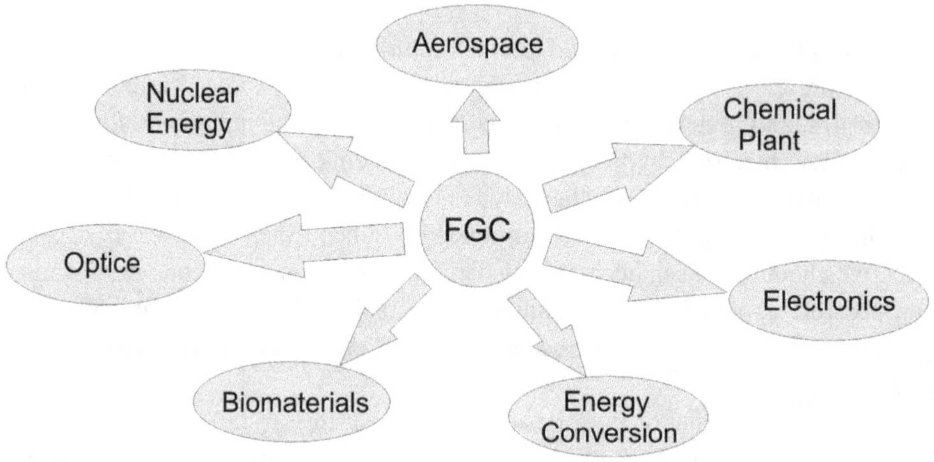

the fabrication of functionally graded coatings into four categories including bulk, layer, pre-form and melt processing and summarized in Figure 2. Though powder metallurgy (PM) is the conventionally applied route and a preferred one as compared to plasma spraying, slip casting, electrophoretic deposition, however, few PM based techniques such as sequential slip casting causes the generation of residual thermal stresses and thermal expansion mismatch between different layers due to presence of sharp interfaces between the successive layers (Moya *et al.*, 1992). In order to eliminate the interfaces successfully and to produce the continuous gradation of material, sedimentation and vibration methods are used in this work.

WEAR

Wear is the progressive loss of material from the operating surface of a solid occurring as a result of relative motion between two surfaces (Budinski, 1988). Wear may broadly be classified into 4 categories, e.g. adhesive, abrasive surface fatigue and erosion. Figure 3 summarizes the different types of wear and the different modes of wear under each category. Abrasive wear occurs when a rough and hard surface, or a soft surface containing hard particles, slides on a softer surface and ploughs a series of grooves in it. The material from the grooves is displaced in the form of wear particles, generally loose ones. It can also arise when hard abrasive particles are introduced between the sliding surfaces and abrade material off each. The mechanism of this form of wear seems to be that an abrasive grain adheres temporarily to one of the sliding surfaces, or else is embedded in it, and plows out

Figure 2. Processing techniques and classification of Fabrication of functionally graded coating

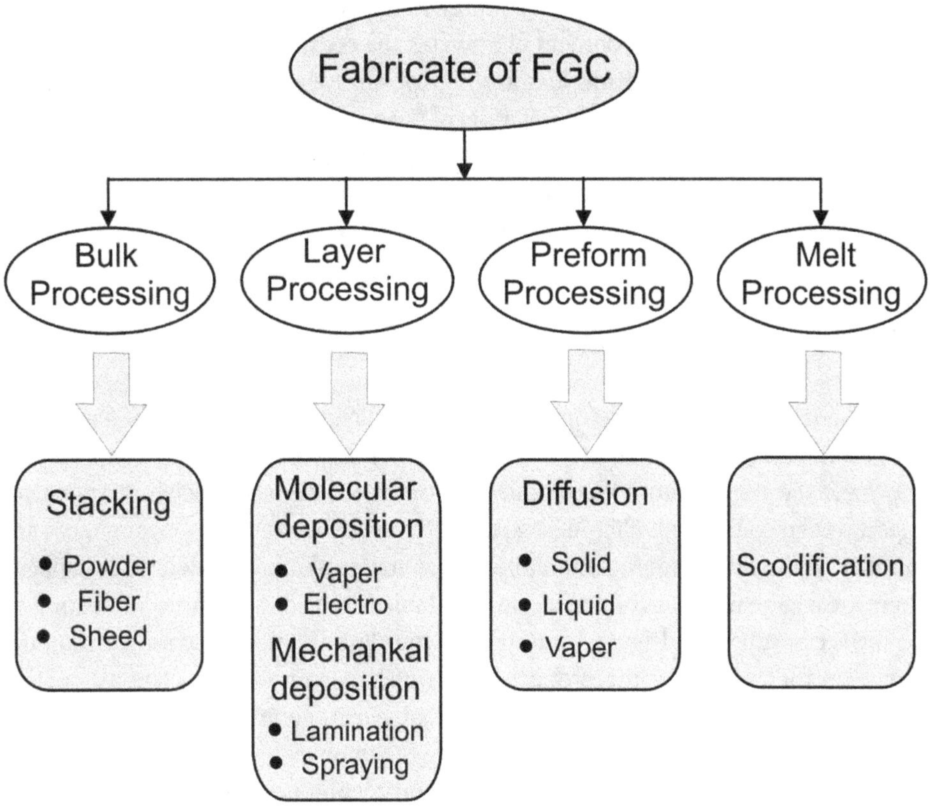

a groove in the other. The two forms of wear, one involving a hard, rough surface and the other hard, abrasive grains, are generally referred to as the two bodies and the three body abrasive wear process, respectively (Kawasaki & Watanabe, 1997). Adhesive wear involves the sliding of two metallic components against each other under an applied load where, no abrasives are present. A strong metallic bond forms between the surface asperities of the contacting materials and hence, the name 'adhesive' is applied (Kawasaki & Watanabe, 1997). In case of stainless steel, there is a thin oxide layer on the surface, which prevents the formation of metallic bonding in between the asperities on the sliding surfaces. Hence, wear rates are low at low load. This type of adhesive wear is known as mild wear or oxidative wear. When the load applied on the surface is high, metallic bonds will form between the surface asperities and the resulting wear rates are high. This kind of wear is more prevalent where a lubricant cannot be used e.g. chain-link conveyer belts, fasteners and sliding components in a valve. (Kawasaki & Watanabe, 1997). Erosive wear may be defined as progressive loss of materials from a solid surface due to the mechanical interaction between that surface and a fluid or impinging fluid stream. The fluids may be multi-component and may contain solids (Kawasaki & Watanabe, 1997). Presence of corrosive environment aggravates the erosion attack due to disruption and removal of the oxide film, leading to exposure of active metal surface to the environment, dissolution or re-passivation of the exposed metal surface, elastic field interaction at asperities in contact with the environment and the interaction between the plastically deformed areas and its environment. Fatigue wear involves the wear caused by the repetitive compressive loads during sliding wear.

SURFACE COATING FOR IMPROVING WEAR RESISTANCE

Surface coating involves development of another layer on the surface of a substrate with a completely different microstructures and/or composition. Coatings are applied to protect the manufactured components from thermal or corrosive degradation, impart wear resistance and hardness to the surface while retaining the toughness and ductility of the bulk component, and enhance the aesthetic and decorative appeal. There are a large numbers of techniques available for the development of coating on the surface of substrate. Figure 4 summarizes the classification of coating techniques employed for improving the properties of finished components. The all coating techniques may broadly be classified under physical, chemical, electro-chemical, thermal and high energy surface engineering techniques.

There is no universal coating material and technique may be applied for all conditions. Selection of coating materials and coating technique depends on the end use of the components. The main factors which influence the selection of coatings for a given application include operating conditions, material compatibility issues, nature and surface preparation of substrate, time or speed of application, cost, safety, environmental effects, coating properties, and structural design. The processes involving the development of coating with a layer of material, not necessarily metallic, to meet the requirements of specific service environments may be categorized into physical, chemical, electro-chemical, thermal and high energy assisted coating.

Physical vapor deposition (PVD) is a family of processes that is used to deposit layers of atoms or molecules from the vapor phase onto a solid substrate in a vacuum chamber. Two very common types of processes used are sputtering and electron beam evaporation. In sputtering, the atoms are ejected from a solid target by bombarding with energetic particles, when its kinetic energy exceeds 1 eV (Mattox, 1964). This process may however, cause a significant erosion of materials during prolonged ion or plasma bombardment. The process is commonly utilized for thin-

Figure 3. General classification of wear

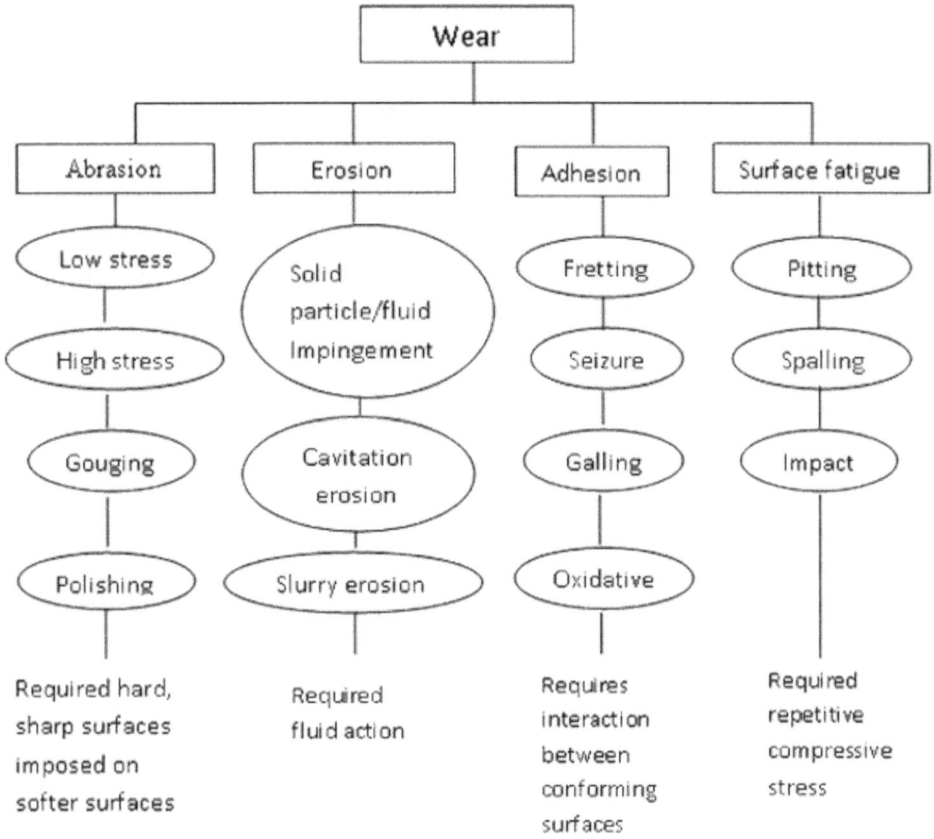

Figure 4. General classification of coating techniques employed for engineering applications

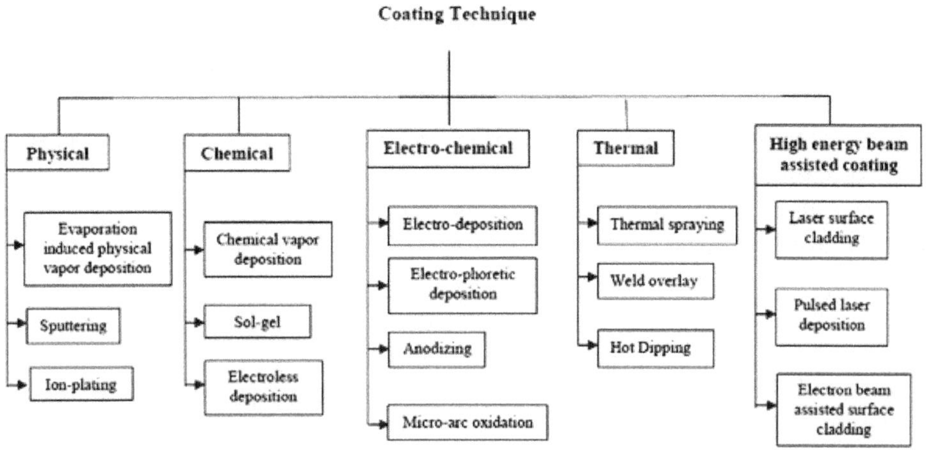

film deposition. Ion plating is a process which causes a periodic bombardment of the substrate and depositing film by transporting atoms or energetic ions and was first described in the technical literature by Donald M. Mattox of Sandia National Laboratories in 1964 (Mattox, 1964). Bombardment prior to deposition is used to clean the substrate and also controls the properties of the depositing film. Sputtering and ion plating may be used to deposit hard coatings of compound materials on tools, adherent metal coatings, optical coatings with high densities, and conformal coatings on complex surfaces.

Chemical vapor deposition (CVD) is a process, where, the substrate is exposed to one or more volatile precursors, which react and/or decompose on the substrate surface to produce the desired deposit and is used to produce high-purity, high-performance solid material and to produce thin film in the semiconductor industry.

The electro-chemical routes applied for development of coatings include electro-deposition, anodizing and micro-arc oxidation techniques. Electro-deposition refers to the technique of deposition of thin film from aqueous solutions, by passing electrical current and may be applied for corrosion resistance, decorative (e.g. gold, rhodium and platinum or electronic/electrical usage or tribological applications. Anodizing is an electrolytic passivation process used to form an oxide layer on the surface of metal parts for corrosion resistance and wear resistance application and provides better adhesion for paint primers and glues. Micro-arc oxidation (MAO), is similar to anodizing, but it employs very high AC voltage so as to cause discharges and the resulting plasma modifies the structure of the oxide layer (Dunleavy, 2009). This process may be used to develop thick oxide coatings on metals such as aluminium, magnesium and titanium which possess high hardness and a act as a continuous barrier, and hence, these coatings can offer protection against wear, corrosion or heat as well as electrical insulation.

Weld overlaying usually involves developing a relatively thick layers (1 mm to several cm) and is applied for improving abrasive wear resistance, such as coating digger teeth, tank tracks and mineral handling equipment. Roll cladding is usually associated with corrosive or mild erosive wear problems, typically those encountered in the chemical, wood pulp, paper and food process industries.

Laser, and electron beams are also used as source of heat for modifying surface microstructure and composition and also for the development of coating. For the modification of microstructure and composition, laser surface melting, laser surface alloying and laser surface cladding are applied. Laser surface alloying is a technique for tailoring the surface microstructure and composition, by melting the alloying elements in the form of powder or wear along with a part of the substrate to develop the alloyed zone confined to a very shallow depth from the substrate. Laser surface cladding is a technique of deposition of a thick metallic/metal matrix composite coating on metallic substrate by melting the clad layer in the form of powders/wires and subsequently, applying the molten materials on the surface of substrate with a minimum dilution at the interface (Dutta Majumdar & Manna, 2012).

THERMAL SPRAY DEPOSITION TECHNIQUE

Thermal spraying refers to a group of coating techniques whereby, droplets of molten or partially molten materials are sprayed onto a solid substrate to develop the coating. Based on the applied heat source and the process characteristics, a large numbers of thermal spray techniques are commercially available, enabling a wide range of materials to be coated. In thermal spraying, the basic mechanism of bond formation at the interface is mechanical interlocking and the bonding between splats can be improved by increasing temperature or particle velocities during particle impact.

Figure 5 summarizes the general classification of thermal spray deposition technique (Tucker, 1994). Thermal spray deposition technique may be classified into four main coating processes: flame spraying, plasma arc spraying, electric arc spraying, and kinetic spraying with many of their subclasses. Each of this process encompasses many more subsets. Each process is having it's own characteristics and hence, develops coating of its own characteristics. The coating characteristics are defined by its bond strength, thickness, microstructure, hardness, compositional homogeneity, porosity and inclusions content. Appropriate coating technique needs to be selected for the desired coating characteristics for the specific application. In the present section, a detailed overview of individual technique would be presented in details.

Flame Spraying

Flame spraying is a thermal spray deposition process which involves combustible gas as a source of heat to melt the coating material (Crawmer, 2004; Chattopadhyay, 2004; Longo, 2005). The advantages of flame spray process include a low equipment and processing costs, high efficiency, low maintenance and easy to operate. However, relatively poor bond strength, higher porosities and a narrower working temperature range as compared to other thermal spray deposition techniques are the disadvantages associated with flame spray deposition process. The coating thickness achieved by flame spray deposition can vary from 100 μm to 2500 μm for single pass and multi pass coating, respectively. The porosity content varies from 10% to 20%. The bond strength determined using tensile adhesion test usually varies from 15 MPa to 30 MPa for metals and alloys.

Figure 5. General classification of thermal spray deposition technique

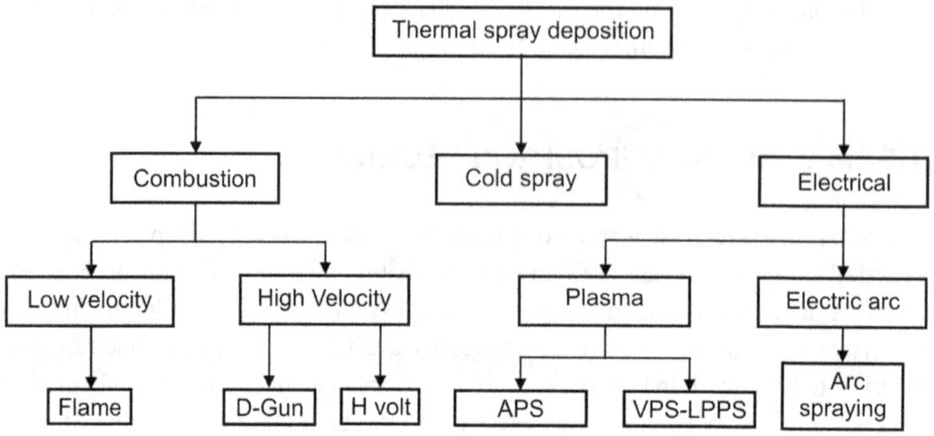

Flame spray deposition is a popular technique for repairing of worn out-parts. Commonly used materials for this application include nickel based alloy, monel, brass, etc. Zinc coating is commonly applied for corrosion resistance application by flame spraying in large structures.

The Electric-Arc (Wire-Arc) Spraying

The electric-arc (wire-arc) spray process uses precursor coating in the form of wire and heating/melting occurs by generating arc between them. Spraying is done by atomization of molten metal and propelling it onto the substrate by a stream of compressed air. Figure 6 shows the schematic of wire arc spraying process. The materials used for coating by wire arc spraying are relatively ductile, electrically conductive wire of pure metal or alloys of about 1.5 mm (0.060 in.) in diameter. However, the recent development of cored wires permits deposition of some composite coatings containing carbides or oxides. Electric-arc coatings are widely used in high-volume, low-cost applications such as deposition of zinc and aluminium for corrosion resistance application or composite coatings for wear resistance application.

Plasma Spraying

In plasma spray deposition, the heat generated by a plasma initiated by high pressure gas (usually argon, but may be any other gas including nitrogen, hydrogen, or helium) flowing through a constrained arc initiated between a tungsten cathode and a water-cooled copper anode using a high-frequency discharge is used to melt

Figure 6. Schematic of wire arc spraying process

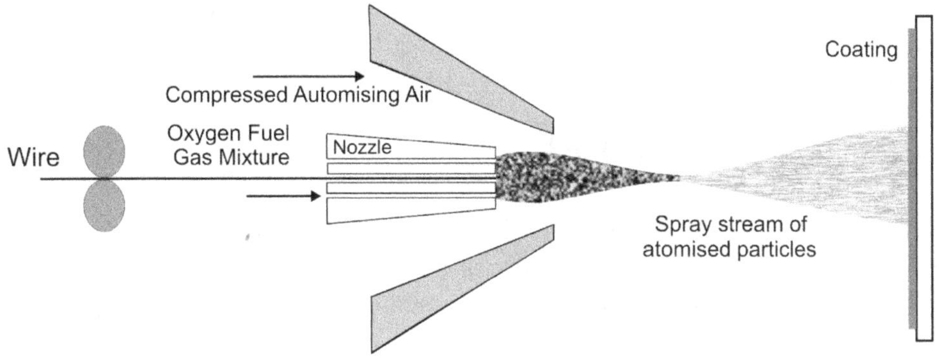

powders introduced into the gas stream either just outside the torch or in the diverging exit region of the nozzle (anode), heated to molten/semi-molten state and accelerated towards the substrate to develop coating. Argon is usually chosen for plasma generation because of its chemical inertness and ionization characteristics. The enthalpy of the gas can be increased by adding the diatomic gases, like hydrogen or nitrogen. Power levels in plasma spray torches are usually in the range of 30 to 80 kW, but they can be as high as 120 kW. The deposition parameters include the gas used for generation of plasma, power level of arc used to generate plasma, the size and morphology of the powder flow, rate of powder flow, carrier gas flow rate, distance from the torch to or from the substrate (stand-off distance) and the angle of deposition. The powder velocity in plasma spray deposition usually varies from 300 to 550 m/s. The density of the deposition is usually in the range of 80 to 95%. The thickness of plasma spray deposited coating usually ranges from about 0.05 to 0.50 mm and the bond strength may vary from 34 MPa (5000 psi) to 69 MPa (10,000 psi).

High Velocity Oxyfuel Flame (HVOF) Spraying

High-velocity, oxy-fuel, (HVOF) coating uses the heat generated by confined combustion of fuel (oxy-acetylene, kerosene of liquid petroleum gas) to heat and accelerate the powdered coating material at a supersonic gas velocities (i.e. greater than MACH 5). The extreme high velocities provide kinetic energy which help in producing dense and very well adhered coatings. Powder is heated to molten/semi-molten stage and accelerated usually axially to a velocity of up to about 550 m/s. As the powder is exposed to the combustible products, they may be melted in either an oxidizing or reducing environment, causing the probability of partial reaction with the environments. The bond strength, density and thickness of coating achieved by

HVOF spray coating usually exceed 69 MPa (10,000 psi), 99% and 0.5 mm (0.002 to 0.020 in.), respectively. HVOF spraying process can be used to develop metallic, ceramic or metal matrix composite coating for the development of wear and corrosion resistance surface.

Detonation Gun Spraying

In detonation gun spraying, ignition of oxy-acetylene fuel is achieved by detonation using a spark into a barrel and a pulse of powder, is introduced into the ignition zone. The powder is heated to molten/semi-molten state by the high-temperature, high-pressure detonation wave and accelerated to a velocity of about 750 m/s. After each detonation, the barrel is purged with nitrogen and the process is repeated at up to about 10 times per second. However, the coating rate is very slow as with each detonation a coating thickness of a few micrometer is achieved and hence, for a thicker deposition, the process needs to be repeated for several times. A uniform coating thickness is achieved by overlapping the circles of coating in many layers.

Detonation gun coating develops coating with a highest bond strengths (usually exceeding the epoxy strength of the test, that is, 69 MPa) and lowest porosities (usually less than 2%). Due to the extremely high velocities and consequent kinetic energy of the particles, the deposited layer usually possesses a residual compressive stress, rather than tensile stress as is typical of most of the other thermal spray coatings, which makes them useful for improved fatigue life. All kind of materials including metallic, ceramic, and cermets for wear and corrosion resistance as well as for many other types of applications may be deposited using detonation gun deposition.

Cold Spraying

Cold Spray (CS) is a process, where, coating is developed by accelerating powder particles the supersonic gas jet at a temperature that is always lower than the melting point of the material, resulting in coating formation from particles in the solid state (Singh, 2012; Schoop, 1917; Turner & Budgen, 1926). Cold spraying is also traditionally known by different names like cold gas dynamic spraying, kinetic spraying, high velocity particle consolidation (HVPC), high velocity powder deposition and supersonic particle/powder deposition (SPD) (Singh, 2012). The carrier gas used in spraying may be compressed air, helium or nitrogen), dry air (79% N_2 - 21% O_2) or a mixture of He and N_2 and. Usually the carrier gasses have good aerodynamic properties. During the process, the gas accelerates particles at a pressure of 1-3 MPa. The particle size of precursor powder ranges between 2 to 150 μm. The distance between the substrate and the nozzle is approximately 25 mm. The particles are heated by HVOF torch and enter the spray nozzle to increase the velocity and ductility which allow easy build up of coating. As the particles impact on the surface, they bond plastically and deposits on substrate. The coated particles are plastically deformed on impact due to the high kinetic energy of the particles and form splats. As coating is carried out at low temperature, and the kinetic energy of the particles are responsible for the bonding of coating with substrate. The advantages of cold spraying over conventional thermal spray deposition include minimization of high-temperature oxidation, retention of chemistry and microstructure of coating, minimum loss of coating materials due to melting and evaporation, reduced residual stress in coating, minimum distortion of substrate, development of high density, high hardness cold worked structure and an increased operation safety because of the absence of high-temperature gas jet, radiation, and explosive gases. The process may be applied for the coating of materials with metastable microstructure including amorphous and nano-structured coatings.

Due to small size of the nozzle (10-15 mm^2) and spray distance (5-25 mm), the footprint of the cold spray beam is very narrow yielding a high-density particle beam, and results in a precise control on the area of deposition over the substrate surface. Due to the high velocity of the particles, the developed coating possesses a large compressive stresses and high bond strength. Due to low temperature, the microstructure of the coatings is usually oxides and other inclusions -free (Schoop, 1917).

Due to the rapid solidification of the molten/semi-molten particles and its flattening due to striking a cold surface at high velocity, the thermal spray coatings posses lenticular or lamellar grain structure. Figure 7 shows the in-flight molten metal transfer sequence and hence, development of microstructure during thermal spraying techniques. The mechanism of bonding during thermal spray processing include (a) mechanical keying or interlocking; (b) diffusion bonding or metallurgical bonding and (c) chemical and physical bonding mechanisms -oxide films, Van der Waals forces, etc. The factors influencing the bond strength and subsequently, build up of the coating are (a) cleanliness, (b) surface area, (c) surface topography or profile, (d) temperature (thermal energy), (e) time (reaction rates & cooling rates etc.), (f) velocity (kinetic energy), (g) physical & chemical reactions.

For improving the adherence of coating, usually, the substrate is subjected to grit blasting for incorporating grooves and also cleaning the substrate. The improvement in bond strength due to grit blasting is due to increase in surface area and also promoting mechanical interlocking due to generation of rough surface. The individual particle cooling rates on impact during thermal spraying can be as high as 1 million °C per second (10^6 Ks^{-1}). Due to the high cooling rate, the thermal interaction due to spraying is very limited. However, increase in thermal and kinetic energy increases chances of metallurgical bonding. However, materials like molybdenum, tungsten, and aluminium/metal composites produce the so called "self bonding" coatings, due to their comparatively high bond strengths (increased metallurgical

Figure 7. In-flight molten metal transfer sequence and hence, subsequent development of microstructure during thermal spray processing

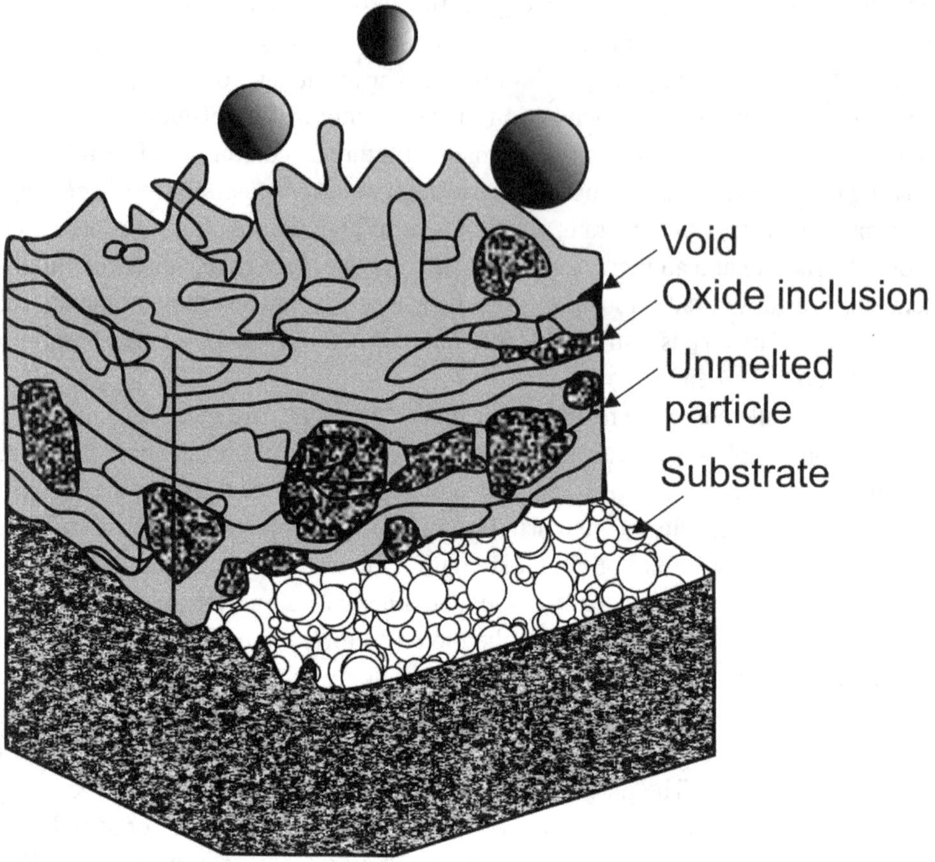

or diffusion bonding). Furthermore, preheating of the substrate increase diffusion bonding activities though increases the probability of oxidation of the substrate. High kinetic energy associated with thermal spraying technique like high velocity oxyfuel coating (HVOF) and cold spraying produce high bond strengths. All conventionally thermally sprayed coatings contain some porosity (0.025% to 50%), which are caused by low impact energy, shadowing effects (un-melted particles / spray angle), and shrinkage and stress relieve effects (Tucker, 1994). The different interactions and thermal effects can change the microstructure and composition of the coatings as compared to that of precursor powder.

RESIDUAL STRESSES IN THERMAL SPRAY DEPOSITION

Residual stresses refer to the internal stresses which originate due to the growth of the coating and depend on the mechanism of growth of coating. Residual stress can arise due to the mismatch in properties between coating and substrate as well as coating microstructure. In many cases, the difference in the thermal expansion coefficients influences the nature and magnitude of the residual stress. The level of stresses varies and is very system specific but, for example, for ceramic coatings on metal substrates the residual stresses are typically in the range of 1–3 GPa. These residual stresses can be either tensile or compressive depending on the nature of the coating and substrate.

Whether the residual stresses are tensile or compressive can also influence the mode of deformation of the coating. Coatings with a residual tensile stress have an increased tendency to surface cracking, while those containing large residual compressive stresses are susceptible to delamination and spalling. Delamination processes can also be assisted by the presence of small flaws in the coating. Excessive levels of residual stress will lead to complete removal of the coating from the substrate. Such coatings will, therefore, offer poor protection to substrates exposed to solid particle erosion, abrasion or adhesion mechanisms. For example, studies of thin hard coatings such as TiN less than 30 mm thick found that compressive residual stresses reduce the tensile stresses induced in the coating during sliding. However, the tendency for spalling and delamination of coatings increased with increased compressive stress or increasing coating thickness. There are typically two sources of residual stress in coated systems: thermal stresses and intrinsic stresses. The total residual stress st is the sum of the two:

$$\sigma_{\tau} = \sigma_{th} + \sigma_{in} \tag{1}$$

Where, σ_{th} is the thermal stress and σ_{in} the intrinsic stress. The thermal stress σth is a consequence of the mismatch between the thermal expansion coefficients of the coating and the substrate. It is usually compressive in nature owing to the greater contraction of most substrates on cooling to room temperature following deposition. It is possible to estimate the thermal stresses σth using the following equations:

$$\sigma_{th} = E_f(\alpha_f - \alpha_s)\Delta T \tag{2}$$

Where, E_f is the biaxial elastic modulus of the coating, α_f and α_s are the thermal expansion coefficients of the coating and substrate respectively and ΔT is the temperature difference, which is typically the difference between the deposition temperature and room temperature. The intrinsic stress σ_{in} in the coating is usually tensile because of a number of microstructural influences during the deposition process. The first source of these stresses lies in the differences between the lattice parameters of the coating and the substrate. The second source is due to the microstructure of the coating and the presence of impurities and structural defects in the coating. These include grain boundaries and structural defects such as dislocations, twins, stacking faults and vacancies. Systems with higher grain boundary densities result in a higher intrinsic stress. Intrinsic stresses are typically in the range 0.1–3.0 GPa. Wheeler106 reported that, for erosion-resistant CVD diamond coatings between 10 and 120 µm thick grown on tungsten substrates, compressive residual stresses between 0.84 and 1.3 GPa are present. Calculations based on equation [1.47] estimate that thermal stresses should be about 3.7 GPa; thus tensile intrinsic stresses of the order of 2.4–2.9 GPa are inferred.

During thermal spray deposition, the particles strike they rapidly cool and solidify, generating a tensile stress within the particle and a compressive stress within the surface of the substrate. As the coatings build up, there are the tensile stresses in the coating, which increases with the increase in thickness. The failure of the coating usually occurs when the tensile stresses in coating exceeds that of the bond strength or cohesive strength. The effect of stress on coated surface is shown on Figure 8. The materials with a high coefficient of thermal expansion like some austenitic stainless steels are prone to high levels of stress build up and thus have low thickness limitations. In general, it may be concluded that thin coatings are usually more durable than thick coatings. Spraying method and coating microstructure also influence the level of stress build up in coatings. Dense coatings are generally more stressed than porous coatings. Combustion powder sprayed coatings generally have greater thickness limitations than plasma coatings.

DEVELOPMENT OF FUNCTIONALLY GRADED METAL-MATRIX COMPOSITE COATING

The development of functionally graded coating for wear resistance applications may be achieved by development of metal matrix composite coating with a gradual change in composition from metal rich at the near substrate region to ceramic rich at the near surface region, or by the development of multilayered composition, such as a combination of a hard and a soft phase, for achieving a maximum hardness and toughness properties in the coating (Lee, 2006; Mellor, 2006). The gradual change in the composition of the coatings may be obtained by a precise control of the process parameters of coating. Graded change in composition also reduces the possible mismatch in chemical or mechanical properties between two different components and hence, improves the durability of coating.

Figure 8. Schematic showing the stress distribution in the coated surface of plasma spray deposited thermal barrier coating

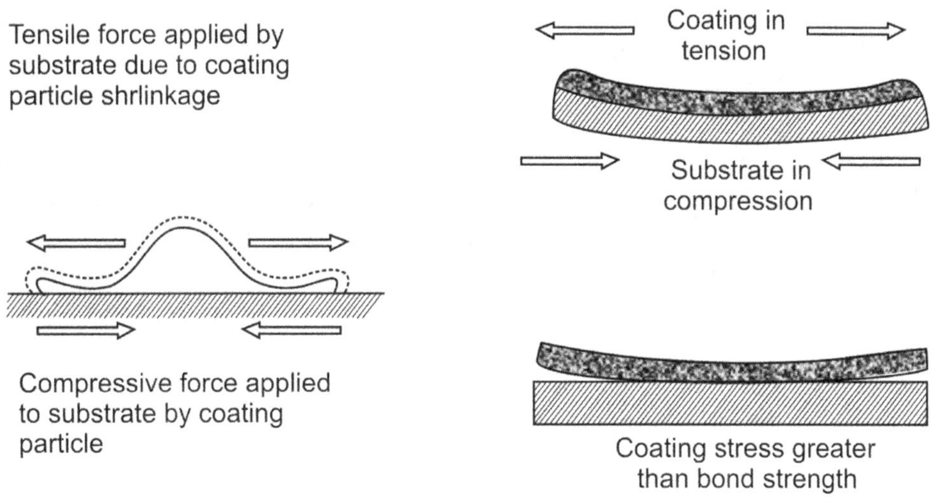

Glass-Alumina Functionally Graded Coatings

Functionally graded glass–alumina coatings were developed on to alumina substrate as multi-layered coatings, with each layer having a mean composition slightly different from the neighboring ones by plasma spraying (Lu et al., 2007). The compositional gradients considered were from 100 vol.% alumina to 100 vol.% glass and from 80–20 vol.% glass to 100 vol.% glass with several heat treatments to improve the substrate-coating interface bonding.

Functionally Graded WC-Co and Mo-Mo_2C Coatings

Functionally graded coatings composed of WC-Co/stainless steel and Mo-Mo_2C/ stainless steel coatings were developed by high velocity oxyfuel coating (Prchlik *et al.*, 2001). A detailed wear studies showed that the wear resistance of the functionally graded WC-Co coating was inferior to those of Mo-Mo_2C coatings with the same steel content (Prchlik *et al.*, 2001). However, with increasing depth/steel content, compositionally graded coating with Mo-Mo_2C/stainless steel exhibited superior behavior. The poor wear behavior of Mo-Mo_2C coating on the surface is attributed to high porosity of pure Mo-19% Mo_2C and limited inter-splat binding. Addition of stainless steel was effective in reducing the coating porosity, increasing fracture toughness and improving friction/wear response. The superior wear behavior of the Mo-Mo_2C/steel composite coating with depth is attributed to toughening of the coatings by a minor addition of stainless steel. Hence, an optimum gradation is essential to achieve the optimum tribological response. However, WC-Co coating is sufficiently dense and addition of stainless steel does not improve its wear response, however, addition of steel increases the resistance against crack propagation.

In an another attempt, a functionally graded WC–Co coating was developed by a novel carburization treatment (Wang et al., 2013). The significantly increased hardness of the surfaces of the FG WC–Co resulted in superior wear resistance of coating. The FG materials showed significantly better fracture toughness when compared to the materials with a comparable wear (which would be of a lower nominal Co content), and the graded materials showed more than a 40% superior wear performance when compared with the conventional materials of comparable fracture toughness. The FG WC–10Co sample showed the highest surface hardness and wear resistance of samples tested, while the FG WC–16Co sample showed a wear resistance nearly that of the "as-received" WC–10Co samples while maintaining fracture toughness. FG WC–Co also exhibited improved impact and fatigue resistance when tested using the dome-shaped geometry of a rock cutting element. The functionally graded materials presented in this study possessed microstructures which exhibited a combination of wear performance and mechanical properties that hold promise for a range of practical wear applications.

In another contribution, the effective damage tolerance of a functionally graded coating (FGC) consisting of 6 layers with a gradual change in composition from 100 vol.% ductile AISI316 stainless steel (bottom layer) to 100 vol.% hard WC–12Co (top layer) deposited by high velocity oxygen fuel (HVOF) spraying was investigated (Valarezo et al., 2010). There was a gradual change in the composition and hardness through the coating thickness which was concluded by scanning electron microscopy and depth-sensing micro-indentation, respectively. The in-situ curvature measurement technique revealed that during the deposition of the coating, compressive stresses was arrested in the metallic layers owing to peening effect of successive impact, which gradually changes to high tensile, in the top layers due to quenching effect. However, the thermal stresses are low because of the gradation. The functionally graded coating structure showed the ability to reduce cracking with increased compliance in the top layer during static and dynamic normal contact loading, while retaining excellent sliding wear resistance (ball-on-disk tests). Results were compared to the coatings with a similar compositions and thicknesses, as well as a thick monolithic WC–12Co sprayed coating and discussed. Further improvements in the processing were proposed to enhance the adhesion strength and avoid coating delamination under high load contact-fatigue conditions.

The average microhardness value of the single layer and each graded layer are compared in Figure 9. From Figure 9 it was concluded that the microhardness of the various layers of the functionally graded coating were not significantly different from those of the corresponding single-layer coatings which indicated that the embedment in a thick graded coating did not affect the mechanical characteristics of the various layers, despite the different residual stress state and peening effect were exerted by the subsequent layers. The peening effect of a layer of the new composition would only modify the top-most portion of the deposited composition and might not have a significant effect on the overall hardness, whereas the residual stress is not significant at the stress levels that the indentation produces from 100 vol.% AISI 316 L stainless steel to 100% WC–Co, was deposited by HVOF-spraying onto a stainless steel substrate. Actual composition of the WC–Co/stainless steel composite layers did not perfectly match with the nominal one due to processing effects. From Figure 9 it may also be noted that there is a smooth gradation in hardness due to the development of functionally graded coating. In addition, the FGC architecture significantly reduced the stress buildup and stress discontinuity arising at the coating/substrate interface during deposition. The tensile quenching stress, which is very large in pure WC–Co, is indeed mitigated by compressive peening stresses in the stainless steel-rich layers close to that interface and hence, minimizing the probability of delamination during deposition of thick coatings.

Due to a smooth gradient in stress distribution from a tensile stress in the cermet top layer to compressive stresses in the stainless steel-rich layers, the FGC therefore acted quite effectively as a damage tolerant structure under severe contact stress conditions. The effectiveness of the FGC structure is particularly manifested when loads are large enough to probe the entire through-thickness behavior of the coating. Both static (high-load indentation) and dynamic (cyclic ball-drop impact) tests showed that the FGC structure has superior ability to control and mitigate the propagation of cracks and defects, in comparison to pure cermet layers of comparable thickness. Cracks may initiate in the top cermet layer due to the presence of a large tensile stress, however, the underlying layers can arrest or deflect them by (a) deflection of cracks along tortuous paths through the interfaces between steel and WC–Co lamellae, (b) arresting of cracks entering the ductile steel lamellae, and (c) reduction of crack-tip stress concentration by the compressive residual stress in the underlying layers. However, when the contact stress is localized in the vicinity of the top surface, as the case for ball on disk wear test, no perceivable difference existed between pure WC–Co and that of the functionally graded coating. A detailed stress distribution analysis showed that there was presence of large residual tensile stress in both monolithic and compositionally graded WC–Co coating due to the large quenching stresses developed by cermet lamellae during the deposition process. However, it was observed that HVOF-sprayed stainless steel suffers from extensive oxidation and phase alteration during spraying, which compromise its mechanical strength and probably impair its adhesion to the substrate, leading to the damage of

Figure 9. Summary of the average microhardness value of the single layer and each graded layer

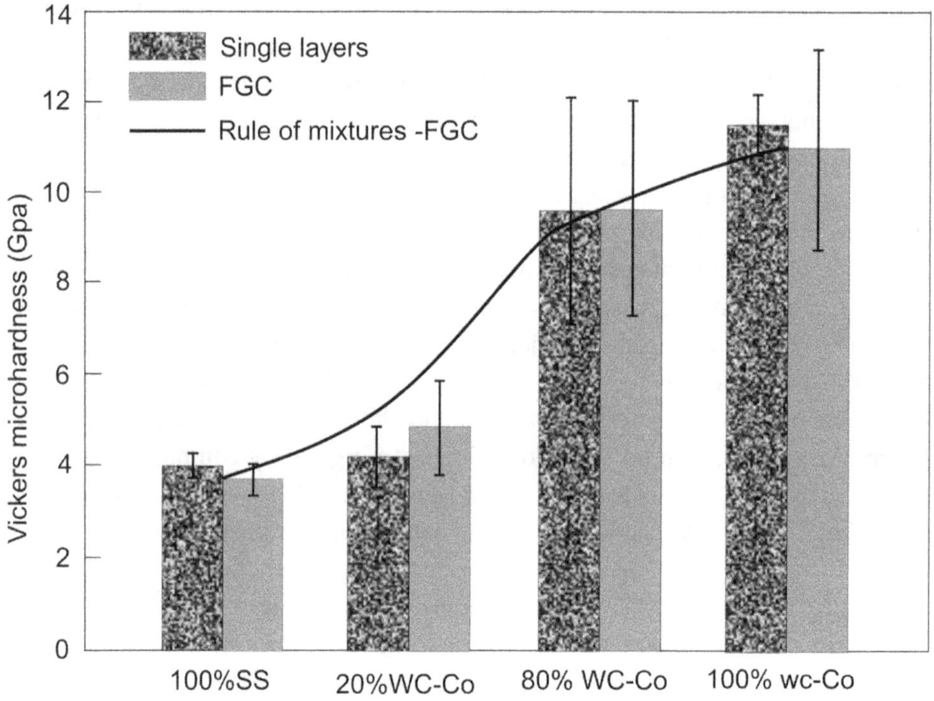

FGC under the ball drop impact test. The use of a fuel-rich flame (oxygen reducing) was proposed as a possible solution.

Functionally Graded Al_2O_3 Coating on Nickel Aluminum Bronze (NAB)

Nickel aluminum bronze (NAB) is known to possess excellent physical, mechanical and tribological properties, and hence, a popular material for mechanical parts used for wear and abrasion-resistance applications both at room and elevated temperature (Li *et al.*, 2006; Yuanyuan *et al.*, 1996). Despite these properties, however, these alloys still have limitations in terms of the design, use, and life of derived components. For example, the turbine impellers and propellers made from nickel aluminum bronze alloys are essentially limited by damage due to wear, cavitation, and erosion. Hence improvements of material performance without degradation of other aspects, particularly general corrosion resistance, will add more promising results under a wider range of service conditions (Lee *et al.*, 2004). Surface modifications can be an effective way of improving material performance without altering the original properties in the remaining parts (Tang et al., 2006). A compositionally graded Al_2O_3 dispersed NAB matrix composite coating were developed on the NAB substrate with a gradual change in distribution of Al_2O_3 phase from the surface towards the interface by high velocity oxyfuel spraying technique (Hypojet – 2700). For the development of coating, naval aluminium bronze (Ni: 5, Al: 10, Fe: 4, and Cu: balance) of dimension: 20 mm x 20 mm x 5 mm were sand blasted (using Al_2O_3 particles of 40-50 μm particle size) and subjected to high velocity oxyfuel (HVOF) spraying techniques (Hypojet – 2700) using a mixture of Al_2O_3 and nickel aluminium bronze (NAB, with the following composition: Ni: 5, Al: 10, Fe: 4, and Cu: balance) as the precursor powders. Figures 10 (a,b) show the scanning electron micrographs of (a) NAB powder and (b) Al_2O_3 powder used as precursor in the present study. From Figure 10 it may be noted that the particle size of Nickle Aluminum (Naval) bronze varies from 10-40 μm (cf. Figure 10a) and the same for Al_2O_3 varies from 30-80 μm (cf. Figure 10b), respectively. Particle size distribution is very important in order to obtain the coatings with the desired properties and deposition efficiency. A detailed energy distribution spectroscopic analysis shows that copper particles are dendritic in shape (labeled as 3 in Figure 10a), nickel particles are rod like in shape (labeled as 4 in Figure 10a), aluminium (labeled as 1) and iron are (labeled as 2) spherical in appearance. A high velocity oxyfuel spraying (Hypojet – 2700) was conducted by melting the precursor powder, subsequently, fragmentation and deposition of the molten droplets at a very high pressure using compressed air. NAB to Al_2O_3 weight ratio was varied from 100% at the substrate surface to 0% at the surface, with a graded distribution at values of 10% Al_2O_3 + 90% NAB, 20% Al_2O_3 + 80% NAB, 30% Al_2O_3 + 70% NAB and 100% Al_2O_3. The process parameters used in the present study is shown in Table 1.

Figures 11(a,b) show the scanning electron micrographs of the cross section of (a) NAB and (b) Al_2O_3 dispersed graded coating deposited by Hypojet universal thermal spraying unit. A double pass deposition was conducted in NAB coating (Figure 11a) and 10 pass deposition was conducted in Al_2O_3 dispersed NAB (Figure 11b). From Figure 11 it may be noted that the microstructures developed by thermal spray deposition is uniform and homogeneous in nature with the presence of repeated splats. A detailed observation of the coating-substrate interface shows that the coating penetrates deep inside the substrate which is attributed to a large energy of impact of the coating materials on the substrate due to a high (500 m/s) velocity of coating. Furthermore, the rate of penetration of the coating was not uniform inside the substrate. The non-uniform penetration of the coating materials within the substrate is attributed to the non-uniform flow velocity of the particles. The individual splat was elongated and irregular in shape, and was clearly visible within the coating where their morphology was quite different from one area to another. A very low area fraction of porosities (2-3 area %) was observed in the

Figures 10. Scanning electron micrographs of (a) NAB powder and (b) Al_2O_3 powder used as precursor in the present study

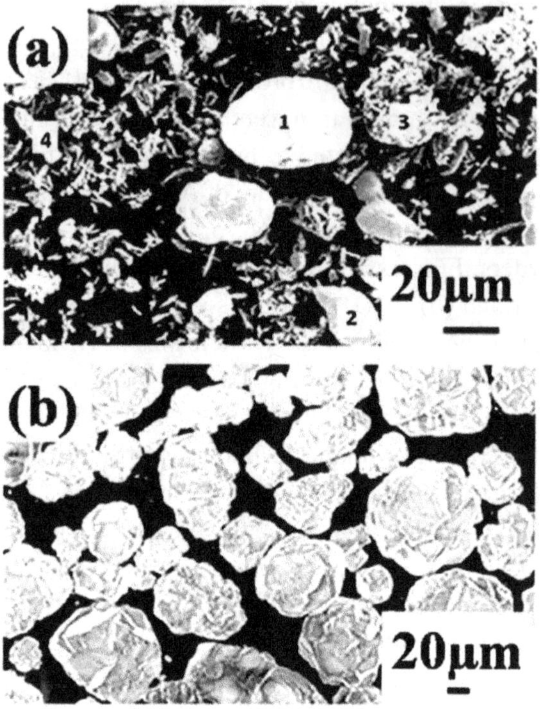

Table 1. Summary of process parameter employed for the development of coating

Sl. No.	Process	Process Parameters		
1	**HVOF Spray**	Flow Rate	Fuel Gas (SLPM*)	70
2			Oxygen (SLPM*)	270
3		Pressure	Oxygen (Bar)	10
4			Fuel Gas (Bar)	8
5		Powder feed rate (gm/min)		30
5		Fuel Gas		LPG

coating. The microstructure of the graded coating on the other hand, consists of four different layers with 100% NAB, 90% NAB+10 wt.% Al_2O_3, 80 wt.% NAB + 20 wt.% Al_2O_3, 70 wt.% NAB + 30 wt.% Al_2O_3 and 100% Al_2O_3 from the interface to surface region. The formation of a flat interface in graded coating as compared to the NAB coating is possibly due to repeated impact of the coating on the previously formed layer causing uniform penetration all throughout. In both the cases, it may be noted that the interface is present deep inside the substrate in contrast to that formed in coating developed by conventional deposition route. The coating splats are interlocked with each other due to its spreading horizontally along the lateral plane of the cross-section. Transverse (perpendicular) cracks usually observed in thermal sprayed coatings are not visible in the microstructure due to the residual compressive stresses imparted onto the coating during the HVOF spraying process.

Figure 12 shows the X-ray diffraction profiles of as received (plot 1) and the graded coating with thermal spray deposition with NAB (plot 2), and the graded coating from the substrate side with 10% Al_2O_3 dispersed in NAB, (plot 3) 20% Al_2O_3 dispersed in NAB, (plot 4) 30% Al_2O_3 dispersed in NAB (plot 5). From Figure 12 it is confirmed that the as received NAB consists of presence of α-Cu, Cu_9Al_4 and NiAl as observed in microstructure. On the other hand, NAB coating by HVOF spraying increased the Cu_9Al_4 peak intensity marginally and also caused a significant broadening of the peaks. The presence of alumina peaks (Al_2O_3) confirms the presence of Al_2O_3 in the alloyed zone. Increased mass fraction of Al_2O_3 in the coating mixture increased the peak intensity of Al_2O_2 peaks along with its extent of broadening in the XRD profiles. From the extent of broadening of XRD peaks the lattice strain developed in the matrix was calculated using Scherrer's formula and are summarized in Table 2. From Table 2 it may be noted that a lattice strain developed in the coating layer was found to vary from .35% to .46% and the intermediate layer (containing 20% Al_2O_3) dispersion experiences a maximum lattice strain. The increasing lattice strain is attributed to the presence of (NiAl in the solid solution) in supersaturated Cu matrix due to rapid quenching associated with thermal spraying. A detailed analysis of residual stress developed on the coating is evaluated by using stress Goniometer attached to XRD Unit (Model No.- D8 Advances, Bruker AXS, Germany) and summarized in the Table 2 .A detailed study of Residual stress developed on the surface and its distribution with depth shows that residual stress of coating is mainly compressive in nature and magnitude is maximum at the top surface which decreases with the depth .The presence of residual stress on the surface is possibly attributed the repetitive impact of the coating on the pre-deposited layer during successive impact.

An Al_2O_3 dispersed Nickel Aluminium Bronze with an improved hardness to as high as 677 VHN was achieved. However, there was a significant scatter in the microhardness value due to the dispersion of hard phase into soft matrix. The hard-

Figure 11. Scanning electron micrographs of the cross section of (a) NAB and (b) Al_2O_3 dispersed graded coating deposited by Hypojet universal thermal spraying unit

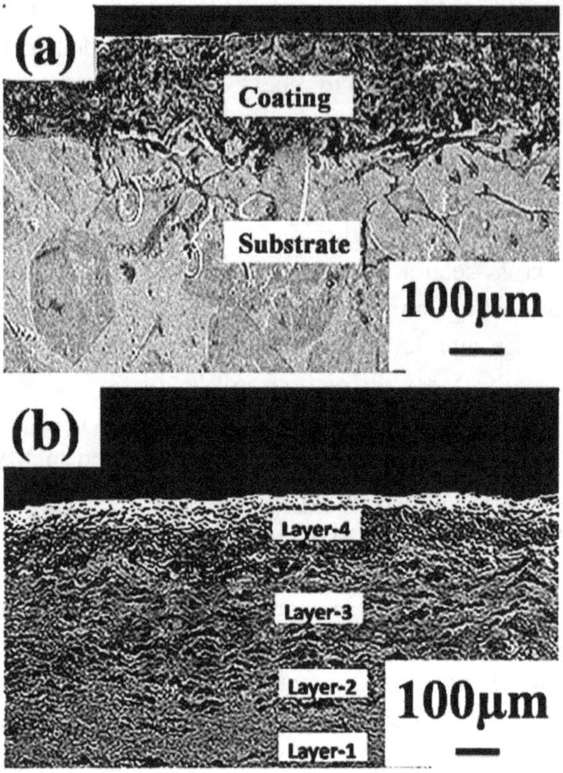

Table 2. Summary of lattice strain and residual stress developed in the graded coating

Sl. No.	Composition	Lattice Strain (in %)	Residual Stress (in MPa)
1.	Cast NAB	0.36175	-220.1±63.8
2.	NAB Coated	0.38725	-261.5±67.9
3.	10% Al_2O_3 (layer-1)	0.3505	-294.2±46.3
4.	10%+20% Al_2O_3 (layer-2)	0.4615	-323.2±54.6
5.	10%+20% +30% Al_2O_3 (layer-3)	0.393	-393.0±153.5
6.	10%+20% +30% +100% Al_2O_3 (layer-4)	0.173	-424.0±103.9

Figure 12. X-ray diffraction profiles of (a) as received and the graded coating with thermal spray deposition with (b) NAB, and the graded coating from the substrate side with (c) 10% Al_2O_3 dispersed in NAB, (d) 20% Al_2O_3 dispersed in NAB, (e) 30% Al_2O_3 dispersed in NAB

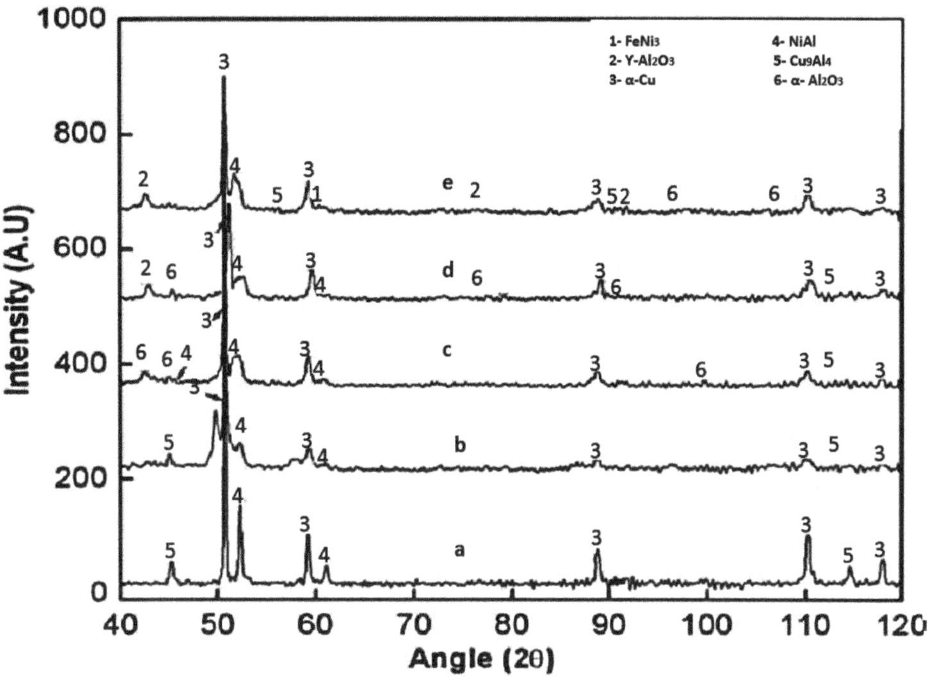

ness was found to increase with increase in Al_2O_3 content in the coating. The microhardness of the composite surfaced region was however, found to decrease with increase in the depth and reached the substrate microhardness at the composite layer-substrate interface (cf. Figure 13). The gradual decrease in the microhardness with increase in depth is attributed to a decrease in area fraction and content of ceramic particles in the matrix. The composite obtained smaller grain size and higher hardness with increasing the content of Al_2O_3 due to the greater strengthening effect of evenly dispersed Al_2O_3 particulates.

Figure 14 shows the fretting wear behavior in terms of cumulative depth of wear as a function of time in as-received nickel aluminium bronze (NAB) (plot 1), and the same coated with NAB (plot 2), and NAB +30% Al_2O_3 (plot 3) against WC ball at an applied load of 10 N at an oscillating frequency of 10 Hz. From Figure 14 it may be noted that the wear kinetics initially increases following which it reaches the steady state. A close comparison of different plots in Figure 14 shows that the

Figure 13. Variation of microhardness with depth beneath the coated surface of Al_2O_3 dispersed Nickel Aluminium Bronze coating

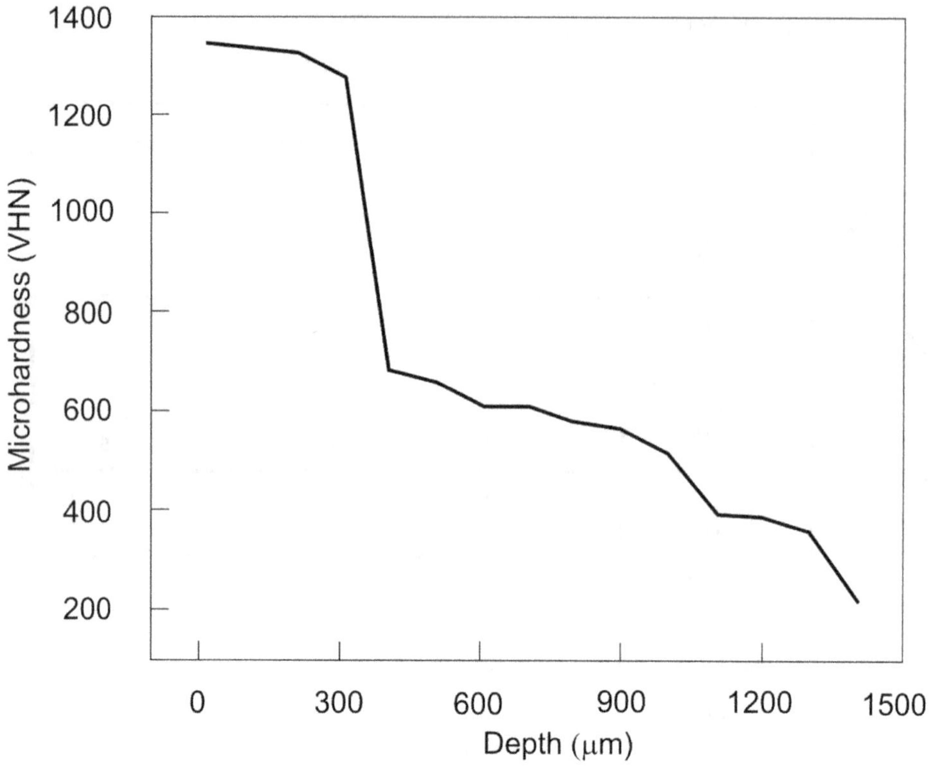

wear kinetics reduces for alumina dispersed composite coating as compared to as received and as deposited nickel aluminium bronze (NAB).. The reduced wear kinetics with addition of Al_2O_3 in the coating is attributed to increased hardness due to dispersion strengthening of the matrix. Figure 15 shows the variation of coefficient of friction of the composite coating with time for as-received nickel aluminium bronze (NAB) (plot 1), and the same coated with NAB (plot 2), and NAB +30% Al_2O_3 (plot 3) against WC ball at an applied load of 10 N at an oscillating frequency of 10 Hz. From Figure 15 it may be noted that the coefficient of friction of composites is higher than as-received and as-coated NAB. It was noted that the hard reinforcing particles resulted in the roughening of worn surfaces during wear, leading to the increase in friction force. Consequently, the addition of Al_2O_3 particles enhanced the friction co-efficient of composites. Additionally, the hard reinforcing Al_2O_3 particles prevented specimens from plastic deformation. There by, the coupling adhesive forces between the ball and the specimen decreased, making it more difficult to press the ball against the specimen and resulting in lowering the wear rate.

Figure 14. Cumulative depth of wear as a function of time in as-received nickel aluminium bronze (NAB) (plot a), and the same coated with NAB (plot b), NAB +10% Al_2O_3 (plot c), NAB +20% Al_2O_3 (plot d), and NAB +30% Al_2O_3 (plot e) against WC ball at an applied load of 10 N at an oscillating frequency of 10 Hz

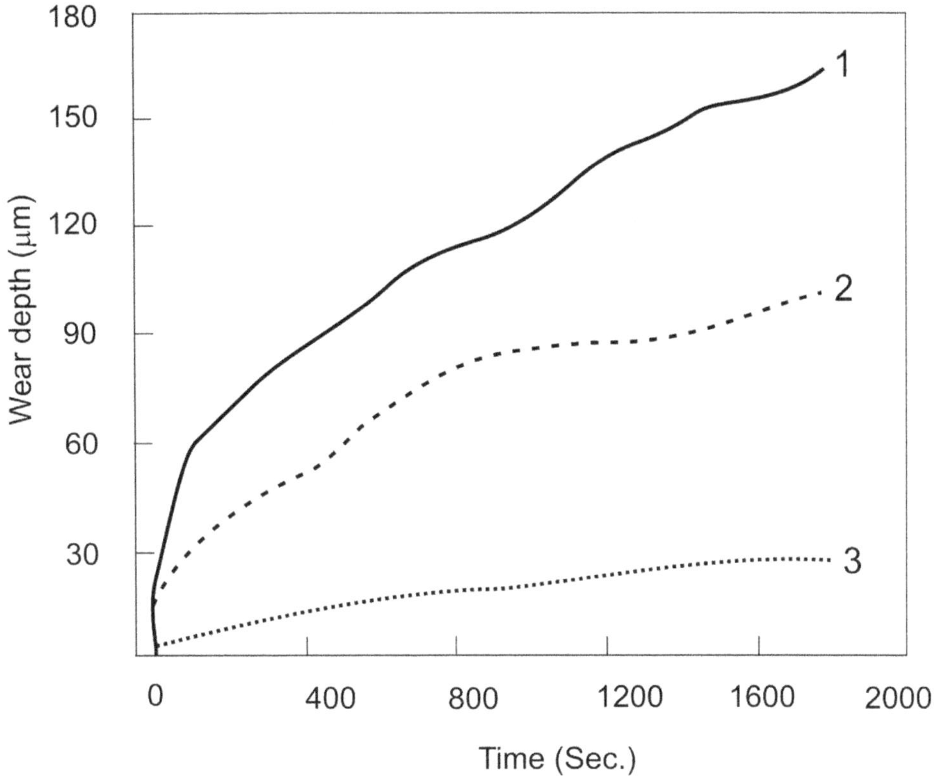

In addition, Al_2O_3 particulates resisted the micro cutting action of abrasives effectively. Accordingly, for the composites, the protrusion of hard Al_2O_3 particles and good bonding between matrix and reinforcing particles also appeared to be responsible for the reduced wear rate.

Table 3 summarizes the results of the fretting wear tests. From Table 3 it may be noted that the matrix material coating shows the highest mass loss. It can be seen that all coatings of the powders show a better wear behavior than the matrix material coating due to their high hardness and fine Al_2O_3 particles in the metal matrix. However, the increase in the wear resistance strongly depended on the Al_2O_3 vol.% in the powder. The coatings of two powders containing 30 vol. % Al_2O_3 were much better than the coatings of two powders containing 10 vol.% Al_2O_3. In addition, it can be seen that there was also a big difference between the coating of powder no.2

Figure 15. The variation of coefficient of friction of the composite coating with time for as-received nickel aluminium bronze (NAB) (plot a), and the same coated with NAB (plot b), NAB +10% Al_2O_3 (plot c), NAB +20% Al_2O_3 (plot d), and NAB +30% Al_2O_3 (plot e) against WC ball at an applied load of 10 N at an oscillating frequency of 10 Hz

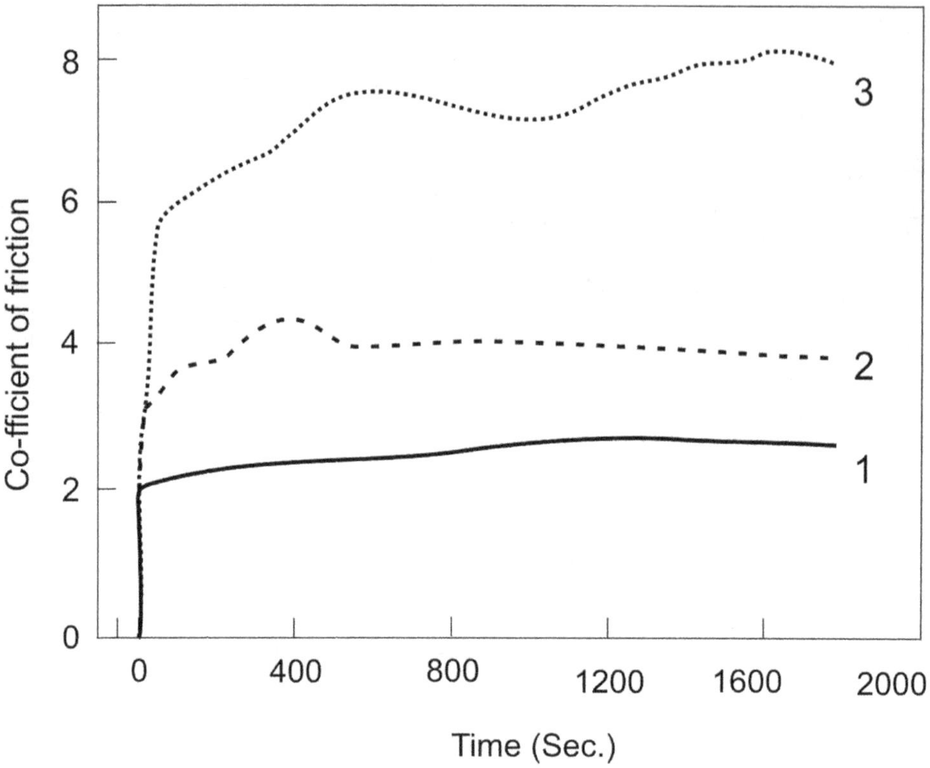

Table 3. Summary of wear behavior of as-received and graded coating

Sl. No.	Ball Used	Load used	Types Sample Used	Time (Min)	Wear Volume (mm^3)
1.	Tungsten carbide	10N	Nickel Aluminium Bronze	30	0.102
2.	Tungsten carbide	10N	100% Nickel Aluminium Bronze coated	30	0.0887
3.	Tungsten carbide	10N	10% Al_2O_3 coated	30	0.071
4.	Tungsten carbide	10N	10% + 20% Al_2O_3 coated	30	0.065
5.	Tungsten carbide	10N	10% + 20%+30% Al_2O_3 coated	30	0.052

and the coating of powder no. 3. The reason is that the higher hardness and more homogenous dispersion structure of powder no. 3 could be represented in the corresponding coating due to change Al_2O_3 volume content.

DEVELOPMENT OF FUNCTIONALLY GRADED THERMAL BARRIER COATING

Thermal barrier coatings are essentially ceramic based coatings applied for providing the thermal insulating barrier and high temperature oxidation resistance properties of the component to be used for high temperature application (Padture, 2002). Typical thermal barrier coatings consists of two layers, the first one is bond coat of MCrAlY type, where 'M' may be replaced with Ni, Co or Fe depending on the material composition of base structure or Pt-aluminide coating and the second one is usually Yttria stabilized Zirconia (YSZ) which provides good thermal resistance. Though thermal barrier coating is extremely important for improving the life time of the component, the major concern in thermal barrier coating is induction of large thermal stress during thermal cycles in an oxidizing environment due to the thermal expansion mismatch between the ceramic top coat and metallic bond coat (Busso *et al.*, 2007). For improving the life time of the thermal barrier coated component at high temperature, ceramic-metal functionally graded material (FGM) is attracting a great attention as thermal barrier coatings for aerospace, turbines, and aircraft engines. The concept of functionally graded thermal barrier coating to reduce the interfacial failure of the coating has been discussed in details by Lee et al. (Lee *et al.*, 1996). Thermal barrier coating may be developed by some processing techniques such as by plasma spraying, powder metallurgy, in-situ synthesis, etc. (Kim *et al.*, 2003]. Plasma spray coatings, involves successive accumulation of molten or semi-molten droplets spreading on the substrate surface and forming thin coating upon solidification remains popular technique for the development of TBC on large structures (Pawlowski, 2008). Though there are several reports on thermal properties of plasma sprayed duplex coating, however, studies on oxidation resistance of functionally graded thermal barrier coatings are limited. Khor & Gu (Khor & Gu, 2000) showed that the thermal cycling resistance of YSZ and NiCoCrAlY functional graded coatings was five times better than that of the duplex coating with a similar thickness. The failure modes of the coatings under thermal cycling conditions were surface cracking and delamination, which were caused by large residual radial/tangential stresses and axial/shear stresses, respectively. Vassen et al. (Vassen et al., 2010) showed that plasma spray deposited thermal barrier coatings made of $La_2Zr_2O_7$ and $Gd_2Zr_2O_7$ pyrochlores and $SrZrO_3$ perovskites with a double layer have a longer cycling lifetime compared to the single layer systems.

In this regard, it is relevant to mention that plasma spray coating technique can be employed to achieve specific properties (low thermal conductivity, longer thermal cycling life etc.) of the top coat and bond coat. However, application of HVOF spraying technique for the deposition of bond coating has been recently reported to form a dense coating with an increased surface roughness which is comparable to the same developed by vacuum plasma spraying technique (Rajasekaran *et al.*, 2012).

In an earlier study, a functionally graded YSZ/ CoNiCrAlY thermal barrier coating consisting of several layers with pre-mixed CoNiCrAlY and YSZ in the weight ratios of 70:30 (of thickness 80 μm in the near bond coat zone as 1st intermediate layer) followed by 50:50 (of thickness 80 μm as 2nd intermediate layer), 30:70 (of thickness 80 μm as 3rd intermediate layer) and 0:100 (of thickness 100 μm as top layer) were developed on Inconel 718 substrate by plasma spray deposition technique. Finally, the non-isothermal oxidation behavior of the graded coating was compared to that of duplex coating to establish the kinetics and mechanism of non-isothermal oxidation for both the conditions (Nath *et al.*, 2013).

A detailed study of the microstructure of the top surface and cross section of the duplex and graded coatings were undertaken. Figure 16 shows the microstructures (scanning electron micrographs) of the cross-section of the (a) duplex coating and (b) graded coating developed in the present study revealing the presence of defect free interface. From Figure 16 (a), it may also be noted that there is a deeper penetration of bond coat inside the substrate due to coating with a very high velocity in HVOF spraying. Figure 16 (b) shows the presence of the graded thermal barrier coating by a continuous change in the composition. From Figure 16b it is also revealed that the gradual change in composition did not create any sharp interface as was in duplex thermal barrier coating and the coating is also continuous and defect free. However, the presence of micro porosities was observed in both the top coatings (duplex and graded) and bond coating with the average area fraction of porosities ranging from 3 to 8% at the top coating and 2-3% at the bond coating, respectively.

Figures 17(a-g) show the distribution of (b) Zr, (c) Y, (d) Co, (e) Ni, (f) Cr, and (g) Al in the selected region of the microstructure of duplex coating (Figure 17a). From Figure 17, the uniform distribution of Zr and Y with a reduced intensity of the later is evident at the top layer and uniform elemental distribution of Co is evident at the bond coat. From Figure 17 it is also evident that the interface is continuous with no signature of diffusion of elements across the interface. Figures 18 (a-g) show the distribution of (b) Zr, (c) Y, (d) Co, (e) Ni, (f) Cr, and (g) Al in the microstructure of the graded coating (Figure 18 a). From Figure 18, it is seen that there is a graded distribution of Zr and Y from the surface to the interface with maximum content at the near surface region. On the other hand Co concentrations are maximum at near interface region and decreases near to the surface. Hence, it may be concluded that the process parameters and coating composition used in the

Figure 16. Scanning electron micrographs of the cross-section of the (a) duplex coating and (b) graded coating developed by plasma spray deposition technique on top of HVOF spray deposited CoNiCrAlY bond coat

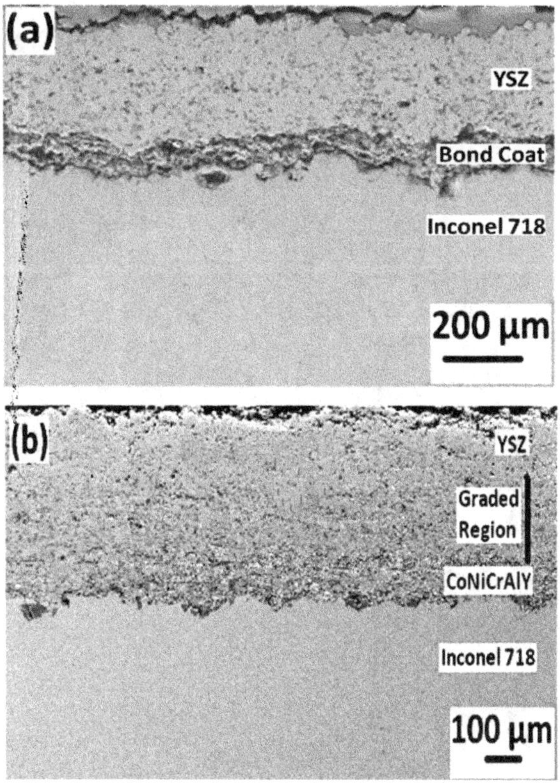

present study was perfect in the development of a thermal barrier coating with a graded distribution of alloying elements.

One of the major aim in development of functionally graded coating is the minimization of residual stress developed in duplex coating. A detailed study of the residual stress in the coated layer shows that on the surface of both the duplex and graded coating there is presence of residual tensile stress of very low magnitude, i.e. 28.3 MPa in duplex coating and 15.5 MPa in graded coating attributed to the thermal quenching effect from the molten state during plasma spraying operations. A variation of residual stress with depth was observed in graded coating. Initially it increases to 34.5 MPa in the 2nd layer (70% YSZ + 30% CoNiCrAlY), 49.8 MPa in the 3rd layer (50% YSZ + 50% CoNiCrAlY), and then decreases with change of its magnitude to a compressive stress at a value of -31.7 MPa (30% YSZ + 70% CoNiCrAlY) at the near interface region. The change in stress from tensile to com-

Figure 17. Elemental mapping showing the distribution of (b) Zr, (c) Y, (d) Co, (e) Ni, (f) Cr, and (g) Al in the selected region of the microstructure of duplex coating (Figure 14a)

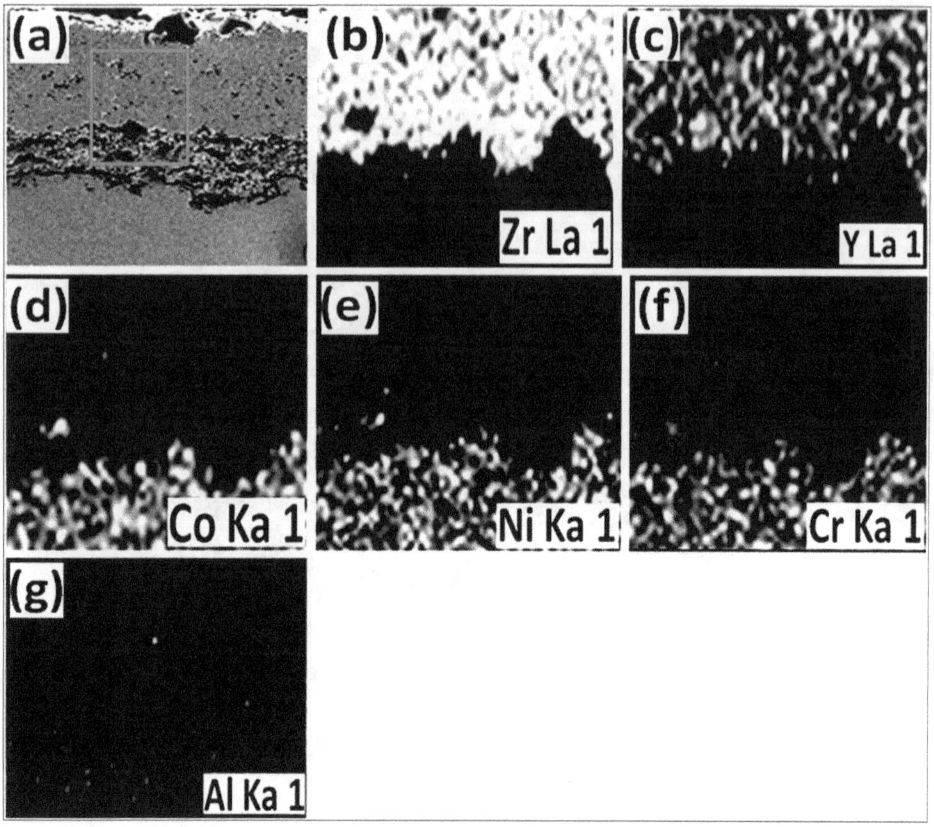

pressive is possibly attributed to the application of large velocity particles during spraying and the presence of transformational stress while transforming from tetragonal to monoclinic zirconia.

A detailed study of kinetics and mechanism of non-isothermal oxidation of both the duplex and graded coating was undertaken. Figure 19 shows the fractional change in weight, α (%) while oxidation from room temperature to 1250 °C at a heating rate of 10 °C/min, 20 °C/min, and 30 °C/min for Inconel 718 during non-isothermal oxidation. Figure 19 concludes that non-isothermal oxidation proceeds in three stages: a very low oxidation in stage I followed by steady oxidation in stage II and a rapid oxidation in stage III. The initial low oxidation at low temperature is attributed to lower diffusivity of ions or oxygen through the protective film (of Cr_2O_3) formed on the surface during oxidation. In stage II, oxidation proceeds at a

Figure 18. Elemental mapping showing the distribution of (b) Zr, (c) Y, (d) Co, (e) Ni, (f) Cr, and (g) Al in the microstructure of the graded coating (Figure 15 a)

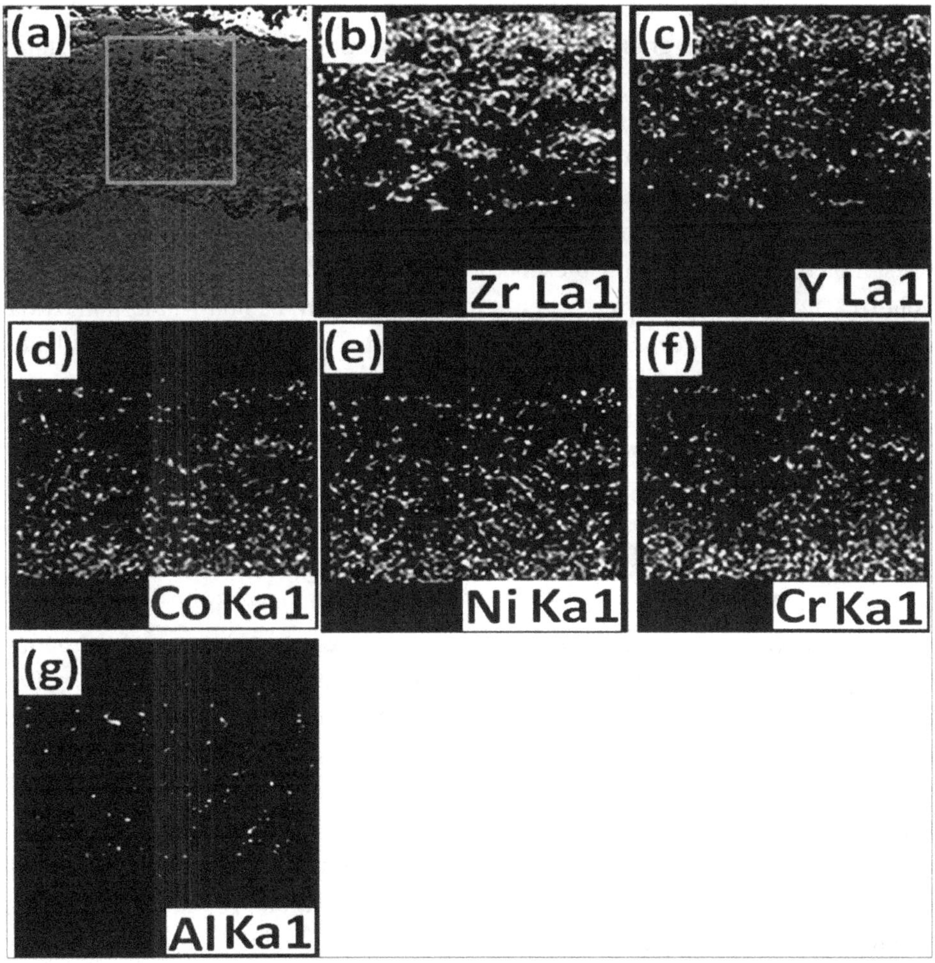

steady state due to the fact the rate of film formation equals to the rate of diffusion. In stage III oxidation, mass gain due to oxidation increases with temperature due to the damage of the protective film formed on the surface. The temperature range for stage I, stage II, and stage III oxidation were however, found to vary with the heating rate. The duration of stage I, stage II, and stage III oxidation for different heating rates are summarized in Table 4. Figure 20 shows the fractional change in weight, α (%) while oxidation from room temperature to 1250 ^{o}C at a heating rate of 10 ^{o}C/min, 20 ^{o}C/min, and 30 ^{o}C/min for duplex thermal barrier coating. Figure 20 reveals that the non-isothermal oxidation behavior of duplex thermal barrier coating revealing that oxidation proceeds in two stages, initial slow oxidation (stage I) and a faster oxidation (stage II). Figure 21 shows the fractional change in weight, α (%) while oxidation from room temperature to 1250 °C at a heating rate of 10 °C/min, 20 °C/min, and 30 °C/min for graded thermal barrier coating. The non-isothermal oxidation behavior of graded thermal barrier coating shows three stages, initial slow oxidation (stage I) followed by a steady oxidation (stage II) and finally, a rapid oxidation (stage III) (cf. Figure 21). The summary of duration of individual stage is shown in Table 4. From Figure 19, Figure 20 and Figure 21, it may be noted that there is an initial decrease in fractional change in weight of all samples with increase in temperature. The initial decrease in the fractional change in weight of the samples is attributed to the presence of moisture in the coating which on heating volatizes and the curve shows decrease in weight. From the non-isothermal oxidation studies of the as-received and as-coated Inconel 718, the kinetics of oxidation was calculated (Nath *et al.*, 2013).

Table 4 summarizes the activation energies of non-isothermal oxidation for Inconel 718, duplex, and compositionally graded thermal barrier coating (Nath *et al.*, 2013). From Table 4, it may be noted that the activation energy for oxidation varies with different stages of oxidation. During stage I oxidation, the activation energy for oxidation of graded thermal barrier coating is lower (94 kJ/mol) than Inconel 718 (162 kJ/mol) attributed to the oxidation of aluminium and hence, formation of alumina at a much faster rate due to its higher free energy of formation as compared to that of other oxides. The activation energy for oxidation increases to 186 kJ/mol in stage II oxidation for Inconel 718 which is due to the presence of Cr_2O_3 layer at the interface between the oxide layers. The activation energy decreases to 163 kJ/mol in the stage III oxidation due to the volatilization of Cr_2O_3 from the surface and porous nature of $NiCr_2O_4$ spinel.

In duplex thermal barrier coating, the activation energy for non-isothermal oxidation is marginally higher during the stage I oxidation (167 kJ/mol) which is due to the heat and oxidation resistance offered by zirconia top coat layer for the formation of alumina scale (Nath *et al.*, 2013). The activation energy increases in stage II, because of formation of an adherent alumina scale on the surface of bond coat. The graded thermal barrier coating shows the highest activation energy (418.9 kJ/mol) for non-isothermal oxidation in stage III as compared to Inconel 718 substrate and duplex thermal barrier coating due to the presence of large quantities of alumina scales distributed throughout the coated layer in an intermittent fashion. Hence, from the activation energy, it may be concluded that even though kinetics of oxidation is fastest at low temperature range, it slows down a maximum extent in graded thermal barrier coating as compared to duplex one.

Figure 22 shows the mechanism of non-isothermal oxidation with time or temperature of exposure derived for the duplex and graded thermal barrier coating based on the detailed thermodynamic analysis followed by the detailed characterization

Figure 19. Non-isothermal oxidation behavior of Inconel 718 showing the fractional change in weight, α (%) with temperature at heating rates of 10 °C/min, 20 °C/min, and 30 °C/min, respectively

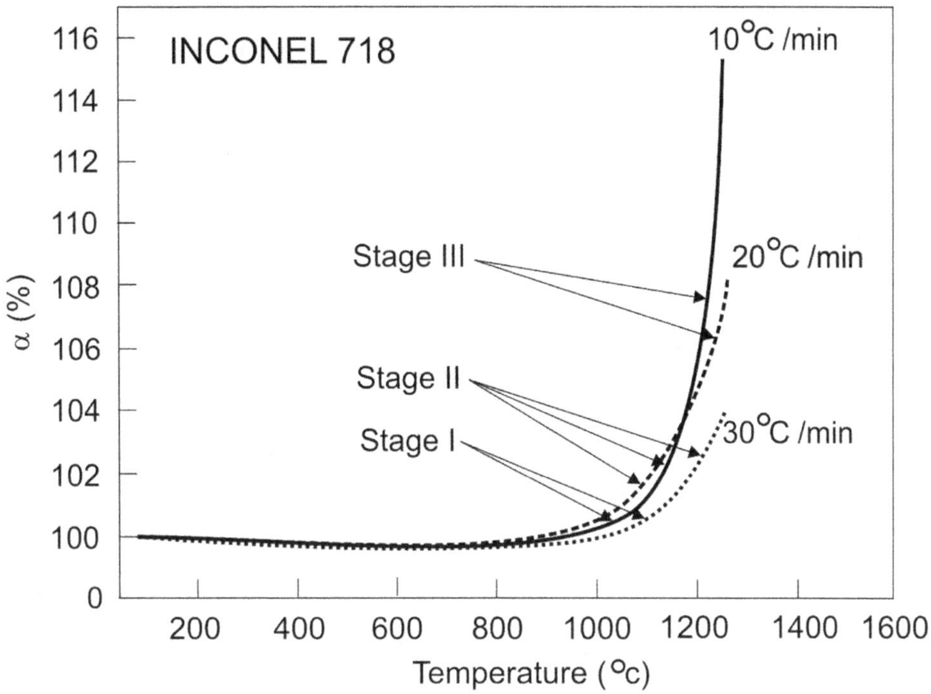

Table 4. Summary of activation energy of non-isothermal oxidation

Inconel 718		
Stage I	**Stage II**	**Stage III**
162 kJ/mol	186 kJ/mol	163 kJ/mol
Duplex TBC		
Stage I	**Stage II**	
167 jK/mol	212 kJ/mol	
Graded TBC		
Stage I	**Stage II**	**Stage III**
94 kJ/mol	187 kJ/mol	419 kJ/mol

Figure 20. Non-isothermal oxidation behavior of duplex thermal barrier coating showing the fractional change in weight, α (%) with temperature at heating rates of 10 °C/min, 20 °C/min, and 30 °C/min, respectively

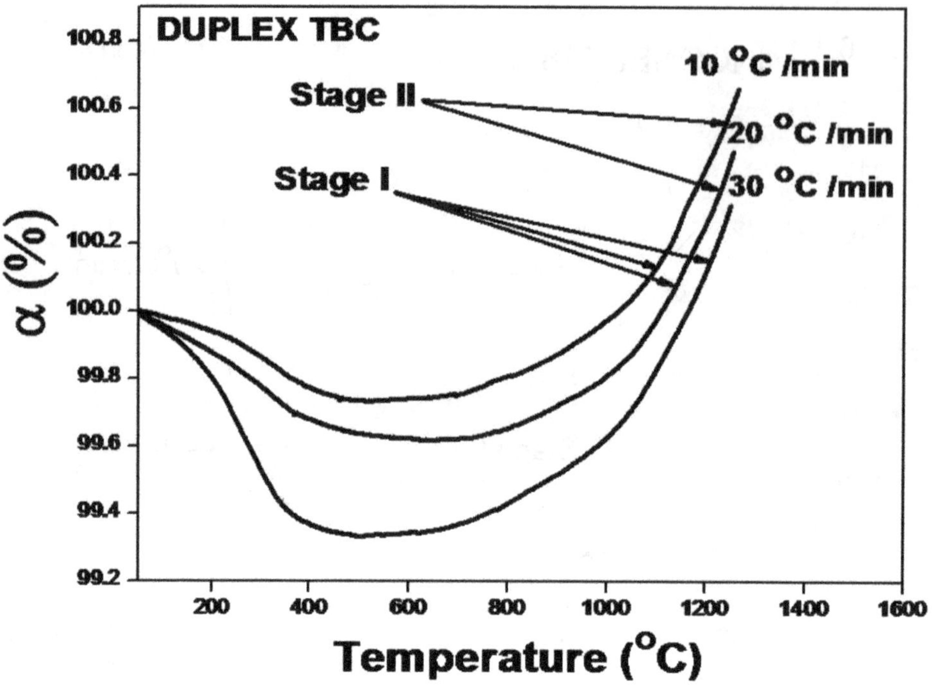

of the oxidized surface (Nath *et al.*, 2012). Figure 22 reveals that in duplex thermal barrier coating, there is the presence of alumina at the interface between the bond coat and top coat which initially, is not so impervious however, as it thickens it becomes protective and the kinetics of the oxidation decreases resulting in increased activation energy for oxidation. On the other hand, in compositionally graded thermal barrier coating, the kinetics of oxidation is initially very low, due to an increased content of aluminium throughout the coating. Furthermore, the alumina layer forms at every interface and as it thickens and becomes protective, the kinetics of oxidation reduces to a maximum extent with maximum activation energy of oxidation. Hence, it may be summarized that presence of Al_2O_3 oxide scale is responsible for improved oxidation resistance in both the duplex and graded coating. The maximum enhancement of oxidation resistance in the graded coating at the later stage of oxidation is attributed to the presence of alumina (Al_2O_3) layer at the interfaces between the graded coatings.

Figure 21. Non-isothermal oxidation behavior of graded thermal barrier coating showing the fractional change in weight, α (%) with temperature at heating rates of 10 °C/min, 20 °C/min, and 30 °C/min, respectively

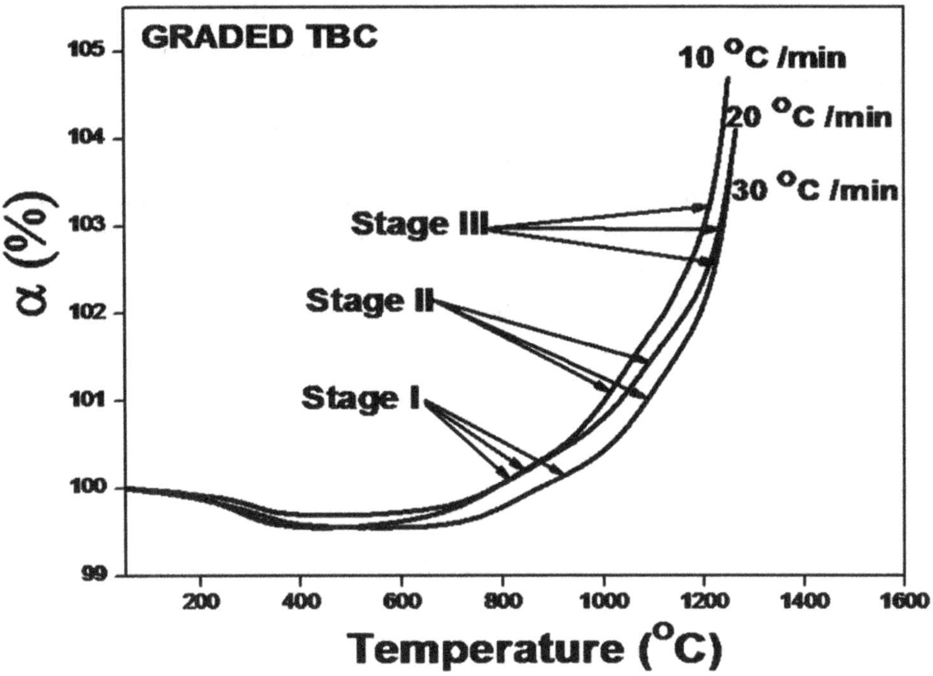

Figure 22. Schematic illustration of the mechanism of oxidation

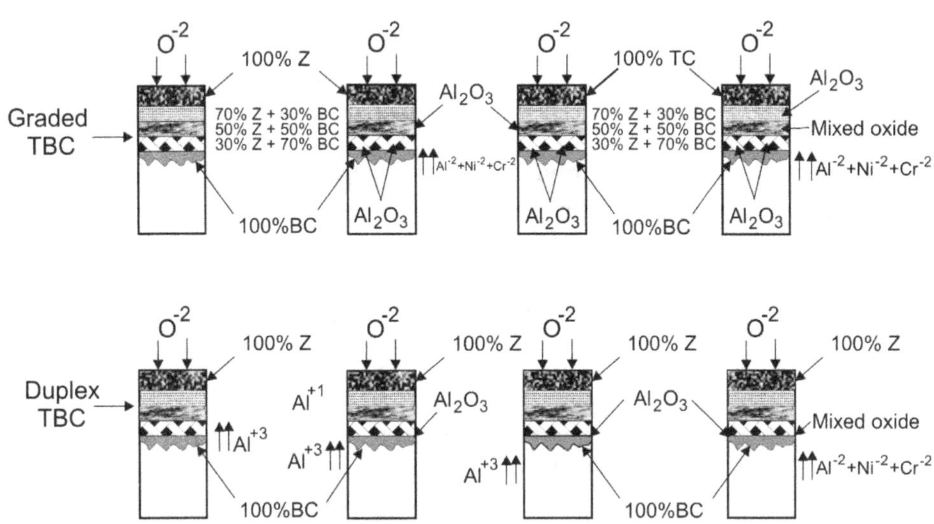

FUTURE SCOPE OF RESEARCH

Compositionally graded coating is the unique approach for the attainment of tailored component with superior properties as compared to the conventional duplex materials. There have been extensive efforts on the development of compositionally graded materials for improving wear, and high temperature oxidation resistance properties of material for reduced stress distribution on the surface and with depth. However, efforts need to be undertaken for the development of compositionally modulated component for functional application leading to the development of (a) hydrophobic surface with improved corrosion resistance, (b) development of multifunctional component with continuously change in coating composition with position for effective utilization of materials and minimum probability of interface for a maximum performance, (c) surface nano-structured graded coating for corrosion and bio-medical application.

REFERENCES

Bever, M. B., & Duwez, P. E. (1972). Gradients in composite materials. *Materials Science and Engineering*, *10*, 1–8. doi:10.1016/0025-5416(72)90059-6

Budinski, K. G. (1988). *Surface engineering for wear resistance* (2nd ed.; Vol. 32). Englewood Cliffs, NJ: Prentice Hall.

Busso, E. P., Wright, L., Evans, H. E., McCartney, L. N., Saunders, S. R. J., Osgerby, S., & Nunn, J. (2007). A physics-based life prediction methodology for thermal barrier coating systems. *Acta Materialia*, *55*(5), 1491–1503. doi:10.1016/j.actamat.2006.10.023

Chattopadhyay, R. (2004). *Advanced thermally assisted surface engineering processes, XII*. Dordrecht, The Netherlands: Kluwer Academic Publishers.

Crawmer, E. D. (2004). Introduction to coatings, equipment, and theory. In Handbook of thermal spray technology. Materials Park, OH: ASM International.

Dunleavy, C. S., Golosnoy, I. O., Curran, J. A., & Clyne, T. W. (2009). Characterization of discharge events during plasma electrolytic oxidation. *Surface and Coatings Technology*, *203*(22), 3410–3419. doi:10.1016/j.surfcoat.2009.05.004

Dutta Majumdar, J., & Manna, I. (2012). Introduction to laser-assisted fabrication of materials. In Laser assisted fabrication-Present status and future scope in laser assisted fabrication of materials. Hiedelberg, Germany: Springer Verlag.

Iwasaki, K. (1997). Production of a functionally graded artificial tooth root by unique sequence of processes. *Materials Research Innovative*, *1*(3), 180–187. doi:10.1007/s100190050038

Kawasaki, A., & Watanabe, R. (1997). Concept and P/M fabrication of functionally gradient materials. *Ceramics International*, *23*(1), 73–83. doi:10.1016/0272-8842(95)00143-3

Khor, K. A., & Gu, Y. W. (2000). Thermal properties of plasma-sprayed functionally graded thermal barrier coatings. *Thin Solid Films*, *372*(1-2), 104–113. doi:10.1016/S0040-6090(00)01024-5

Kieback, B., Neubrand, A., & Riedel, H. (2003). Processing techniques for functionally graded materials. *Materials Science and Engineering A*, *362*(1-2), 81–106. doi:10.1016/S0921-5093(03)00578-1

Kim, J. H., Kim, M. C., & Park, C. G. (2003). Evaluation of functionally graded thermal barrier coatings fabricated by detonation gun spray technique. *Surface and Coatings Technology*, *168*(2), 275–280. doi:10.1016/S0257-8972(03)00011-2

Koizumi, M., & Niino, M. (1995). Overview of FGM research in Japan. *MRS Bulletin*, *20*(1), 19–26.

Krumova, M., Klingshirn, C., Haupert, F., & Friedrich, K. (2001). Microhardness studies on functionally graded polymer composites. *Composites Science and Technology*, *61*(4), 557–563. doi:10.1016/S0266-3538(00)00228-1

Kumar Roop, R., & Wang, M. (2002). Functionally graded bioactive coatings of hydroxyapatite/titanium oxide composite system. *Materials Letters*, *55*(3), 133–137. doi:10.1016/S0167-577X(01)00635-8

Lee, C. S., Ahn, S. H., DeJonghe, L. C., & Thomas, G. (2006). Effect of functionally graded material (FGM) layers on the residual stress of polytypoidally joined Si_3N_4 –Al_2O_3. *Materials Science and Engineering A*, *434*(1-2), 160–165. doi:10.1016/j.msea.2006.06.139

Lee, M. K., Hong, S. M., Kim, G. H., Kim, K. H., & Kim, W. W. (2004). Structural properties in flame quenched Cu-9Al-4.5Ni-4.5Fe alloy. *Metals and Materials International*, *10*(4), 313–319. doi:10.1007/BF03185979

Lee, W. Y., Stinton, D. P., Berndt, C. C., Erdogan, F., Lee, Y. D., & Mutasim, Z. (1996). Concept of functionally graded materials for advanced thermal barrier coating applications. *Journal of the American Ceramic Society*, *79*(12), 3003–3012. doi:10.1111/j.1151-2916.1996.tb08070.x

Li, W. S., Wang, Z. P., Lu, Y., Jin, Y. H., Yuan, L. H., & Wang, F. (2006). Mechanical and tribological properties of a novel aluminum bronze material for drawing dies. *Wear, 261*(2), 155–163. doi:10.1016/j.wear.2005.09.032

Longo, F. N., & Davis, J. R. (Eds.). (2005). *Handbook of thermal spray technology*. Materials Park, OH: ASM International.

Lu, F., Li, H., Ji, Q., Zeng, R., Wang, S., Chi, J., & Sola, A. et al. (2007). Characterization of glass-alumina functionally graded coatings obtained by plasma spraying. *Journal of the European Ceramic Society, 27*(4), 1935–1943. doi:10.1016/j.jeurceramsoc.2006.05.105

Mattox, D. M. (1964). Film deposition using accelerated ions. *Electrochemical Technology, 2*(9-10), 295–298.

Mellor, B. G. (Ed.). (2006). Surface coatings for protection against wear. Cambridge, UK: Woodhead Publishing Limited. doi:10.1201/9781439823422

Miyamoto, Y., Kaysser, W., Rabin, B., Kawasaki, A., & Ford, R. G. (1999). Functionally graded materials: Design, processing and applications. Dordrecht, The Netherlands: Kluwer Academic Publishers.

Moro, A., Kuroda, Y., & Kusaka, K. (2002). Development status of the reusable high-performance engines with functionally graded materials. *Acta Astronautica, 50*(7), 427–432. doi:10.1016/S0094-5765(01)00174-6

Mortensen, A., & Suresh, S. (1995). Functionally graded metals and metal-ceramic composites: Part 1. *International Materials Reviews, 40*(6), 239–265. doi:10.1179/imr.1995.40.6.239

Moya, J. S., Sanchezherencia, A. J., Requena, J., & Moreno, R. (1992). Functionally gradient ceramics by sequential slip casting. *Materials Letters, 14*(5-6), 333–335. doi:10.1016/0167-577X(92)90048-O

Nath, S., Manna, I., & Dutta Majumdar, J. (2013). Compositionally graded thermal barrier coating by hybrid thermal spraying route and its non-isothermal oxidation behavior. *Journal of Thermal Spray Technology, 22*(6), 901–917. doi:10.1007/s11666-013-9937-4

Padture, N. P., Gell, M., & Jordan, E. H. (2002). Thermal barrier coatings for gas-turbine engine applications. *Science, 296*(5566), 280–284. doi:10.1126/science.1068609 PMID:11951028

Pawlowski, L. (2008). *The science and engineering of thermal spray coatings* (2nd ed.). Chichester, UK: John Wiley and Sons Ltd. doi:10.1002/9780470754085

Prchlik, L., Sampath, S., Gutleber, J., Bancke, G., & Ruff, A. W. (2001). Friction and wear properties of WC-Co and Mo-Mo_2C based functionally graded materials. *Wear*, *249*(12), 1103–1115. doi:10.1016/S0043-1648(01)00839-0

Rajasekaran, B., Mauer, G., & Vassen, R. (2012). Enhanced characteristics of HVOF sprayed MCrAlY bond coats for TBC applications. *Journal of Thermal Spray Technology*, *20*(6), 1209–1216. doi:10.1007/s11666-011-9668-3

Schoop, M. U., Guenther, H. & Das. (1917). *Schoopsche, Metallspritz-Verfahren* [The Schoop Metal Spraying Technique]. Stuttgart, Germany: Franckh Publishers.

Schulz, U., Peters, M., Bach, F.-W., & Tegeder, G. (2003). Graded coatings for thermal, wear and corrosion barriers. *Materials Science and Engineering A*, *362*(1-2), 61–80. doi:10.1016/S0921-5093(03)00579-3

Shen, M., & Bever, M. B. (1972). Gradients in polymeric materials. *Journal of Materials Science*, *7*(7), 741–746. doi:10.1007/BF00549902

Singh, H., Sidhu, T. S., & Kalsi, S. B. S. (2012). Cold spray technology: Future of coating deposition processes. *Integrità Strutturale Frattura*, *22*, 69–84.

Tampieri, A., Celotti, G., Sprio, S., Delcogliano, A., & Franzese, S. (2001). Porosity-graded hydroxyapatite ceramics to replace natural bone. *Biomaterials*, *22*(11), 1365–1370. doi:10.1016/S0142-9612(00)00290-8 PMID:11336309

Tang, C. H., Cheng, F. T., & Man, T. H. (2006). Laser surface alloying of a marine propeller bronze using aluminium powder part I: Microstructural analysis and cavitation erosion study. *Surface and Coatings Technology*, *200*(8), 2602–2609. doi:10.1016/j.surfcoat.2004.12.021

Tucker, R. C. (1994). Thermal sprays coatings. In ASM handbook, surface engineering (vol. 5, pp. 497-509). Materials Park, OH: ASM International.

Turner, T. H., & Budgen, N. E. (1926). *The origin, development and applications of the metal spray process of metallization*. London: Charles Griffin & Co.

Valarezo, A., Giovanni, B., Wanhuk, B. C., Sanjay, S., Valeria, C., Luca, L., & Roberto, R. (2010). Damage tolerant functionally graded WC–Co/stainless steel HVOF coatings. *Surface and Coatings Technology*, *205*(7), 2197–2208. doi:10.1016/j.surfcoat.2010.08.148

Vassen, R., Kassner, H., Stuke, A., Mack, D. E., Maria, J. O., & Stöver, D. (2010). Functionally graded thermal barrier coatings with improved reflectivity and high-temperature capability. *Materials Science Forum*, *631/632*, 73–78. doi:10.4028/www.scientific.net/MSF.631-632.73

Wang, X., Hwang, K. S., Koopman, M., Fang, Z., & Zhang, L. (2013). Mechanical properties and wear resistance of functionally graded WC-Co. *International Journal of Refractory Metals & Hard Materials*, *36*, 46–55. doi:10.1016/j.ijrmhm.2012.04.011

Watari, F., Yokoyama, A., Saso, F., Uo, M., & Kawasaki, T. (1995). Functionally gradient dental implant composed of titanium and hydroxyapatite. In *Proc. 3rd Int. Symp. Structural & Functional Gradient Materials* (pp. 703–8). Lausanne, Switzerland: Polytech. Univ. Romand.

Watari, F., Yokoyama, A., Saso, F., Uo, M., & Kawasaki, T. (1997). Fabrication and properties of functionally graded dental implant. *Composites. Part B, Engineering*, *28B*(1-2), 5–11. doi:10.1016/S1359-8368(96)00021-2

Yuanyuan, L., Tungwai, L. N., & Wei, X. (1996). Mechanical, friction and wear behaviors of a novel high-strength wear-resisting aluminum bronze. *Wear*, *197*(1-2), 130–136. doi:10.1016/0043-1648(95)06890-2

KEY TERMS AND DEFINITIONS

Functionally Graded Coatings: Coating consisting of continuously changing composition from substrate to the surface to achieve combinations of toughness and bond strength.

Residual Stresses: Residual stresses refer to the internal stresses which originate due to the growth of the coating and depend on the mechanism of growth of coating.

Thermal Barrier Coating: Thermal Barrier Coating: are essentially multilayered coating structures that consist of a bond coat that is applied to the substrate, a thermally grown oxide (TGO) and a ceramic topcoat, typically an yttria stabilized zirconia.

Thermal Spray Coatings: Coating obtained by a group of processes that apply a consumable powders or wires in the form of finely divided molten and semi molten droplets.

Tribology: Science of friction.

Chapter 6
High Temperature Sliding Wear of Thermal Sprayed Coatings

A. Pauschitz
AC2T Research GmbH., Austria

Manish Roy
Defence Metallurgical Research Laboratory, India

ABSTRACT

In this chapter, various layers that are formed during sliding wear of thermal sprayed coatings at elevated temperature are discussed. Glazed layers are formed on the worn surfaces during elevated temperature sliding wear of thermal sprayed coatings. These layers have a characteristic physical appearance, mechanical properties, chemical compositions, and failure mechanisms. Wearing conditions, wearing material, and mating material influence formation and characteristics of glazed layers. Among these parameters, wearing material and mating material are most important. These glazed layers are divided into different types of layers, namely Transfer Layer (TL), Mechanically Mixed Layer (MML), Reaction Layer (RL), and Composite Layer (CL). The recent results on friction of thermal sprayed coatings at elevated temperature are rationalised in the light of different types of glazed layer formation.

DOI: 10.4018/978-1-4666-7489-9.ch006

INTRODUCTION

Advancement of technology has enhanced the demand for tribosystem capable of running under severe conditions such as at higher relative motions, at higher applied loads, at higher temperatures, etc. Sliding wear at elevated temperature is a serious problem in a large number of engineering components in various industrial environment such as power generation, transport, materials processing, high temperature bearing, impeller bearing of slurry pumps operated in waste tank, etc. (Taktak, 2006; Roy *et al.*, 2004; Jiang *et al.*, 2005; Mateos *et al.*, 2001). In order to improve the sliding wear performance of materials at elevated temperature for enhanced elevated temperature performance, sometimes it is imperative to modify the surfaces or use coatings with superior oxidation resistance and mechanical properties without compromising bulk properties. Thermal sprayed coating in this respect is one of the most widely used coating technique which offers several advantages.

Thermal sprayed coatings are used extensively in several engineering components requiring resistance to wear at high temperature, such as fuel rod mandrel, forming tools, gas turbines, etc. (Lai, 1979, Watreme *et al.*, 1996; Li, 1980). Li & Lai (Li & Lai, 1979) noted that at elevated temperature the friction and wear of thermal sprayed Cr_3C_2-NiCr coating is initially high due to stick-slip phenomenon and as the sliding distance increases these values decrease. Further, friction coefficient increases with decrease of sliding speed. An extensive work on room temperature sliding wear of this coating by Mohanty *et al.* (Mohanty *et al., 1996)* indicated decohesion at intersplat controls the wear rate at room temperature. Wear rate is proportional to applied load but friction coefficient decreases with increase of sliding velocity. Mateos *et al.* (Mateos *et al.,* 2001) reported presence of two distinct zones on the wear versus sliding distance plot. Guilemany *et al.* (Guilemany *et al.*, 2002) demonstrated an improvement of the sliding wear behaviour of similar coating by heat treatment through formation of Cr_3C_2 precipitates. However, these studies are confined to the understanding of the parametric effect of wearing condition on the wear rate and carried out primarily at ambient condition. So far, no attempt has been made to compile the state of art on the influence of test temperature on the wear behaviour of thermal sprayed coatings.

The objective of the present work is to give a systematic and brief survey of the current status and future trends of sliding wear of thermal sprayed coatings at elevated temperature. For the purpose of convenience this paper is divided into six sections. After introduction of the elevated-temperature wear in Section 1, background of elevated-temperature wear of metallic materials and thermal sprayed coatings is discussed in Section 2. Formation and mechanisms of formation of wear protective glazed layer is the subject matter in section 3. Classification of various types of glazed layers is made in section 4. The results of recent research are rationalised in section 5. Summary and direction of future research is the subject matter of section 6.

BACKGROUND

The oxidation of the materials in air environment plays a significant role during elevated temperature (300-1000°C) wear of materials. It changes the overall wear rate under sliding contact when operating temperature of the bulk material is high. The importance of oxidation during wear of metallic materials was first identified by Fink (Fink, 1930) Archard & Hirst (Archard & Hirst, 1956) proposed the classification of mild and severe wear based on measurement of contact resistance, wear debris analysis and microscopic examination. The importance of oxide scale for ambient temperature wear was discussed extensively by Quinn (Quinn *et al.* 1984) and Lim and Ashby (Lim &d Ashby, 1987). Subsequently a large number of investigations were carried out on the elevated temperature sliding wear of metallic materials (Stott, 2002; Radu & Li, 2007; Roy *et al.*, 2002, 2004, 2008; Inman *et al.*, 2006a, 2006b; Pauschitz *et al.*, 2003a, 2003b, 2006; Tu *et al.* 1998; Wang & Li, 2003). Most of the studies indicate the formation of glazed layers on the substrate under certain conditions of load, temperature and sliding speed. A few investigations examined the structure of glazed layer on the substrate and the mechanisms of the formation of the layer in the tribosystem containing Nimonic against Stellite 6 (Inman *et al.*, 2006a, 2006b). Fontalvo & Mitterer (Fontalvo & Mitterrer, 2005) reported the influence of Si and Al addition on high-temperature wear of hot working steel X38CrMoV5-1. In a series of experiments by Li and co-workers (Radu & Li, 2007; Wang & Li, 2003) demonstrated improvement of the mechanical properties of the oxide scale and consequently the wear behaviour of these materials at elevated temperature by addition of yttrium (Y) in small quantities to various Stellite alloys.

In recent years, ZrO_2 ceramic coatings as thermal barrier coatings in advanced engines and gas turbines, and wear resistant coatings in lubrication and seal systems has received increased attention (Ahn *et al.*, 1997; Sliney, 1982; Wang *et al.*, 1998). The tribological behavior of ZrO_2 coatings at elevated temperatures have been extensively investigated (Ahn *et al.*, 1997; Wang *et al.*, 1988, 1998). However, the tribological characteristics of ZrO_2 coatings with the lowest friction and wear under dry sliding conditions are still far from satisfactory for many high temperature applications. Thus, it is imperative to effectively lubricate ZrO_2 coatings if they are to be used extensively in tribological applications at high temperatures. To date, only a limited amount of experiments have been done on the lubrication of

ZrO_2 coatings. These studies utilized several kinds of solid lubricants (CaF_2, BaF_2 or silver) as additives or well adhering soft metallic coatings, as boundary films (Wang *et al.*, 1998; Della-Corte, 1996; Ouyang *et al.*, 2001). Improved tribological characteristics of ZrO_2 coatings at an elevated temperature were reported through solid lubrication methods (Wang *et al.*, 1998; Ouyang *et al.*, 2001). The friction and wear coefficients of ZrO_2-CaF_2 coatings doped with and without silver additives showed that both coating combinations had fairly high wear rates at room temperature, but wear rates were much lower for the ZrO_2-CaF_2 coating at 650°C as compared to ZrO_2 coatings without any solid lubricant (Wang *et al.*, 1998). The addition of silver did not have a beneficial effect in improving friction and wear. Plasma-sprayed ZrO_2-CaF_2 coatings also exhibit a low friction and wear in sliding against a sintered Al_2O_3 ball at 600–700°C (Ouyang *et al.*, 2001). The fluoride provides a low shear strength film between the sliding surfaces to mitigate friction and wear. Plasma-sprayed solid lubricant films having combinations of CaF_2, BaF_2 and Cr_2O_3 powders also illustrated excellent friction properties in a wide temperature range of 300 – 900°C. The low friction coefficient was considered to be due to the coexistence of CaF_2 and $BaCrO_4$, which are formed by the tribo-chemical reaction on the worn surfaces at high temperature.

The tribological behavior of Cr_2O_3 coating at elevated temperatures has been extensively investigated (Wang *et al.*, 1988, Wang, 1993, Richard *et al.*, 1995). However, it is still necessary to find ways of effectively lubricating Cr_2O_3 ceramic coating if they are to be used extensively in tribological applications at extreme temperatures. Only a limited amount of experiments has been done on lubrication of Cr_2O_3 ceramic coating. These studies utilized several kinds of solid lubricants (CaF_2, BaF_2 or silver) as additives or a well-adhering soft metallic coating as a boundary film [Wang, 1993; Della-Corte, 1996; Liu, 1993; Vos *et al.*, 1998; Li & Lai, 1979; Mohanty *et al., 1996*). Improved tribological characteristics of Cr_2O_3 ceramic coatings at elevated temperature were reported through solid lubrication method (Wang, 1993). Spray drying of a Cr_2O_3–CaF_2 powder blend allows obtaining a homogeneous distribution of the lubricant in the composite coatings plasma-sprayed with these agglomerates (Vos *et al.*, 1998). It was proven that optimized parameters for plasma-spraying the pure Cr_2O_3 are also suitable to spray the Cr_2O_3–CaF_2 composite coatings. At room temperature, both the pure Cr_2O_3 and the composite coatings reach a steady-state friction coefficient of approximately 0.40 after 10,000 cycles. A friction coefficient of 0.25 and a wear volume of 37×10^3 mm^3 for a coating containing 16 vol.% of CaF_2 are obtained under the wear condition of 400°C temperature, 5N normal load and 1Hz frequency against the corundum ball using reciprocating wear test. At 400°C, the CaF_2 obtains a much lower shear strength compared to the shear strength at room temperature and the CaF_2 particles will be spread out, creating a lubricating layer in the wear track that reduces the friction and wear (Vos *et al.*, 1998). The tribological results of plasma-sprayed Cr_2O_3 composite coating with different concentrations of CaF_2 and BaF_2 as solid lubricants at 425°C (Liu, 1993) demonstrated that the optimal solid lubricant content was about 14–21 vol.% for Cr_2O_3–CaF_2 coatings and 20–31 vol.% for Cr_2O_3–BaF_2 coatings. The mean friction coefficient at 425°C had been reduced from about 0.40 for Cr_2O_3/Cr_2O_3 sliding to 0.20–0.25 for Cr_2O_3/Cr_2O_3-solid lubricant sliding. The friction and wear coefficients of plasma-sprayed nichrome bonded Cr_2O_3 coating with Ag and BaF_2/CaF_2 lubricant additives against Al_2O_3 or Inconel alloy counterfaces in a pin-on-disk tribometer (Della-Corte, 1996) showed that, compared to the Al_2O_3, the Inconel counterface generally exhibited lower friction and wear at 25°C but higher friction and wear at 650°C. The performance of each counterface was affected by the ability of solid lubricant additives to form an adequate transfer film. But no further data were obtained for higher temperature test or against other different counterface materials.

It was noted that the wear loss of yttria stabilized zirconia coatings increased with the increase in the test temperature and reaching its maximum loss at approximately 400°C. This unique behavior was explained by the tetragonal to monoclinic transformation (Kim & Lim, 1994). Others found the presence of $Y(OH)_3$ crystallites and suggested that dissolution of Y elements from the surface into H_2O was responsible for the low temperature transformation (Li & Watanabe, 1996). The influence of annealing process on the high temperature tribological performance of plasma-sprayed 3 mol% yttria stabilized zirconia coatings with and without Fe_2O_3 addition was investigated by Shin *et al.* (Shin *et al.*, 2000). The wear experiments were carried out on a ring-on-plate type reciprocating wear tester at selected temperatures within the range of 25-600°C. In addition, plasma-sprayed zirconia-based coatings were annealed at 500°C. The results show that annealing treatment decreases the wear rate due to the release of tensile residual stress. The stress relaxation improves the tribological properties by reducing the influence of low temperature degradation. The addition of Fe_2O_3 to the yttria stabilized zirconia coating leads to a reduction of wear and friction. This decrease is attributed to the stabilization of the tetragonal phase and the increase of microhardness.

Yang *et al.* (Yang *et al.*, 2006) noted that the friction coefficient of WC-Co coating does not alter significantly with change of temperature or carbide grain size up to 400°C. The coatings show higher sliding wear resistance at moderately elevated temperatures, contrary to other materials such as ceramics (Yang *et al.*, 2004), which have shown much lower wear resistance at moderately elevated temperatures.

To study the effect of temperature on a Ni-based alloy reinforced with small quantities of WC, wear tests of plasma sprayed WC-NiCrBSi coatings were carried out in a friction machine with alumina cylinders as counterbody by Martin *et al.* (Martin *et al.*, 2001). Considerable differences were found between the wear rates of the counterbody of diverse tests. Change of the bulk temperature changes the characteristics of the counterbody-specimen contact. Some layers of specimen material are transferred to the counterbody, reducing the severity of the original contact, and consequently the wear rate is lowered. The tungsten carbide particles play an abrasive role which greatly increases the wear rates of the counterbody. Nevertheless, the percentage of reinforcement particles does not enhance the wear resistance of the specimens tested. When the bulk temperature is increased up to 500°C, the tungsten carbide particles hinder the transfer of material from the specimen to the counterbody. Thus, the wear rates are found to be very similar at 20 and 500°C for this reinforced alloy.

Thus, above reports clearly show that elevated temperature wear of thermal sprayed coating has been studied to a considerable extent. As the test is carried out over a number of varieties of conditions, it is not possible to compare the test results. Wear rates and friction coefficients of various coatings change with temperature in different ways. Most widely investigated coatings are WC-Co and Cr_3C_2-NiCr coating. There are several other coatings which are used at elevated temperatures namely oxide variety coatings such as Cr_2O_3, TiO_2, TiO_2–Cr_2O_3 etc. cerment variety coatings such as NiCrFeAlTi-TiC, NiCr-(Ti, Ta)C and NiCrMo-(Ti, Ta)C, (Ti,Mo)(C,N)-Ni(Co) etc. and self fluxing coatings such as WC-NiCrBSi.

GLAZED LAYER FORMATION

A glazed layer is formed on the worn surfaces during elevated temperature sliding wear of thermal sprayed coatings. The mechanical properties and microstructural features of the glazed layer are governed by the wearing and test conditions and the materials involved in sliding process. Detailed literatures have shown that this layer is formed by developing compacted layers of oxide and partially oxidized alloy particles on the sliding surfaces. The detail discussion on the development of this layer from the agglomeration and compaction of the wear debris particles is carried out elsewhere (Jiang et al., 1998). After the generation of debris particles, some are lost from between the surfaces, but most are retained and they involve in development of the compact, load-bearing layers. These particles are subjected to deformation, fragmentation and comminution and break down into smaller particles. A fraction of these finer particles is lost from the wear track while others are agglomerated on the wear surfaces, particularly in grooves formed in the early stages of sliding. As sliding continues, these agglomerated clusters of particles are subjected to thermo-elastic stresses and compaction and their surrounding areas of contact are subjected to wear. They come into contact with the counter face and become load-bearing areas. At the same time, the fine particles are sintered together, to some extent, to form more solid layers. Studies of sintering of fine alloy particles (40 nm in diameter) have found that significant interactions occurred as soon as the temperature exceeded 20°C (Zhou *et al.*, 1989). In the presence of the high compressive pressures during wear, sintering is enhanced (Burke, 1962). Moreover, the particles in the outer surfaces of the layers of wear debris are very small and heavily deformed. As the temperature is increased, the rate of sintering increases, leading to the establishment of very solid, smoothly burnished, hard surfaces, termed 'glazes'. These layers can provide wear-protection for long periods since breakdown results in formation of further oxide debris that can be re-compacted into the 'glaze' surface once again. Although the 'glaze' surfaces are generally not observed at temperatures below 150°C, less well-sintered wear-debris layers are able to develop, particularly after longer periods of sliding, leading to the transition to mild wear (Archard & Hirst, 1956). Usually these layers are accompanied by positive contact resistances, although intermittent periods of zero resistance indicate that the layers tend to break up and then re-form elsewhere on the contacting surfaces (Jiang *et al.*, 1994). This is consistent with the smaller adhesion forces between the particles in the layers at these lower temperatures. As the wear debris particle layer is established, two competitive processes occur during sliding. These are breakdown of the layers, resulting in formation of further debris particles and consolidation of the layers by further sintering/cold welding of the particles. The effect of increase in temperature is to increase the rate of sintering and consolidation and to increase the rate of oxidation of the residual metal in the particles. If the surfaces become solid before breakdown occurs, the 'glaze' is established and the wear rate becomes very low. Conversely, if the particles are not well compacted and sintered, loose particles are removed more easily and the wear rate is higher. This can explain the observed critical temperature, above which the wear rate decreases considerably, due to establishment of a 'glaze' surface under the sliding conditions.

As mentioned earlier, the development and formation of the glazed layer depends on the wearing and test conditions. When the counter face material is soft, shallow grooves are formed. The shallow groves are not able to retain oxide debris to form the glazed layer. In contrast, if the counter face is relatively hard, deep grooves are generated which lead easily to glazed layer formation. These deep grooves are capable of retaining oxide debris to form the wear protective layer. The relative sliding speed between the wearing material and the counter face material can play an important role in the formation of wear protective layers. At high sliding speed abrasive wear debris is formed. Under high sliding speed, formation of glazed layers is less favourable due to their low residence time and the debris acts as abrasives leading to higher wear rate. The formation of the wear protective layer is influenced by the partial pressure of the oxygen during sliding. At partial pressure of oxygen lower than 0.1 atm and at low test temperature and, the glaze layer fails to form in some tribosystem. In general, glaze layer forms at a temperature higher than 423

K. A significant role in controlling the nature of the glaze layers is played by the geometry of the tribosystem. A conformable geometry helps in retaining wear debris and in turns favours formation of glazed layers. When the geometry is nonconformable it fails to retain the debris and fails to form glazed layer. Composition of the oxide debris also decides formation of glaze layers. Oxides based on Co and FeCr alloy can easily be sintered and hence glaze layer formation is easy. If the debris is oxide of Ni or Cr, it cannot be sintered all that easily. Thus, they act as abrasive particles enhancing the wear rate.

CLASSIFICATION OF GLAZED LAYER

Pauschitz *et al.* (Pauschitz *et al.*, 2003, 2004) for the first time carried out a series of systematic experiments. On the basis of experimental results, they classified the glazed layers under different heads based on their nature and characteristics. They deposited Cr_3C_2-NiCr coatings on 253 MA alloy procured from Avesta Polariod, (Sweden). The nominal composition of the steel is given in Table 1. The composition of the steel is nearly same to that of ASTM 309S with addition of N_2 and Ce. The microstructure of the as received test material is given in Figure 1. The microstructure consists of equiaxed grains of around 40 μm size. Presence of some twin boundary can also be seen. This indicates that the material is in annealed condition. Thermal spraying of the coating was carried out at Terolab Services Austria GmbH. A JP-5000 HVOF system was employed for that purpose. The schematic diagram of the system is presented in Figure 2. The spraying parameters are summarised in Table 2. The low magnification optical micrograph of the coating is presented in Figure 3. A good adhesion between the coating and the substrate can be seen. The coating thickness is around 200 μm. The micrograph shows that the coating is build up by depositing flat plate like lamella, layers after layers. The high magnification back scattered image of the coating is given in Figure 4. This figure shows presence of three zones. The 1st zone is dark area, which is essentially Cr_3C_2. The second zone is white representing the NiCr matrix. The third zone is grey, which contains both the phases, NiCr and Cr_3C_2. The presence of three zones corresponding to three different phases and the mixture of phases are confirmed by EDAX analysis. Wear tests were carried out using a self made high temperature wear test rig as shown in Figure 5. The rig is essentially a disc-on-disc configuration and modified for the present work as 3 pin-on-disc configuration. The 3 pins are the entire test time interacting with the ring, thus the access of oxygen to the contact surface is limited. In contrast, within the ring sliding track oxidation can occur during the time between two pin contacts (which is a function of the sliding track diameter, the rotation frequency and the pin cross section). Wear test conditions are listed in Table 3. Counterbody materials are PM 1000 alloy and 100Cr6 steel. The nominal compositions of these materials are given in Table 4. Although 100Cr6 steel is not used at elevated temperature, it is used to develop various types of glazed layers clearly.

Wearing conditions, wearing material and mating material influence formation and characteristics of glazed layers. Among these parameters wearing material and mating material are most important. Wearing material and mating material combination can be metal against metal, metal against ceramic, ceramic against metal and ceramic against ceramic. These layers determine the friction coefficient and the wearing rate. These layers have characteristics physical appearance, mechanical properties, chemical compositions and failure mechanisms. These layers are:

Table 1. Nominal composition of the substrate material

Elements	C	N	Cr	Ni	Si	Others
Wt %	0.09	0.17	21.0	11.0	1,7	Ce

Figure 1. Optical micrograph of the substrate material 754 MA (Roy et al., 2002)

Figure 2. Schematic diagram of the thermal spraying system (Roy et al., 2002)

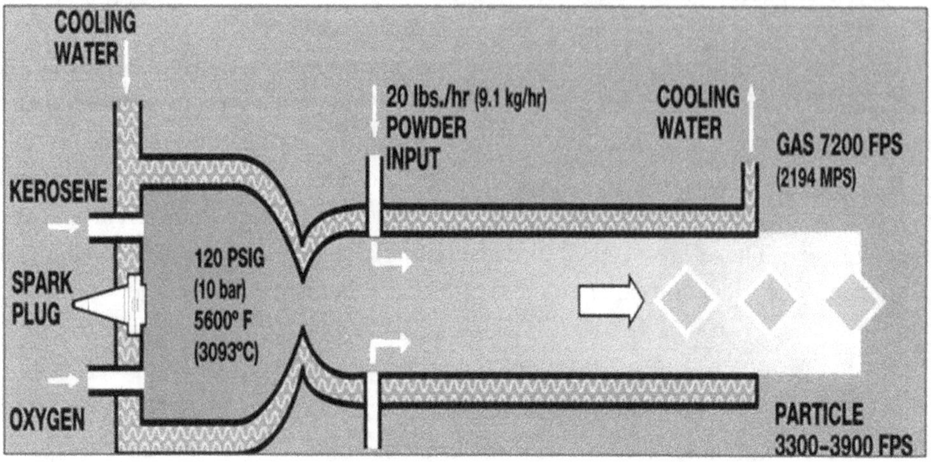

Figure 3. Low magnification optical image of Cr3C2-NiCr coating (Roy et al., 2004)

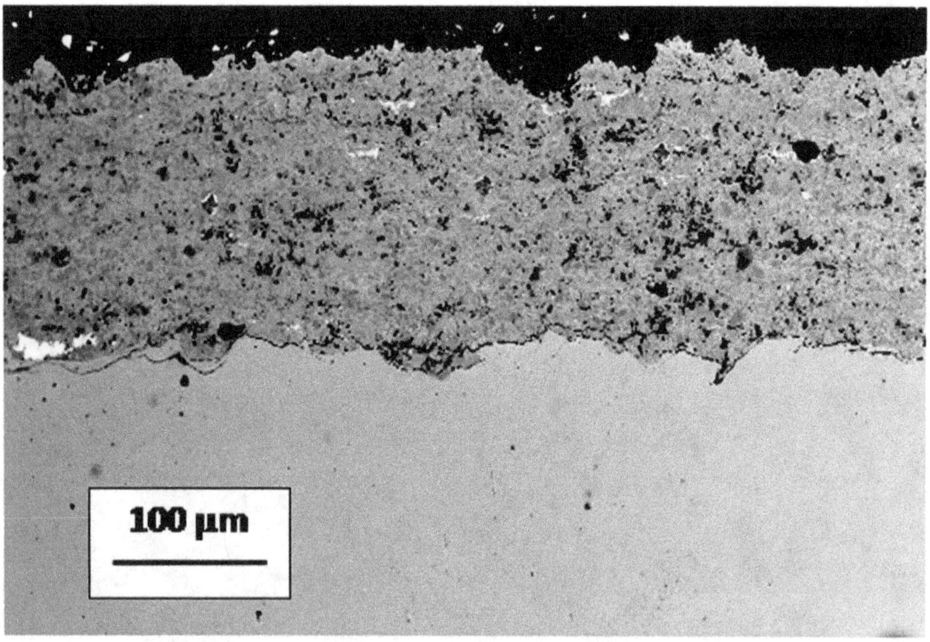

Table 2. HVOF spraying parameters

Para-meters	Liquid Fuel	Gaseous Fuel	Fuel Flow Rate	Fuel Flow Pressure	Powder Carrier Gas	Gas Flow Rate	Gas Pressure	Powder Feed Rate	Spray Distance
			m^3/h	bar		m^3/h	bar	Kg/h	mm
Values/ names	Kerosene	Oxygen	23	22-25	Air	1	7	2	260

Figure 4. High magnification SEM image of the investigated Cr3C2-NiCr coating (Roy et al., 2004)

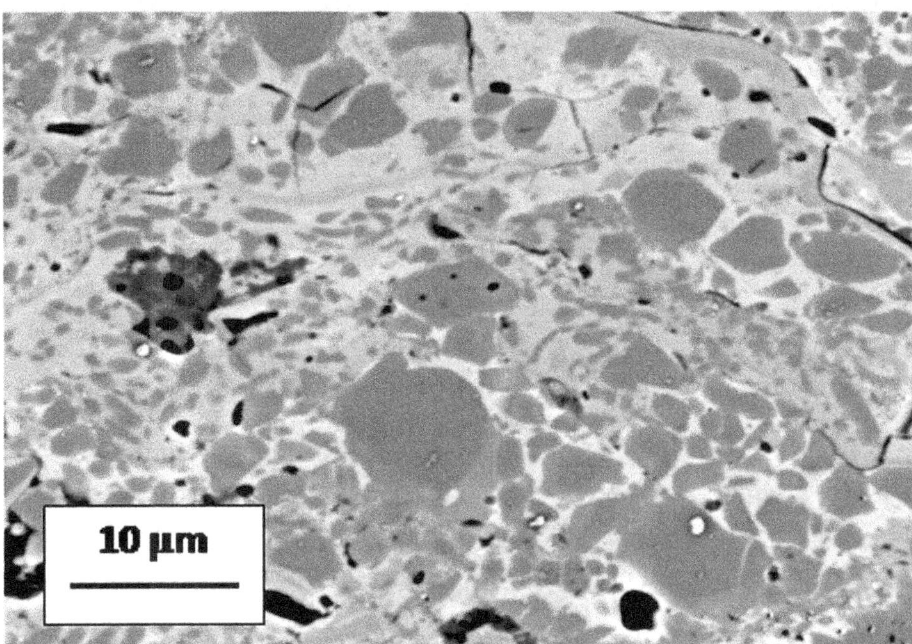

Table 3. Wear Test Conditions

Parameter/Materials	Pin	Mating Material	Load	Sliding Speed	Test Temperature
Values/names	NiCr-Cr_3C_2 coating on 253 MA alloy	100Cr6 Steel, PM 1000 alloy	70 N	0.25 m/s	Ambient Temperature to 1073 K

Figure 5. High temperature ring-ring wear test rig (Roy et al., 2004)

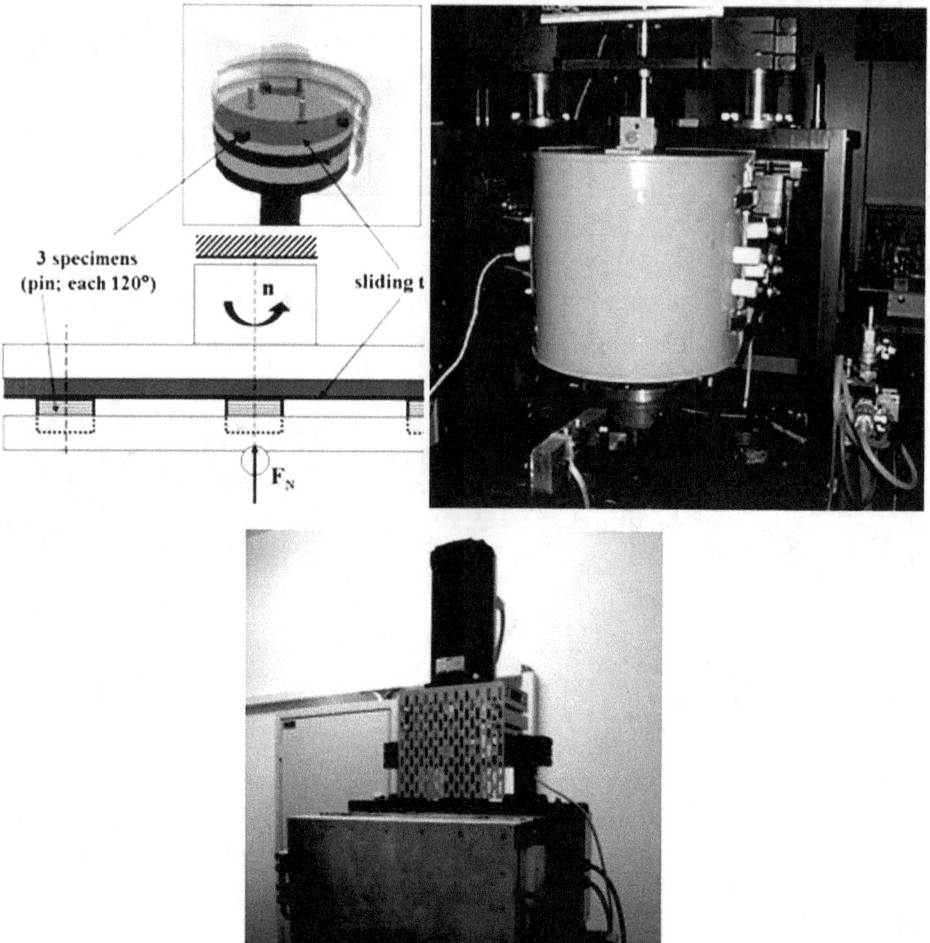

Table 4. Nominal composition of the counterbody material

PM-1000 Alloy	**Elements**	**Cr**	**Fe**	**T_2O_3**	**Ti**	**Al**	**Ni**
	Wt %	20.0	3.0	0.6	0.5	0.3	bal
100Cr6 Steel	**Elements**	**Cr**	**C**	**Mn**	**Si**	**Cu**	**Fe**
	Wt %	1.5	1.3	0.35	0.25	0.25	bal

1. Wear is by pure delamination without any layer formation.
2. Transfer layer (TL) formation,
3. Mechanically mixed layer (MML) formation
4. Reaction layer formation and
5. Composite layer (CL) formation.

At near ambient condition, no layer is formed. Sometimes formation of transfer layer can take place. When the hardness of the mating surface is significantly higher than the hardness of the coating, no layer is formed. Under this condition, the near surface region and the debris have a composition similar to the test specimen. In other words composition of the worn surface and wear debris similar to the specimen surface confirms pure delamination. The worn surface under this condition has high average surface roughness R_a. This situation is characterised by reasonably high friction coefficient. Delamination wear is possible with metal-metal, metal-ceramic system. Morphology of the coating surface when delamination wear takes place is shown in Figure 6a. Delamination cracks can be seen on the surface. SEM image of transverse section of delamination wear obtained for the investigated coating rubbed against 100Cr6 steel is shown in Figure 7a.

In contrast, if the mating surface is softer than coating material, transfer layer (TL) is formed. TL formation takes place at near ambient temperature. As the transfer layer is primarily from the mating surface, the chemical composition of that layer and the composition of the wear debris resemble closely to that of the mating surface. Thus when the compositions of the worn surface and the debris are similar to the mating surface wear mechanism is governed by formation of TL. The surface roughness under such circumstances is low but it increases with increase in temperature. Transfer layer is formed during metal metal and ceramic metal contact. SEM images showing the topography and the transverse section of the transfer layer formed by rubbing the investigated coating against 100Cr6 steel is shown in Figure 6b and Figure 7b respectively. X-ray dot mapping showing the distribution of Cr, Ni and O in the transfer layer is provided in Figure 8a. Although uniform distribution of oxygen can be noted this layer is depleted with Cr and enriched with Ni.

At intermediate temperature, sometimes, reaction layer is formed. The reaction layer is formed when there is intermixing between the rubbing surfaces. The reaction layer (RL) is characterised with a composition in between the composition of the wearing surface and the mating surface. However, the reaction layer has a distinct interface. The layer can fail at the interface. The friction coefficient generally decreases when RL forms. RL forms during metal ceramic contact. Morphology of the reaction layer can be seen in Figure 6c. SEM images showing transverse section of reaction layer as result of rubbing of the investigated coating against 100Cr6 steel

Figure 6. SEM images showing the morphology of the worn surfaces of Cr3C2-NiCr coating at a) 298 K against 100Cr6 steel, b) 473 K against PM 1000 alloy, c) 873 K against PM 1000 alloy and d) 1073 K against 100Cr6 steel (Roy et al., 2004)

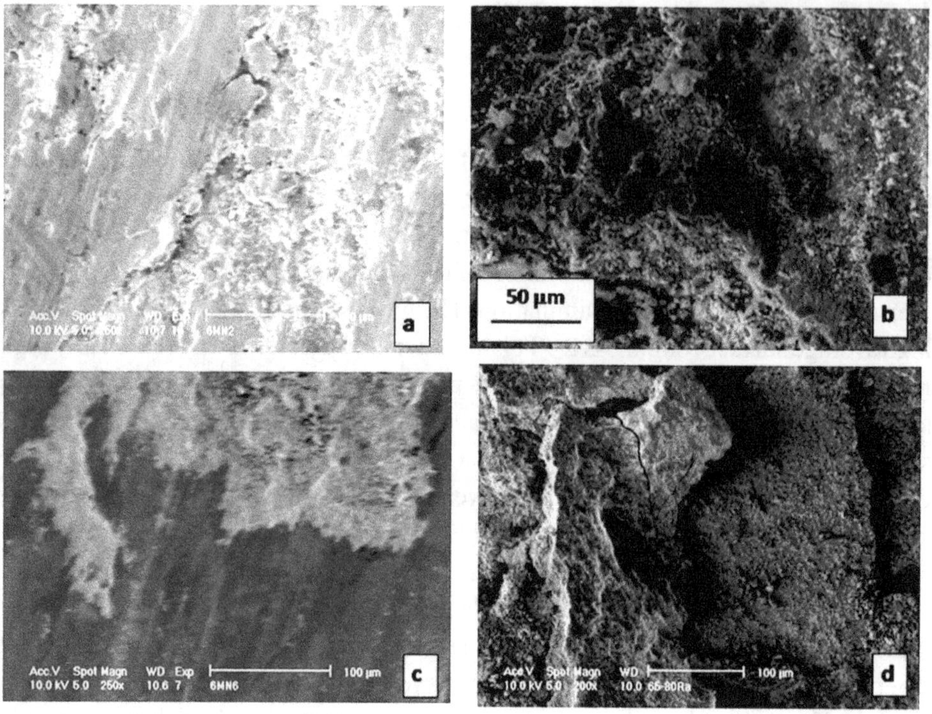

and PM 1000 alloys at 873 K are presented in Figure 7c and Figure 7d respectively. A distinct interface between the RL and the coating is evident. The corresponding dot mapping of Cr, Ni and O is illustrated in Figure 8b. Similar to TL, RL is also characterised by depletion of Cr and enrichment of Ni.

At high temperature, composite layer (CL) is found. The composition of the worn surface with composite layer and the composition of the wear debris will be in between the composition of the wearing sample and the composition of the mating surface. In addition, CL contains high amount of oxygen. This makes the layer hard and brittle. This leads to higher friction coefficient and lower wear rate. The interface between the CL and the substrate is very week giving rise to detachment of the layer at the interface. The strength of this interface improves with increase in temperature. Further, the roughness of this layer is quite high and this roughness decreases with increase of temperature. Thus, composite layer formation is characterized with high roughness and high friction coefficient, particularly at low temperature and decrease of roughness and friction coefficient with increase of tem-

Figure 7. SEM images of the transverse section of various glazed layers (Roy et al., 2008)

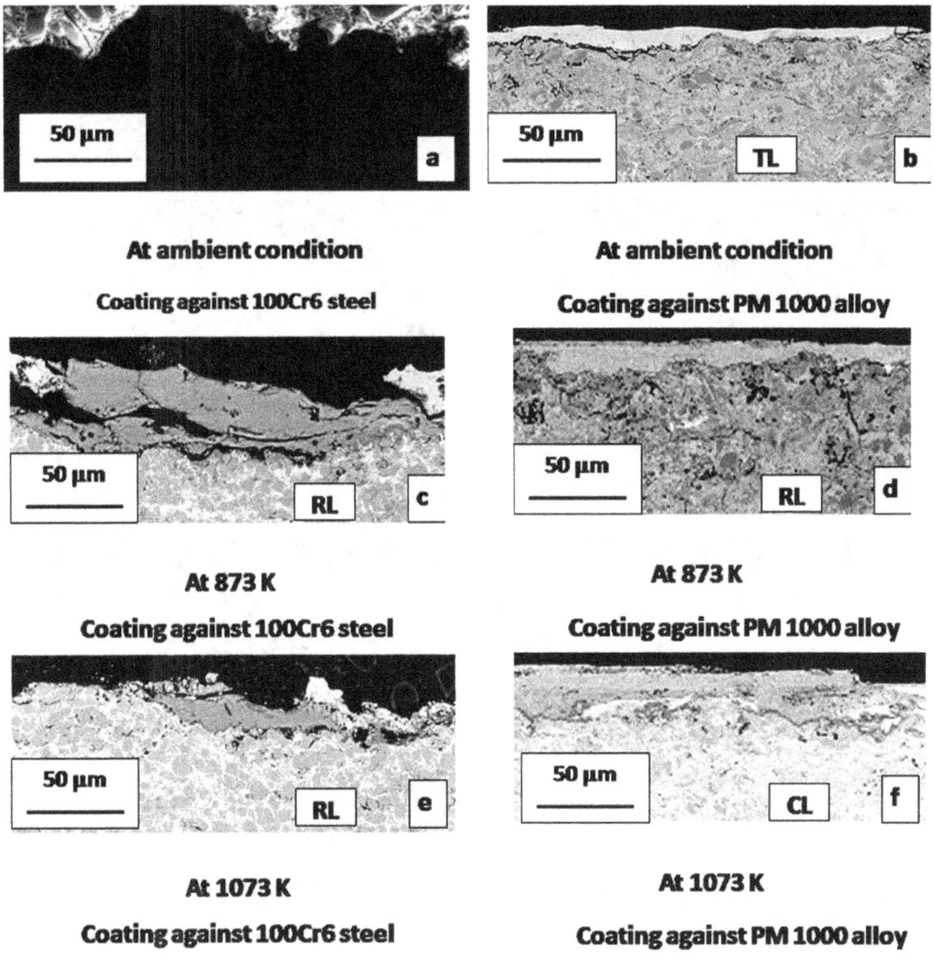

perature. This decrease of roughness of composite layer with increase of temperature is different from what is observed for transfer layer where the roughness increases with increase of temperature. CL can form by metal metal or metal ceramic rubbing. SEM images of the morphology of the composite layer are presented in Figure 6d and the corresponding transverse sections formed during rubbing of Cr_3C_2-NiCr coatings against 100Cr6 steel and PM 1000 alloys at 1073 K are given in Figure 7e and Figure 7f respectively. It is clear from Figure 6d that the composite layer is failing at the interface. The dot mapping of Cr, Ni and O of the image of Figure 7e is shown in Figure 8c.

Figure 8. X-ray dot mapping of the transverse section of various glazed layers (Roy et al., 2008)

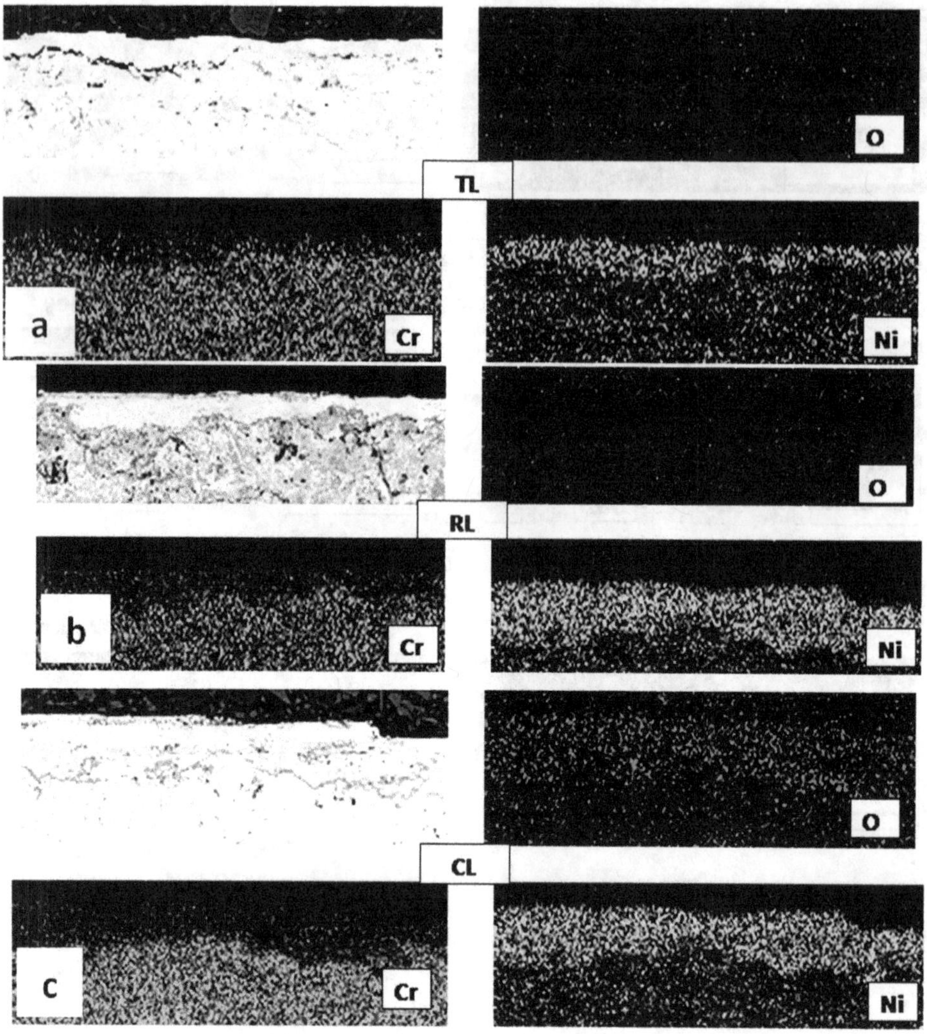

At relatively higher temperature, or under condition of relatively soft but comparable hardness of the tribosystem, mechanically mixed layer (MML) is formed. The chemical state of such layer is characterized by low oxygen content and composition in between the composition of the wearing sample and the composition of the mating material. Similar compositional feature can be seen for wear debris as well. This hard but ductile layer formation is responsible for increase in surface roughness and other corresponding parameters. Interestingly, the friction coefficient exhibits very low value under this condition. The roughness parameter during MML layer formation decreases with increase in temperature. MML layer formation is accompanied by a diffused adhesive interface between the MML and the substrate. MML layer is different from CL and RL in the fact that there is no well defined interface between the layer and the wearing material as case of CL and RL. Unlike composite layer this layer contains practically no oxygen. At this stage it is to be stated that MML can be formed only for metal-metal contact. The SEM image of the transverse section of MML formed by rubbing 754 MA alloy against 100Cr6 steel and PM 1000 alloys at 1073 K are given in Figure 9 along with x-ray dot mapping of Cr, Ni and O. Uniform distribution of various alloying elements and depletion of oxygen is the main characteristics of the layer.

WEAR AT ELEVATED TEMPERATURE FOR THERMAL SPRAYED COATINGS

As mentioned earlier Pauschitz *et al.* (Pauschitz *et al.*, 2003, 2004) carried out a series of systematic experiments on high temperature wear of Cr_3C_2-NiCr coatings on 253 MA alloy against 100Cr6 steel and PM 1000 alloys at different temperatures (test chamber and bulk temperature from 293-1073 K). The result of their work is presented in Figure 10. The variation of friction coefficient of Cr_3C_2-NiCr coatings on 253 MA alloy as function of sliding distance for Cr_3C_2-NiCr coatings rubbed against 100Cr 6 steel at various temperature at a load of 70 N and sliding speed of 0.25 m/s is illustrated in Figure 10. Friction coefficient of thermal sprayed coatings

Figure 9. SEM image and X-ray dot mapping of the transverse section of mechanically mixed layers (Roy et al., 2006)

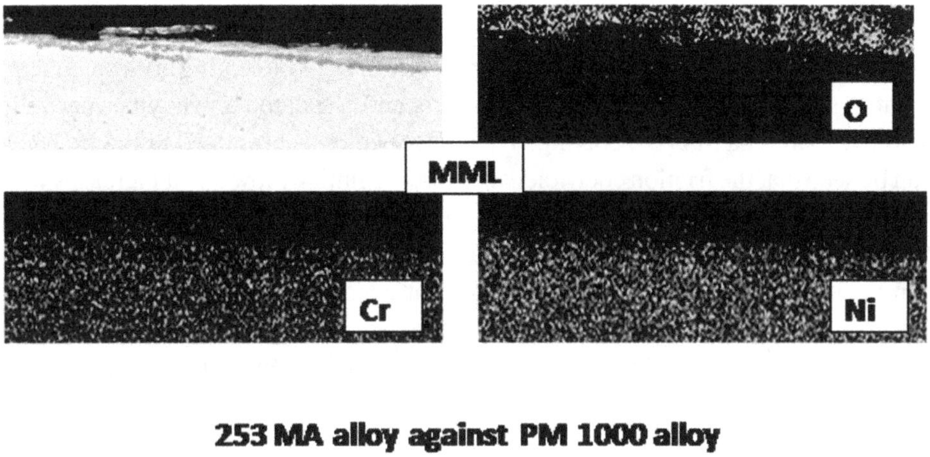

in ambient and also at elevated temperature attains steady state value after an initial run in period. Friction coefficient at ambient condition is the highest. With increase in temperature friction coefficient generally decreases. It can be seen from the Figure 10 that the friction coefficient is high at room temperature and then it decreases. It increases again on increase of test temperature and decreases subsequently on further increase of temperature. At ambient temperature, high friction coefficient is due to delamination wear which takes place without formation of any glazed layer. As the temperature is increased glazed layer in the form of reaction layer is formed and friction coefficient decreases. Finally at 873 K friction increases again due to formation of composite layer and with further increase of temperature friction coefficient decreases as the composite layer gets smoothened. The variation of friction coefficient with sliding distance for Cr_3C_2-NiCr coatings rubbed against PM 1000 alloy at various temperatures, at a load of 70 N and sliding speed of 0.25 m/s is presented in Figure 11. At 873 K friction coefficient is slightly higher than that at other elevated temperature. The friction coefficient is high at ambient temperature due to delamination wear. At intermediate temperature, reaction layer is formed leading to decrease of friction coefficient. As the temperature is increased further, composite layer is formed and the friction coefficient increases.

High temperature wear of thermal sprayed coatings are reported extensively. Cermet variety of coatings such as WC-Co, Cr_3C_2-NiCr are well known for their excellent wear behaviour. Among this cermet variety of coatings, WC-Co and Cr_3C_2-NiCr coatings find application in large variety of industrial usages (Lai, 1978; Li, 1980; Taylor *et al.* 1983). However, WC-Co coatings decarburise to W_2C, or even to elemental W on exposure to high temperature (Russo & M. Dorfmann, 1995). Decarburisation degrades the mechanical properties of the coatings. Such degradation will result in poor performances under deformation-controlled wear such as mild wear. In contrast, Cr_3C_2-NiCr coating exhibits improved performance even up to a temperature of 1173 K (Mateos *et al.*, 2001).

Ouyang *et al.* (Ouyang *et al.*, 2002) studied the friction and wear characteristics of low-pressure plasma-sprayed (LPPS) ZrO_2-$BaCrO_4$ (ZB) coating at elevated temperatures using a high-temperature wear tester and compared it with yttria partially stabilised zirconia (YPAZ) coating. Their observation is presented in Figure 12. It can be seen that the friction coefficient of YPAZ is initially low and it increases with increase of test temperature. In contrast, the friction coefficient of ZB is initially high but it decreases with temperature and stabilise at that value. This behaviour can be attributed to the fact that near ambient condition a transfer layer forms on YPSZ and at elevated temperature composite layer is formed. This results in low friction coefficient at ambient temperature and high friction coefficient at elevated

Figure 10. Variation of friction coefficient with sliding distance for HVOF sprayed Cr3C2-NiCr coatings against 100Cr6 steel as counter body at different test temperatures (Roy et al., 2004, Roy et al., 2008)

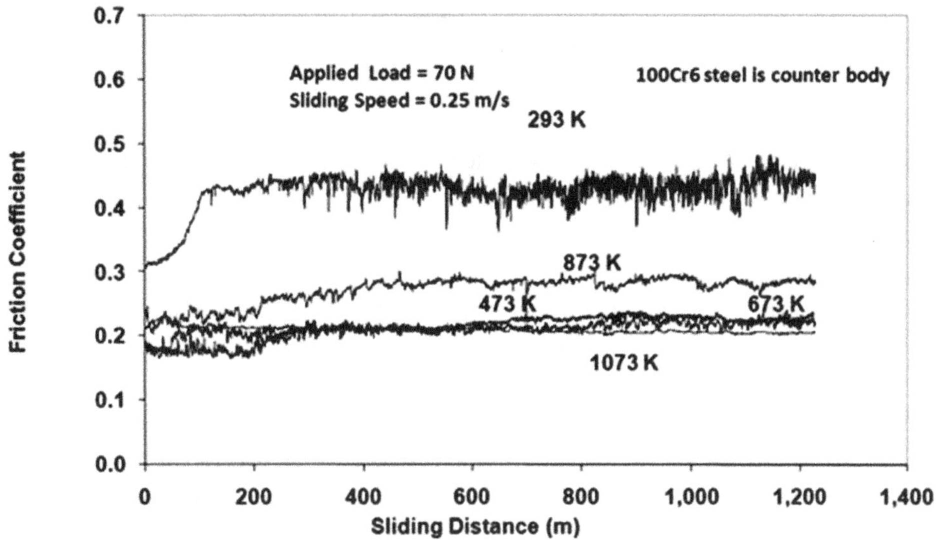

Figure 11. Variation of friction coefficient with sliding distance for HVOF sprayed Cr3C2-NiCr coatings against PM 1000 alloy as counter body at different test temperatures (Pauschitz et al. 2008, Roy et al., 2008)

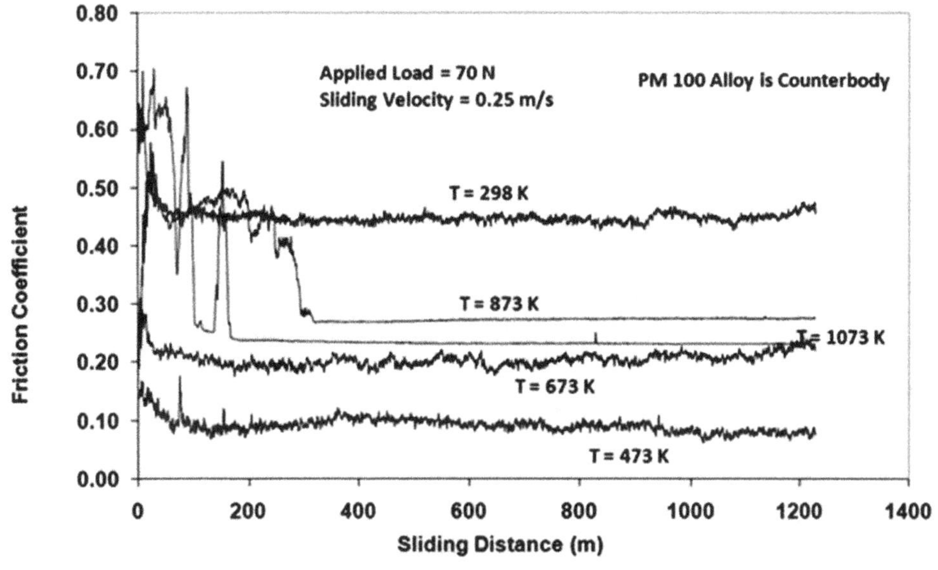

temperature. In case of ZB coating the friction is governed by delamination at ambient condition and subsequently at elevated temperature stable reaction layer is formed leading to low and stable friction coefficient.

As discussed earlier, formation of mechanically mixed layer results in significant decrease in friction coefficient. Mechanically mixed layer is generally formed when both the rubbing materials are metallic. The work due to Hager *et al.* (Hager *et al.*, 2009) supports such hypothesis. They investigated the effect of temperature on inter-metallic wear between Ti6Al4V and cold-sprayed, commercially pure nickel coatings. Their results are illustrated in Figure 13. It can be seen that near ambient temperature friction coefficient is considerably high due to delamination wear of thermal sprayed metallic coating pure Ni. However, as the temperature increases, pure Ni forms mechanically mixed layer with Ti-6Al-4V. This leads to reduction of friction coefficient which is significantly stable.

Alumina-titania coatings are used where a high wear and oxidation resistances are required. Rico *et al.* (Rico *et al.*, 2013) compared the high temperature wear behaviour of Al_2O_3-13%TiO_2 nanostructured coatings with that of the conventional ones. Wear rates and friction coefficients were measured by means of a pin on disc

Figure 12. Variation of friction coefficient with temperature for low-pressure plasma-sprayed (LPPS) ZrO2-BaCrO4 (ZB) coating and yttria partially stabilised zirconia (YPAZ) coating (Ouyang et al., 2002)

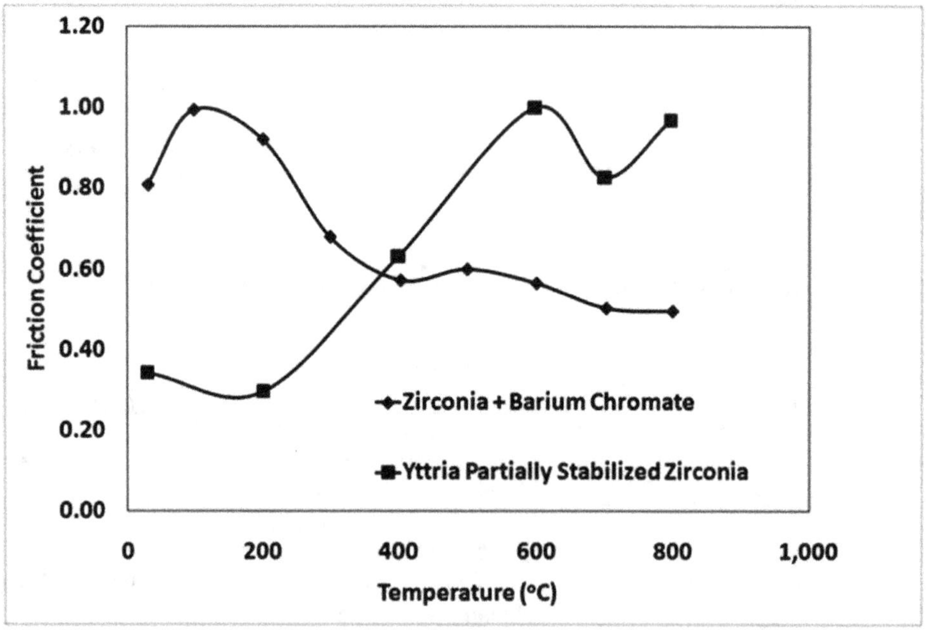

Figure 13. Variation of friction coefficient as function of temperature for cold sprayed commercially pure Ni coating (Hager et al., 2009)

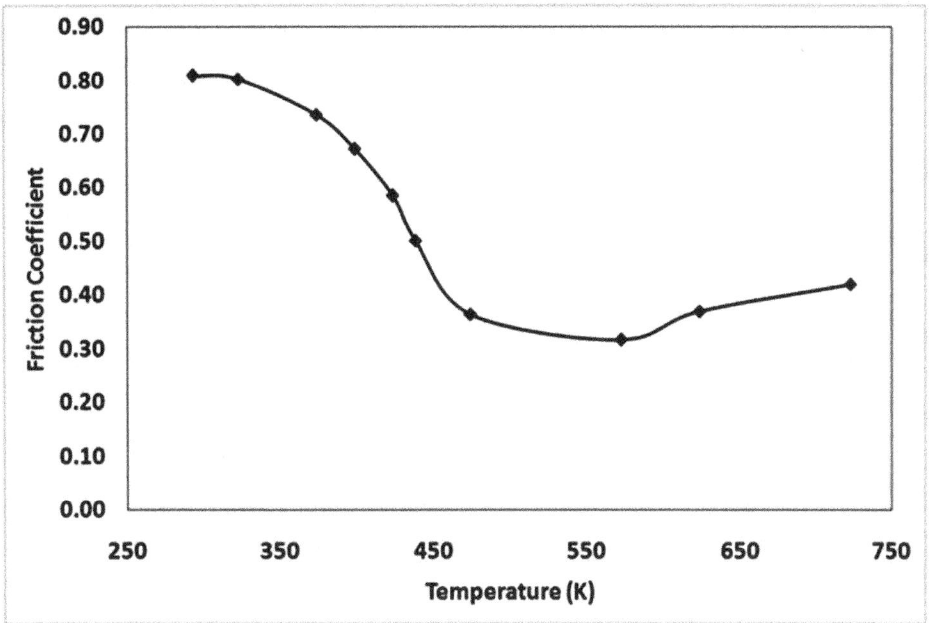

machine in which both counterparts were made of Al_2O_3-13%TiO_2. Although the mechanical properties are slightly higher in the nanostructured coatings, they exhibit a much better wear behaviour under all the testing conditions. Their result is illustrated in Figure 14. It can be noted that at ambient condition at low applied load friction coefficient is less but with increase in load friction coefficient suddenly increases for conventional coating. In contrast, such increase in friction coefficient for nanostructure coating is gradual. This can be attributed to the fact that at ambient condition and at low applied load transfer layer is formed. As the load increases there is increase in temperature at locale contact point and composite layer is formed and friction coefficient increases. In case of nanostructured coating, such transformation takes place gradually and hence a gradual increase of friction coefficient is noticed. However, at elevated temperature and at low load reaction layer is formed and this layer transforms to composite layer at higher load characterised by sudden increase of friction coefficient. This transformation from reaction layer to composite layer takes place at around 30 to 40 MPa for coating with conventional grain and at 40 to 50 MPa for nanostructure coating.

Economou *et al.* (Economou *et al.* 2000) investigated the tribological behaviour of plasma sprayed cermet coatings. Low pressure plasma sprayed NiCrFeAlTi-TiC,

Figure 14. Variation of friction coefficient with applied load at ambient temperature and at 773 K for thermal sprayed conventional and nanostructure Al2O3-13%TiO2 coating (Rico et al., 2013)

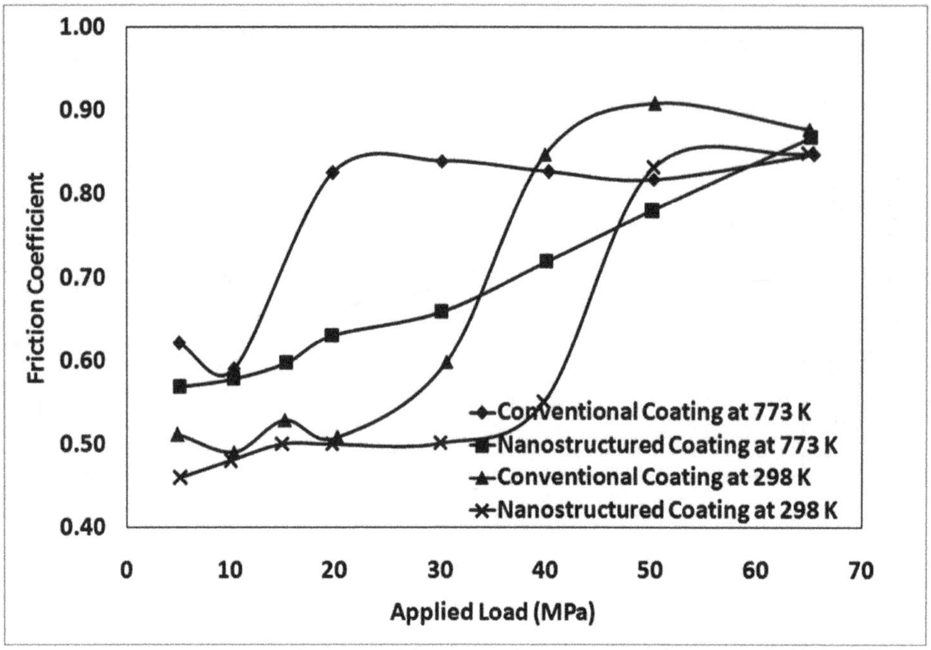

NiCr-(Ti, Ta)C and NiCrMo-(Ti, Ta)C coatings were investigated at room temperature and at 550°C, against sapphire ball. Air plasma sprayed NiCr and NiCr-TiC coatings, as well as the high velocity oxy-fuel sprayed WC-Co coating, were investigated for comparison. The wear behaviour of the coatings is shown in Figure 15. The matrix and the carbide modifications of the NiCr-TiC coating influenced positively coating wear and friction behaviour. The wear resistance of all the coatings decreased at high temperature. The NiCrFeAlTi-TiC coating showed the best wear resistance among the TiC-based coatings. At room temperature and at 550°C, the WC-Co coating was more wear resistant than the TiC-based coatings, even though the coating cracked in the high temperature tests. The coefficients of friction of the WC-Co coating, tested against sapphire, were lower than the coefficients of friction of the TiC-based coatings at room temperature, but not at 550°C. The observed wear rate can be rationalised with respect to various layer formation as follows. Since counter body is sapphire ball only NiCr coating conforms to metal ceramic system. Consequently for NiCr coating, transfer layer governs wear at ambient condition and reaction layer governs wear at elevated temperature. This results in comparable wear rate for the coating at both temperatures. Although reaction layer formation results in lower wear rate for ceramic materials, it is not so for metallic material when the counterbody is ceramic. For all other coatings it is composite layer formation during sliding against ceramic. Thus, reaction layer is not formed. This resulted in no layer formation at ambient condition and composite layer formation at elevated temperature. Hence all coatings exhibit higher wear rate at elevated temperature than that at room temperature.

SUMMARY AND DIRECTION OF FUTURE RESEARCH

During sliding wear of thermal sprayed coatings at elevated temperature glazed layers are formed. These glazed layers can be divided into different types of layers, namely, transfer layer (TL), mechanically mixed layer (MML), reaction layer (RL) and composite layer (CL). The friction coefficient is always low with MML. Formation of MML requires extensive strain hardening of the near surface region of the wearing materials. It also needs good intermixing of the wear debris and worn surfaces. Formation of MML is favoured by wearing surfaces having higher oxidation resistance. MML is formed during metal metal contact. In the event these

Figure 15. Bar diagram showing the wear rate of various thermal sprayed coatings at ambient temperature and at 550oC (Economou et al. 2000)

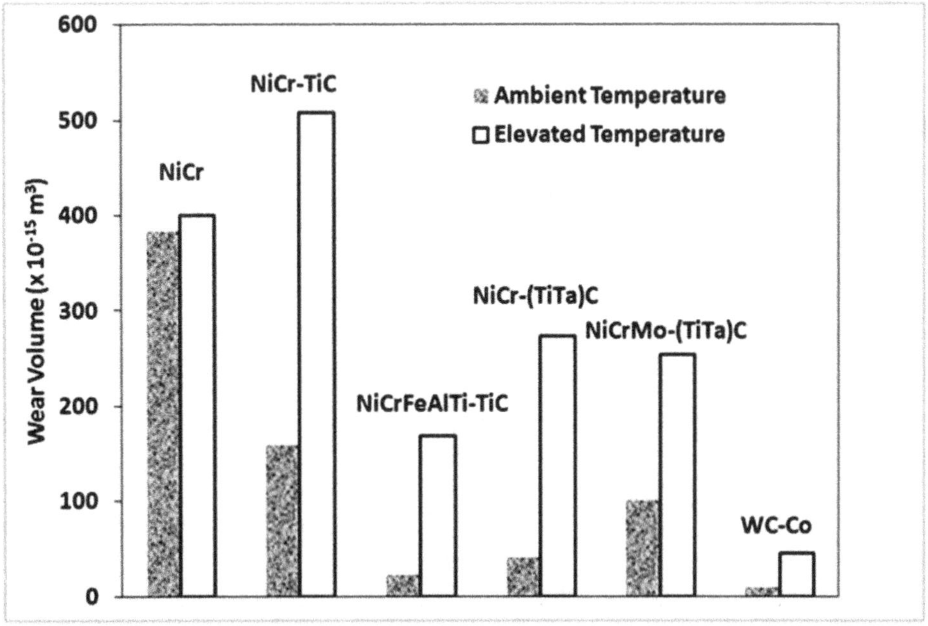

conditions are not met, reaction layer is formed. Friction coefficient is high when CL is in its initial stage of formation. At this stage, CL has tendency to get detached from the wearing surfaces leading to higher roughness, higher friction coefficient and higher wear rate. With increased temperature composite layer is consolidated as a thin adherent layer giving rise to decrease in friction coefficient. The absolute value of friction coefficient is dependent on the mechanical properties of layers formed and the tribosystem involved. The recent results on friction of thermal sprayed coatings at elevated temperature has been rationalised in the light of different types of glazed layer formation.

In spite of advancement of the understanding of glazed layers, more work needs to be done. The mechanical properties of these glazed layers are to be evaluated using nanoindentation and nanoscratch tester. Microstructural features of these layers are also important. Thus, it is imperative to examine the microstructural features of these layers employing transmission electron microscopy and electron back scattered diffraction techniques. Roughness is always an influential parameter concerning wear and friction. Thus a comprehensive understanding of the roughness parameters of these layers as function of various wearing condition and friction response is needed. Electrochemical responses of these layers are crucial and should be studied extensively in order to understand characteristics features and tribocorrosion responses of these layers. As new and new thermal spraying techniques are developed and new and new coating materials are developed, it has become mandatory to develop data base with the coating developed by new technique and new materials. Last but not least is the generation and documentation of more extensive data base of these layers involving wider materials combination and wearing conditions.

REFERENCES

Ahn, H., Kim, J., & Lim, D. (1997). Tribological behaviour of plasma-sprayed zirconia coatings. *Wear*, *203–204*, 77–87. doi:10.1016/S0043-1648(96)07395-4

Archard, J. F., & Hirst, W. (1956). The wear of metals under unlubricated conditions. *Proceedings of the Royal Society of London. Series A, Mathematical and Physical Sciences*, *236*(1206), 397–410. doi:10.1098/rspa.1956.0144

Burke, J. E. (1962). The science and technology of sintering. In C. Klingsberg (Ed.), *Physics and chemistry of ceramics* (pp. 167–178). New York: Gordon Beach.

Della-Corte, C. (1996). The effect of counterface on the tribological performance of a high temperature solid lubricant composite from 25 to 650°C. *Surface and Coatings Technology*, *86–87*, 486–492. doi:10.1016/S0257-8972(96)02959-3

Economou, S., De Bonte, M., Celis, J. P., Smith, R. W., & Lugscheider, E. (2000). Tribological behaviour at room temperature and at 550°C of TiC-based plasma sprayed coatings in fretting gross slip conditions. *Wear, 244*(1-2), 165–179. doi:10.1016/S0043-1648(00)00455-5

Fink, M. (1930). Wear oxidation, a new component of wear. *Transaction ofAmerican Society of Steel, 18*, 1026–1034.

Fontalvo, G. A., & Mitterrer, C. (2005). The effect of oxide-forming alloying elements on the high temperature wear of a hot work steel. *Wear, 258*(10), 1491–1499. doi:10.1016/j.wear.2004.04.014

Guilemany, J. M., Miguel, J. M., Vizcaino, S., Lorenzana, C., Delgado, J., & Sanchez, J. (2002). Role of heat treatments in the improvement of the sliding wear properties of Cr_3C_2–NiCr coatings. *Surface and Coatings Technology, 157*(2-3), 207–213. doi:10.1016/S0257-8972(02)00148-2

Hager, C. H. Jr, Sanders, J., Sharma, S., Voevodin, A., & Segall, A. (2009). The effect of temperature on gross slip fretting wear of cold-sprayed nickel coatings on Ti6Al4V interfaces. *Tribology International, 42*(3), 491–502. doi:10.1016/j.triboint.2008.08.009

Inman, I. A., Rose, S. R., & Dutta, P. K. (2006). Studies of high temperature sliding wear of metallic dissimilar interfaces II: Incoloy MA956 versus Stellite. *Tribology International, 39*(11), 1361–1375. doi:10.1016/j.triboint.2005.12.001

Inman, I. A., Rose, S. R., & Dutta, P. K. (2006). Development of a simple 'temperature versus sliding speed' wear map for the sliding wear behaviour of dissimilar metallic interfaces. *Wear, 260*(9-10), 919–932. doi:10.1016/j.wear.2005.06.008

Jiang, J., Stott, F. H., & Stack, M. M. (1994). Some frictional features associated with the sliding wear of the nickel-base alloy N80A at temperatures to 250 °C. *Wear, 176*(2), 185–194. doi:10.1016/0043-1648(94)90146-5

Jiang, J., Stott, F. H., & Stack, M. M. (1998). The role of triboparticulates in dry sliding wear. *Tribology International, 31*(5), 245–256. doi:10.1016/S0301-679X(98)00027-9

JiangX.LiuW.DongS.YXuB.S., (2005);

Jiang, X., Liu, W., Dong, S. Y., & Xu, B. S. (2005, April). High temperature tribology behaviors of brush plated Ni–W–Co/SiC composite coating. *Surface and Coatings Technology, 194*(1), 10–15. doi:10.1016/j.surfcoat.2004.04.095

Kim, J. Y., & Lim, D. S. (1994). The effect of annealing temperature and environment on room temperature wear behaviour of plasma sprayed partially stabilised zirconia coating. *Journal of Korean Ceramic Society*, *31*(10), 1176–1180.

Lai, G. Y. (1978). Evaluation of sprayed chromium carbide coatings for gas cooled reactor application. *Thin Solid Films*, *53*(3), 343–351. doi:10.1016/0040-6090(78)90227-4

Lai, G. Y. (1979). Factors affecting the performances of sprayed chromium carbide coatings for gas cooled reactor heat exchangers. *Thin Solid Films*, *64*(2), 271–280. doi:10.1016/0040-6090(79)90520-0

Li, C. C. (1980). Characterisation of thermally sprayed coatings for high temperature wear protection. *Thin Solid Films*, *73*(1), 59–77. doi:10.1016/0040-6090(80)90329-6

Li, C. C., & Lai, G. Y. (1979), Sliding wear studies of sprayed chromium carbide-nichrome coatings for gas-cooled reactor applications. In *Proc. Int. Conf. on Wear of Materials*. Dearborn, MI: ASME Newyork.

Li, J. F., & Watanabe, R. (1996). $BaFe_{12-2x}Co_xTi_xO_{19}$ (x=0, 0.6, 0.9) films synthesized by the spray-ICP technique. *Materials Transactions*, *37*, 1171–1181. doi:10.2320/matertrans1989.37.1171

Lim, S. C., & Ashby, M. F. (1987). Wear-mechanism maps. *Acta Materialia*, *30*(1), 1–24. doi:10.1016/0001-6160(87)90209-4

Liu, G. H., Robbevalloire, F., Gras, R., & Blouet, J. (1993). Improvement in tribological properties of chromium oxide coating at high temperature by solid lubricants. *Wear*, *160*(1), 181–189. doi:10.1016/0043-1648(93)90419-M

Martın, A., Rodr'ıguez, J., Fernández, J. E., & Vijande, R. (2001). Sliding wear behaviour of plasma sprayed WC-NiCrBSi coatings at different temperatures. *Wear*, *251*(1-12), 1017–1022. doi:10.1016/S0043-1648(01)00703-7

Mateos, J., Cuetos, J. M., Vijande, R., & Fernandej, E. (2001). Tribological properties of plasma sprayed and laser remelted 75/25 Cr_3C_2/NiCr coatings. *Tribology International*, *34*(5), 345–351. doi:10.1016/S0301-679X(01)00023-8

Mohanty, M., Smith, R. W., De Bonte, M., Celis, J. P., & Lugscheider, E. (1996). Sliding wear behavior of thermally sprayed 75/25 Cr_3C_2/NiCr wear resistant coatings. *Wear*, *198*(1-2), 251–266. doi:10.1016/0043-1648(96)06983-9

Ouyang, J. H., Sasaki, S., & Umeda, K. (2001). Low-pressure plasma-sprayed ZrO_2–CaF_2 composite coating for high temperature tribological applications. *Surface and Coatings Technology*, *137*(1), 21–30. doi:10.1016/S0257-8972(00)00918-X

Ouyang, J. H., Sasaki, S., & Umeda, K. (2002). The friction and wear characteristics of low-pressure plasma-sprayed ZrO_2-$BaCrO_4$ composite coating at elevated temperatures. *Surface and Coatings Technology*, *154*(2-3), 131–139. doi:10.1016/S0257-8972(02)00024-5

Pauschitz, A., Roy, M., & Franek, F. (2003). On the chemical composition of the layers formed during sliding of metallic alloys at high temperature. *Tribologia*, *188*, 127–143.

Pauschitz, A., Roy, M., & Franek, F. (2003). Comparism of high temperature wear behaviour of chromia forming Fe-based and Ni-based alloy. *Tribol Schmierungstech*, *50*, 40–49.

Pauschitz, A., Roy, M., & Franek, F. (2004). Identification of the mechanisms of wear during sliding of metallic materials at elevated temperature using an optical interferometer. In D. Dowson (Ed.), *Transient phenomena in tribological processes* (pp. 721–730). Elsevier.

Pauschitz, A., Roy, M., & Franek, F. (2008). Mechanisms of sliding wear of metals and alloys at elevated temperatures. *Tribology International*, *41*(7), 584–602. doi:10.1016/j.triboint.2007.10.003

Quinn, T. F. J., Sullivan, J. L., & Rowson, D. M. (1984). Origins and development of oxidational wear at low ambient temperatures. *Wear*, *94*(2), 175–191. doi:10.1016/0043-1648(84)90053-X

Radu, I., & Li, D. Y. (2007). The wear performance of yttrium-modified Stellite 712 at elevated temperatures. *Tribology International*, *40*(2), 259–265. doi:10.1016/j.triboint.2005.09.027

Richard, C. S., Lu, J., Beranger, G., & Decomps, F. (1995). Study of Cr_2O_3 coatings: Part I: Microstructure and modulus. *Journal of Thermal Spraying Technology (Elmsford, N.Y.)*, *4*(4), 342–346.

Rico, A., Poza, P., & Rodríguez, J. (2013). High temperature tribological behavior of nanostructured and conventional plasma sprayed alumina-titania coatings. *Vacuum*, *88*, 149–154. doi:10.1016/j.vacuum.2012.01.008

Roy, M., Pauschitz, A., & Franek, F. (2002), Elevated temperature wear of 253 MA alloy. In *Proceedings of the Third Aimeta International Tribology Conference*. Vietri sul Mare, Italy: Academic Press.

Roy, M., Pauschitz, A., & Franek, F. (2004). The influence of temperature on the wear of Cr_3C_2-25(Ni20Cr) coating – comparism between nanocrystalline grains and conventional grains. *Wear, 257*(7-8), 799–811. doi:10.1016/j.wear.2004.05.001

Roy, M., Pauschitz, A., Polak, R., & Franek, F. (2008), High temperature wear of thermal sprayed Cr3C2-25(Ni-20Cr) coating against Fe based and Ni based mating surfaces. In *Proc. 16th Intl. Colloquium Tribology*. Stuttgart, Germany: Academic Press.

Roy, M., Pauschitz, A., Wernisch, J., & Franek, F. (2004). Influence of mating surface on elevated temperature wear of 253 MA alloy. *Materials & Corrosion, 55*(4), 259–273. doi:10.1002/maco.200303715

Russo, L., & Dorfmann, M. (1995). Structure evaluation of HVOF sprayed NiCr-Cr_3C_2 coatings. In A. Ohmori (Ed.), *Thermal spraying: Current status and future trends* (pp. 681–686). Osaka, Japan: High Temperature Society of Japan.

Shin, L. D. S., Lim, D.-S., & Ahn, H.-S. (2000). Effect of annealing and Fe O addition on the high temperature tribological behavior of the plasma sprayed yttria-stabilized zirconia coating. *Surface and Coatings Technology, 133-134*, 403–410. doi:10.1016/S0257-8972(00)00965-8

Sliney, H. E. (1982). Solid lubricant materials for high temperatures—A review. *Tribology International, 15*(5), 303–315. doi:10.1016/0301-679X(82)90089-5

Stott, F. H. (2002). High-temperature sliding wear of metals. *Tribology International, 35*(8), 489–495. doi:10.1016/S0301-679X(02)00041-5

Taktak S., (2006), Tribological behaviour of borided bearing steels at elevated temperatures. *Surface & Coatings Technology, 201*, 2230-2239.

Taylor, T. A., Overs, M. P., Quets, J. M., & Tucker, R. C. Jr. (1983). Development of several new nickel alluminide and chromium carbide coatings for use in high temperature nuclear reactors. *Thin Solid Films, 107*(4), 427–435. doi:10.1016/0040-6090(83)90304-8

Tu, J. P., Jie, X. H., Mao, Z. Y., & Matsumara, M. (1998). The effect of temperature on the unlubricated sliding wear of 5 CrNiMo steel against 40 MnB steel in the range 400–600°C. *Tribology International, 31*(7), 347–353. doi:10.1016/S0301-679X(98)00039-5

Vos, F., Delaey, L., De Bonte, M., & Froyen, L. (1998), Plasma-sprayed self-lubricatinng Cr2O3–CaF2 coatings: Friction and wear properties. In *Proceedings of the 15th International Thermal Spray Conference*. Nice, France: Academic Press.

Wang, L., & Li, D. Y. (2003). Effects of yttrium on microstructure, mechanical properties and high-temperature wear behavior of cast Stellite 6 alloy. *Wear*, *255*(1-6), 535–544. doi:10.1016/S0043-1648(03)00057-7

Wang, Y. (1993). Friction and wear performances of detonation-gun and plasma-sprayed ceramic and cermet hard coatings under dry friction. *Wear*, *161*(1-2), 69–78. doi:10.1016/0043-1648(93)90454-T

Wang, Y., Hsu, S. M., & Jones, P. (1998). Evaluation of thermally-sprayed ceramic coatings using a novel ball-on-inclined plane scratch method. *Wear*, *218*(1), 96–102. doi:10.1016/S0043-1648(98)00182-3

Wang, Y., Jin, Y., & Wen, S. (1988). The analysis of the friction and wear mechanisms of plasma-sprayed ceramic coatings at 450 °C. *Wear*, *128*(3), 265–276. doi:10.1016/0043-1648(88)90063-4

Watreme, M., Bricout, J. P., Marguet, B., & Qudin, J. (1996). Friction, temperature, and wear analysis for ceramic coated brake disks. *Journal of Tribology*, *118*(3), 457–465. doi:10.1115/1.2831558

Yang, Q., Senda, T., & Hirose, A. (2004). Sliding wear behavior and tribofilm formation of ceramics at high temperatures. *Surface and Coatings Technology*, *184*(2-3), 270–277. doi:10.1016/j.surfcoat.2003.10.157

Yang, Q., Senda, T., & Hirose, A. (2006). Sliding wear behavior of WC–12% Co coatings at elevated temperatures. *Surface and Coatings Technology*, *200*(14-15), 4208–4212. doi:10.1016/j.surfcoat.2004.12.032

Zhou, Y. H., Hermelin, M., & Bigot, J. (1989). Sintering behaviour of ultra-fine Fe, Ni and Fe-25wt%Ni powders. *Scripta Metalia*, *23*(8), 1391–1396. doi:10.1016/0036-9748(89)90065-3

KEY TERMS AND DEFINITIONS

Composite Layer: Layer formed due to mixing and oxidation of debri from two contacting surfaces in relative motion.

Mechanically Mixed Layer: Layer formed due to mechanical mixing of debri from two contacting surfaces in relative motion.

Oxidation: Degradation due to reaction with the environment at elevated temperature.

Reaction Layer: Layer formed due to mixing and reaction of debri from two contacting surfaces in relative motion.

Sliding Wear: Degradation due to relative motion between two contacting surfaces.

Thermal Spraying Coatings: Coating obtained by a group of processes that apply a consumable powders or wires in the form of finely divided molten and semi molten droplets.

Transfer Layer: Layer formed due to transfer of debri from softer surface to harder surface during relative motion of two contacting surfaces.

Tribology: Science of friction.

Wear Oxidation: Synergy between wear and oxidation.

Chapter 7
Solid Particle Erosion of Thermal Sprayed Coatings

S. G. Sapate
Visvesvaraya National Institute of Technology, India

Manish Roy
Defence Metallurgical Research Laboratory, India

ABSTRACT

Solid particle erosion is an important material degradation mechanism. Although various methods of coating are tried and used for protection against erosion, thermal sprayed coating for such purpose is the most widely used method. In this chapter, evolution of thermal sprayed coating, erosion testing methods, and erosive wear of thermal sprayed coatings are discussed extensively with emphasis on recent developments. It is generally found that erosion of thermal sprayed coatings depends on erosion test conditions, microstructural features, and mechanical properties of the coating materials. Most thermal sprayed coatings respond in brittle manner having maximum erosion rate at oblique impact and velocity exponent in excess of 3.0. Erosion rate is also dependent on thermal spraying techniques and post coating treatment. However, little work is done on dependence of erosion rate on coating techniques and coating conditions. Future direction of work is also reported.

DOI: 10.4018/978-1-4666-7489-9.ch007

1 INTRODUCTION

Solid particle erosion, more popularly known as erosion or erosive wear is essentially degradation of materials due to impact of moving particles. Erosion is mechanistically different from liquid erosion, cavitation erosion etc. Erosion is a common occurrence and particularly a matter of considerable technical and economical importance. Erosion of choke valves used in oil and gas industry, erosion of steam pipes by entrained iron oxide particles in steam turbines, erosion in fluidized bed combustion systems are some typical examples of erosive wear. Although erosion is generally regarded as a material degradation process, in some cases erosion is a useful phenomenon, as in sand blasting and high-speed abrasive water jet cutting. (Reddy & Sundarrajan,1986; Kosel,1992). The inconsistency in prediction of erosion resistance of engineering materials based on improvement of mechanical properties by various strengthening mechanisms (Roy, Subramaniyan & Sundarrajan, 1992; Roy, Tirupataiah & Sundarrajan, 1993; Roy, Tirupataiah & Sundarrajan, 1995), alteration in microstructure by thermal / mechanical treatment has been the major impetus to the development of erosion resistant surface coatings. Design modification and change of service parameters are often unsuitable solutions to reduce erosion severity. Therefore in many cases especially in industries where service parameters and design are fixed, the most practical route is based on use of alternative material or surface coatings. Since the properties required for wear control are often specific to the surface, use of surface coatings for erosion mitigation had been widely practiced in the past. Several coating techniques ranging from surface hardening, weld hardfacing, chemical vapour deposition, and physical vapour deposition to ion implantation are currently in practice for erosive wear protection. Table 1 gives some of the manufacturing techniques to produce wear resistant surface (Schmidt & Steinhauser, 1996). Weld hardfacing, although considered sometimes as a 'black art', offers several economic advantages. The use of hard coatings offers several advantages over bulk material since surface properties can be engineered to be independent of the bulk material to produce a composite system and use of cheaper substrate material can significantly reduce the cost of the product. The selection of a particular coating technique is dictated by several factors such as coating thickness, mechanical properties, morphology and microstructure of the coating as influenced by the manufacturing techniques and its cost, used for coating deposition, coating defects and most importantly the operating conditions such as erosion mode, properties of erodent particles, fluid flow variables such as impact angle and the impact velocity. In this context thermal spray coatings have been widely used for protection of industrial components against solid particle erosion (SPE). Thermal spray coatings offer attractive mechanical properties and are being used for wear, abrasion, erosion and corrosion protection of engineering components in automotive, power plants and mining industries. The erosion resistance of thermal spray coatings relies on use of hard ceramic second phase particles such as carbides and borides in a relatively softer matrix; a binder phase. Figure 1 shows use of different reinforcement phases in thermal spray coatings metals and metalloids. The dispersion of a ceramic phase in a more ductile matrix to improve hardness and toughness has been successfully used for improving and abrasion and erosion resistance of dual phase materials such as tool steels, high chromium cast irons, sintered WC-Co alloys and composite materials. In view of the above, an attempt has been made to examine the performance of solid particle erosion of thermal spray coatings and highlight the development in recent past. Present chapter has been divided under seven sections. After introducing solid particle erosion of thermal sprayed coatings in the first section, evolution of thermal spraying and various testing techniques are discussed in section two and three respectively. Erosion behavior of coatings is elaborated in section four. Applications of thermal sprayed coating for erosion resistant applications are provided in section five. Finally, direction of future research and concluding remark is summarized in section six and seven.

2 EVOLUTION OF THERMAL SPRAY COATINGS

Thermal spraying is a process where consumable powders or wires in the form of finely divided molten and semi molten droplets are propelled to a substrate to produce coatings. Thermal spraying dates back in 1911 when Dr. Schoop for the first time atomized molten metal by high pressure gas and propelled them on to a surface. In 1912 a device to spray metal wires known as flame spraying was produced. Later on new technique for thermal spraying by using electricity to melt feedstock materials was introduced. These methods are known as arc spraying. Usage of powder for the

Table 1. Manufacturing methods for wear resistant surfaces (Schmidt & Steinhauser, 1996)

Surface Coating Methods	Surface Treatment Methods
Thermal spraying	Thermo chemical treatment
Deposit welding	Thermal surface hardening
Electrochemical deposition	Mechanical surface hardening
Chemical deposition	Surface remelting
Physical vapour deposition	Anodic oxidation
Chemical vapour deposition	Ion implantation
Plating	

Figure 1. Use of different reinforcement phases, metals and metalloids in thermal spray coating (GFE Schmalkalden, http://www.fh-schmalkalden.de/schmalkaldenmedia/-p-16916. 389-394)

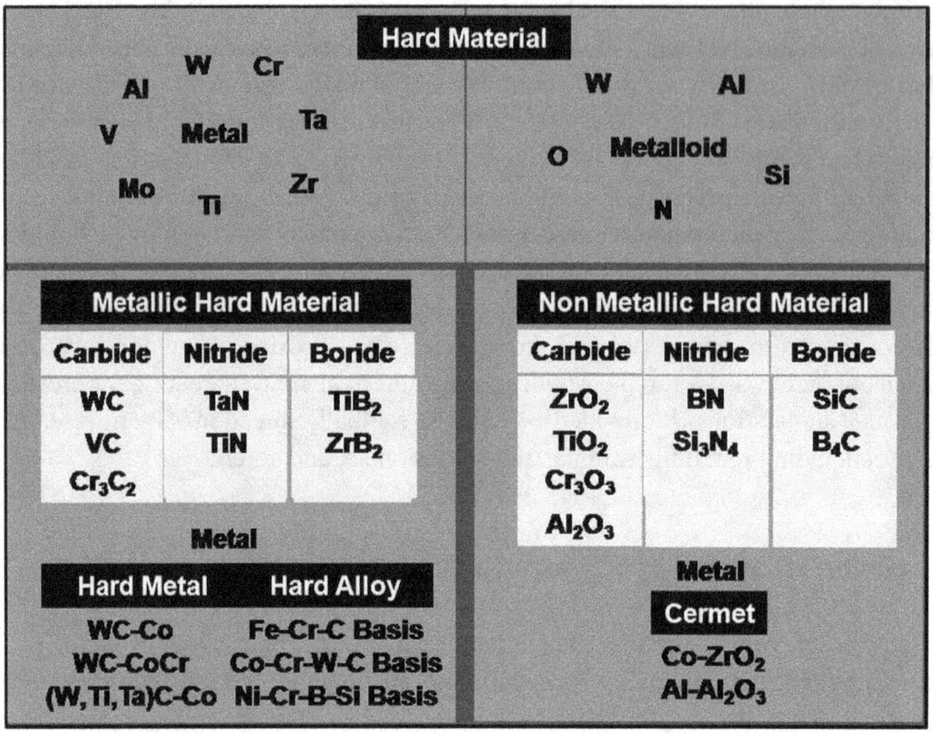

flame spraying process was developed for the first time by F. Schori in 1930. These processes can use only low melting metallic powders. Slowly people started using powders with high melting temperature and oxidation resistance. This resulted in development of detonation spraying in 1958 by Union Carbide and atmospheric plasma spraying in 1960. This was followed by development of vacuum plasma spraying (VPS) and low-pressure plasma spraying (LPPS) in late 1970 and 1980. However, the main development of thermal spraying occurred in 1980 when a novel technique for spraying powder employing high oxygen, known as high velocity oxy fuel (HVOF) technique was introduced. James A Browning developed liquid fuel based coating gun 'Jet-Kote' in 1982. The most important technology cold spraying was developed in 1990. This process has the ability to produce unique coating that is not possible by other coating method. (Li, et al., 2008; Melendez et al., 2013). Over the last several decades, thermal spraying has evolved from experimental stage to commercial techniques. Figure 2 depicts evaluation of usage of thermal sprayed coatings. It took 35 five years to get first published paper on tribology since development of detonation gun and it was nearly 50 years before HVOF sprayed WC –Co coating found application in turbine blades.

Over the years different thermal sprayed coatings are developed. Different techniques have different features, different advantages and limitations. Different techniques are suitable for spraying different types and grades of powder. The process temperatures and particle velocities obtained using different spraying gun can vary over a wide range. Comparison of characteristics (temperature and feed velocity) of different thermal spray processes and cold spray process is made in Figure 3.

3 EXPERIMENTAL METHODS IN EROSION TESTING

The selection of an "appropriate" erosion test is very often difficult and is dictated by several parameters. Very often it is difficult to simulate the exact service conditions in actual laboratory test. The analysis of the service conditions such as nature of wear process responsible for material removal and degradation, abrasive medium and its properties (hardness, size and shape), and environmental conditions, forms an integral part in the selection of a most suitable erosion test in the laboratory. The extent to which the laboratory erosion test data can be extrapolated into service decreases from the field test to the very simplified model test. Very often the shortening of test times by increased loading in a model test can result in different erosion rates, different erosion mechanism making transferability of test results from model test to service, uncertain. However, model tests are advantageous for scientific investigations due to high reproducibility of the test parameters and ease and precision with which operating conditions /test parameters can be controlled. Model tests are very useful in preliminary selection of materials or ranking of

Figure 2. Evolution of WC coatings and their industrial applications (Wood& Robert, 2010)

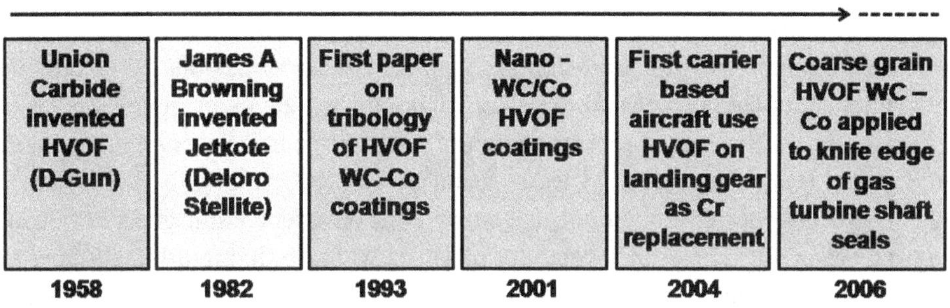

Figure 3. Comparison of characteristics (temperature and feed velocity) of different thermal spray and cold spray process (GFE Schmalkalden, http://www.fh-schmalkalden.de/schmalkaldenmedia/-p-16916. 389-394)

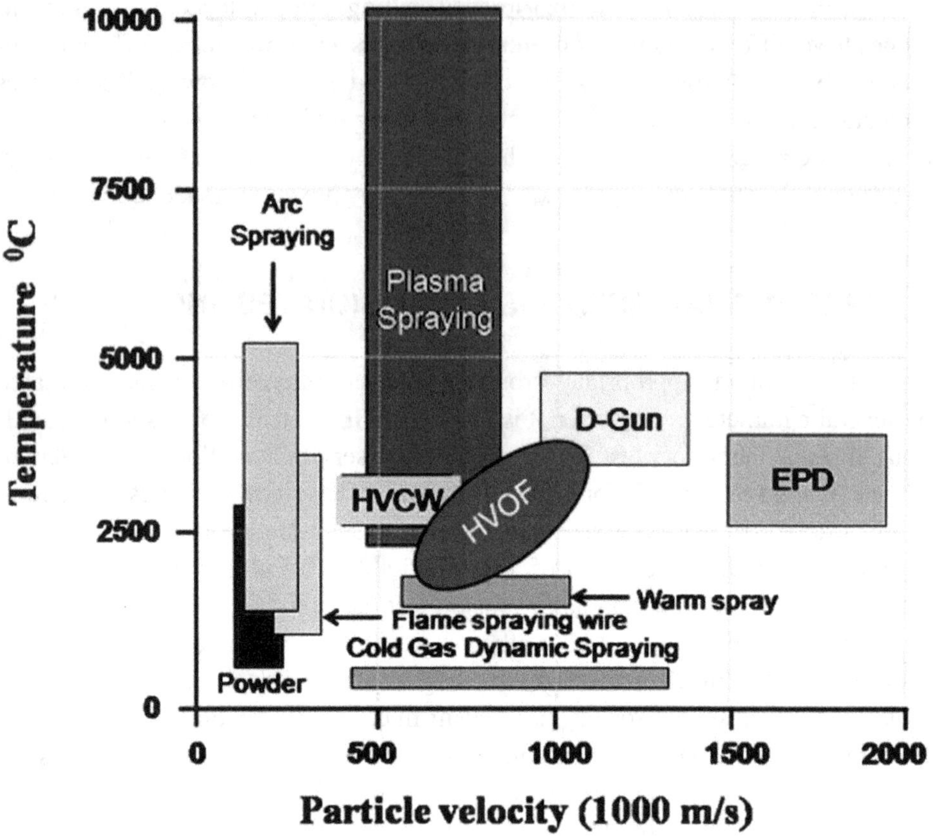

materials for a specific application and in investigation of the mechanisms and /or the influence of operating variables e.g. pressure, velocity and environment (Gahr, 1987; Hammitt, 1976).

To study the erosive wear behavior of materials, several types of erosion test apparatus have been used by the investigators. The most common designs being gas blast test rig and centrifugal erosion accelerator. In case of gas blast type erosion test rig, the erodent particles are accelerated along a nozzle in a flowing stream of gas, usually air. Schematic representation of air jet erosion test apparatus is made in Figure 4 (Roy et al., 1992). Various techniques for controlling the particle feed rate have been reported e.g. vibratory feeding, belt conveyer feeding and screw feeding. The centrifugal erosion accelerator has the advantage that multiple number of specimens can be tested simultaneously. In this system the erodent particles are fed into the center of a rotor rotating in a horizontal plane and are accelerated by the rotation producing a stream of particles to strike specimens positioned at various angles around the perimeter of the enclosure. Both the methods produce divergent stream of particles, resulting in variation of particle flux over the surface of the specimen (Hutchings, 1997). The details regarding standard test conditions, test procedure and reporting of test results are given in ASTM G-76 (ASTM G-76, 1995).

The whirling arm rigs are developed to enable tests to be carried out at precisely controlled velocity over a range of impact conditions. The target specimens are attached to the tips of the rotor arm and whirled through a certain or a narrow band of erosive particles. These rigs simulate the degradation conditions prevalent in fluidised bed combustor. The whirling arm erosion rig was initially developed by Tilly and Sage (Tilly & Sage, 1967). One of the advantages of this type of rig is that the erodent velocity can be controlled precisely as the velocity is governed by the rotating speed of the arms.

Particle velocity is one of the important parameters influencing the erosion behavior of materials. Due to strong dependence of erosion rate on velocity, (E ∝

Figure 4. Schematic diagram of air jet erosion test apparatus (Roy et al., 1992)

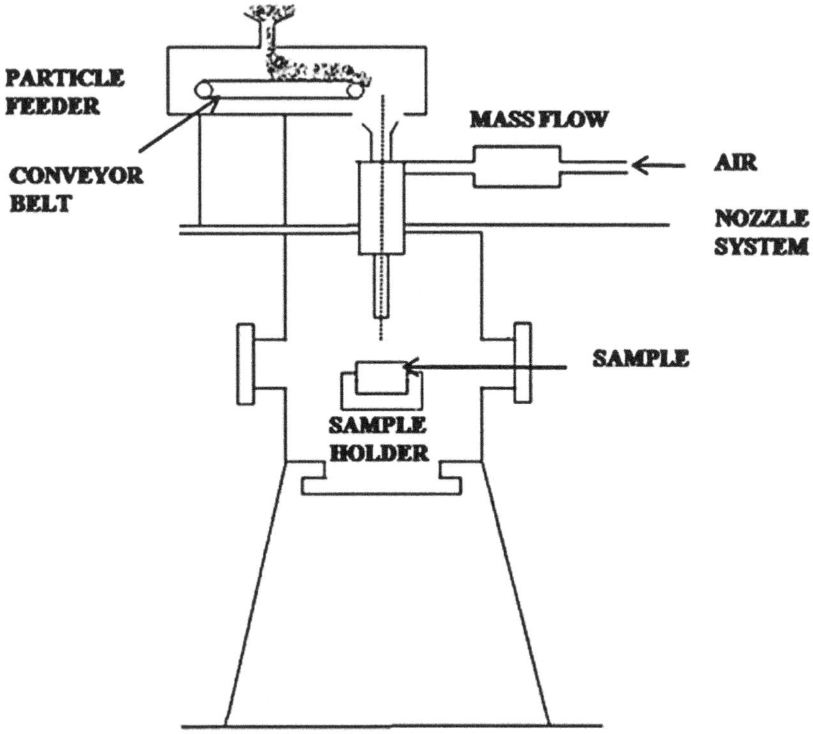

$V^{n)}$ in case of metals and ceramics an accurate knowledge of particle velocity is essential for reliability of erosion data and for determination of velocity exponents in erosion studies. Several methods of velocity calibrations such as Double Rotating Disk, Multiple Flash Photography, High Speed Camera, Laser Doppler Velocimeter and Opto-electronic Velocimeter etc. have been reported. Ruff and Ives (Ruff and Ives, 1975) devised relatively simple method for velocity measurement by using double disk apparatus, in which two circular discs with a known spacing are fixed on the same rotating shaft. A single radial slit in the upper disc permits particles to incident and eventually erode a mark on the lower disc. Two erosion exposures are made, one with the disks stationary and the other with the disks rotating at a known, constant speed. Measurement of the angular displacement between these marks along with measure of the time of flight of the particles as they cross the space between the discs gives impact velocity of the erodent. Though this method is useful, several shortcomings have been reported (Kosel & Anand, 1989). Hovis (Hovis, Anand, Conard & Scattergood, 1985) modified double disk method, still keeping the basic principle similar to that used in the original method, called as paddle wheel method in which the particles were allowed to pass through a multiple slit arrangement to strike radial paddles mounted on a shaft parallel to the direction of flight. Since the paddles are at different distances from their respective slits, impact patterns are produced at different positions along the paddles. The impact pattern will be formed at a distance from the disk. A linear least square fit of the distances of the spot patterns from the ends of the paddles is used to obtain the velocity. This method offers an advantage in that; it uses least square analysis which helps to reduce the statistical errors in the calibrations. However this method is still subject to some of the problems mentioned above.

The accurate determination of velocity of smaller particles 20-40 μm is considerably difficult since small particles are considerably influenced by drag force due to the difference between particle velocity and fluid velocity. The use of photodiode and high speed photography is thought to be inadequate in this context since particle stream is nearly invisible. Kagimato (Kagimato., 1987) used laser doppler velocimeter (LDV) to measure the particle velocity of small ash particles and to determine the effect of impact velocity on erosion rate in boilers. The method requires some numerical calculation to determine the air flow pattern and to estimate particle deceleration rate. Ninham and Hutchings (Ninham and Hutchings, 1983) developed a computer program to calculate the particle velocity, which is based on standard fluid mechanics equations, for compressible flow in pipes with friction and this defines the gas pressure, density and velocity at all points along the tube. The model predicts the velocity of spherical particles and requires an empirical correction for measurement of velocity for angular particles. Stevenson and Hutchings (Stevenson and Hutchings, 1995) used opto-electronic velocimeter which uses two photo diodes to measure the time of flight of the particles. Multiple flash photography method is based on identification of a particle at two different positions on a double exposed photograph. The time interval between the two exposures presents a means of calculating the particle velocity. The technique is relatively simple and is best suited for single particle impact studies with a particle size about 25 μm. However an uncertainty of 10% in the measured velocity values has been reported. The high speed camera technique is similar to multiple flash photography except that single exposures on consecutive films are made instead of multiple exposures on the same film. The method suffers from same constraints and uncertainty as that of multiple flash photography. Further details of erosion testing can be found elsewhere (Roy, 2000).

4 EROSION BEHAVIOR

In this section solid particle erosion (SPE) of thermal sprayed coatings is discussed. Since most thermal sprayed coatings which are used for protection against erosion are composite in nature, mostly cermet variety, erosion of multiphase material, in general, is discussed. However, to understand erosion of multiphase materials it is necessary to understand erosion of single phase materials and hence erosion of metals and alloys has briefly been introduced.

4.1 Solid Particle Erosion of Metallic and Ceramic Materials

Wear by erosion is the loss of material due to mechanical interaction between the component surfaces and the high velocity particles entrained in a fluid stream. Until 1950s most of the erosion studies were restricted to devising practical solutions to the erosion problems, until 1958 when a pioneering effort was made by Ian Finnie (Finnie, 1995) to understand the fundamental mechanism of material removal from surface by erosion. His model was based on cutting mechanism of erosion which predicted accurately the variation of the weight loss with angle of impingement and maximum erosion rate was observed at 20-30° impact angle for ductile metals (Finnie, 1979). Subsequently it was shown that the material removal mechanism involved, material raised in the form of a lip which is vulnerable to be removed in subsequent impacts. Hutchings (Winter & Hutchings, 1974; Hutchings &Winter, 1975; Hutchings, 1977; Hutchings, 1979) made significant contribution by way of his single particle erosion studies on ductile metals and proposed ploughing and cutting mechanisms. A considerable body of evidence exists to show that erosion mechanism is a strong function of erodent particle properties like particle hardness, particle size, shape and its orientation and the material properties. Observations on erosion debris further confirmed that ploughing and cutting are the dominant mechanisms at lower angles (Kosel, *1984*). Various erosion mechanisms of material removal have been proposed at normal impingement angle. The widely accepted erosion mechanisms at normal impact are, deformation (Bitter, 1963; Nelson & Gilchrist. 1968), platelet formation (Bellamn & Levy, 1981; Levy, 1986), fragmentation of the particle (Tilly, 1969; Tilly,1973), low cycle fatigue (Hutchings, 1981) and localization of plastic flow (Sundararajan & Shwemon,1983; Sundarrajan, Roy & Venkataraman, 1990; Sundararjan & Roy,1997). However no clear agreement exists on any single mechanism which is operative at all the angles. Although it is suggested that platelet mechanism may be the predominant mechanism of material removal for metals at 90° and micromachining may predominate at low angles of incidence; plastic deformation is still considered as the dominant mechanism of erosion (Finnie, 1995; Kosel, 1992).

A different mechanism termed 'adiabatic shear induced spalling' involves the formation of intersecting adiabatic shear bands at the base of the crater and subsequent removal of chunks of material is relevant for hard and brittle metallic material. This mode of weight loss, which is highly efficient in terms of energy expended per unit volume of target material removed, is important only at normal impact where the constrained of deformation is maximum. The erosion response under such circumstances will be similar to that observed for ceramic materials (Quadir and Schewmon 1981; Chrisman and Schewmon, 1979). But the underlying mechanism is entirely different from the way materials are removed from ceramic materials.

The mechanism of material removal for brittle solids particularly ceramics involves generation and propagation of subsurface lateral cracks driven by the residual stresses produced in an elastic-plastic contact zone. Although the mechanism of erosion in brittle materials involves brittle fracture, it is considered that plastic deformation might be operative at lower angle of impingement (Finnie, 1995). The erosion behavior of brittle materials is distinctly different from that of ductile metals. The peak in erosion is observed at 90° impingement angle. In contrast to ductile metals, brittle materials show stronger dependence on velocity. The models proposed by Evans, 1978; Evans & Wilshaw, 1977; Lawn and Wilshaw, 1975; Ruff and Wiederhorn, 1979) based on elastic plastic interaction, predicted reasonably accurately the dependence of erosion on velocity, particle size and material properties like hardness and fracture toughness and this could not be established in the Hertzian fracture model (Sheldon and Finnie, 1966).

4.2 Erosion Behavior of Multiphase Materials

The ceramic materials exhibit superior abrasion and erosion resistance as compared to metals, especially when the abrasive particles are softer than the target material.

The dispersion of a ceramic in a more ductile matrix to improve toughness has been successful in abrasion but with a little success in erosion except for polymer matrix composites or carbon composites (Roy, Vishwanathan &Sundararajan 1994; Sarkar, Sekharan, Mitra & Roy 2009). For example, WC-Co alloys and high chromium iron alloys, containing a dispersion of hard carbides, exhibit excellent abrasion resistance. The literature suggests that beneficial effect of high volume fraction of hard ceramic particles could only be observed with softer erodent particles and under benign erosion conditions. The important factors influencing erosion behavior of white irons were observed to be volume fraction of carbides, relative hardness of erodent particles in relation to carbide hardness, relative size of erodent particles with respect to size of carbides, mean free path, impact velocity, impact angle and the environmental conditions. (Sapate & RamaRao, 2004)

Hansen (Hansen, 1979) compared the erosion rates of a large number of metallic alloys, ceramic materials and cermets when eroded with 27 μm alumina particles at a velocity of 170 ms^{-1} at 90° impingement angle. The WC-Co alloys exhibited relative erosion factor (with respect to stellite 6B) in the range of 0.1-1.6 and relative erosion factor increased with binder content. Most of the ceramic materials had relative erosion factor values in the range of 0.3-0.6. Ninham and Levy (Ninham & Levy, 1988) performed erosion tests on high chrome iron, cermets and stellite alloys containing carbide volume fraction from 10 to 90%, with 75-200 μm angular quartz particles travelling at velocity of 60 ms^{-1}. The erosion rate decreased at high volume fractions for WC and TiC cermets, when the carbide was the major constituent. In tristellite containing large primary carbides the localization of deformation in the overlying matrix and subsequent fracture of unsupported carbides resulted in higher erosion rates. The brittle signature was favored at higher binder levels, since binder phase controlled the erosion but was severely constrained by the carbides particles and therefore cermets behaved in a brittle manner. Anand (Anand, 1985) studied scaling effects in erosion of multi phase materials. They performed erosion studies on WC-Co alloys with different sizes of WC particles (2-16μm) and a composite consisting of alumina rods (500 μm diameter) in a stainless steel matrix with alumina particles (40-270 μm) at a velocity of 50 ms^{-1}. An increase in erosion rate with increase in particle size of the erodent was observed. The decrease in grain size of WC resulted in decrease in erosion rate. The increase in velocity from 50 ms^{-1} to 93 ms^{-1} resulted in shifting the peak in erosion to angles less than 90°. They suggested that the relative differences in erosion rates of WC-Co alloys under different erosion conditions could be rationalized on the basis of the scale of the damage event relative to the scale of the microstructure due to variation in erodent particle size and /or velocity. When the damage event is small as compared to the WC grain size, there is no apparent constraint provided by the binder phase since cracks can extend to their full extension with single grain and the overall erosion rate is dominated by the erosion rate of WC. The growth of the lateral cracks in small grains is constrained by the Co binder phase and ductile type erosion is favoured when the size of the damage event is large as compared to WC grain size.

Anand and Konard (Anand & Konard, 1988; Anand & Konard, 1989) further observed microstructural effects in the erosion of WC-Co cemented carbides. The erosion response of WC-Co alloys was investigated as a function of binder content (10-70% by volume) and WC grain size (0.42-3.25 μm) with 63-405 μm alumina particles in the velocity range of 35 to 93 ms^{-1}. The change in erosion behavior from brittle to ductile was favored under severe erosion conditions of increasing particle size and velocity. In the ductile erosion regime the erosion rate of cemented carbides varied linearly with square root of binder mean free path. In the ductile regime the flow stress of the constrained binder controlled the erosion rate. In brittle regime, WC grains undergo brittle fracture however crack propagation was restricted to WC grain due to surrounding binder phase. They proposed a mechanism involving crushing, corner chipping and cracking of WC grains lying within and along the edge of the impact crater in the brittle erosion mode. Ball and Patterson (Ball & Peterson, 1985) investigated the erosion response of number of WC cermets with 100 μm silicon carbide particles at a velocity of 40 ms^{-1} and at impinging angle of 45°. The erosive wear volume loss vs. binder curve showed a maximum at about 17% binder and a minimum was observed at 34 volume % binder. They suggested that cermets with fine carbide grains suffered a greater volume loss as compared to coarse-grained cermets. It was pointed out that for binder volume fractions of less than 20%, the erosion is controlled by the deformation and fracture of the grains of WC skeleton while ductile deformation of the binder phase controlled the erosion of cermets containing more than 20 volume % binder. The results of erosion studies by Pennefather (Pennefather, 1988) on WC cermets with 150 μm silicon carbide particles at a velocity of 40 ms^{-1} showed peak in erosion at 10 weight % cobalt. The peak in erosion was found to be more pronounced at higher impact angles. For impact angles less than 45 ^{o}, the erosion rate increased linearly with binder content. They considered that for low cobalt content, the erosion is controlled by the rigidity and toughness of the carbide skeleton while deformation of binder phase controls the erosion of cermets with higher cobalt content. They put forward similar arguments as proposed by Patterson and Ball (Patterson & Ball, 1985) to rationalize the observed erosion rates of WC-Co cermets.

4.3 Erosion Behavior of Thermal Spray Coatings

As mentioned earlier, thermal sprayed coatings especially WC-M coatings are promising layers for protection against solid particle erosion. WC-M (Co, N, Ni-Cr) thermal spray coatings offer attractive properties such as high hardness imparted by carbide particle and good toughness due to the relatively softer binder phase. The softer metal matrix provides good adhesion and wetting characteristics resulting in good toughness. The improved mechanical properties of WC-Co thermal spray coatings in relation to percentage of hard phase are provided in Figure 5. The erosion resistance of WC-M coatings is influenced by the volume fraction of carbides and binder phase, mechanical properties of the carbides and binder phase, manufacturing technique and porosity of the coating and the operating conditions such as impact angle, velocity of erodent particles, properties of erodent particles and the environment; temperature/corrosive media (Karimi, Verdonm Martin & Schimd, 1995; Kullu, Hussainova, & Veinthal 2005; Machio, Akdogan, Witcomb & Luyckx 2005; Mathews, James & Hyland 2007). The erosion rates of detonation gun sprayed WC –Co coatings were found to increase with increasing hardness of erodent particles (Babu, Basu & Sundarrajan, 2011).

Erosion of thermal spray coatings are influenced by the complex splat network and phase variation inherent in their formation. In general, coatings with low inter-splat adhesive strength are prone to fracture and crack propagation along splat boundaries during erosion (Kingswell, Rickerby, Bull & Scot, 1991; Lathabai, Ottmuller &, Fernandez 1998). This results in removal of entire splats or groups of splats. Sometimes cracks are linked up and form chips of material that are readily ejected upon subsequent impact. Increased inter-splat adhesion reduces the magnitude of damage generated by each impact. Fracture becomes more localised within splats and results from brittle fracture and fatigue around impact sites (Kingswell *et al,* 1991; Lathabai et al., 1998). Preferential mass loss occurs along splat boundaries via chipping or plastic deformation (Wang, 1996; Wang, Geng & Levy, 1990; Wang and Lee, 1997). Link up of vertical cracking with subsurface splat boundaries is an additional erosion mechanism (Wood, Mellor & Binfield, 1997). Splat structural effects are the most important parameters for erosive loss as noted for a range of deposition techniques, coating compositions and testing conditions (Kingswell et al., 1991; Lathabai et al., 1998; Wayne, Baldoni & Buljan 1990; Xia, Zhang & Song, 1999). Contributing factors have included splat size (in relation to the erodent) (Levy, 1988; Levy & Wang, 1988.), degree of intra-splat microcracking upon solidification (Westergard, Axen, Wiklund & Hogmark 2000), high concentrations of oxide stringers (Kingswell *et al.,* 1991) and inter-splat porosity (Wayne & Sampath, 1992).

The contribution of sub-surface crack propagation along regions of microstructural weakness to erosion damage of thermally sprayed coatings is demonstrated by several studies. The SEM images showing the morphology of the eroded surface and the region beneath the eroded surface of detonation sprayed WC-Co coating is presented in Figure 6 due to Roy (Roy, 2014). The images correspond to normal impact and impact velocity of 150 m/s. Propagation of cracks along the splat boundary

Figure 5. Advantages offered by cermets and ceramic coatings as a function of hard phases (Kulu 2005)

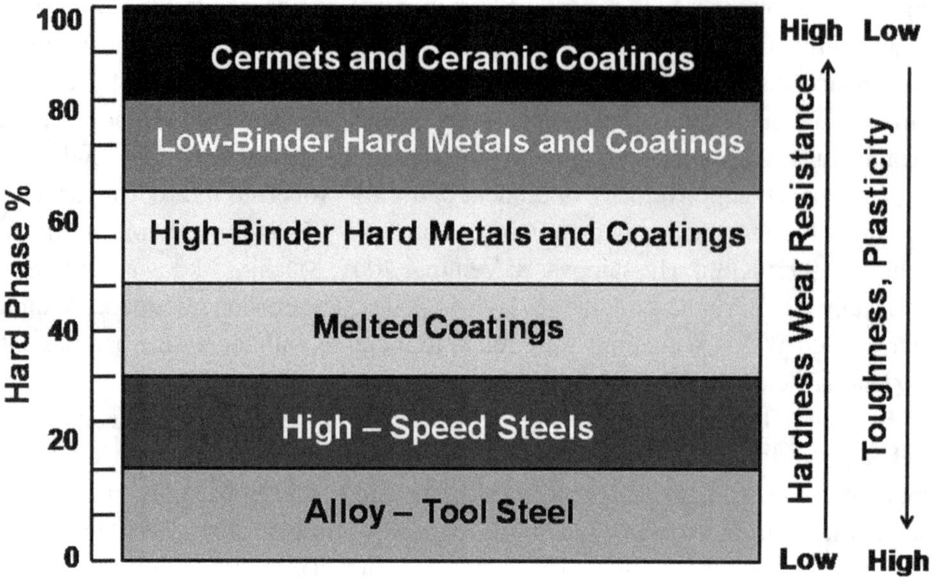

is evident. Verdon Karimi and Martin (Verdonm Karimi & Martin, 1997) deduced that the relative contribution of plastic deformation, cracking and delamination mechanism of material removal in erosion of thermally sprayed WC-Co coatings is decided by impact velocity and the ease of crack propagation along splat boundaries. The detachment of subsplat and splat was observed at an impact velocity of 70 ms^{-1} during erosion of plasma sprayed ceramic coating (Westergard et al., 2000). Similar observation were noted by Lofti, Shipway, McCartney and Eldris (Lofti, Shipway, McCartney & Eldris, 2003) regarding extensive subsurface cracking in erosion of NiCr-TiB_2 coatings and the presence of intersplat oxide phases thought to be responsible for cracking around splat boundaries.

Hawthorne (Hawthorne, Arsenault., Immarigeon, Legoux & Parmeswaran, 1999) conducted dry sand and slurry erosion test on ten different HVOF WC coating with Co, Co-Cr-Ni, Ni—Cr, Ni-Cr-Si-B. The dry sand erosion tests were carried out at a velocity of 84 ms^{-1} at impact angles of 20° and 90°. Figure 7 presents comparison of dry erosion with 50 μm alumina particles at 20 and 90° impact angles for various thermal sprayed coatings. It was observed that dry erosion rates were three times higher than slurry erosion rates. No direct correlation between hardness and erosion rate was observed for thermal spray coatings and at normal impact erosion rate exhibited tendency to decrease with increase of hardness as observed in Figure 7. It was concluded that coating composition and microstructural integrity were the

Figure 6. The SEM images showing the morphology of the eroded surface and the region beneath the eroded surface of detonation sprayed WC-Co coating. (Roy, 2014)

dominant factors in deciding erosion behavior. The addition of Cr and VC proved to be beneficial for erosion resistance of HVOF WC-Co coatings (Karimi et al., 1995; Machio et al., 2005). It was observed that addition of Cr to HVOF sprayed WC-Co coatings improved the erosion resistance by a factor 4-6 times due to improvement in toughness and binding of carbides to the matrix.

The studies due to Li, Yang & Ohmori (Li, Yang & Ohmori, 2006) indicate that mean lamella thickness and mean bonding ratio of plasma sprayed coatings is influenced by plasma power and spray distance as shown in Table 2 and Table 3. The

Figure 7. Dry sand erosion test of several thermally spayed HVOF coatings at 20 and 90° impact angles with 50 μm alumina particles. (Hawthorne et al., 1999)

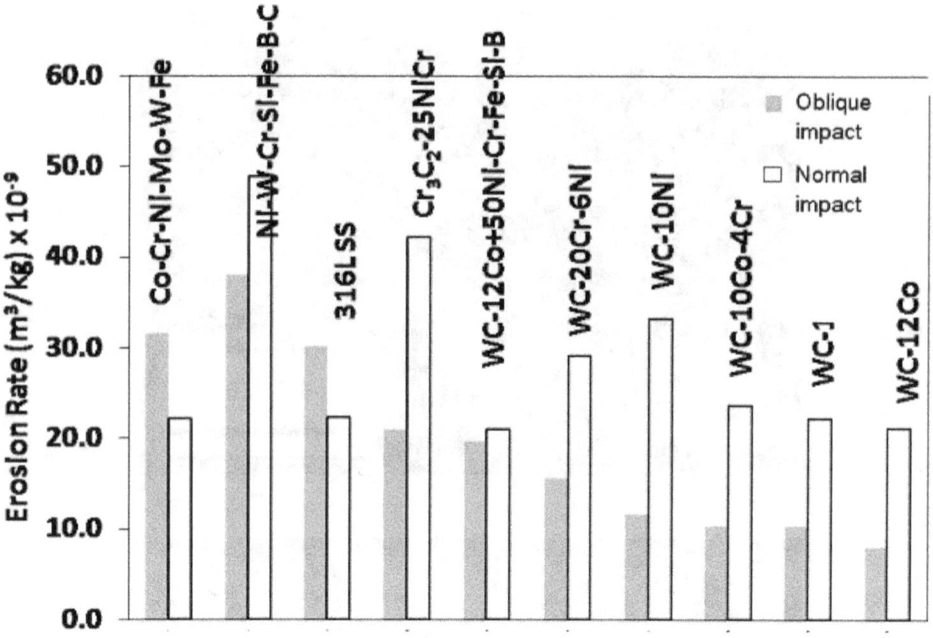

lamellae thickness, δ and mean bonding ratio, α increased with plasma power whereas with increase in spray distance mean bonding ratio increased and lamellae thickness exhibited an increasing trend. The erosion rate of Al_2O_3 coating at an impact angle of 90° and velocity of 63 m s^{-1} and with 330 μm alumina erodent particles was inversely proportional to lamellar interface bonding ratio and is directly proportional to lamellae thickness. The dependence of the erosion resistance was expressed in terms of a microstructural parameter involving mean bonding ratio between lamellae and thickness of the lamellae by the equation 1

$$\frac{1}{E} = \frac{2\gamma_c \alpha}{E_{eff}\delta} \quad (1)$$

where γ_c is the effective surface energy of the lamella material, α is the bonding ratio of the interface between lamellae, δ is the thickness of the lamella, *1/E* is the erosion resistance and E_{eff} is the fraction of kinetic energy per unit mass of impacting particle utilized for cracking of the bonded lamellar interface.

Erosion rate of thermal sprayed coatings are dependent on the angle of impact. Hearley, Little & Sturgeon (Hearley, Little & Sturgeon, 1999) observed classical

Table 2. Mean lamella thickness and mean bonding ratio of Al_2O_3 coatings plasma sprayed under different powers (Li et al., 2006)

Power (kW)	Lamella Thickness (μm)	Mean Bonding Ratio (%)
21.0	1.85	23.5
24.5	2.1	32
28.0	1.58	32
31.5	1.75	32

Table 3. Mean lamella thickness and mean bonding ratio of Al_2O_3 coatings plasma sprayed at different spray distance (Li et al., 2006)

Spray Distance(mm)	Lamella Thickness (μm)	Mean Bonding Ratio (%)
80	1.89	31.9
100	1.58	32
150	1.82	17.5
200	3.44	16.8

brittle erosion response of HVOF sprayed NiAl coating as can be seen from Figure 8. It is to be stated that the eroded surfaces indicates ductile materials removal mechanism governed by deformation rather than fracture even though erosion response is brittle. This apparent contradictory observation may be attributed to the nature of erodent, their size and shape. Similar observation is also made by Kulu et al. (Kulu et al., 2005) for several variety of thermal sprayed coatings. As noted by Babu (Babu et al., 2011) the ratio of erosion rate at impact angles of 90° and 30° in the case of thermal sprayed coatings, increases with decreasing impact velocity and with increasing hardness of the erodent. With SiO_2 erodent particles the ratio (i.e.E_{90}/E_{30}) for WC-Co coatings was marginally lower than mild steel whereas with alumina and SiC erodent particles the ratio was 1.5-2 times higher than mild steel at a velocity of 25 m s^{-1} (Figure 9). In contrast, at 45 ms^{-1} the ratio for coatings was nearly 1.5-2 times lower than mild steel as shown in Figure 10. In all cases thermal sprayed coatings generally exhibited brittle erosion response. Guilemany, Cincam Fenandez & Sampath (Guilemany, Cincam, Fenandez & Sampath, 2008), however, demonstrated that near stoichometric Fe-40Al HVOF coating iron aluminides exhibited a ductile signature under the conditions of normal impact. Similarly Hawthrone (Hawthorne et al., 1999) reported ductile erosion for several varieties of coatings (Figure 7).

Figure 8. Erosion rate of Ni-Al coatings as a function of impact angle and impact velocity (Hearley et al., 1999)

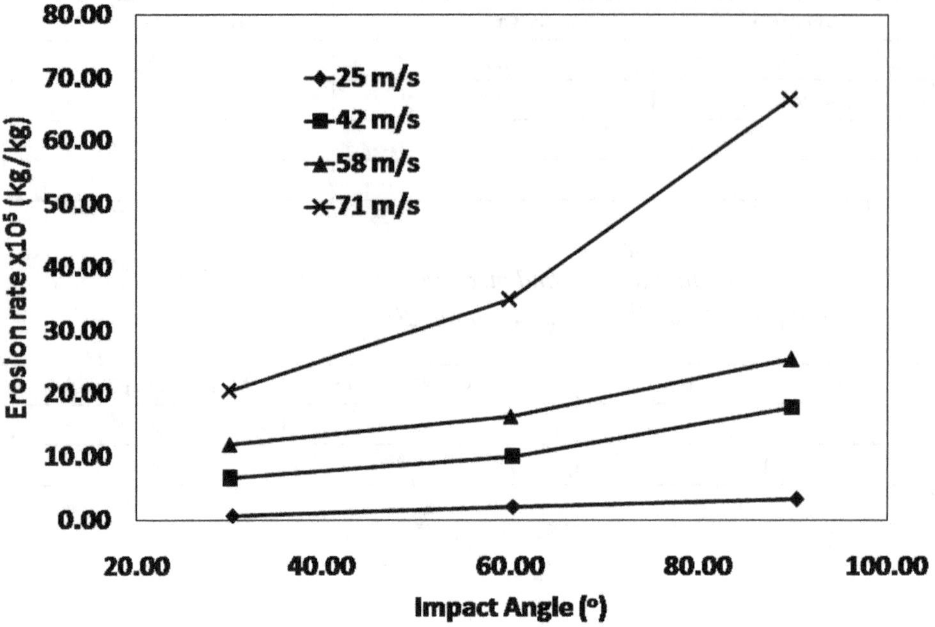

Erosion rate strongly depends on the impact velocity. As shown in Figure 11, erosion rate increased with impact velocity; the relation between erosion rate and impact velocity is given in Equation 2.

$$E \propto V^{3.2} r^{3.7} \rho^{1.3} H^{-1.25} \qquad (2)$$

where V is the impact velocity, r is the radius of the erodent; ρ and H are density and hardness of the target material. Similar observations were made by Levy and Wang (Levy & Wang 1988) for plasma sprayed WC-Co. It is observed by Wood et al. (Wood et al., 1997) that for WC-Co coatings, erosion rate was independent of impact angle at lower impact velocity whereas erosion rate was dependent on impact angle at higher impact velocity. Most coatings, in general, exhibit velocity exponents in access of 3.0.

The post coating treatment also alters the erosion performances of coatings. According to Matthews, James & Hyland (Matthews James & Hyland, 2009) heat treatment resulted in increased ductility of the NiCr phase of high velocity air fuel (HVAF) sprayed and HVOF sprayed Cr_3C_2-NiCr coating. Heat treatment resulted

Figure 9. Erosion rate ratio (E_{90}/E_{30}) for mild steel, bulk WC-Co coating and detonation gun sprayed WC-Co coatings at a velocity of 25 ms^{-1} with different erodent particles (Babu et al., 2011)

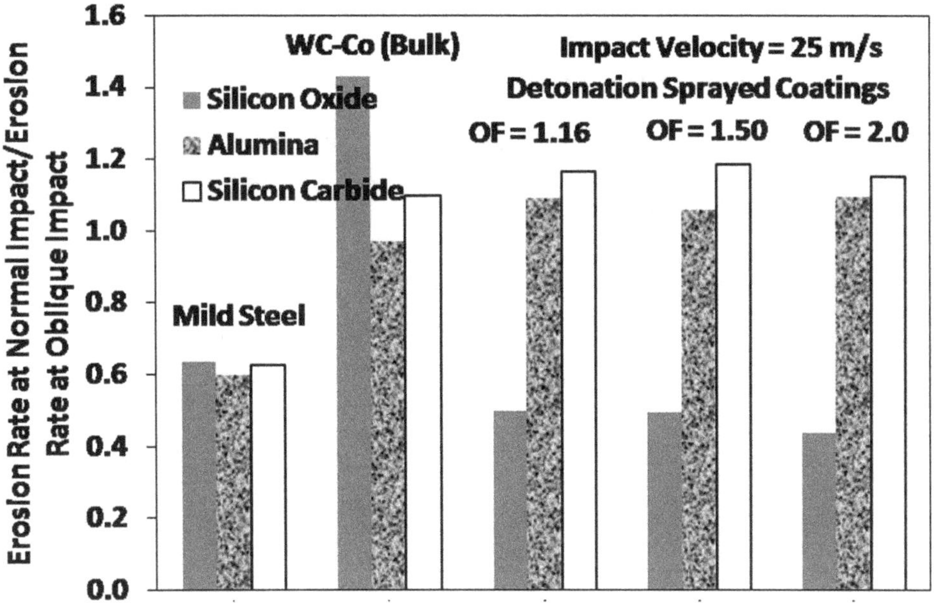

Figure 10. Erosion rate ratio (E_{90}/E_{30}) for mild steel, bulk WC-Co coating and detonation gun sprayed WC-Co coatings at a velocity of 45ms^{-1} with different erodent particles (Babu et al., 2011)

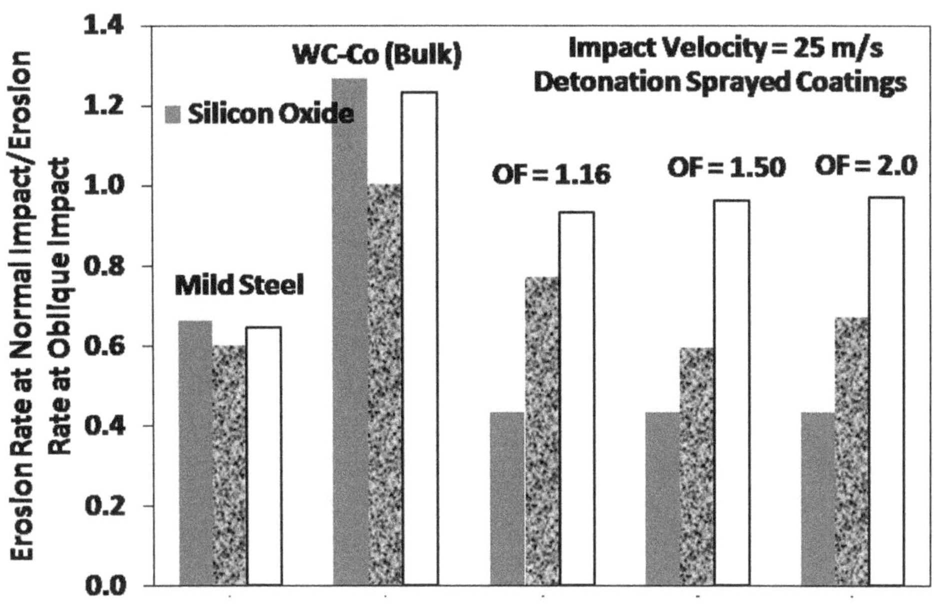

Figure 11. Erosion rate a s function of impact velocity for different impact angles (Hearley,1999)

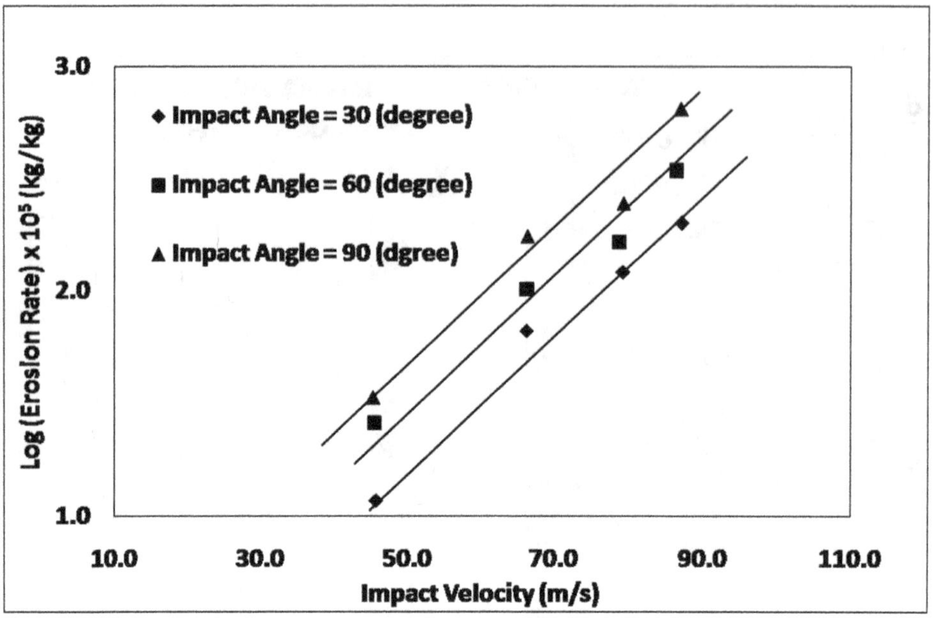

in the reduced contribution of splat based erosion mechanisms by intersplat sintering and this leads to improved erosion resistance of heat treated coatings (Matthews *et al.,* 2009). The heat treatment of HVOF and HVAF Cr_3C_2-NiCr coatings proved beneficial for erosion resistance as a result of improvement in toughness of the coating due to re-precipitation of the carbide phase in network form. This restricted relative movement of carbides during impact and applied load being shared by relatively larger volume of the coating. Heat treatment also resulted in ductile erosion in contrast to untreated coatings which exhibited brittle erosion signature (Matthews et al., 2007). Similar improvement for chromium carbide based coating has been reported by several other investigators (Walsh& Tabakoff, 1992; Wlodek, 1985; Nerz et al., 1991). Effect of heat treatment on the erosion of tungsten carbide based coatings is found to be less prominent (Nerz et al., 1991; Iron *et al.,* 1994).

Nerz, Kushner, Kaufold and Rotolico (Nerz, Kushner, Kaufold and Rotolico, 1991) reported increase in erosion rate for blended powder based HVOF Cr_3C_2–NiCr coatings following heat treatment. In contrast, coatings based on sintered/crushed powder showed negligible variation with heat treatment. However, heat treatment significantly improved the erosion performance of a blended powder based

Cr_3C_2–FeCrAlY sprayed by the same technique. Irons, Kratochivil, Bullock, Roy & Berndt (Irons, Kratochivil, Bullock, Roy & Berndt, 1994) noted an improvement in erosion resistance at 20° impact angle but higher erosion rate at 90° impact after heat treatment. He correlated the erosion response with microhardness as a function of heat treatment. However, erosion rate and hardness have mixed response to different heat treatment conditions (time, temperature etc) (Wlodek, 1985; Nerz et al,. 1991). D-Gun sprayed Cr_3C_2 based coatings have tendecy to harden with heat treatment (Walsh, 1992). For HVOF coatings the response was mixed. Nerz et al. (Nerz et al., 1991) reported increases in hardness for treatment of blended and sintered/crushed powder based coatings, while Irons et al. (Irons et al., 1994) noted a softening with heat treatment. Wlodek (Wlodek, 1985) and Nerz (Nerz et al., 1991) attributed the increased erosion resistance with heat treatment to the development of Cr_7C_3 phases. This phase was postulated to form as a result of carbon loss from Cr_3C_2 during spraying. It was considered that the decarburised phase to be harder than the original starting carbide, thereby improving the erosion resistance. The development of an ultrafine grain structure was also linked to the increased hardness and improvement in performance following heat treatment. Tsai, Lee and Chang (Tsai, Lee and Chang, 2007) investigated the improvement of erosion resistance of plasma sprayed thermal barrier coatings on laser glazing. Several erosion tests were conducted at room temperature using 50 μm silica erodent particles with impact velocity of 50 m s^{-1}. The erosion rate increased as the impingement angle increased for both plasma-sprayed and laser-glazed TBC. Laser glazing enhanced the erosion resistance of plasma-sprayed TBC by about 1.5 to 3 times with the impingement angle ranging between 30° and 75°, while the erosion resistance did not significantly improve when the impingement angle reached 90°. Erosion morphology analysis clearly indicates that the erosion of the plasma-sprayed TBC is deemed to be the erosion of the protrusions and the sprayed splats. The erosion of the laser-glazed TBC is proven to be the spallation of the glazed layer. Spallation occurred in the laser-glazed layer/plasma-sprayed splats interface.

Erosion behavior of thermal sprayed coatings is dependent on the coating process. The observations due to Roy, Narkhede & Paul (Roy, Narkhede & Paul, 1999) indicated that that the erosion resistance of detonation sprayed WC-Co coating is better than plasma sprayed and HVOF sprayed coatings at normal impact and comparable at oblique impact. Erosion resistance of HVAF sprayed Cr_3C_2-NiCr coating is found to be better than HVOF sprayed Cr_3C_2-NiCr coating as large scale splat boundaries contributed to increased erosion rate. The preferential path for crack propagation in HVOF sprayed Cr_3C_2-NiCr coating was coating splat structure although but less significant in the HVAF sprayed Cr_3C_2-NiCr coating. Walsh and Tabakoff (Walsh and Tabakoff, 1992; Walsh, 1992) noted that the plasma coatings generated poorer erosion resistance with 85% carbide than the D-Gun coatings with 65% carbide. An increasing erosion resistance with increasing carbide content for bulk sintered samples and HVOF sprayed coatings, but the opposite trend for plasma sprayed coatings was reported by Wayne and Sampath (Wayne and Sampath, 1992). In each deposition techniques (HVOF, HVAF, D-Gun) variation in erosion response has been noted with the starting powder size. It has been reported that for Cr_3C_2 the erosion resistance of D-Gun sprayed coatings increased with decreasing starting powder size. Interestingly, the opposite was true for HVOF sprayed coatings (Fukuda & Kumon, 1995; Irons et al., 1994). Similar results are reported for tungsten carbide coatings (Sue and Tucker, 1987).

Erosion of nanocomposite thermal sprayed coatings exhibits some interesting features. Nano size WC-Co coatings manufactured by HVOF process offered better erosion resistance as compared to conventional WC-Co coatings which was attributed to decreased mean size of WC grain and improved hardness and toughness. However the problems of increased decarburization due to increased surface area of nano coatings have been reported. Similar mechanisms of material removal e.g. chipping, deformation of the binder phase and carbide pullout have also been observed in erosion of nano WC- Co coatings (Thakur, Arora, Jayagnathan & Sood, 2011). Work due to Dent, Depalo and Sampath. (Dent, Depalo and Sampath 2002) indicates that erosion resistance of nanocomposite coating decreases as compared to conventional coatings.

In this context it is pertinent to mention the material removal mechanisms in case of erosion of sintered WC-Co alloys. This involves plastic deformation of the binder phase followed by accumulation of plastic strain in WC grain, cracking and detachment of unsupported WC grains (Kulu et al., 2005). The subsurface cracking also contributed to erosion damage of sintered WC-Co alloys as observed by Ninham and Levy (Ninham & Levy, 1988). The mechanism of ploughing and microcutting of the binder phase and propagation of subsurface cracks by fatigue mechanism in slurry erosion of detonation gun sprayed WC-Co coating was observed by Wood et al. (Wood et al., 2010). These mechanisms of erosion can be compared with erosion of high chromium cast irons which are although different on microstructural scale. Small scale chipping of carbides at lower impact angle and lower velocities induced by Hertzian cracking and brittle fracture of chromium carbides with subsurface cracking at normal impact angle at relatively higher velocities were predominant mechanisms of material removal from carbides in erosion of high chromium white irons with silica sand erodent particles. With harder erodent particles even at shallow impact angles brittle fracture of carbides was observed. (Sapate & RamaRao, 2004; Sapate & RamaRao, 2006).

5 APPLICATION OF THERMAL SPRAY COATINGS FOR PROTECTION AGAINST EROSIVE WEAR

WC-M thermal spray coatings are used in several industrial applications like aerospace, automotive, power generation industries and in extraction, conveying, sand collection machinery, and piping systems in mining and mineral processing, fluidised beds and turbines for protection against erosive wear. Table 4 gives application of thermal spray coatings where erosion is considered to be important problem. Erosion is very important in aero engine operating in dusty terrain. The vanes of aero engines are subjected to erosive wear over wide range of temperature including room temperature during operation in dusty terrain. Similar problem can be encountered on the engine shaft coated with thermal sprayed WC-Co coating. Erosion of thermal sprayed abradable coatings such as Al-Si-Polyester, Al-Si-Hexagonal BN, Ni-Graphite etc is a serious problem. Attempt has been made to coat helicopter blades with thermal sprayed coatings to reduce erosion. Several parts of land based gas turbines are also coated with thermal sprayed coatings to improve their performances. Figure 12a illustrates damage around the wear ring and perforation into the standard duplex stainless steel impellers used in water injection applications after 7 months service. When super duplex wear ring coated with WC-based thermal spray coating as in Figure 12b is used, the integrity of the component is retained (Souza & Neville, 2005) even after nine and half months of operation under the same conditions.

6 DIRECTION OF FUTURE STUDY

In spite of above discussion, a lot of areas need comprehensive attention. Erosion mechanisms of various coatings should be studied in an extensive way. A database of mechanical properties of various thermal sprayed coatings should be compiled and a correlation between mechanical properties of the coatings and their erosion resistance should be established. Cold spraying is an area which is least explored for solid particle erosion. Further, significant research needs to be carried out to examine the influence of thermal spraying technique on erosive wear. It is, in particular, imperative to understand whether various plasma spraying techniques such as air plasma spraying, vacuum plasma spraying, low pressure plasma spraying etc. has any influence on mechanical properties and erosion rates of coatings. Similar effort should be made for various HVOF processes. The effect of spaying condition of each thermal spraying technique on the erosive wear should also be addressed.

Table 4. Application of Thermal spray coating (GFE Schmalkalden,http://www.fh-schmalkalden.de/schmalkaldenmedia/-p-16916. 389-394)

Industry	Application Areas
Automotive	Pistons, cylinders
Chemical industry	Valves, punps, storage reservoir
Electrotechnical industry	Isolators, conductive layers, capacitor, fuel cells
Consumables	Electric irons, fry pan
Engineering	Compressors, pumps, cylinders,
Medicine	Coated implants
Aerospace industry	Run-in layer, thermal barrier coatings, blades, combustion chamber,
Textile industry	Thread eye, thread brake

A large variety of cermet based coatings reinforced with various particles such as TiC, TiMo (CN) etc. should also be studied extensively. Erosion micromechanism and erosion efficiency (Sundararajan et al., 1990) of various coatings has to be compiled and compared.

7 CONCLUDING REMARKS

The chapter presents solid particle erosion behavior of thermal spray coatings. The similarities and marked differences amongst erosion behavior of monolithic metallic, ceramic, multiphase materials and thermal spray coatings can be observed. The complexity in prediction of erosion behavior increases from metallic, ceramic to multiphase materials manufactured by conventional techniques (casting, powder metallurgy) to thermal spray coatings. This complexity in prediction of erosion behavior of thermal spray coatings arises primarily due to characteristics of different thermal spray coating techniques. The erosion behavior of thermal spray coatings is greatly influenced by resulting microstructure and morphology of phases, phase transformation of reinforcement phases, defects such as porosity and nature of splat boundaries, anisotropy of mechanical properties, in particular fracture toughness variation, properties of interfaces between binder and reinforcement particles. Erosion conditions such as impact velocity and impact angle greatly affect erosion rate. Most thermal sprayed coating exhibit brittle erosion response with velocity exponent in access of 3.0. Erosion rate of thermal sprayed coatings are greatly governed by

Figure 12. (a) Standard duplex stainless steel impeller after 7 months (b) Super duplex stainless steel impeller and eye wear ring. Wear ring coated with HVOF tungsten carbide (50%) coating after 9 and 1/2 months service. (Souza & Neville, 2005)

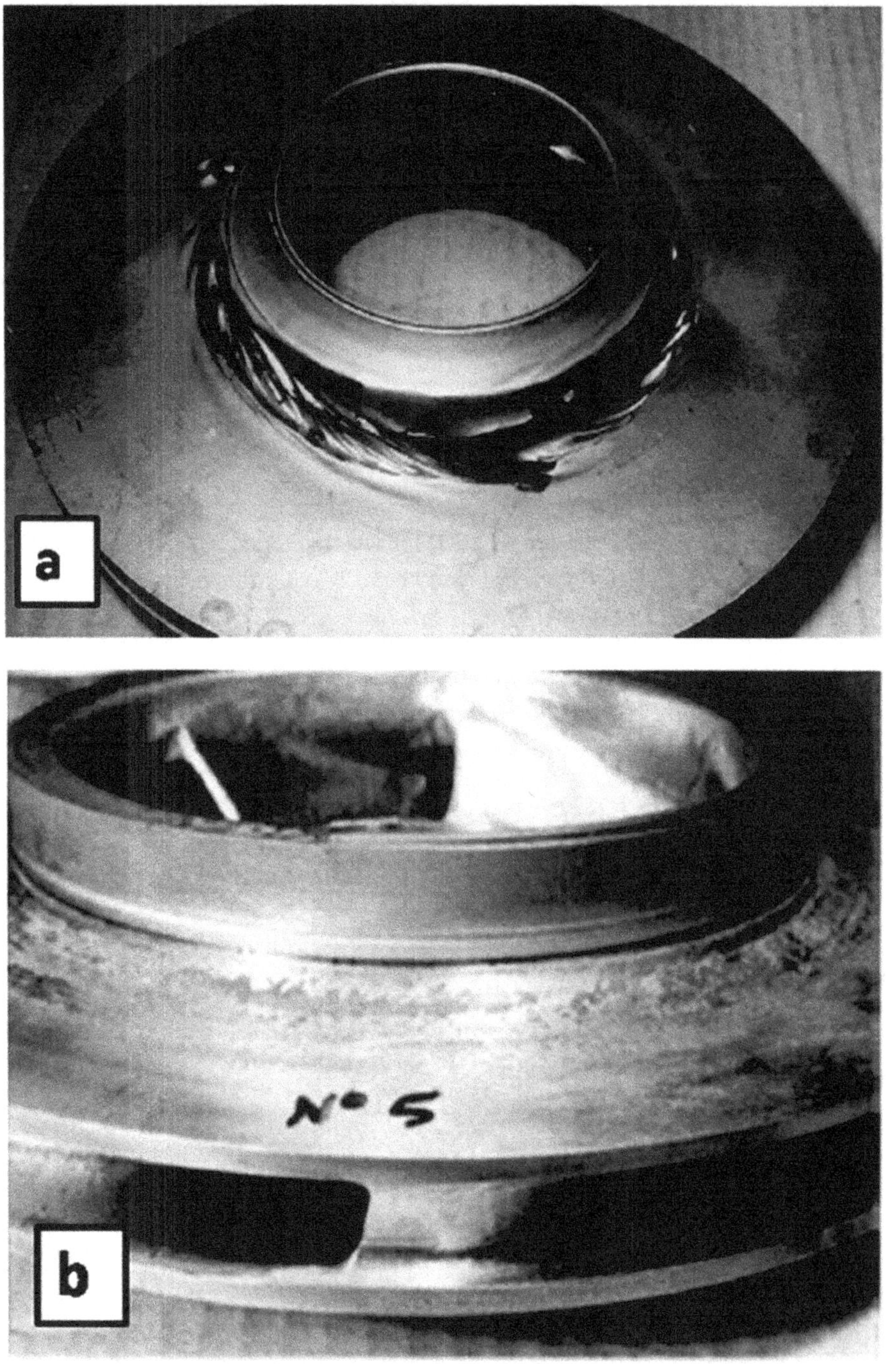

coating technique and post coating treatment. It is to be mentioned that in many respect published results are contradictory. The emergence of cold spray technique has reduced some of the problems associated with phase transformation and it is expected to improve their erosion resistance.

REFERENCES

Anand, K. (1985). Erosion of multiphase materials. In *Proceedings of International Conference on Science of hard Materials*. Rhodes, Greece: Academic Press.

Anand, K., & Conard, H. (1988). Microstructure and scaling effects in the damage of WC-Co alloys by single impacts of hard particles. *Materials Science and Engineering*, *23*, 2931–2942.

Anand, K., & Conard, H. (1989). Microstructural effect sin the erosion of cemented carbides. In *Proceedings of International Conference on Wear of Materials* (pp. 135- 142). Academic Press.

Babu, P., Basu, B., & Sundararajan, G. (2011). The influence of erodent hardness on the erosion behavior of detonation sprayed WC-12Co coatings. *Wear*, *270*(11–12), 903–913. doi:10.1016/j.wear.2011.02.019

Ball, A., & Patterson, A. W. (1985, May). Microstructural design of erosion resistant hard materials. In *Proc. 11th Int. Plausee Seminar on New Applications, Recycling and Technology of Refractory Metals and Hard Metals*. Reutte Tirol, Austria: Academic Press.

Bellman, R. Jr, & Levy, A. V. (1981). Erosion mechanism in ductile metals. *Wear*, *70*(1), 1–27. doi:10.1016/0043-1648(81)90268-4

Bitter, J. G. A. (1963). A study of erosion phenomena: Part I. *Wear*, *6*(1), 5–21. doi:10.1016/0043-1648(63)90003-6

Bitter, J. G. A. (1963). A study of erosion phenomena: Part II. *Wear*, *6*(3), 169–190. doi:10.1016/0043-1648(63)90073-5

Christman, T., & Shewmon, P. G. (1979). Erosion of strong Al alloy. *Wear*, *51*(1), 57–70. doi:10.1016/0043-1648(79)90196-0

Dent, A. H., DePalo, S., & Sampath, J. (2002). Examination of the wear properties of HVOF sprayed nanostructured and conventional WC-Co cermets with different binder phase contents. *Journal of Thermal Spray Technology*, *11*(4), 551–558. doi:10.1361/105996302770348691

Evans, A. G., Gulden, M. E., & Rosenblatt, M. (1978). Impact damage in brittle materials in the elastic-plastic response regime. *Proceedings of the Royal Society of London. Series A, Mathematical and Physical Sciences*, *361*(1706), 343–365. doi:10.1098/rspa.1978.0106

Evans, A. G., & Wilshaw, T. R. (1977). Dynamic solid particle damage in brittle materials: An appraisal. *Journal of Materials Science*, *12*(1), 97–116. doi:10.1007/BF00738475

Finnie, I. (1979). Fundamental mechanism of the erosive wear of ductile metals by solid particles. In W. F. Adler (Ed.), Erosion, prevention, and useful applications, ASTM STP 664 (pp. 36-58). American Society for Testing and Materials. doi:10.1520/STP35794S

Finnie, I. (1995). Some reflections on the past and future of erosion. *Wear*, *186-187*, 1–10. doi:10.1016/0043-1648(95)07188-1

Fukuda, Y., & Kumon, M. (Eds.). (1995). Thermal spraying current status and future trends. High Temperature Society of Japan.

GFE Schmalkalden E. V. (n.d.). *Principles of coating technology*. Retrieved on 27th February 2014, from http://www.fh-schmalkalden.de/schmalkaldenmedia/-p-16916

Guilemany, J. M., Cinca, N., Ferna'ndez, J., & Sampath, S. (2008). Erosion, abrasive, and friction wear behavior of iron aluminide coatings sprayed by HVOF. *Journal of Thermal Spray Technology*, *17*(5-6), 762–773. doi:10.1007/s11666-008-9252-7

Hammitt, F. G. (1976). Erosive wear testing. In R. G. Bayer (Ed.), Selection and use of wear tests for metals, ASTM STP 615 (pp. 45-67). American Society for Testing and Materials. doi:10.1520/STP38612S

Hansen, J. S. (1979). Relative erosion resistance of several materials. In W. F. Adler (Ed.), Erosion: Prevention and useful Applications ASTM STP 664 (pp. 148-162). American Society for Testing and Materials. doi:10.1520/STP35800S

Hawthrone, H. M., Arsenault, B., Immarigeon, J. P., Legoux, J. G., & Parameswaran, V. R. (1999). Comparison of slurry and dry erosion behaviour of some HVOF thermal sprayed coatings. *Wear*, *225-229*, 835–834.

Hearley, A., Little, J. A., & Sturgeon, A. J. (1999). The erosion behaviour of NiAl intermetallic coatings produced by high velocity oxy-fuel thermal spraying. *Wear*, *328*, 233–235.

Hovis, S. K., Anand, K., Conrad, H., & Scattergood, R. O. (1985). A new method of velocity calibration for erosion testing. *Wear*, *101*(1), 69–76. doi:10.1016/0043-1648(85)90211-X

Hutchings, I. M. (1977). Deformation of metal surfaces by the oblique impact of square plates. *International Journal of Mechanical Sciences*, *19*(1), 45–52. doi:10.1016/0020-7403(77)90015-7

Hutchings, I. M. (1979). Mechanism of the erosion of metals by solid particles. In W. F. Adler (Ed.), Erosion: Prevention and useful applications, ASTM STP 664 (pp. 59-76). American Society for Testing and Materials. doi:10.1520/STP35795S

Hutchings, I. M. (1981). A model for the erosion of metals by spherical particles at normal incidence. *Wear*, *70*(3), 269–281. doi:10.1016/0043-1648(81)90347-1

Hutchings, I. M. (1997). Abrasive and erosive wear tests for thin coatings: A unified approach, new directions in tribology. Mechanical Engineering Publications.

Hutchings, I. M., & Winter, R. E. (1975). The erosion of ductile metals by spherical particles, *Journal of Physics (D) Applied Physics (Berlin)*, *8*, 9–14.

Irons, G., Kratochvil, W. R., Bullock, W. R., & Berndt, R. A. (1994). Thermal spray industrial applications, investigation of erosion resistant chromium carbide coatings sprayed with the high pressure HVOF process. ASM International.

Kagimato, Y. (1987). Experimental study on the relation between erosion wear rate and particle impact velocity measured by LDV. In *Proceedings of 7th International Conference on Erosion by Liquid and Solid Impact.* Cambridge, UK: Academic Press.

Karimi, A., Verdon, Ch., Martin, J. L., & Schimd, R. K. (1995). Slurry erosion behavior of thermally sprayed WC-M coatings. *Wear*, *186-187*, 480–486. doi:10.1016/0043-1648(95)07154-7

Kingswell, R., Rickerby, D. S., Bull, S. J., & Scot, K. T. (1991). Erosive wear of thermally sprayed tungsten coatings. *Thin Solid Films*, *198*(1-2), 139–148. doi:10.1016/0040-6090(91)90332-R

Kosel, T. H. (1992). Solid particle erosion, friction, lubrication and wear technology. In ASM handbook (Vol. 18). ASM International.

Kosel, T. H., & Anand, K. (1989). An optoelectronic erodent particle velocitimeter. In Proceedings of Corrosion and Particle Erosion at High Temperatures. The Minerals, Metals & Materials Society.

Kosel, T. H., Mao, Z. Y., & Prasad, S. V. (1984). Erosion debris particle observations and the micromachining mechanism of erosion. *ASLE Transactions*, *28*(2), 268–276. doi:10.1080/05698198508981621

Kulu, P., Hussainova, I., & Veinthal, R. (2005). Solid particle erosion of thermal sprayed coatings. *Wear*, *258*(1-4), 488–496. doi:10.1016/j.wear.2004.03.021

Lathabai, S., Ottmuller, M., & Fernandez, I. (1998). Solid particle erosion behavior of thermal sprayed ceramic, metallic and polymer coatings. *Wear*, *221*(2), 93–108. doi:10.1016/S0043-1648(98)00267-1

Lawn, B., & Wilshaw, R. (1975). Review Indentation fracture: Principles and applications. *Journal ofMaterials Science*, *10*(6), 1049–1081. doi:10.1007/BF00823224

Leibhard, M., & Levy, A. (1991). The effect of erodent particle characteristics on the erosion of steels. *Wear*, *151*(2), 381–390. doi:10.1016/0043-1648(91)90263-T

Levy, A. (1986). The platelet mechanism of erosion of ductile metals. *Wear*, *108*(1), 1–21. doi:10.1016/0043-1648(86)90085-2

Levy, A. (1988). The erosion-corrosion behavior of protective coating. *Surface and Coatings Technology*, *36*(1-2), 387–406. doi:10.1016/0257-8972(88)90168-5

Levy, A., & Wang, B. Q. (1988). Erosion of hard material coating systems. *Wear*, *121*(3), 325–346. doi:10.1016/0043-1648(88)90209-8

Li, C. J., Yang, G. J., & Ohmori, A. (2006). Relationship between particle erosion and lamellar microstructure of plasma sprayed alumina coatings. *Wear*, *260*(11-12), 1166–1172. doi:10.1016/j.wear.2005.07.006

Li, W.-Y., Zhang, C., Guo, X. P., Zhang, G., Liao, H. L., Li, C.-J., & Coddet, C. (2008). Effect of standoff distance on coating deposition characteristics in cold spraying. *Materials & Design*, *29*(2), 297–304. doi:10.1016/j.matdes.2007.02.005

Lofti, B., Shipway, P. H., McCartney, D. G., & Eldris, H. (2003). Abrasive wear behaviour of NiCr-TiB2 coatings deposited by HVOF spraying of SHS-derived cermet powders. *Wear*, *254*(3-4), 340–349. doi:10.1016/S0043-1648(03)00014-0

Roy, M. (2014). Solid particle erosion behaviour of WC coating obtained by electro spark technique and detonation spraying. *Communicated to Tribology Transactions.*

Machio, C. N., Akdogan, G., Witcomb, M. J., & Luyckx, S. (2005). Performance of WC-VC-Co thermal spray coatings in abrasion and slurry erosion tests. *Wear*, *258*(1-4), 434–442. doi:10.1016/j.wear.2004.09.033

Manish, R. (2000). Wear testing: An assessment of test methodology. *Transactions of the Indian Institute of Metals, 62*, 623–638.

Mathews, S. J., James, B. J., & Hyland, M. M. (2007). Microstructural influence on erosion behavior of thermal spray coatings. *Materials Characterization, 58*(1), 59–64. doi:10.1016/j.matchar.2006.03.014

Matthews, S., James, B., & Hyland, M. (2009). The role of microstructure in the mechanism of high velocity erosion of Cr3C2–NiCrthermal spray coatings: Part 2 - Heat treated coatings. *Surface and Coatings Technology, 203*(8), 1094–1100. doi:10.1016/j.surfcoat.2008.10.013

Melendez, N. M., Narulkar, V. V., Fisher, G. A., & McDonald, V. (2013). Effect of reinforcing particles on the wear rate of low-pressure cold-sprayed WC-based MMC coatings. *Wear, 306*(1-2), 185–195. doi:10.1016/j.wear.2013.08.006

Neilson, J. H., & Gilchrist, A. (1968). Erosion by a stream of solid particles. *Wear, 11*(2), 111–122. doi:10.1016/0043-1648(68)90591-7

Nerz, J., Kushner, B., Kaufold, R., & Rotolico, A. (1991). Thermal spray coatings: Properties, processes and applications. In Reduction of solid particle erosion by using HVOF and HEP coating deposition methods. Pittsburgh, PA: ASM International.

Ninham, A. J., & Hutchings, I. M. (1983). A computer model for particle velocity calculation in erosion testing. In *Proceedings of 6th International Conference On Erosion by Liquid and Solid Impact.* Cavendish Laboratory.

Ninham, A. J., & Levy, A. V. (1988). The erosion of carbide metal matrix composites. *Wear, 121*(3), 347–361. doi:10.1016/0043-1648(88)90210-4

Pennefather, R.C. (1988). Recent observations of the erosion of hard materials. *Materials Science and Engineering, A*(105/106).

Qadir, T., & Shewmon, P. (1981). Solid particle erosion mechanism in copper and two copper alloys. *Metallurgical Transactions, 12A*(7), 1163–1176. doi:10.1007/BF02642330

Roy, M., Narkhede, B. E., & Paul, S. N. (Eds.). (1999). *Proceedings of 2nd international conference on industrial tribology*. Hyderabad, India: Academic Press.

Roy, M., Subramaniyam, M., & Sundararajan, G. (1992). Room temperature erosion behavior of a precipitation hardened stainless steel. *Tribology International, 25*(4), 271–280. doi:10.1016/0301-679X(92)90064-T

Roy, M., Tirupataiah, Y., & Sundararajan, G. (1993). Effect of particle shape on the erosion of Cu and its alloys. *Materials Science and Engineering A*, *165*(1), 51–63. doi:10.1016/0921-5093(93)90626-P

Roy, M., Tirupataiah, Y., & Sundararajan, G. (1995). Influence of solid solution and dispersion strengthening mechanism on room temperature erosion behavior of nickel. *Materials Science and Technology*, *11*(8), 791–797. doi:10.1179/mst.1995.11.8.791

Roy, M., Vishwanathan, B., & Sundararajan, G. (1994). The solid particle erosion of polymer matrix composites. *Wear*, *171*(1-2), 149–161. doi:10.1016/0043-1648(94)90358-1

Ruff, A. W., & Ives, L. K. (1975). Measurement of solid particle velocity in erosive wear. *WEA*, *35*(1), 195–199. doi:10.1016/0043-1648(75)90154-4

Ruff, A. W., & Wiederhorn, S. M. (1979). Erosion by solid particle impact. In C. M. Preece (Ed.), *Treatise on material science and technology* (Vol. 16, pp. 69–123). Academic Press.

Sapate, S. G., & Rama Rao, A. V. (2004). Effect of carbide volume fraction on erosive wear behaviour of hardfacing cast irons. *Wear*, *256*(7-8), 774–786. doi:10.1016/S0043-1648(03)00527-1

Sapate, S. G., & RamaRao, A. V. (2006). Erosive wear behaviour of weld hardfacing high chromium cast irons: Effect of erodent particles. *Tribology International*, *39*(3), 206–212. doi:10.1016/j.triboint.2004.10.013

Sarkar, S., Sekharan, V., Mitra, R., & Roy, M. (2009). The solid particle erosion behavior of carbon/carbon and carbon/phenolic composite used in re-entry vehicles. *Tribology Transactions*, *52*(6), 777–787. doi:10.1080/10402000903097411

Schimdt, G., & Steinhauser, S. (1996). Characterzation of wear protective coatings. *Tribology International*, *29*(3), 207–214. doi:10.1016/0301-679X(95)00090-Q

Sheldon, G. L., & Finnie, I. (1966). The mechanism of material removal in the erosive cutting of brittle materials. *Journal of Engineering for Industry*, *88*(4), 393–401. doi:10.1115/1.3672667

Souza, V. A. D., & Neville, A. (2005). Corrosion and synergy in a WC-Co-Cr HVOF thermal spray coating-understanding their role in erosion–corrosion degradation. *Wear*, *259*(1-6), 171–180. doi:10.1016/j.wear.2004.12.003

Standard Test Method for Conducting Erosion Tests by Solid Particles Impingement Using Gas Jets, G 76-95. (1995). *Annual book of ASTM standards* (vol. 3.02). Wear and Erosion, Metal Corrosion ASTM.

Stevenson, A. N. J., & Hutchings, I. M. (1995). Scaling laws for particle velocity in the gas blast erosion test. *Wear, 181-183*, 56–62. doi:10.1016/0043-1648(95)90008-X

Sue, J. A., & Tucker, R. C. Jr. (1987). High temperature erosion behavior tungsten- and chromium-carbide-based coatings. *Surface and Coatings Technology*, *32*(1-4), 237–248. doi:10.1016/0257-8972(87)90110-1

Sundararajan, G., & Roy, M. (1997). Solid particle erosion behavior of metallic materials at room and elevated temperatures. *Tribology International*, *30*(5), 339–359. doi:10.1016/S0301-679X(96)00064-3

Sundararajan, G., Roy, M., & Venkataraman, B. (1990). Erosion efficiency-A new parameter to characterize the dominant erosion micromechanism. *Wear*, *140*(2), 369–381. doi:10.1016/0043-1648(90)90096-S

Sundararajan, G., & Shewmon, P. G. (1983). A new model for the erosion of metals a normal incidence. *Wear*, *84*(2), 237–258. doi:10.1016/0043-1648(83)90266-1

Thakur, L., Arora, N., Jayaganthan, R., & Sood, R. (2011). R. An investigation on erosion behavior of HVOF sprayed WC–CoCr coatings. *Applied Surface Science*, *258*(3), 1225–1234. doi:10.1016/j.apsusc.2011.09.079

Tilly, G. P. (1969). Erosion caused by airborne particles. *Wear*, *14*(1), 63–79. doi:10.1016/0043-1648(69)90035-0

Tilly, G. P. (1973). A two stage mechanism of ductile erosion. *Wear*, *23*(1), 87–96. doi:10.1016/0043-1648(73)90044-6

Tsai, P. C., Lee, J. H., & Chang, C. L. (2007). Improving erosion resistance of plasma sprayed zirconia thermal barrier coatings by laser glazing. *Surface and Coatings Technology*, *201*(4-7), 719–724. doi:10.1016/j.surfcoat.2007.07.005

Venugopal Reddy, A., & Sundarajan, G. (1986). Erosion behavior of ductile materials with a spherical non-friable erodent. *Wear*, *111*(3), 313–323. doi:10.1016/0043-1648(86)90190-0

Venugopal Reddy, A., & Sundararajan, G. (1986). Selection of steels for erosion resistant applications. *Tool and Alloy Steels*, *20*(12), 379–388.

Verdon, C., Karimi, A., & Martin, J. L. (1997). Microstructural and analytical study of themally sprayed WC-C0 coatings in connection with their wear resistance. *Materials Science and Engineering A*, *234-236*, 731–734. doi:10.1016/S0921-5093(97)00377-8

Walsh, P. (1992). Erosion resistant coatings at steam turbine temperatures. *PWR-Steam Turbine Generator Developments for the Power Generation Industry, 18*, 123–128.

Walsh, P., & Tabakoff, W. (1990). Comparative erosion resistance of coatings intended for steam turbine components. In *Proceedings of International Joint Power Generation Conference*. Boston: Academic Press.

Wang, B. (1996). The dependence of erosion-corrosion wastage on carbide/metal binder proportion for HVOF carbide-metal cerment coatings. *Wear, 196*(1-2), 141–146. doi:10.1016/0043-1648(95)06883-X

Wang, B. Q., Geng, G. Q., & Levy, A. V. (1990). Article. *Erosion-Corrosion of Thermal Spray Coatings Surface and Coating Technology, 43-44*, 859–874.

Wang, B. Q., & Lee, S. W. (1997). Elevated temperature erosion of several thermal-sprayed coatings under the simulated erosion conditions of in-bed tubes in a fluidized bed combustor. *Wear, 203–204*, 580–587. doi:10.1016/S0043-1648(96)07381-4

Wayne, S. F., Baldoni, J. G., & Buljan, S. T. (1990). Abrasion and erosion of WC-Co with controlled microstructures. *Tribology Transactions, 33*(4), 611–617. doi:10.1080/10402009008981996

Wayne, S. F., & Sampath, S. (1992). Structure/property relationships in sintered and thermally sprayed WC-Co. *Journal of Thermal Spray Technology, 1*(4), 307–315. doi:10.1007/BF02647158

Westergard, R., Axen, N., Wiklund, U., & Hogmark, S. (2000). An evaluation of plasma sprayed ceramic coatings by erosion, abrasion and bend testing. *Wear, 246*(1-2), 12–19. doi:10.1016/S0043-1648(00)00506-8

Winter, R. E., & Hutchings, I. M. (1974). Solid particle erosion studies using single angular particles. *Wear, 29*(2), 181–194. doi:10.1016/0043-1648(74)90069-6

Wlodek, S. T. (1985). Development and testing plasma sprayed coatings for solid particle erosion resistance. In *Proceedings of SPE of Utility Steam Turbines*. EPRI CS4683.

Wood, R. J. K. (2010). Tribology of thermal sprayed WC–Co coatings. *International Journal of Refractory Metals & Hard Materials, 28*(1), 82–94. doi:10.1016/j.ijrmhm.2009.07.011

Wood, R. J. K., Mellor, B. G., & Binfield, M. L. (1997). Sand erosion performance of detonation gun applied tungsten carbide/cobalt-chromium coatings. *Wear*, *211*(1), 70–83. doi:10.1016/S0043-1648(97)00071-9

Xia, Z., Zhang, X., & Song, J. (1999). Erosion resistance of plasma sprayed coatings. *Journal of Materials Engineering and Performance*, *8*(6), 716–718. doi:10.1361/105994999770346495

Zum Gahr, H. (1987). Microstructure and wear of materials. Elsevier.

KEY TERMS AND DEFINITIONS

Ceramic: An inorganic, non-metallic material produced by the application of heating and cooling under the action of pressure.

Coating: A layer of a substance spread over a surface for protection against degradation by erosion or corrosion.

Cracking: A mechanism of fracture of material occurring on microscopic or macroscopic level.

Erosion: A process of material removal from the surface by repeated impingement of solid particles carried along with a fluid medium.

Fracture: Separation of a surface under the action of heat, load or presence of chemical species.

Metal: An element prepared from its oxide or found in its native form having good heat and electrically conducting property.

Microstructure: An image showing structure / phases present in an engineering material.

Multiphase: Presence of more than one, usually two or more in a microstructure on microscopic or macroscopic level.

Thermal Spray: A surface coating technique in which melted particles of metallic and or ceramic materials are deposited on to a surface.

Chapter 8
Tribology of Thermally Sprayed Coatings in the Al_2O_3–Cr_2O_3–TiO_2 System

Lutz-Michael Berger
Fraunhofer Institute IWS, Germany

ABSTRACT

With the exception of ZrO_2, the individual oxides and binary compositions in the system Al_2O_3-Cr_2O_3-TiO_2 are the most important oxide materials for thermally sprayed coating solutions. Traditionally, these coatings are prepared by Atmospheric Plasma Spraying (APS), but processes such as Detonation Gun Spraying (DGS) and High Velocity Oxy-Fuel (HVOF) spraying can produce coatings with lower porosity and higher wear resistance. Traditionally, feedstock powders have been used for coating preparation. Recent developments have seen the emergence of suspensions as a new feedstock, but tribological properties of coatings prepared using suspensions have not yet been studied in detail. This chapter summarizes some important issues regarding wear protection applications of coatings in the Al_2O_3-Cr_2O_3-TiO_2 system, the advantage of alloying the individual oxides, and the influence of different feedstocks and spray processes.

INTRODUCTION

Thermal spray processes represent an important and rapidly growing group of surface modification technologies, which use a very wide range of solid feedstock materials, including metals and alloys, hardmetals, ceramics and polymers. Ceramics have an outstanding role as a group of materials processed by thermal spraying into coatings

DOI: 10.4018/978-1-4666-7489-9.ch008

for a broad variety of applications. Many ceramic coatings show multifunctional properties, they can serve as wear resistant coatings, but also as electrically isolating and conductive coatings as well as thermal barriers. A general overview of thermal spray technologies can be found in several textbooks (Pawlowski, 2008; Mathesius & Krömmer, 2009; Tucker, Jr. (ed.), 2013; Fauchais, Heberlein, & Boulos, 2014) and book chapters (Vuoristo, 2014).

The process of ceramic coating preparation by thermal spraying is illustrated in Figure 1 (Toma, Berger, Langner, & Naumann, 2010). This figure shows that the coating properties are dependent upon both from the properties of the feedstock and the characteristics of the spray process. Conventionally, ceramic coatings are prepared from feedstock powders, commonly in the size range 10–45 µm (see Figure 1, upper part), or with adapted particle size depending on the characteristics of the spray process. Thermal spraying with conventional feedstock powders can be considered as a two-step shaping technology for the preparation of ceramic materials: first step is the preparation of a processable feedstock powder, the second step is the spray process (Berger, 2014). The use of suspensions as the feedstock (see Figure 1, lower part) is a recent trend in coating preparation by thermal spraying, e.g. (Bolelli, Cannillo *et al.*, 2009; Bolelli, Rauch *et al.*, 2009; Toma, Berger, Langner, & Naumann, 2010; Berger, Toma, & Potthoff, 2013). It allows the direct use of finely dispersed ceramic powders with a particle size of a few µm as applied in ceramic technologies for sintered components. Thus, coating manufacturing is transformed into a one-step shaping technology without the additional production step of feedstock powder preparation.

For ceramic materials, full or partial melting of the feedstock material (conventional feedstock powder or finely dispersed powder when using suspensions) occurs during the thermal spray process, while the substrate remains unmelted, i.e. the coating is primarily mechanically bonded to the substrate. The coating thickness typically lies within the range 100-500 µm. The process of coating formation is characterised by high cooling rates, which also, in the case of ceramic coatings, leads to the existence of high-temperature and non-equilibrium phases as well as nanocrystalline and amorphous structures. Thus, heat treatment at high temperatures (including high temperature applications) leads to changes in the microstructure, and, possibly, phase composition. In the case of substoichiometry, a heat treatment in air will also lead to oxidation.

Both bulk ceramics and hardmetals are traditionally manufactured by the principles of powder metallurgy (sintering). At the same time they are used as thermally sprayed coatings. The production of hardmetal (Berger, 2014) and ceramic parts by sintering is limited in size by technical and economical reasons. Thus for larger

Figure 1. Illustration of the preparation of ceramic coatings by thermal spraying (HVOF process with Al_2O_3 feedstocks), using conventional feedstock powders (top) and suspensions (bottom) as feedstock materials. Adapted from (Toma, Berger, Langner, & Naumann, 2010)

components thermal spray can be considered as a technology which allows the functionality of these materials to be realized by applying them as a coating on the surface of the component.

An important difference between thermally sprayed coatings and conventionally manufactured sintered materials, including ceramics, is that thermal spray coatings exhibit an anisotropic behavior and a characteristic splat structure (Ang & Berndt, 2014). Another difference is that the properties of the coatings are less exactly defined than for sintered parts. The range of values for one specific property of a coating is significantly larger, as schematically illustrated in Figure 2. This results from the use of different feedstock materials, which are applied with different spray processes and sprayed with different parameters. In addition different test procedures and conditions contribute to a large scattering of values for the mechanical and other coating properties. The yellow part in the highest area in Figure 2 demonstrates that the properties of some of the coatings of a given composition will fulfill the requirements of service in a certain application, while the remaining nominally identical coatings are not suitable. Developments in thermal spray technology and feedstock materials have already narrowed the coating property ranges of individual compositions and this tendency is expected to progress further (Berger, 2014).

An analysis of the oxide materials used for thermal spray coating applications showed that, with the exception of materials based on zirconia, a significant share is prepared from the individual oxides Al_2O_3, Cr_2O_3 and TiO_2, and their binary compounds or solid solutions (Berger, Stahr, Toma, Stehr, & Beyer, 2007; Berger & Stahr, 2008). It is a common goal when these materials are applied as wear re-

Figure 2. Factors influencing the property range of a thermal spray coating for a certain nominal composition (Berger, 2014). The yellow part in the highest area demonstrates that some of the coatings of this composition will fulfill the requirements of service in a certain application, while the remaining coatings are not suitable. (© 2014, Elsevier. used with permission.)

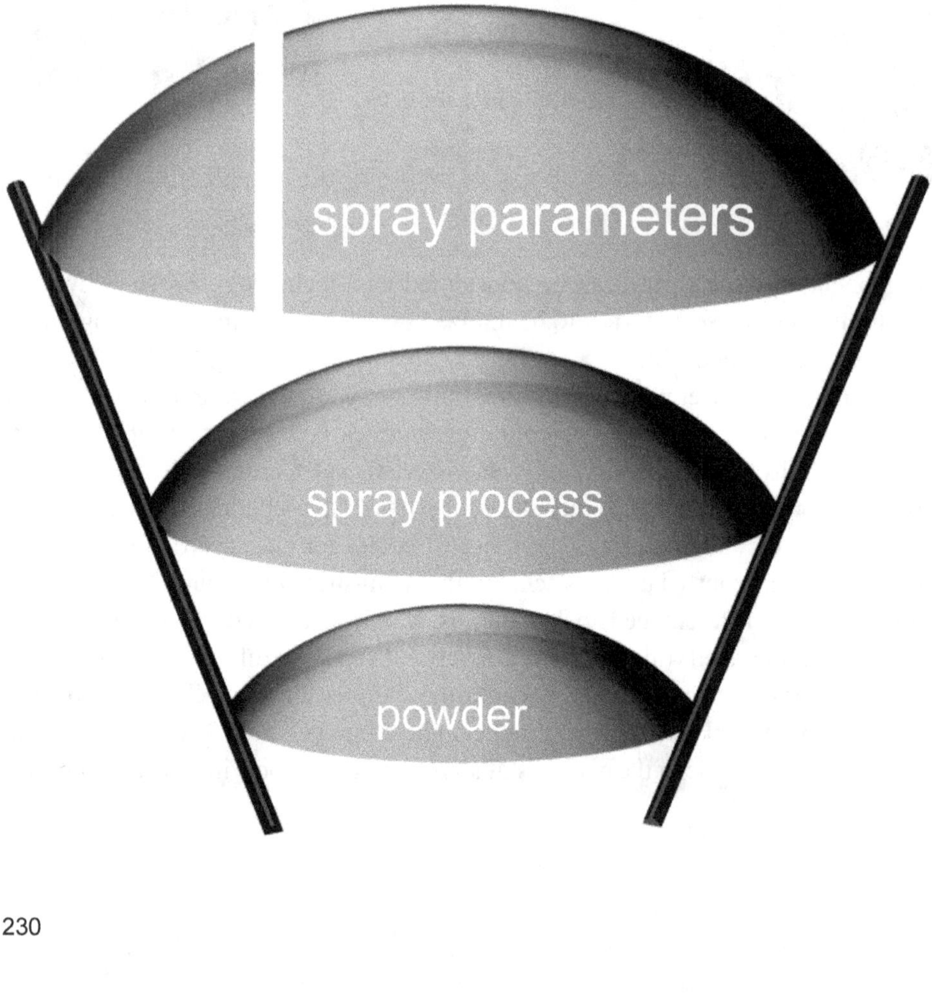

sistant coatings they have to be manufactured with the minimum of defects. Figure 3 shows a schematic presentation of the important features of the Al_2O_3-Cr_2O_3-TiO_2 system with relevance to coating applications for wear protection. A detailed review of studies focusing on the coating microstructures as well as electrical and corrosion properties is given elsewhere (Berger, Toma, Scheitz, Trache, & Börner, 2014).

Each of the individual oxides in the Al_2O_3-Cr_2O_3-TiO_2 system shows significant disadvantages in thermal spray processing. In the case of Al_2O_3 the occurrence of phase transformations and formation of metastable phases is detrimental. Cr_2O_3 is characterized by a high volatility in the spray process, and subsequently a low deposition efficiency. TiO_2 shows an easy loss of oxygen, or substoichiometric titania can easily change its oxygen content. Some of these disadvantages can be at least partially eliminated by use of suspension feedstocks or conventional feedstock powders consisting of solid solutions or compounds of the corresponding binary systems.

Figure 3. Schematic presentation of the Al_2O_3-Cr_2O_3-TiO_2 system with relevance to coatings application for wear resistance

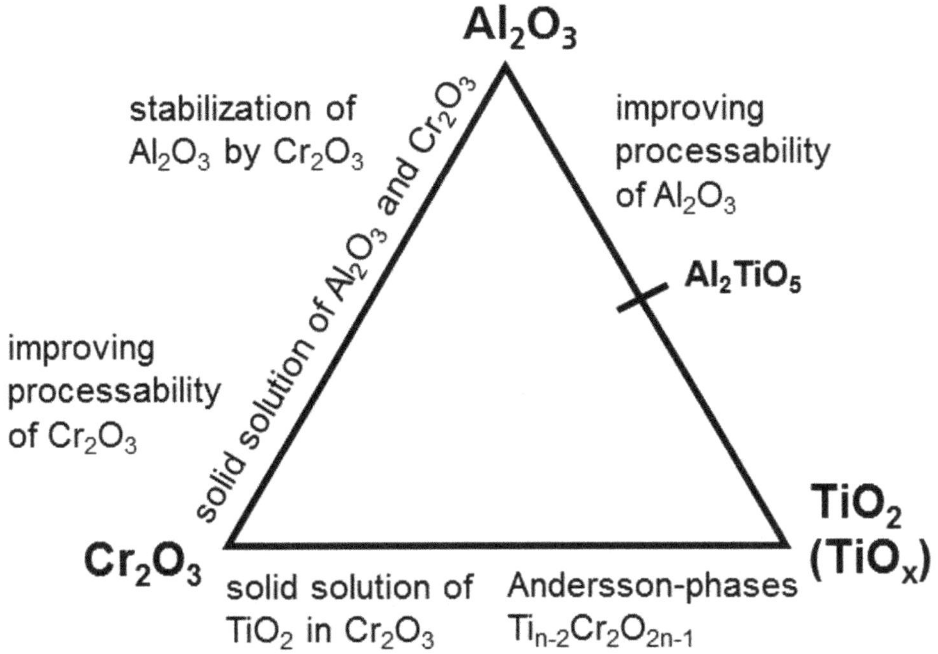

THERMAL SPRAY PROCESSES FOR PREPARATION OF CERAMIC COATINGS OF THE Al_2O_3-Cr_2O_3-TiO_2 SYSTEM

New coating solutions from the Al_2O_3-Cr_2O_3-TiO_2 system are based both on new technological and materials developments. In the past, the improvement of coating properties was mainly achieved by progress in the thermal spray process itself. Comparatively little progress has been made by improvement of the feedstock materials. The main thermal spray processes are presented in Figure 4 schematically as a function of particle velocity and temperature. In addition to these two technical parameters, which have a crucial influence on the coating properties, the economic parameters spray rate and deposition efficiency must also to be considered in the selection of the spray process in order to achieve efficient coating solutions. Atmospheric plasma spraying (APS) is traditionally the most common spray process for preparation of coatings of the Al_2O_3-Cr_2O_3-TiO_2 system due to the high melting temperatures. Besides conventional plasma spray processes, recently developed new processes with four electrodes (three cathodes or three anodes) have became more popular. They enable the use of higher powder feed rates and result in higher deposition efficiencies, and thus the economic efficiency of coating manufacturing is raised, e.g. (Müller & Kreye, 2001). Another variation of plasma gun devices are spray guns with axial feed, also reported to be applied for preparation of coatings of the Al_2O_3-Cr_2O_3-TiO_2 system, e.g. (Westergård, Axén, Wiklund, & Hogmark, 1998; Westergård, Erickson, Axén, Hawthorne, & Hogmark, 2000). Still another alternative process for oxide feedstock materials is the water stabilized plasma spraying (WSP), e.g. (Chráska, Dubsky, Neufuss, & Písacka, 1997; Ilavsky, Berndt, Herman, Chraska, & Dubsky, 1997; Stahr *et al.*, 2007; Ageorges & Ctibor, 2008). In addition to a high productivity, this process also shows a unique behavior to the materials, e.g. enabling the stabilization of alumina by mechanically blended chromia (Stahr *et al.*, 2007), but the excess of energy also creates high stress in the material.

Detonation gun spraying (DGS), which is based on the combustion of hydrocarbon fuel, was the first high velocity process employed for preparation of coatings of the Al_2O_3-Cr_2O_3-TiO_2 system, e.g. (Niemi, Vuoristo, & Mäntylä, 1991; Vuoristo, Niemi, & Mäntylä, 1992; Niemi *et al.*, 1995; Sundararajan, Sen, & Sivakumar, 2005; Niemi, 2009). Due to the wide use of acetylene as the fuel gas, this process supplies enough energy to the oxide materials to melt them in order to form a coating. However, this technology has a rather limited distribution as a spray process. Since the invention of high velocity oxy-fuel (HVOF) spraying in the 1980s it has become a very popular spray process in the industry, due to the significantly lower process temperatures and higher particle velocities generated. Three generations of HVOF spray gun devices can be distinguished (Gärtner, Kreye, & Richter, 2006; Krömmer, Heinrich, & Kreye, 2009). The whole development of this spray process

Figure 4. Schematic presentation of the combination of process temperature and velocity for the different spray processes. (FS - flame spraying, ARC - arc spraying, APS/VPS - atmospheric/vacuum plasma spraying, DGS - detonation gun spraying, HVOF/HVAF - high velocity oxy-fuel / air-fuel spraying) (Berger, 2014). (© 2014, Elsevier. used with permission.)

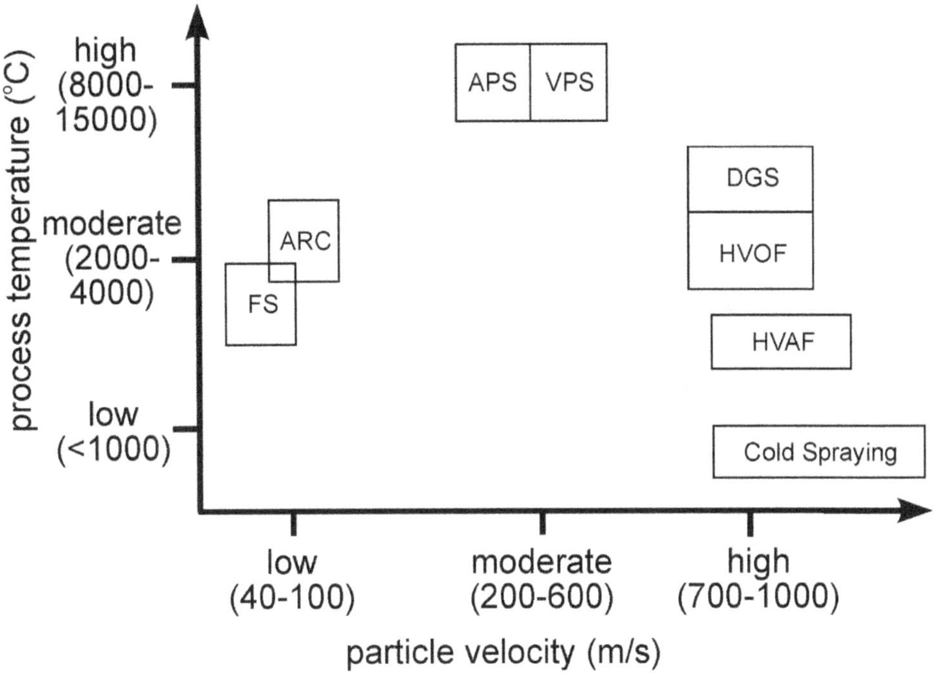

was directed to a decrease of the thermal load of the particles. Thus it is currently predominantly applied for the manufacturing of metallic and hardmetal coatings. However, second generation HVOF spray guns, which generate high process temperatures, are also suitable for the deposition of oxide coatings. Thus, many efforts have been made to show the capabilities of HVOF spraying of oxide materials, e.g. (Müller & Kreye, 2001; Lima & Marple, 2006; Turunen *et al.*, 2006; Berger, Stahr *et al.*, 2007; Niemi, 2009; Berger, *et al.*, 2014). In general, oxide coatings prepared by HVOF show denser microstructures, lower porosity, higher bond strength, and lower roughness, e.g. (Berger, Stahr *et al.*, 2007; Niemi, 2009). The main concern regarding the application of HVOF for the manufacturing of oxide coatings, is due to the significantly lower powder feeding rates, e.g. (Müller & Kreye, 2001). Normally, the deposition efficiencies are also lower.

Occasionally the use of flame spraying has been reported for preparation of coatings of the Al_2O_3-Cr_2O_3-TiO_2 system, e.g. (Habib *et al.,* 2006). There exist also special flame spray processes using ceramic rods (Lathabai, Ottmüller, & Fernandez, 1998; Tucker, Jr. (ed.), 2013) or ceramic-filled flexible cords (Vargas, Ageorges, Fournier, Fauchais, et al., 2014). It should be mentioned that in industrial practice there existed and exist many individual technical solutions regarding the spray equipment.

BASIC PROPERTIES OF THE INDIVIDUAL OXIDES AND THE BINARY SYSTEMS OF THE Al_2O_3-Cr_2O_3-TiO_2 SYSTEM

Al_2O_3 and Al_2O_3-Rich Compositions

Sintered alumina materials consist of the only thermodynamically stable modification of alumina: α-Al_2O_3 (corundum). The typical properties of sintered alumina (high melting point of 2050 °C, good mechanical properties, high chemical stability, and high electrical resistance up to high temperatures) are also associated with the properties of alumina coatings, e.g. (Stahr *et al.*, 2007; Berger & Stahr, 2008). However, sintered alumina and alumina coatings differ in the modifications of Al_2O_3 found in them. Regardless of the spray process used for their preparation, the coatings predominantly consist of other thermodynamically unstable modifications, mostly γ-Al_2O_3. This atypical behavior of Al_2O_3 has been known to exist for a long time and was studied in detail previously (McPherson, 1973; McPherson, 1980).

Thermally sprayed coatings are currently produced with this phase transformation being either ignored or accepted. Despite this transformation, coating properties are still acceptable for many applications, but problems with the long-term stability of the coating properties are often discussed. High-temperature treatments, e.g. (Lima & Bergmann, 1996), do not present an acceptable solution for many coated substrates. Thus, there is a strong need for the deposition of thermally sprayed coatings consisting of α-Al_2O_3 or containing α-Al_2O_3 as the main phase. By the selection of an appropriate grade of fine alumina powder, and the use of aqueous suspensions as feedstock, coatings containing α-Al_2O_3 as the main phase can be prepared without additional alloying elements (Toma, Berger, Stahr, Naumann, & Langner, 2008).

The possibility of stabilization of the α-Al_2O_3 by addition of isostructural Cr_2O_3 (eskolaite) is based on the formation of solid solutions (Chráska *et al.*, 1997; Stahr *et al.*, 2007). At high temperatures full dissolution of both oxides in each other occurs. Below 1250 °C a miscibility gap exists (Stahr *et al.*, 2007).

Additions in the range of 3-40 mass % TiO_2 (commonly 3, 13, and 40 mass %) are used in commercially available standard materials. All these three commercial standard compositions are located in the Al_2O_3-Al_2TiO_5 partial system. The addition of TiO_2 to Al_2O_3 leads to the formation of the liquid phase at the eutectic temperature of 1840 °C. With increasing TiO_2-content, the amount of liquid phase formed at 1840 °C increases, resulting in an improved sprayability of the feedstock powders. At a mole ratio of 1:1 (44 mass % TiO_2) in the Al_2O_3-TiO_2 system the compound Al_2TiO_5 (aluminium titanate) is formed (see Figure 3).

As Al_2TiO_5 is stable only in a certain temperature range, the conditions of feedstock powder preparation and in the thermal spray process will determine the appearance of this phase. In particular, the compound Al_2TiO_5 has very specific properties that are different from Al_2O_3 and TiO_2 properties. This relates also to its very low Young's modulus (13 or 37 GPa depending on sintering temperature) (Thomas & Stevens, 1989), which can be assumed to have a significant influence on the tribological properties. Under reducing conditions, there is also a tendency for Al_2TiO_5 to lose oxygen. The resulting shift of the peaks in the x-ray diffraction pattern, is caused by the formation and dissolution of Ti_3O_5 in Al_2TiO_5, with the resulting phase can than described by the formula $Al_{2-x}Ti_{1+x}O_5$ (Herrmann, Toma, Berger, Kaiser, & Stahr, 2014) It should be mentioned that quite recently other binary oxides, such as $Al_6Ti_2O_{13}$ and $Al_6Ti_7O_{15}$ have also been discovered (Herrmann *et al.*, 2014).

Cr_2O_3 and Cr_2O_3-Rich Compositions

The high temperature behaviour of this oxide is dependent upon the environment, and, therefore, contradictory information exists about the melting point of this oxide. However, it is above 2300 °C, making it the highest of the individual oxides in the Al_2O_3-Cr_2O_3-TiO_2 system. A possible small substoichiometry of Cr_2O_3 leads to a change of color from green to black (Eschnauer, 1980). In oxidizing atmospheres, due to the oxidation from Cr^{3+} to Cr^{6+}, the gaseous CrO_3 is formed which causes a high volatility. An immediate reconversion occurs when cooling down. Additions of Al_2O_3 to Cr_2O_3 and the formation of a solid solution on the chromia-rich side of the Al_2O_3-Cr_2O_3 binary system can decrease the material loss by evaporation and the formation of hexavalent chromium (Yu & Wallar, 2002). The addition of TiO_2 to Cr_2O_3 (eskolaite) has a similar effect, as up to a concentration of 50 mass % TiO_2 (depending on the temperature) a solid solution with the crystallographic structure of eskolaite is formed (Werner, 1974; Sōmiya, Hirano, & Kamiya, 1978). A commercially available feedstock powder exists with the composition Cr_2O_3 - 25 mass % TiO_2.

TiO_2 and TiO_2-Rich Compositions

There exist several modifications of TiO_2, the most important of which are rutile and anatase. At higher temperatures the latter transforms irreversibly into rutile.

This phase transformation is mostly discussed in connection with photocatalytic applications of these coatings, e.g. (Toma, Berger, Stahr, Naumann, & Langner, 2010). Under reducing conditions titanium dioxide (TiO_2) loses oxygen easily. Furthermore, substoichiometric titania can easily change its oxygen content. For this reason titanium oxide coatings are better described by the formula TiO_x rather than by TiO_2. TiO_2 has a melting point of 1857 °C. According to the Ti-O phase diagram there is a eutectic at 1679 °C in the suboxide composition range, this supports their sprayability. A characteristic feature of TiO_x (titanium suboxide) is the ability of the defects in the oxygen sublattice to form crystallographic shear planes. These ordered defect structures are called Magnéli-phases and are described by the formula Ti_nO_{2n-1} (Berger, 2004).

So-called Andersson-phases with the general formula $Ti_{n-2}Cr_2O_{2n-1}$ appear on the TiO_2-rich side of the Cr_2O_3-TiO_2 phase diagram (Werner, 1974; Sōmiya *et al.*, 1978). These are isostructural with the Magnéli-phases of TiO_x. They are of special interest as they can be synthesized from the oxides in air, and are partially stable in air at high temperatures. While the phases $Ti_{n-2}Cr_2O_{2n-1}$ with n = 6-8 have no homogeneity range and are only stable in a certain temperature range, the compound $Ti_2Cr_2O_7$ (phase E) has a wide range of homogeneity over a large temperature range (Werner, 1974; Sōmiya *et al.*, 1978). However, the high-temperature range of the Cr_2O_3-TiO_2 phase diagram is poorly investigated (Berger, Stahr, Saaro, Thiele, & Woydt, 2009a; Berger *et al.*, 2009b; Berger, Woydt, Saaro, Stahr, & Thiele, 2010); for this reason the behavior of these phases during the spray process is particularly difficult to describe.

Additions of Al_2O_3 to TiO_2 have not been studied yet.

FEEDSTOCK MATERIALS

As thermally sprayed coatings of the Al_2O_3-Cr_2O_3-TiO_2 system are prepared from solid feedstock materials, their characteristics will have a significant influence on the resulting coating properties. The commercially available binary compositions in the Al_2O_3-Cr_2O_3-TiO_2 system have not been changed since the beginning of their use. As a rule, the coatings are prepared from feedstock powders with a particle size distribution adapted to the corresponding spray process.

Various technologies existing for feedstock powder preparation result in different particle morphologies and purities. The technology "fusing and crushing" is predominatly used for the production of oxide powders. An SEM micrograph of an Al_2O_3 powder showing the typical morphology of feedstocks prepared by this technology is included in Figure 1. One advantage of this technology is the flexibility of preparation of different binary compositions, as jointly fused feedstock powders, e.g. in the case of Al_2O_3 and TiO_2 (Beczkowiak, Keller, & Schwier, 1996). However, fused and crushed oxide feedstock powders are also often sprayed as mechanical blends, e.g. Al_2O_3 and TiO_2, e.g. (Beczkowiak *et al.*, 1996; Normand, Fervel, Coddet, & Nikitine, 2000; Ananthapadmanabhan *et al.*, 2003), or Al_2O_3 and Cr_2O_3, e.g. (Chráska *et al.*, 1997; Stahr *et al.*, 2007; Dubsky, Chraska, Kolman, Stahr, & Berger, 2011; Yang, Feng, Zhou, & Tao, 2012). One of the disadvantages of mechanical blends is that only one of the components can be sprayed with optimum parameters, and coatings made from these powders will always be heterogeneous. The formation of solid solutions or compounds from mechanically blended powders during the spray process is rather unlikely, e.g. (Normand *et al.*, 2000), since these processes are based on diffusion. For preparation of coatings with homogeneous microstructures of binary compositions, prealloyed feedstock powders have to be used. Also, the reducing action of the graphite electrodes in the powder production process can lead easily to an oxygen deficiency, which is particularly important in the case of titania and titania-rich compositions.

Due to the specific properties of chromia these feedstock powders are often prepared by "sintering and crushing", see Figure 5. For advanced coating applications, oxide feedstock powders can also be prepared by "agglomeration and sintering", e.g. (Normand *et al.*, 2000). Spray drying is normally used as the agglomeration process. Spray drying of nanoparticles with subsequent mild consolidation (sintering) is a way to produce nanostructured coatings from nanostructured feedstocks using conventional thermal spray equipment, e.g. (Luo, Goberman, Shaw, & Gell, 2003; Sánchez *et al.*, 2010; Fauchais, Montavon, Lima, & Marple, 2011). For illustration, Figure 6 shows the morphology of a spherical feedstock powder particle consisting of nanostructured primary particles. Such feedstock powders are also characterized by high chemical purity.

Ceramic rods (Lathabai *et al.*, 1998; Tucker, Jr. (ed.), 2013; Fauchais *et al.*, 2014) or flexible ceramic cords (plastic tubes filled with ceramic particles) (Vargas *et al.*, 2010; Fauchais *et al.*, 2014) are other forms of ceramic feedstocks but are only used for special flame spray processes.

In recent years, the number of studies and developments for the use of suspensions as feedstocks has increased enormously, e.g. (Bolelli, Cannillo *et al.*, 2009; Bolelli, Rauch *et al.*, 2009; Toma, Berger, Langner, & Naumann, 2010; Toma, Berger, Stahr *et al.*, 2010; Fauchais *et al.*, 2011; Berger, *et al.*, 2013). This is a way of directly processing nanosized powders into nanostructured coatings, e.g. (Fauchais *et al.*, 2011). However, it appears as still more important that conventional ceramic oxide powders, which are traditionally used in the preparation of sintered technical ceramics, can be directly used in thermal spray technologies, such as APS and HVOF (Berger, *et al.*, 2013). Water and different alcohols are predominatly used for preparing suspensions. It is recommended to abbreviate the processes as "APS-S"

Figure 5. SEM micrograph of a sintered and crushed chromia feedstock powder

Figure 6. SEM micrograph of an agglomerated and sintered titania feedstock powder

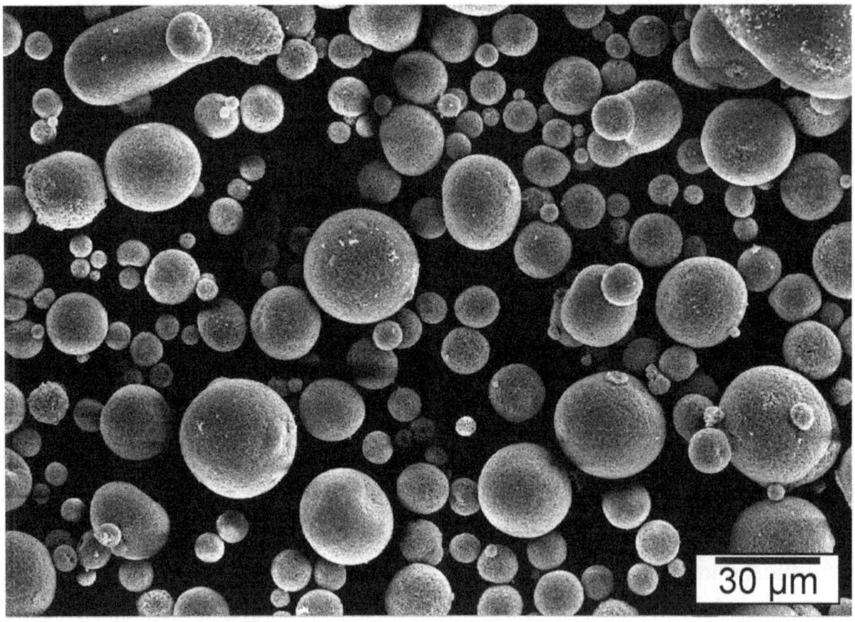

and "HVOF-S" to identify them as processes using suspensions as feedstocks. The economic efficiency of the use of suspensions depends on the availability of suitable additional system components to existing thermal spray booths, e.g., suspension feeders and injectors, and special highly concentrated, stable aqueous suspensions in order to decrease the amount of water to be evaporated and to reduce the spray times (Berger *et al.*, 2013). In order to achieve economic efficiency, significant efforts have been made to increase the solids content of the suspensions. Depending on the nature of the material, a mass fraction up to 70% has been reached (Berger *et al.*, 2013). The use of suspensions has been concentrated so far on the individual oxides (Toma, Berger, Stahr *et al.*, 2010; Berger, *et al.*, 2013). From the individual oxides of the Al_2O_3-Cr_2O_3-TiO_2 system mostly Al_2O_3 and TiO_2 have been sprayed, e.g. (Bolelli, Cannillo *et al.*, 2009; Bolelli, Rauch *et al.*, 2009; Toma, Berger, Stahr *et al.*, 2010). Cr_2O_3 is more difficult to spray (Gadow, Killinger, Rempp, & Manzat, 2010) and is less intensively studied so far.

METHODS OF COATING CHARACTERIZATION

General

Methods of coating characterisation have been described in detail in textbooks (Pawlowski, 2008; Mathesius & Krömmer, 2009; Tucker, Jr. (ed.), 2013; Fauchais *et al.*, 2014) and in a recent review paper (Ang & Berndt, 2014). These methods include standardized procedures, customized tests developed and applied in single companies or research laboratories, or highly sophisticated research methods, accessible only with special equipment (Berger, 2014). In the majority of cases, coating properties are determined in the as-sprayed state only. Changes in properties occurring during service are most often unknown but can be significant, in particular at elevated temperatures.

Coating adhesion is a property often discussed for thermal spray coatings but is more critical for processes with lower particle velocities, such as plasma spraying. It is predominantly determined by tensile adhesion testing according to standards DIN EN 582 and ASTM C 633. However, these test conditions do not reflect typical service conditions of the coatings.

Metallographically prepared cross sections have a high importance for investigation of the coating microstructure including the determination of porosity. These investigations are most often made by optical microscopy and scanning electron microscopy (SEM). They have an additional high importance, because many investigation methods, such as measurements of hardness, Young's modulus by indentation methods as well as indentation fracture toughness determinations are using also the cross sections. For Young's modulus measurements many alternative methods exist. The measurement using laser-acoustic waves is a fast and non-destructive method (Berger, Schneider, Barbosa, & Puschmann, 2012).

Coating properties can be divided into intrinsic properties (porosity and residual stress state) and extrinsic properties (adhesion, hardness, Young's modulus, fracture toughness, Poisson ratio) (Ang & Berndt, 2014) as well as system properties (corrosion, wear resistance, etc.). As coating hardness can be measured according to the standards EN ISO 14923 and ASTM E384 relatively easily and quickly, it is one of the properties most frequently given. Most often the Vickers hardness test is applied, the most common load being 2.94 N (300 g). A dependence of the wear resistance is often proposed and discussed, however, this appears not to be of relevance for coatings with high hardness. Different from hardness measurements, for measurements of the Young's modulus and fracture toughness there are no standardized procedures, but there are many reports in the literature, e.g. (Luo *et al.*, 2003; Bolelli, Cannillo, Lusvarghi, & Manfredini, 2006, Habib *et al.*, 2006; Lima & Marple, 2006; Bolelli *et al.*, 2007a; Vargas *et al.*, 2010; Berger *et al.*, 2012). An extension of the determination of mechanical properties is necessary in order better to describe system properties of the coatings, such as wear resistance. For quantitatively comparable values of the indentation fracture toughness more systematic work on the measurement procedure and the selection of the model and the corresponding equation is still required.

Oxide coatings are frequently used in corrosive environments, even when their major purpose is to provide wear resistance (Ashary & Tucker, Jr., 1989). The corrosion resistance of a coating can be understood as the ability to protect the substrate from corrosion (Toma, Stahr *et al.*, 2010; Herrmann *et al.*, 2014). Thus, also the corrosion properties of the substrate material are of importance (Ashary & Tucker, Jr., 1989). One of the many test methods widely applied is electrochemical testing in aqueous media (e.g. according to ASTM G5). Exposure testing is mostly customized to the different corrosive media, such as corrosive liquids like acids or bases.

Tribological Characterisation

There are numerous tribological test methods applicable to the coatings of the Al_2O_3-Cr_2O_3-TiO_2 system, including both standardized and customized methods. These methods reflect the different wear mechanisms and failure modes (Holmberg & Matthews, 2009). As a rule, real service conditions are not fully reflected by these test methods. Some of these tribological test methods (e.g. dry abrasion wear test) are often used for coating development and comparative purposes.

Abrasion and sliding wear are the most widely studied wear mechanisms for the coatings of the Al_2O_3-Cr_2O_3-TiO_2 system. The most important abrasion wear test is the dry sand rubber wheel test according to ASTM G65, e.g. (Niemi *et al.*, 1991; Niemi *et al.*, 1995; Niemi, 2009; Börner, Berger, Saaro, & Thiele, 2010). Modifications from the standard relate most often to the abrasive material, or the number of samples simultaneously tested. Occasionally, several two-body abrasion wear tests, e.g. (Niemi *et al.*, 1995; Müller & Kreye, 2001; Habib *et al.*, 2006) and a wet abrasion wear test (ASTM G75) are employed, e.g. (Knuuttila, Ahmaniemi, & Mäntylä, 1999; Houdková, Zahálka, Kašparová, & Berger, 2011).

Sliding wear is tested both in dry (including high temperature) and lubricated conditions, e.g. (Storz, Gasthuber, & Woydt, 2001; Skopp, Kelling, Woydt, & Berger, 2007; Landa *et al.*, 2007; Bolelli *et al.*, 2007b; Houdková *et al.*, 2011). Motion in sliding testing can be unidirectional, oscillating or reciprocating. There are several standards like ASTM G99 for pin-on-disk or ASTM G77 for block-on-ring testing. Testing can be performed against different counterpart materials, such as alumina, silicon carbide or hardmetals, different sliding speeds or frequencies and loads.

Erosion wear is often studied using centrifugal accelerator devices, e.g. (Kleis & Kulu, 2008). Sand is most often applied as erodent. Impact angles between 15 and 90° can be studied but are often limited to one or two different angles. The most important parameters are the hardness, size and morphology of the erodent, its velocity and the temperature. For slurries pot-type erosion tests are applicable, e.g. (Knuuttila, Ahmaniemi, & Mäntylä, 1999).

COATING MICROSTRUCTURES AND PROPERTIES

General

Although coatings of the single oxides Al_2O_3 and TiO_2 also contain several phases, for binary compositions of the Al_2O_3-Cr_2O_3-TiO_2 system the homogeneity of the coating microstructures is in particular often a question of concern. In this case, preference is given to the use of prealloyed instead of mechanically blended feedstock powders. In general, oxide coatings sprayed by high-velocity processes, namely DGS and HVOF, show denser microstructures, lower porosity, higher bond strength, higher hardness, and lower roughness than coatings sprayed by APS, e.g. (Berger, Stahr *et al.*, 2007; Niemi, 2009). However, the appearance of microcracks due to relaxation of residual stresses is a common feature of all ceramic coatings. As shown so far for the individual oxides, with the use of suspensions significantly smaller and more homogeneous microstructures can be obtained, these microstructures differ both from those of APS- and HVOF-sprayed coatings (Toma, Berger, Stahr *et al.*, 2010).

Al_2O_3 and Al_2O_3-Rich Compositions

Figures 7-10 show optical and SEM micrographs of Al_2O_3-coatings sprayed by APS and HVOF, respectively. Although HVOF-sprayed coatings appear denser in optical micrographs than APS-sprayed coatings (see Figures 7 and 9), the SEM micrographs of coatings sprayed by both processes show many similarities (see Figures 8 and 10). The main difference is the finer microstructure of the HVOF coating, due to the finer feedstock powder used for this process. Typical microhardness values are around 800-1200 HV0.3. The structure of Al_2O_3-coatings prepared from suspensions is discussed elsewhere, e.g. (Toma, Berger, Stahr *et al.*, 2010).

As a rule, only minor amounts of α-Al_2O_3 exist in the coatings. The content of α-Al_2O_3 can be increased and can be the main phase in the coating without any additions by using suspensions as the feedstock material and by the selection of an appropriate grade of fine alumina powder and corresponding spray conditions (Toma *et al.*, 2008).

A second way of stabilization of alumina is realized by addition of chromia (Chráska *et al.*, 1997). A full stabilization of the α-phase is possible, as was demonstrated through spraying of a solid solution alumina-chromia powder containing 33 mass % Cr_2O_3, but this was demonstrated with the WSP process only (Stahr *et al.*, 2007; Dubsky *et al.*, 2011). Information from the literature (Müller & Kreye, 2001; Marple, Voyer, & Béchard, 2001) suggests that with a content of about 30 mass % Cr_2O_3 the α-phase may be stabilized, regardless of the spray process used. However, a number of experimental series summarized elsewhere (Berger *et al.*, 2014) have shown that the mechanism of α-phase stabilization is much more complex than assumed earlier and still requires a lot of experimental efforts in order to be clarified. The change of coating hardness of the Al_2O_3-rich compositions of the Al_2O_3-Cr_2O_3-system has yet to be studied in detail. So far different tendencies have been observed depending on the spray process (Müller & Kreye, 2001).

Many studies have been performed with the compositions of the Al_2O_3-rich side of the Al_2O_3-TiO_2 system, for which a broad variety of feedstock materials containing 3, 13, and 40 mass % TiO_2 are commercially available. The preparation of nanostructured coatings from nanostructured feedstock powders has particularly been studied for the Al_2O_3-13%TiO_2 composition, e.g. (Gell *et al.*, 2001; Jordan *et al.*, 2001; Luo *et al.*, 2003; Lima, Moreau, & Marple, 2007; Sánchez *et al.*, 2010). Except for Al_2O_3- and TiO_2-phases, binary oxides are already found in these feedstock powders, e.g. (Herrmann *et al.*, 2014; Toma, Stahr *et al.*, 2010; Vargas *et al.*, 2010). As there are so many different types of feedstock materials and spray processes applied, e.g. (Niemi, Vuoristo, & Mäntylä, 1994; Habib *et al.*, 2006; Niemi, 2009; Vargas *et al.*, 2010; Börner *et al.*, 2010), the results for the coating properties and their interpretations are very often contradictory, e.g. (Vargas *et al.*, 2010).

Figure 7. Optical micrograph of Al_2O_3-coating sprayed by APS

Figure 8. SEM micrograph of Al_2O_3-coating sprayed by APS

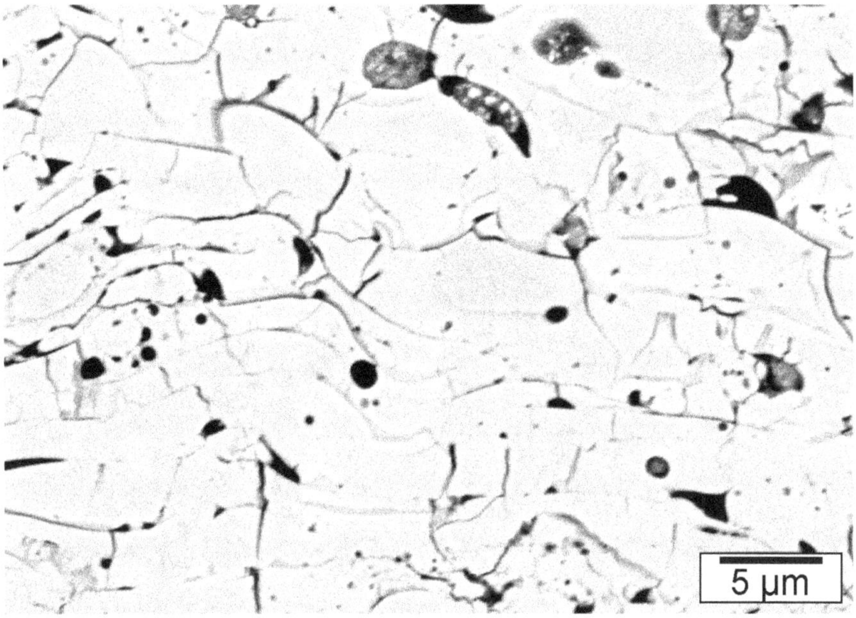

Figure 9. Optical micrograph of Al_2O_3-coating sprayed by HVOF

Figure 10. SEM micrograph of Al_2O_3-coating sprayed by HVOF

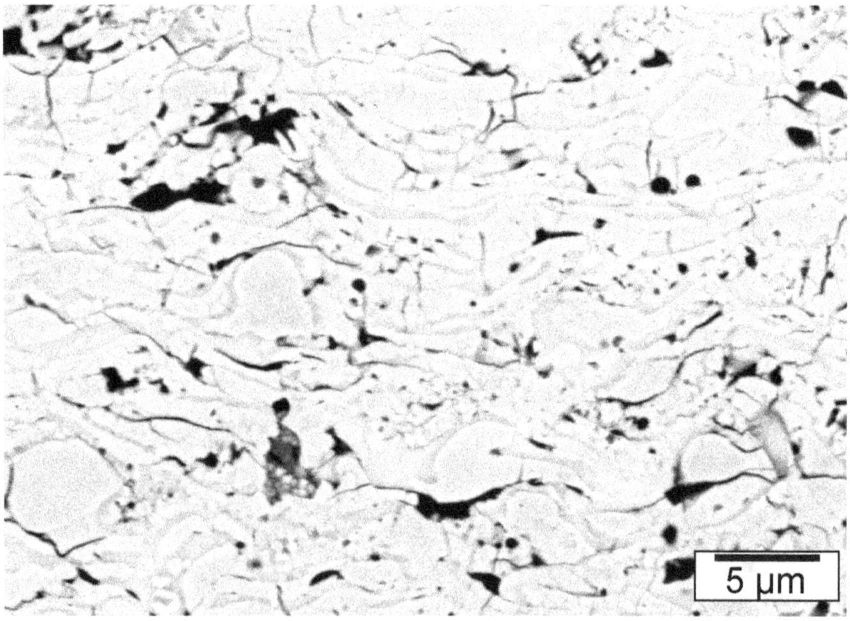

For illustration, the microstructures of an HVOF-sprayed Al_2O_3-13%TiO_2 coating and of an APS-sprayed Al_2O_3-40%TiO_2 coating are shown in Figure 11 and Figure 12, respectively. Both coatings have been prepared from a jointly fused and crushed powder and show the typical multiphase microstructure. As a rule, the microstructures of HVOF-sprayed coatings show always finer structures, e.g. (Niemi, 2009; Herrmann *et al.*, 2014).

Although the phase compositions of the coatings are rather complex, these coatings are often designated as Al_2O_3/TiO_2 coatings only. The majority of studies have been performed with prealloyed feedstock powders. As in plain Al_2O_3-coatings, in Al_2O_3-3% TiO_2 coatings basically γ- and α-Al_2O_3 are found. One possible reason is the insensitivity of x-ray diffraction to small contents of a phase in a material. For Al_2O_3-13% TiO_2 coatings TiO_2 was found to be in solid solution with γ-Al_2O_3 (Gell *et al.*, 2001; Jordan *et al.*, 2001). For nanostructured Al_2O_3-13% TiO_2 coatings the existence of a non-equilibrium χ-$Al_2O_{3*}TiO_2$ phase was postulated (Jordan *et al.*, 2001). With increasing TiO_2-content the amount of binary phases in the coatings increases, in particular for the Al_2O_3-40% TiO_2 composition, which is already close to the composition of Al_2TiO_5. However, it strongly depends on the structure and the phase composition of the feedstock powder, if individual Al_2O_3-phases, substoichiometric titania phases, or binary phases are found in the coatings. The binary phases $Al_{2-x}Ti_{1+x}O_5$ (Herrmann *et al.*, 2014) and $Al_6Ti_2O_{13}$ (Vargas *et al.*, 2010) have been identified in the coatings containing $\geq$ 40% TiO_2. Surprisingly, for these coatings significantly higher values of the Young's modulus were measured than were expected from the low Young's modulus of Al_2TiO_5 (Vargas *et al.*, 2010; Trache, Berger, & Leyens, 2014).

As the appearance and the amount of the liquid phase during spraying are connected with the content of TiO_2, the coating porosity usually decreases with increasing TiO_2 content (Habib *et al.*, 2006; Niemi, 2009). There is no clear dependence but a tendency that coating hardness decreases with increasing TiO_2 content (Niemi *et al.*, 1994; Habib *et al.*, 2006; Niemi, 2009; Börner *et al.*, 2010; Herrmann *et al.*, 2014). As a rule, for coatings of the composition Al_2O_3-40%TiO_2 deposited by APS, DGS and HVOF a lower hardness was reported than for the other binary compositions, sprayed with the same processes, respectively (Niemi *et al.*, 1994; Znamirowski, Pawlowski, Cichy, & Czarczynski, 2004; Habib *et al.*, 2006; Niemi, 2009; Börner *et al.*, 2010; Vargas *et al.*, 2010; Herrmann *et al.*, 2014). Regarding the fracture toughness it can be stated that the data existing in the literature is still insufficient in order to draw general conclusions on the influence of coating microstructure, chemical and phase composition.

The corrosion resistance of Al_2O_3 and Al_2O_3-TiO_2 bulk ceramics against solutions of 1 N NaOH and H_2SO_4 was superior to that of the corresponding thermally sprayed coatings (Herrmann *et al.*, 2014). This is due to the specific microstructures of the

Figure 11. Optical micrograph of an HVOF-sprayed Al_2O_3-13% TiO_2 coating

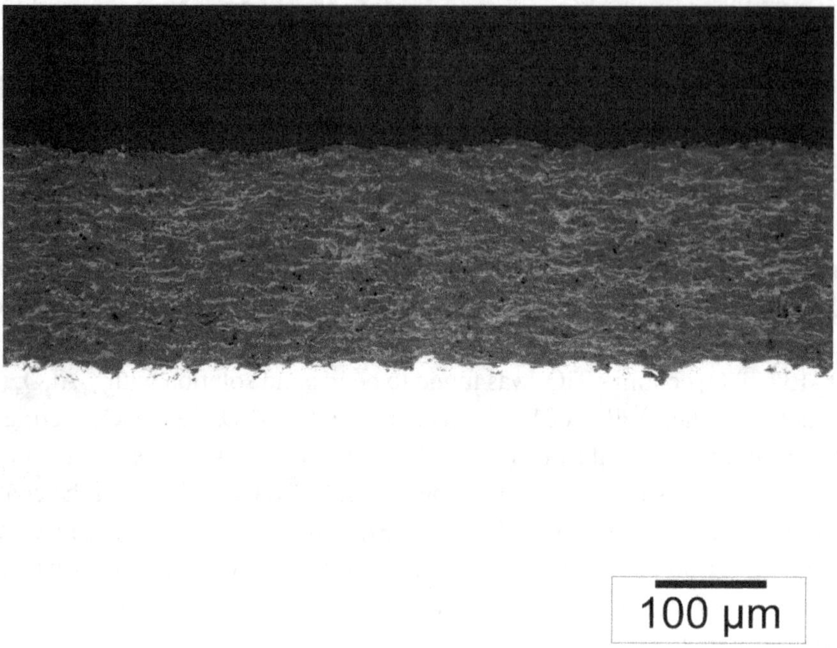

Figure 12. Optical micrographs of an APS-sprayed Al_2O_3-40% TiO_2 coating

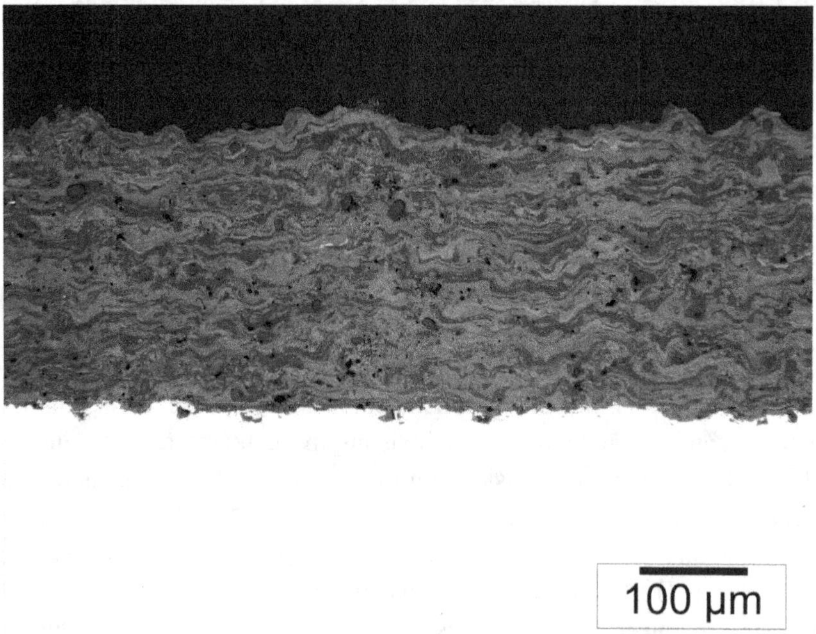

thermally sprayed coatings and the content of amorphous and/or metastable phases (Herrmann *et al.*, 2014). In contrast to the expectations, the corrosion resistance of the APS-sprayed coatings was higher than that of the denser HVOF-sprayed coatings in nearly all cases (Toma, Stahr *et al.*, 2010; Herrmann *et al.*, 2014). The main factor in the enhancement of the corrosion resistance with increasing content of TiO_2, was the denser microstructure of both the APS- and the HVOF-sprayed coatings (Toma, Stahr *et al.*, 2010; Herrmann *et al.*, 2014).

Cr_2O_3 and Cr_2O_3-Rich Compositions

Due to its high melting point and the specific properties of Cr_2O_3 as described above, the deposition efficiency in conventional plasma spray processes is only about 40%, and reaches a maximum of 60%. This is also the reason why spraying of Cr_2O_3 by DGS (Niemi *et al.*, 1991; Niemi *et al.*, 1995) or by HVOF, e.g. (Bolelli *et al.*, 2007a; Bolelli *et al.*, 2007b) is seldom reported in the literature.

For illustration, Figure 13 shows an optical and Figure 14 an SEM micrograph of an APS-sprayed Cr_2O_3 coating. Cr_2O_3 is characterized by high phase stability. However, low purity of the Cr_2O_3 starting material, conditions of feedstock powder preparation and processes occurring during spraying can lead to the incorporation of metallic chromium into the coating. This results in inhomogeneous coating properties, which can be avoided through use of the corresponding quality of the starting material as well as by control of the process conditions. The microstructure of plain chromia coatings is characterized by a large number of microcracks.

Typical hardness values of APS-sprayed coatings are in the range 1000-1300 HV0.3. The high brittleness of the coating results in a high scatter of published hardness values.

Among a broad variety of oxide coatings, Cr_2O_3 coatings were found together with TiO_x coatings as the most corrosion resistant (Berger *et al.*, 2009c). Due to its brittleness, a metallic bond coat is often applied together with the Cr_2O_3 coatings, which increases the bond strength and is also beneficial for the protection of the substrate against corrosion.

With increasing content of alumina, the number of microcracks appearing in SEM micrographs decreases (Börner *et al.*, 2010). Although commercial feedstock powders of Cr_2O_3-rich compositions both with alumina and titania additions are available, the number of studies found in the literature for these compositions is rather limited. These studies include both mechanically blended powders (Yang *et al.*, 2012) but predominantly jointly heat-treated powders, e.g. (Harju, Levänen, & Mäntylä, 2004; Berger *et al.*, 2009a; Berger *et al.*, 2010; Börner *et al.*, 2010). On the base of the currently available information, for alumina additions, a limit where coatings contain only one phase cannot be given. In the case of titania additions, all

Figure 13. Optical micrograph of an APS-sprayed Cr_2O_3 coating

Figure 14. SEM micrograph of an APS-sprayed Cr_2O_3 coating

coatings of Cr_2O_3-rich compositions consist of the eskolaite lattice only (Harju *et al.*, 2004; Berger *et al.*, 2009a). The decrease in hardness of APS-sprayed coatings due to alumina additions is less than for titania additions (Börner *et al.*, 2010). The dependence of the coating hardness on the titania content is different for APS- and HVOF-sprayed coatings (Berger *et al.*, 2009a). Titania prevents effectively the formation of metallic chromium, as it is reduced faster.

TiO_2 and TiO_2-Rich Compositions

The low melting point makes TiO_2 suitable for processing by HVOF. As oxygen deficiency decreases the melting temperature the processability by HVOF is further improved. A detailed review of the material behavior during the spray process, the coating properties and their applications for TiO_2 and TiO_2-rich compositions is given elsewhere (Berger, 2004). There is an enhancement of the mechanical properties due to use of nanostructured feedstock powders (Lima & Marple, 2006).

Titanium oxide coatings are characterized by substoichiometry and defect structures. The O/Ti ratio of the substoichiometric feedstock powders is changed in nearly all spray process conditions. In particular, use of an Ar/H_2 mixture as a plasma-forming gas in the APS process results in oxygen loss by reduction with hydrogen. This change predominantly occurs in the outer regions of the spray powder particles and leads to locally different O/Ti ratios throughout the coatings, leading to the appearance of different gray scales in the optical micrographs of the coatings (Buchmann & Gadow, 2001; Berger, 2004). This is illustrated by the optical micrograph in Figure 15. As shown in Figure 16, spraying with HVOF leads only to minor changes of the O/Ti ratio and less pronounced differences in the gray scales of the optical micrograph. Application of TiO_x-coatings above 380 °C in air leads to oxidation (Berger *et al.*, 2009b).

Typical coating hardness values are in the range 700-900 HV0.3 (Berger, 2004). TiO_x coatings were found to be more corrosion resistant than plain alumina and Al_2O_3-TiO_2 coatings (Toma, Stahr *et al.*, 2010), and also showed better corrosion resistance than any other sprayed ceramic coating, including chromia (Berger *et al.*, 2009c).

For the preparation of coatings, which should contain Andersson-phases with the general formula $Ti_{n-2}Cr_2O_{2n-1}$ (compositions of the TiO_2-rich side of the Cr_2O_3-TiO_2 phase diagram) experimental feedstock powders have been developed. As summarized elsewhere (Berger, *et al.*, 2014), agglomerated and sintered, as well as fused and crushed feedstock powders can be used for the deposition of coatings with low porosity. The hardness for both APS- and HVOF- coatings increases with increasing Cr_2O_3-content (Berger *et al.*, 2009a). Similar to plain titania coatings,

Figure 15. Optical micrograph of an APS-sprayed coating using an Ar-H_2 plasma from an experimental agglomerated and sintered Ti_5O_9-Ti_6O_{11} powder

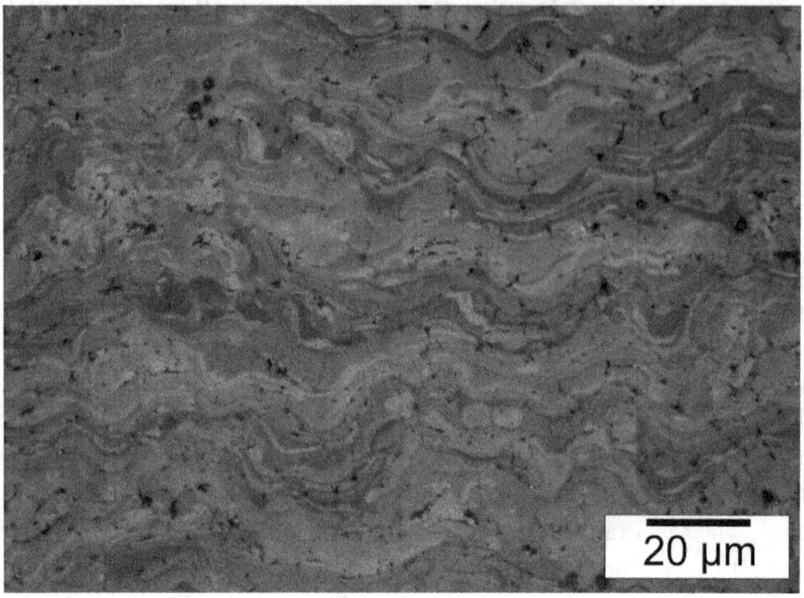

Figure 16. Optical micrograph of an HVOF-sprayed coating using hydrogen as a fuel gas from an experimental fused and crushed Ti_5O_9-Ti_6O_{11} powder, prepared by reduction with hydrogen from a commercial powder. Sample sprayed at Tampere University of Technology, Finland

oxygen deficiency is observed in particular for APS-sprayed coatings, prepared with Ar/H_2 plasma forming gases (Berger *et al.*, 2009a). Thus, a reoxidation will occur during service in air at high temperatures.

TRIBOLOGICAL PROPERTIES

General

With the use of adapted feedstock powders, spraying by HVOF or DGS produces coatings with denser microstructures, lower porosity, higher hardness, higher bond strength, and lower roughness can compared to those sprayed by APS. Finer microstructures also decrease the costs for surface finishing. In order to be applied in tribological applications in industrial practice, economically competitive spray rates and deposition efficiencies are required for both spray processes.

Corrosion resistance of the coatings of the Al_2O_3-Cr_2O_3-TiO_2 system is of importance as service conditions for wear protection are often accompanied by corrosive attack. The coating microstructure, in particular the appearance of interconnected open porosity and cracks, can be more important than the corrosion resistance of the coating itself, e.g. (Ashary & Tucker, Jr., 1989; Ashary & Tucker, Jr., 1991). In order to prevent any corrosion attack of the substrate in service, as a rule, ceramic coatings are sealed. The influence of the sealant on wear resistance is reported in the literature, e.g. (Leivo *et al.*, 1997; Knuuttila, Ahmaniemi, & Mäntylä, 1999; Knuuttila, Sorsa, & Mäntylä, 1999; Ahmaniemi *et al.*, 2002; Niemi, 2009). On the other hand, the corrosion properties of the coating material are detrimentally influenced by the formation of amorphous phases as a result of the spray process, e.g. (Herrmann *et al.*, 2014; Toma, Stahr *et al.*, 2010; Berger, *et al.*, 2014).

When tribological test results of different studies are compared, often a disagreement of the results is found. The special character of thermal spray coating properties and the often very different test conditions as described in the introduction have to be taken into account. Unfortunately, many studies describing the coating microstructures, phase compositions and mechanical properties in detail are limited to one selected tribological test only (most often abrasive or sliding wear).

The abrasion wear resistance of the individual oxides and the binary compositions of the Al_2O_3-Cr_2O_3-TiO_2 system has been systematically studied (Börner *et al.*, 2010), as well as studies of parts of this system, e.g. (Niemi *et al.*, 1995; Niemi, 2009). There is a generally good agreement that DGS- and HVOF-sprayed coatings show a better abrasion wear resistance than APS-sprayed coatings due to their more homogeneous microstructures, lower porosity and (in the case they are investigated) improved mechanical properties, e.g. (Niemi *et al.*, 1995; Bolelli *et al.*, 2007b; Niemi, 2009; Börner *et al.*, 2010).

Bulk ceramic components are considered as useful for sliding wear applications if the specific wear rate is $< 10^{-6}$ mm^3/Nm, which is the upper limit for mild wear (Rainforth, 2004). This specific wear rate is considered as the border to distinguish a mild wear regime from a severe wear regime for thermally sprayed coatings (Bolelli *et al.*, 2007b). Total wear rates below this threshold value (sum of wear rates of the coating and the counterpart material) are typically measured under mixed/boundary lubrication conditions (Berger, Saaro, & Woydt, 2007). If this condition is met under dry sliding conditions, it characterizes the special application potential of a tribological pair to serve under these conditions. These criteria are met at lower loads for most of the coatings independent of the spray process. For higher loads the HVOF-sprayed coatings show better performance (Bolelli *et al.*, 2007b).

Based on a comparative investigation of Al_2O_3, Al_2O_3-13TiO_2 and Cr_2O_3 coatings by scratch testing it is proposed that these coatings exhibit a sharp transition from a mild wear regime to a severe wear regime when the contact load exceeds the critical value, with extensive cracking and material removal by lateral cracking (Xie & Hawthorne, 1999).

The tribological properties of coatings sprayed from suspensions have only rarely been studied so far. Some results have been reported for dry sliding wear resistance of Al_2O_3 and TiO_2 coatings against sintered alumina (Bolelli, Cannillo *et al.*, 2009; Bolelli, Rauch *et al.*, 2009). However, these coatings have been produced from suspensions with low solids content, thus representing uneconomic conditions of coating deposition compared to conventional coatings. Thus, the tribological characterization of high solids content and more economical suspension-sprayed coatings remains as a future task.

Tribological testing at high temperatures, e.g. 800 °C, can lead to a change of the phase composition and the microstructure, and can increase the hardness, as has been found for APS-sprayed coatings in the Cr_2O_3-TiO_2 system (Berger *et al.*, 2010).

Al_2O_3 and Al_2O_3-Rich Compositions

Wear protection is one of the most important applications of alumina coatings. Compared to other wear resistant materials it is also a cost-efficient solution, as for tribological applications no special requirements regarding the purity exist. It has been shown in several studies, that the wear resistance of plain alumina coatings can be improved by use of DGS or HVOF processes, e.g. (Müller & Kreye, 2001; Niemi, 2009). The abrasion wear resistance of DGS- and HVOF-sprayed alumina coatings is higher than that of bulk alumina (Niemi, 2009). Contrary to this, bulk alumina shows a higher erosion wear resistance for impingement angles of 30° and 90° than all thermally sprayed coatings. As typical for ceramic coatings, the mass loss is always higher for an impingement angle of 90 ° (Niemi, 2009).

There are only very few studies concerning the wear resistance of coatings with Cr_2O_3 additions, e.g. (Müller & Kreye, 2001). Thus, the information available in the literature is not sufficient for drawing general conclusions on the effect of chromia additions.

For different spray processes (APS, DGS, HVOF) different tendencies on the influence of the amount of titania addition on the abrasion wear resistance have been observed (Niemi, 2009; Börner *et al.*, 2010). Remarkably, high abrasion wear resistance was found for Al_2O_3-40%TiO_2 coatings (Niemi, 2009; Börner *et al.*, 2010), but also for Al_2TiO_5 coatings (Niemi, 2009). The abrasion wear resistance of APS-sprayed plain Al_2O_3 coatings and coatings of the Al_2O_3-rich side of the Al_2O_3-TiO_2 system is strongly influenced by sealers (Knuuttila, Sorsa, & Mäntylä, 1999; Niemi, 2009).

For flame sprayed coatings, due to the reduction in coating hardness with increasing content of TiO_2, the wear resistance decreases as well (Habib *et al.*, 2006).

In dry sliding wear tests (pin-on-disk) at room temperature, both alumina and Al_2O_3-3%TiO_2 coatings showed unfavorable coating and sample wear rates against 100Cr6 and alumina, as well as high coefficients of friction (Bolelli *et al.*, 2006). Wear rate and the coefficient of friction depend on the load. For nanostructured Al_2O_3-3%TiO_2 coatings the transition from mild to severe wear regimes appeared at higher loads (Rico, Rodríguez, Otero, Zeng, & Rainforth, 2009).

Independent of the spray process, the erosion wear resistance for most of the Al_2O_3-3%TiO_2 coatings was slightly better than that of the plain alumina coatings. Further increase of the titania content decreases the erosion wear resistance (Niemi, 2009). DGS-sprayed Al_2O_3-13%TiO_2 coatings showed a erosion wear resistance comparable to that of chromia coatings (Niemi *et al.*, 1995).

Cr_2O_3 and Cr_2O_3-Rich Compositions

Chromia coatings are mainly used in the production of rotogravure rolls for printing machines. Besides a high wear resistance, the advantageous material behavior in the laser engraving process is important for this application (Pawlowski, 1996). Due to their high hardness chromia coatings are also often used for general applications of wear protection. The wear resistance of APS-sprayed chromia coatings is often compared with that of metallic or hardmetal coatings, e.g. (Ashary & Tucker, Jr., 1991; Bolelli *et al.*, 2006; Berger, Saaro, & Woydt, 2007; Houdková *et al.*, 2011). The dry abrasion wear resistance of APS-sprayed chromia coatings is comparable with metallic self-fluxing alloy or stainless steel coatings. In the case of wet abrasion wear resistance, the performance is close to that of the group of hardmetal coatings (Houdková *et al.*, 2011). The abrasion wear resistance of the DGS-sprayed coatings was higher than that of the APS-sprayed coating (Niemi *et al.*, 1991; Niemi *et al.*, 1995), but a coating sprayed by HPPS (high power plasma spraying) showed the best performance (Niemi *et al.*, 1995).

In general, APS-sprayed Cr_2O_3-coatings show a high resistance in sliding wear conditions at room temperature (Bolelli *et al.*, 2006; Berger, Saaro, & Woydt, 2007; Berger *et al.*, 2009a; Berger *et al.*, 2009b; Houdková *et al.*, 2011).

The erosion wear resistance of the APS-, HPPS- and DGS-sprayed coatings is very high and shows relative small differences (Niemi *et al.*, 1995). The mass loss of an APS-sprayed chromia coating in an erosion wear test increased with the impingement angle and was highest at 90 ° (Houdková *et al.*, 2011; Niemi *et al.*, 1995).

Despite a high coating hardness, coatings of Cr_2O_3-rich compositions with Al_2O_3 additions showed a lower abrasion wear resistance than coatings of Cr_2O_3-rich compositions with TiO_2 additions (Börner *et al.*, 2010).

The study of the dry sliding wear of APS-sprayed chromia-rich coatings of the Cr_2O_3-TiO_2 system against sintered alumina has shown that, at room temperature, the coatings are characterized by high wear rates and decreasing coefficients of friction with increasing sliding speeds. Low wear rates ($< 10^{-6}$ mm^3/Nm) in combination with low coefficients of friction have been measured at 800 °C only (Berger *et al.*, 2009a; Berger *et al.*, 2009b).

TiO_2 and TiO_2-Rich Compositions

In comparison with other oxide coatings, APS sprayed titanium oxide coatings are generally characterized by an intermediate but satisfactory wear resistance, which has led to their application in mechanical engineering (light bearings, pump seals, etc.) (Berger, 2004). The electrical conductivity of these coatings can be advantageous for several tribological applications.

The abrasion wear resistance of APS-sprayed titanium oxide coatings (Vuoristo, Määttä, Mäntylä, Berger, & Thiele, 2002) is inside the group of APS-sprayed Al_2O_3/TiO_2 coatings (Vuoristo *et al.*, 1992). A similar result is found, if the abrasion wear resistance of HVOF-sprayed coatings (Vuoristo *et al.*, 2002) is compared with DGS-sprayed coatings (Vuoristo *et al.*, 1992). There are also reports on the increase of the abrasion wear resistance, due to the use of nanostructured titania feedstock powders (Lima & Marple, 2006).

The term 'lubricious oxides' has been used to describe the expected low coefficients of friction and wear rates of titanium suboxide coatings in unlubricated dry sliding conditions (Gardos, 1988), (Woydt, 2005). This suggestion led to numerous tribological studies being carried out on titanium suboxides, including thermal spray coatings, e.g. (Storz *et al.*, 2001; Landa *et al.*, 2007). For potential engine applications the coatings have been studied in dry and lubricated sliding conditions, e.g. (Storz *et al.*, 2001; Skopp *et al.*, 2007; Landa *et al.*, 2007). They were found to be characterized by good tribological behavior, but the predicted low coefficients of friction were not found experimentally (Skopp *et al.*, 2007).

The study of the dry sliding wear of APS-sprayed titania-rich coatings of the Cr_2O_3-TiO_2 system against sintered alumina has shown that, at room temperature, the coatings are characterized by high wear rates and decreasing coefficients of friction with increasing sliding speeds (Berger *et al.*, 2009a; Berger *et al.*, 2009b). Low wear rates (< 10^{-6} mm^3/Nm) in combination with low coefficients of friction have been measured at 800 °C only. As for pure TiO_x-coatings, the functional benefits of these materials are different from the meaning "lubricious oxide" (Berger *et al.*, 2009b).

These tribological studies at 800 °C (maximum duration of 14 h) have also surprisingly shown that under these conditions significant changes of the microstructure, a strong increase of coating hardness, and changes of the phase compositions occur (Berger *et al.*, 2010). For illustration, Figure 17 shows the optical micrograph of a 23.5 mass % Cr_2O_3 -76.5 mass % TiO_2 coating cross section of a tribological specimen after testing at room temperature. A bond coat was applied in order to decrease the stresses due to different coefficients of thermal expansion. The ceramic coating was sprayed from an experimental agglomerated and sintered feedstock powder consisting of the Andersson phases $Cr_2Ti_6O_{15}$ and $Cr_2Ti_7O_{17}$. As a result of the spray process the so-called high-temperature Andersson phases were formed. Significant changes occurred after the dry sliding wear test at 800 °C with a sliding speed of 0.1 m/s (test duration of about 14 h). As evidenced by Figure 18 the microstructure changed. Also the phase composition changed, with rutile being the only phase identified by XRD. An increase of hardness from 814 HV0.3 to 1125 HV0.3 after testing at 800 °C for 14 h was observed.

SUMMARY AND OUTLOOK

Thermally sprayed coatings of the individual oxides and binary compositions in the Al_2O_3-Cr_2O_3-TiO_2 system are widely manufactured and used, in particular, for tribological applications. There is a wide variety of spray processes (APS, DGS, HVOF, and others) available, the choice depends both on the required coating quality and the economic parameters (deposition efficiency, powder feedrates). APS is the most common process, however, high velocity processes, such as DGS and HVOF, have been found to improve the coating properties, in particular the performance in tribological applications. Spray powders with a particle size 10-45 µm (or variations depending on the spray process characteristics) are commonly used as feedstock.

Figure 17. Optical micrograph of the coating cross section of an APS-sprayed 23.5 mass % Cr_2O_3 -76.5 mass % TiO_2 coating with a NiCrAlY bond coat after sliding wear tests at room temperature at a speed of 0.1 m/s, test duration of about 14 h (Berger et al., 2010)

Figure 18. Optical micrographs of the coating cross section of an APS-sprayed 23.5 mass % Cr_2O_3 -76.5 mass % TiO_2 coating with a NiCrAlY bond coat after sliding wear tests at 800 °C at a speed of 0.1 m/s, test duration of about 14 h (Berger et al., 2010)

The use of suspensions as feedstock is emerging and allows direct coating manufacture from common ceramic powders of only a few micrometers in size. These coatings offer a variety of technical advantages, such as lower roughness, possible thinner coating thickness, and others. However, the potential of coatings sprayed with economically reasonable parameters in tribological applications has not yet been studied sufficiently.

Current coating solutions have been predominantly developed on an empirical base. Although they have been applied for decades, the chemical and physical processes occurring during coating deposition are still not fully understood, e.g. the phase transformations of alumina in the spray process. Efforts need to be made to overcome this situation as the base for future knowledge-based developments. They will lead to a better understanding of the process-properties relationships of the coatings. Thus, more defined and reliable coating properties relative to those currently produced, will be obtained.

REFERENCES

Ageorges, H., & Ctibor, P. (2008). Comparison of the structure and wear resistance of Al_2O_3 – 13 wt.% TiO_2 coatings made by GSP and WSP plasma process with two different powders. *Surface and Coatings Technology*, *202*(18), 4362–4368. doi:10.1016/j.surfcoat.2008.04.010

Ahmaniemi, S., Vippola, M., Vuoristo, P., Mäntylä, T., Buchmann, M., & Gadow, R. (2002). Residual stresses in aluminium phosphate sealed plasma sprayed oxide coatings and their effect on abrasive wear. *Wear*, *252*(7-8), 614–623. doi:10.1016/S0043-1648(02)00020-0

Ananthapadmanabhan, P. V., Thiyagarajan, T. K., Sreekumar, K. P., Satpute, R. U., Venkatramani, N., & Ramachandran, K. (2003). Co-spraying of alumina-titania: Correlation of coating composition and properties with particle behaviour in the plasma jet. *Surface and Coatings Technology*, *168*(2-3), 231–240. doi:10.1016/S0257-8972(03)00006-9

Ang, A. S. M., & Berndt, C. C. (2014). A review of testing methods for thermal spray coatings. *International Materials Reviews*, *59*(4), 179–223. doi:10.1179/1743280414Y.0000000029

Ashary, A. A., & Tucker, R. C. Jr. (1989). Electrochemical corrosion studies of alloys plasma sprayed with Cr_2O_3. *Surface and Coatings Technology*, *39-40*(2), 701–709. doi:10.1016/S0257-8972(89)80032-5

Ashary, A. A., & Tucker, R. C. Jr. (1991). Corrosion characteristics of several thermal spray cermet-coating/alloy systems. *Surface and Coatings Technology*, *49*(1-3), 78–82. doi:10.1016/0257-8972(91)90035-U

Beczkowiak, J., Keller, H., & Schwier, G. (1996). Al_2O_3-TiO_2-Schichten - eine Alternative zu Cr_2O_3. In E. Lugscheider (Ed.), *Thermische Spritzkonferenz 1996, DVS-Berichte 175* (pp. 68–75). Düsseldorf: DVS-Verlag.

Berger, L.-M. (2004). Titanium oxide – New opportunities for an established coating material, In *Thermal Spray 2004, Advances in Technology and Application, Proc. Int. Thermal Spray Conf. ITSC 2004*. Düsseldorf: DVS-Verlag.

Berger, L.-M. (2014). Coatings by thermal spray. In Comprehensive hard materials (Vol. 1, pp. 471-506). Elsevier.

Berger, L.-M., Saaro, S., & Woydt, M. (2007). Reib-/Gleitverschleiß von thermisch gespritzten Hartmetallschichten, In R. Suchentrunk (Ed.), Jahrbuch Oberflächentechnik 2007 (vol. 63, pp. 242–267). Bad Saulgau: Eugen G. Leuze Verlag.

Berger, L.-M., Schneider, D., Barbosa, M., & Puschmann, R. (2012). Laser acoustic surface waves for the non-destructive characterisation of thermally sprayed coatings. *Thermal Spray Bulletin*, *5*(1), 56–64.

Berger, L.-M., & Stahr, C. C. (2008). State and perspectives of thermally sprayed ceramic coatings in the Al2O3-Cr2O3-TiO2 system, In T.S. Sudarshan & M. Jeandin (Eds.), *Proc. 21st Int. Conf. Surf. Modification Technol.* (pp. 469–478). Valar Docs.

Berger, L.-M., Stahr, C. C., Saaro, S., Thiele, S., & Woydt, M. (2009a). Development of ceramic coatings in the system Cr_2O_3-TiO_2. *Thermal Spray Bulletin*, *2*(1), 64–77.

Berger, L.-M., Stahr, C. C., Saaro, S., Thiele, S., Woydt, M., & Kelling, N. (2009b). Dry sliding up to 7.5 m/s and 800°C of thermally sprayed coatings of the TiO_2-Cr_2O_3 system and (Ti,Mo)(C,N)-Ni(Co). *Wear*, *267*(5-8), 954–964. doi:10.1016/j.wear.2008.12.105

Berger, L.-M., Stahr, C. C., Toma, F.-L., Saaro, S., Herrmann, M., Deska, D., & Thiele, S. et al. (2009c). Corrosion of thermally sprayed oxide ceramic coatings. *Thermal Spray Bulletin*, *2*(1), 40–56.

Berger, L.-M., Stahr, C. C., Toma, F.-L., Stehr, G. C., & Beyer, E. (2007). Ausgewählte Entwicklungstendenzen bei der Herstellung thermisch gespritzter keramischer Schichten. In R. Suchentrunk (Ed.), Jahrbuch Oberflächentechnik 2007 (vol. 63, pp. 71–84). Bad Saulgau: Eugen G. Leuze Verlag.

Berger, L.-M., Toma, F.-L., & Potthoff, A. (2013). Thermal spraying with suspensions – An economic spray process. *Thermal Spray Bulletin*, *6*(2), 98–101.

Berger, L.-M., Toma, F.-L., Scheitz, S., Trache, R., & Börner, T. (2014). Thermisch gespritzte Schichten im System Al_2O_3-Cr_2O_3-TiO_2 – ein Update. *Materialwissenschaft und Werkstofftechnik*, *45*(6), 465–475. doi:10.1002/mawe.201400260

Berger, L.-M., Woydt, M., Saaro, S., Stahr, C. C., & Thiele, S. (2010). High-temperature sliding wear of APS-sprayed coatings of the Cr2O3-TiO2 system. In *Proceedings 14th Nordic Symposium on Tribology*. Luleå, Sweden: Luleå University of Technology.

Bolelli, G., Cannillo, V., Gadow, R., Killinger, A., Lusvarghi, L., & Rauch, J. (2009). Properties of high velocity suspension flame sprayed (HVSFS) TiO_2 coatings. *Surface and Coatings Technology*, *203*(12), 1722–1732. doi:10.1016/j.surfcoat.2009.01.006

Bolelli, G., Cannillo, V., Lusvarghi, L., & Manfredini, T. (2006). Wear behaviour of thermally sprayed ceramic oxide coatings. *Wear*, *261*(11-12), 1298–1315. doi:10.1016/j.wear.2006.03.023

Bolelli, G., Lusvarghi, L., Manfredini, T., Pighetti Mantini, F., Polini, R., Turunen, E., & Hannula, S.-P. et al. (2007a). Comparison between plasma- and HVOF-sprayed ceramic coatings. Part I: Microstructure and mechanical properties. *International Journal of Surface Science and Engineering*, *1*(1), 38–61. doi:10.1504/IJSURFSE.2007.013620

Bolelli, G., Lusvarghi, L., Manfredini, T., Pighetti Mantini, F., Turunen, E., Varis, T., & Hannula, S.-P. (2007b). Comparison between plasma- and HVOF-sprayed ceramic coatings. Part II: Tribological behaviour. *International Journal of Surface Science and Engineering*, *1*(1), 62–79. doi:10.1504/IJSURFSE.2007.013621

Bolelli, G., Rauch, J., Cannillo, V., Killinger, A., Lusvarghi, L. V., & Gadow, R. (2009). Microstructural and tribological investigation of high-velocity suspension flame sprayed (HVSFS) Al_2O_3 coatings. *Journal of Thermal Spray Technology*, *18*(1), 35–49. doi:10.1007/s11666-008-9279-9

Börner, T., Berger, L.-M., Saaro, S., & Thiele, S. (2010). Systematische Untersuchung der Abrasionsbeständigkeit thermisch gespritzter Schichten aus dem Werkstoffsystem Al_2O_3-TiO_2-Cr_2O_3. In B. Wielage (Ed.), *Tagungsband zum 13. Werkstofftechnischen Kolloquium in Chemnitz, 30. September - 01. Oktober 2010, Werkstoffe und werkstofftechnische Anwendungen* (Vol. 37, pp. 178–187). Chemnitz: TU Chemnitz.

Buchmann, M., & Gadow, R. (2001). Mechanical and tribological characterization of APS and HVOF sprayed TiO2 coatings on light metals, In C.C. Berndt, K.A. Khor, & E.F. Lugscheider (Eds.), *Thermal Spray 2001: New Surfaces for A New Millennium, Proc. Int. Thermal Spray Conf.* Materials Park, OH: ASM International.

Chráska, P., Dubsky, J., Neufuss, K., & Písacka, J. (1997). Alumina-base plasma-sprayed materials, Part I: Phase stability of alumina and alumina-chromia. *Journal of Thermal Spray Technology*, *6*(3), 320–326. doi:10.1007/s11666-997-0066-9

Dubsky, J., Chraska, P., Kolman, B., Stahr, C. C., & Berger, L.-M. (2011). Phase formation control in plasma sprayed alumina-chromia coatings. *Ceramics-Silikáty*, *55*(3), 95–301.

Eschnauer, H. (1980). Hard material powders and hard alloy powders for plasma surface coating. *Thin Solid Films*, *73*(1), 1–17. doi:10.1016/0040-6090(80)90323-5

Fauchais, P., Montavon, G., Lima, R. S., & Marple, B. R. (2011). Engineering of new class of thermal spray nano-based microstructures from agglomerated nano-structured particles, suspensions and solutions: An invited review. Journal of Physics D: Applied Physics, 44, 093001.

Fauchais, P. L., Heberlein, J. V. R., & Boulos, M. I. (2014). Thermal spray fundamentals: From powder to part. New York, NY: Springer. doi:10.1007/978-0-387-68991-3

Gadow, R., Killinger, A., Rempp, A., & Manzat, A. (2010). Advanced ceramic tribological layers by thermal spray routes. *Advances in Science and Technology*, *66*, 106–119. doi:10.4028/www.scientific.net/AST.66.106

Gardos, M. N. (1988). The effect of anion vacancies on the tribological properties of rutile (TiO_{2-x}). *Tribology Transactions*, *31*(4), 427–436. doi:10.1080/10402008808981844

Gärtner, F., Kreye, H., & Richter, H. J. (2006). HVOF Spraying with powder and wire. In Proceedings 7th Kolloquium Hochgeschwindigkeitsflammspritzen (pp. 39-56). Unterschleißheim, Gemeinschaft Thermisches Spritzen.

Gell, M., Jordan, E. H., Sohn, Y. H., Goberman, D., Shaw, L., & Xiao, T. D. (2001). Development and implementation of plasma sprayed nanostructured ceramic coatings. *Surface and Coatings Technology*, *146-147*, 48–54. doi:10.1016/S0257-8972(01)01470-0

Habib, K. A., Saura, J. J., Ferrer, C., Damra, M. S., Gimenez, E., & Cabedo, L. (2006). Comparison of flame sprayed Al_2O_3/TiO_2 coatings: Their microstructure, mechanical properties and tribology behavior. *Surface and Coatings Technology*, *201*(3-4), 1436–1443. doi:10.1016/j.surfcoat.2006.02.011

Harju, M., Levänen, E., & Mäntylä, T. (2004). Surface characterization of plasma sprayed oxide materials: Estimation of surface acidity using mass titration. *Journal of Colloid and Interface Science*, *276*(2), 346–353. doi:10.1016/j.jcis.2004.03.058 PMID:15271562

Herrmann, M., Toma, F.-L., Berger, L.-M., Kaiser, G., & Stahr, C. C. (2014). Comparative study of the corrosion resistance of thermally sprayed ceramic coatings and their bulk ceramic counterparts. *Journal of the European Ceramic Society*, *34*(2), 493–504. doi:10.1016/j.jeurceramsoc.2013.08.033

Holmberg, K., & Matthews, A. (2009). Coatings tribology: Properties, mechanisms, techniques and applications in surface engineering (2nd ed.). Amsterdam: Elsevier.

Houdková, Š., Zahálka, F., Kašparová, M., & Berger, L.-M. (2011). Comparative study of thermally sprayed coatings under different types of wear conditions for hard chromium replacement. *Tribology Letters*, *43*(2), 139–154. doi:10.1007/s11249-011-9791-9

Ilavsky, J., Berndt, C. C., Herman, H., Chraska, P., & Dubsky, J. (1997). Alumina-base plasma-sprayed materials, Part II: Phase transformations in aluminas. *Journal of Thermal Spray Technology*, *6*(4), 439–444. doi:10.1007/s11666-997-0028-2

Jordan, E. H., Gell, M., Sohn, Y. H., Goberman, D., Shaw, L., Jiang, S., & Strutt, P. et al. (2001). Fabrication and evaluation of plasma sprayed nanostructured alumina-titania Coatings with superior properties. *Materials Science and Engineering A*, *301*(1), 80–89. doi:10.1016/S0921-5093(00)01382-4

Knuuttila, J., Ahmaniemi, S., & Mäntylä, T. (1999). Wet abrasion and slurry erosion resistance of thermally sprayed oxide coatings. *Wear*, *232*(2), 207–212. doi:10.1016/S0043-1648(99)00147-7

Knuuttila, J., Sorsa, P., Mäntylä, T., Knuuttila, J., & Sorsa, P. (1999). Sealing of thermal spray coatings by impregnation. *Journal of Thermal Spray Technology*, *8*(2), 249–257. doi:10.1007/s11666-999-0002-2

Krömmer, W., Heinrich, P., & Kreye, H. (2009). High velocity oxy-fuel flame spraying: Past – present – future. In Proceedings 8th Kolloquium Hochgeschwindigkeitsflammspritzen (pp. 9–16). Unterschleißheim: Gemeinschaft Thermisches Spritzen.

Landa, J., Illaramendi, I., Kelling, N., Woydt, M., Skopp, A., & Hartelt, M. (2007). Potential of thermal sprayed Ti_nO_{2n-1}-coatings for substituting molybdenum-based ring coatings. *Industrial Lubrication and Tribology*, *59*(5), 217–229. doi:10.1108/00368790710776801

Lathabai, S., Ottmüller, M., & Fernandez, I. (1998). Solid particle erosion behaviour of thermal sprayed ceramic, metallic and polymer coatings. *Wear*, *221*(2), 93–108. doi:10.1016/S0043-1648(98)00267-1

Leivo, E. M., Vippola, M. S., Sorsa, P. P. A., Vuoristo, P. M. J., & Mäntylä, T. A. (1997). Wear and corrosion properties of plasma sprayed Al_2O_3 and Cr_2O_3 coatings sealed by aluminum phosphates. *Journal of Thermal Spray Technology*, *6*(2), 205–210. doi:10.1007/s11666-997-0014-8

Lima, R. S., & Bergmann, C. P. (1996). Phase transformations on flame sprayed alumina. In C.C. Berndt (Ed.), *Proceedings of the National Thermal Spray Conference*. (pp.765-771) Cincinnati, OH: ASM International.

Lima, R. S., & Marple, B. R. (2006). From APS to HVOF spraying of conventional and nanostructured titania feedstock powders: A study on the enhancement of the mechanical properties. *Surface and Coatings Technology*, *200*(11), 3428–3437. doi:10.1016/j.surfcoat.2004.10.137

Lima, R. S., Moreau, C., & Marple, B. R. (2007). HVOF-sprayed coatings engineered from mixtures of nanostructured and submicron Al_2O_3-TiO_2 powders: An enhanced wear performance. *Journal of Thermal Spray Technology*, *16*(5-6), 866–872. doi:10.1007/s11666-007-9092-x

Luo, H., Goberman, D., Shaw, L., & Gell, M. (2003). Indentation fracture behavior of plasma-sprayed nanostructured Al_2O_3-13wt.%TiO_2 coatings. *Materials Science and Engineering A*, *346*(1-2), 237–245. doi:10.1016/S0921-5093(02)00523-3

Marple, B. R., Voyer, J., & Béchard, P. (2001). Sol infiltration and heat treatment of alumina-chromia plasma-sprayed coatings. *Journal of the European Ceramic Society*, *21*(7), 861–868. doi:10.1016/S0955-2219(00)00276-4

Mathesius, S., & Krömmer, W. (2009). *Praxis des thermischen Spritzens: Anleitung für das Fachpersonal*. Düsseldorf: DVS-Media.

McPherson, R. (1973). Formation of metastable phases in flame- and plasma-prepared alumina. *Journal of Materials Science, 8*(6), 851–858. doi:10.1007/BF02397914

McPherson, R. (1980). On the formation of thermally sprayed alumina coatings. *Journal of Materials Science, 15*(12), 3141–3149. doi:10.1007/BF00550387

Müller, J.-H., & Kreye, H. (2001). Mikrostruktur und Eigenschaften von thermisch gespritzten Aluminiumoxidschichten. *Schweissen und Schneiden, 53*(6), 336–345.

Niemi, K. (2009). *Abrasion wear characteristics of thermally sprayed alumina based coatings*. Doctoral Thesis, Publication 820. Tampere: Tampere University of Technology.

Niemi, K., Vuoristo, P., & Mäntylä, T. (1991). Chromium oxide coatings deposited by plasma spraying and detonation gun spraying. In S. Blum-Sandmeier, H. Eschnauer, P. Huber, & A.R. Nicoll (Eds.), Proceedings 2nd Plasma-Technik-Symposium (pp. 311-322). Wohlen, Switzerland: Plasma-Technik AG.

Niemi, K., Vuoristo, P., & Mäntylä, T. (1994). Properties of alumina-based coatings deposited by plasma spray and detonation gun spray processes. *Journal of Thermal Spray Technology, 3*(2), 199–203.

Niemi, K., Vuoristo, P., Mäntylä, T., Lugscheider, E., Knuuttila, J., & Jungklaus, H. (1995). Wear characteristics of oxide coatings deposited by plasma spraying, high power plasma spraying and detonation gun spraying. In C.C. Berndt & S. Sampath (Eds.), *Advances of Thermal Spray Science and Technology, Proceedings of the 8th National Thermal Spray Conference* (pp. 645-650). Materials Park, OH: ASM International.

Normand, B., Fervel., V., Coddet, C., Nikitine, V. (2000). Tribological properties of plasma sprayed alumina-titania coatings: Role and control of the microstructure. *Surface and Coatings Technology, 123*(2-3), 278-287.

Pawlowski, L. (1996). Technology of thermally sprayed Anilox rolls: State of the art, problems, and perspectives. *Journal of Thermal Spray Technology, 5*(3), 317–334. doi:10.1007/BF02645884

Pawlowski, L. (2008). *The science and engineering of thermal spray coatings* (2nd ed.). Chichester, UK: John Wiley & Sons. doi:10.1002/9780470754085

Rainforth, W. M. (2004). The wear behaviour of oxide ceramics - A review. *Journal of Materials Science, 32*(22), 6705–6721. doi:10.1023/B:JMSC.0000045601.49480.79

Rico, A., Rodriguez, J., Otero, E., Zeng, P., & Rainforth, W. M. (2009). Wear behaviour of nanostructured alumina-titania coatings deposited by atmospheric plasma spray. *Wear*, *267*(5-8), 1191–1197. doi:10.1016/j.wear.2009.01.022

Sánchez, E., Moreno, A., Vicent, M., Salvador, M. D., Bonache, V., Klyatskina, E., & Moreno, R. et al. (2010). Preparation and spray drying of Al_2O_3-TiO_2 nanoparticle suspensions to obtain nanostructured coatings by APS. *Surface and Coatings Technology*, *205*(4), 987–992. doi:10.1016/j.surfcoat.2010.06.002

Skopp, A., Kelling, N., Woydt, M., & Berger, L.-M. (2007). Thermally sprayed titanium suboxide coatings for piston ring/cylinder liners under mixed lubrication and dry-running conditions. *Wear*, *262*(9-10), 1061–1070. doi:10.1016/j.wear.2006.11.012

Sōmiya, S., Hirano, S., & Kamiya, S. (1978). Phase relations of the Cr_2O_3-TiO_2 system. *Journal of Solid State Chemistry*, *25*(3), 273–284. doi:10.1016/0022-4596(78)90112-3

Stahr, C. C., Saaro, S., Berger, L.-M., Dubsky, J., Neufuss, K., & Herrmann, M. (2007). Dependance of the stabilization of α-alumina on the spray process. *Journal of Thermal Spray Technology*, *16*(5-6), 822–830. doi:10.1007/s11666-007-9107-7

Storz, O., Gasthuber, H., & Woydt, M. (2001). Tribological properties of thermal-sprayed Magnéli-type coatings with different stoichiometries (Ti_nO_{2n-1}). *Surface and Coatings Technology*, *140*(2), 76–81. doi:10.1016/S0257-8972(01)01024-6

Sundararajan, G., Sen, D., & Sivakumar, G. (2005). The tribological behaviour of detonation sprayed coatings: The importance of coating process properties. *Wear*, *258*(1-4), 377–391. doi:10.1016/j.wear.2004.03.022

Thomas, H. A. J., & Stevens, R. (1989). Aluminium titanate – A literature review part 2: Engineering properties and thermal stability. *British Ceramic. Transactions and Journal*, *88*(5), 184–190.

Toma, F.-L., Berger, L.-M., Langner, S., & Naumann, T. (2010). Suspension spraying – The potential of a new spray technology. *Thermal Spray Bulletin*, *3*(1), 24–29.

Toma, F.-L., Berger, L.-M., Stahr, C.C., Naumann, T., & Langner, S. (n.d.). Thermisch gespritzte Al_2O_3-Schichten mit einem hohen Korundgehalt ohne eigenschaftsmindernde Zusätze und Verfahren zu ihrer Herstellung, DE 10 2008 026 101 B4, filed: 30.5.2008, granted: 18.2.2010; EP 2300630 A1; WO 2009/146832 A1; US 8,318,261, B2; CA 2726434 A1; JP 2011-522115 A.

Toma, F.-L., Berger, L.-M., Stahr, C. C., Naumann, T., & Langner, S. (2010). Microstructures and functional properties of suspension-sprayed Al_2O_3 and TiO_2 coatings: An overview. *Journal of Thermal Spray Technology, 19*(1-2), 262–274. doi:10.1007/s11666-009-9417-z

Toma, F.-L., Stahr, C. C., Berger, L.-M., Saaro, S., Herrmann, M., Deska, D., & Michael, G. (2010). Corrosion resistance of APS- and HVOF-sprayed coatings in the Al_2O_3-TiO_2 system. *Journal of Thermal Spray Technology, 19*(1-2), 262–274. doi:10.1007/s11666-009-9417-z

Trache, R., Berger, L.-M., & Leyens, C. (n.d.). Comparison of alumina-titania coatings deposited by plasma and HVOF spray techniques. In *Proceedings of the International Thermal Spray Conference* (pp. 368-372). Düsseldorf: DVS Media.

Tucker, R. C., Jr. (Ed.). (2013). ASM Handbook: Thermal spray technology (vol. 5A). Materials Park, OH: ASM International.

Turunen, E., Varis, T., Hannula, S.-P., Vaidya, A., Kulkarni, A., Gutleber, J., & Herman, H. et al. (2006). On the role of particle state and deposition procedure on mechanical, tribological and dielectric response of high velocity oxy-fuel sprayed alumina coatings. *Materials Science and Engineering A, 415*(1-2), 1–11. doi:10.1016/j.msea.2005.08.226

Vargas, F., Ageorges, H., Fournier, P., Fauchais, P., & López, M. E. (2010). Mechanical and tribological performance of Al_2O_3-TiO_2 coatings elaborated by flame and plasma spraying. *Surface and Coatings Technology, 205*(4), 1132–1136. doi:10.1016/j.surfcoat.2010.07.061

Vuoristo, P. (2014). Thermal spray coating processes. In D. Cameron (Ed.), *Comprehensive materials processing* (Vol. 4, pp. 229–276). Elsevier. doi:10.1016/B978-0-08-096532-1.00407-6

Vuoristo, P., Määttä, A., Mäntylä, T., Berger, L.-M., & Thiele, S. (2002). Properties of ceramic coatings prepared by HVOF and plasma spraying of titanium suboxide powders. In E. Lugscheider (Ed.), *Proceedings of the International Thermal Spray Conference 2002* (pp. 488-493). Düsseldorf: DVS-Verlag.

Vuoristo, P. M. J., Niemi, K. J., & Mäntylä, T. A. (1992). On the properties of detonation gun sprayed and plasma sprayed ceramic coatings. In C. C. Berndt (Ed.), *Thermal spray: International advances in coatings technology* (pp. 171–175). Materials Park, OH: ASM International.

Werner, H.-D. (1974). Phasenbeziehungen im system TiO_2-Cr_2O_3. *Neues Jahrbuch für Mineralogie Monatshefte*, 218–234.

Westergård, R., Axén, N., Wiklund, U., & Hogmark, S. (1998). The erosion and abrasion characteristics of alumina coatings plasma sprayed under different spraying conditions. *Tribology International, 31*(5), 271–279. doi:10.1016/S0301-679X(98)00033-4

Westergård, R., Erickson, L. C., Axén, N., Hawthorne, H. M., & Hogmark, S. (2000). An evaluation of plasma sprayed ceramic coatings by erosion, abrasion and bend testing. *Wear, 246*(1-2), 12–19. doi:10.1016/S0043-1648(00)00506-8

Woydt, M. (2005). Triboaktive Werkstoffe für hohe Gleitgeschwindigkeiten bis 800°C. *Tribologie und Schmierungstechnik, 52*(4), 5–12.

Xie, Y., & Hawthorne, H. M. (1999). The damage mechanisms of several-sprayed ceramic coatings in controlled scratching. *Wear, 233-235*, 293–305. doi:10.1016/S0043-1648(99)00211-2

Yang, K., Feng, J. W., Zhou, X. M., & Tao, S. Y. (2012). Microstructural characterization and strengthening-toughening mechanism of plasma-sprayed Al_2O_3-Cr_2O_3 composite coatings. *Journal of Thermal Spray Technology, 21*(5), 1011–1024. doi:10.1007/s11666-012-9796-4

Yu, S.H., & Wallar, H. (n.d.). Chromia spray powders and a process for making the same, US 6,774,076, B2, filed: 3.09.2002, granted: 10.08.2004, WO 03/086974 A1.

Znamirowski, Z., Pawlowski, L., Cichy, T., & Czarczynski, W. (2004). Low macroscopic field electron emission from surface of plasma sprayed and laser engraved TiO_2, Al_2O_3+13TiO_2 and Al_2O_3+40TiO_2 coatings. *Surface and Coatings Technology, 187*(1), 37–46. doi:10.1016/j.surfcoat.2004.01.001

KEY TERMS AND DEFINITIONS

Alumina (Al_2O_3): Oxide material, widely used for the preparation of thermal spray coatings.

Atmospheric Plasma Spraying (APS): One of the most widely applied thermal spray processes using a plasma as an energy source.

Chromia (Cr_2O_3): Oxide material, widely used for the preparation of thermal spray coatings.

Detonation Gun Spraying (DGS): Spray process using a flame, produced by controlled detonations of a hydrocarbon-oxygen mixture.

High Velocity Oxy-Fuel (HVOF) Spraying: One of the most widely applied thermal spray processes using a flame, produced by combustion of gaseous or liquid hydrocarbon in a pressurized chamber as an energy source.

Thermal Spray Coatings: Coatings with a typical thickness in the range 100-500 µm, which are prepared by one of the thermal spray processes and are primarily mechanically bonded to the substrate.

Titania (TiO_2): Oxide material, widely used for the preparation of thermal spray coatings.

Chapter 9
Slurry Erosion Behavior of Thermal Spray Coatings

Harpreet Singh Grewal
Indian Institute of Technology Ropar, India

Harpreet Singh
Indian Institute of Technology Ropar, India

ABSTRACT

Slurry erosion is a degrading phenomenon usually observed in machineries dealing with particle-laden fluid such as hydro power plants, ship propellers, pump impellers, valves, and connecting pipes. The low erosion resistance of commonly employed structural materials prompts the use of different surface modification techniques. Among several types of surface modification techniques, thermal spraying has achieved a significant recognition worldwide due to its versatile nature. In this chapter, slurry erosion behavior of thermal sprayed coatings has been discussed with special emphasis on the contribution of different coating related parameter. It has been observed that microstructure play an important role in determining the slurry erosion performance of thermal spray coatings. Different microstructural features such as splat boundaries, pores, un-melted particles, and cracks are detrimental for the thermal spray coatings exposed to erosive environment. A parameter useful for identification of primary erosion mechanism for thermal sprayed coatings is also discussed.

DOI: 10.4018/978-1-4666-7489-9.ch009

INTRODUCTION

Slurry erosion is a type of wear phenomenon in which the degradation of the target surface takes place due to impact of solid particles entrained in liquid medium. During impact, significant amount of complex stresses accompanied by high strain rates originate in the target surface (Hutchings & Winter, 1974). This result in significant deformation of the target surface eventually leading to loss of material. The problem caused by slurry erosion is highly severe and this results in the loss of performance of many fluid machineries and components such as pumps, hydroturbines, propeller, control values and connecting pipelines to name a few (Mann & Arya, 2002). The direct and indirect economic costs involved in repair and maintenance of the slurry affected components could be enormous (Grewal, Bhandari, & Singh, 2012; Mann, 2000; Mann & Arya, 2001).

Various factors that affect the slurry erosion performance of a target material can be categorized into following three categories(Finnie, 1995)

- Factors dependent upon the operating parameters such as impact velocity, impingement angle, concentration of erodent particles, viscosity, density and temperature of the carrier fluid.
- Factors dependent upon the erodent particles such as size, shape, composition, hardness.
- Factors dependent upon the target material such as yield and ultimate strength, ductility, hardness, fracture toughness, microstructure and composition.

Other than these factors, it has been observed that restitution coefficient, which is a complex function of different mechanical properties of target and erodent materials and the operating parameters such as impact velocity and angle (Hussein & Tabakoff, 1974) has been observed to play an important role (Grewal, Agrawal, & Singh, 2013a; Papini, 2003).

It is understood that with an increase in velocity of the erodent particles, the slurry erosion rate also increases (Kleis & Kulu, 2008). It has been observed that slurry erosion rate exhibits a power-law relation with velocity as shown in Equation (1)

$$E = KV^n \tag{1}$$

Here 'E' represents the slurry erosion rate, 'K' is the proportionality constant, 'V' is the erodent velocity and 'n' the velocity exponent. It has been normally observed that the value of 'n' varies in between 2 to 5 (Kleis & Kulu, 2008). However, value below two has also been reported (Grewal, Agrawal, & Singh, 2013a; Shivamurthy, Kamaraj, Nagarajan, Shariff, & Padmanabham, 2009).

Impingement angle of the erodent particles has also been observed to significantly influence the slurry erosion performance of the materials. It has been generally observed that materials exhibit maximum slurry erosion either at the normal impingement angle or at bleak angles typically between 20° to 30°. Materials showing maximum erosion rate at bleak angles are said to exhibit ductile mode of erosion (Finnie, 1995; A.V. Levy, 1995). However, materials for which maximum erosion takes place at 90° are termed to exhibit brittle mode of erosion. This type of subtle categorization based upon the dependence on impingement angle can be related to the erosion mechanism (Levy, 1995; Stachowiak & Batchelor, 2005).

Further, the concentration of the erodent particles in the carrier medium has been observed to exhibit a non-linear relation with slurry erosion rates (Kleis & Kulu, 2008). This type of relationship has been observed by number of investigators (Anand, Hovis, Conrad, & Scattergood, 1987; Arnold & Hutchings, 1989; Deng, Chaudhry, Patel, Hutchings, & Bradley, 2005; Prasad, Jha, Modi, & Yegneswaran, 2004; Turenne, Fiset, & Masounave, 1989). However, some investigators have also reported a linear relation between the slurry erosion rates and the erodent concentration (Bhandari, Singh, Kansal, & Rastogi, 2012; Bhandari, Singh, Kumar, & Rastogi, 2012; Grewal, Bhandari, & Singh, 2012; Krishnamoorthy, Seetharamu, & Sampathkumaran, 1993; Kumar Goyal, Singh, Kumar, & Sahni, 2012; Padhy & Saini, 2009; Stack & Abd El-Badia, 2008; Sun, Wharton, & Wood, 2009; Telfer, Stack, & Jana, 2012). Generally, with an increase in concentration of the erodent particles initially the slurry erosion rates increases, mainly because of increase in the number of impact events. However, further increase in concentration beyond a certain threshold value would result in interaction between the incoming and the rebounded particles. This interaction commonly known as shielding phenomenon leads to the lowering of erosion rates due to reduction of the incident velocity of the particles (Anand,Hovis, Conrad, & Scattergood, 1987; Uuemo & Kleis, 1975). With further increase in concentration beyond threshold value it is expected that a steady state is attained due to the balance formed between the impact events and the shielding phenomenon (Kleis & Kulu, 2008). Other operating parameters such as viscosity and density of the carrier fluid have also been observed to play a crucial role in the slurry erosion process (Clark, 1992; Wong & Clark, 1993). Both of these factors have been observed to affect significantly the particle trajectories and properties of the squeeze film formed inside the impact zone.

Various different types of test rigs have been used for slurry erosion testing of materials and coatings. A detailed review along with their advantages and limitation could be found elsewhere (Grewal, Agrawal & Singh, 2013b). Among various designs, pot-type and jet-type rigs have been prominently used. Pot-type rigs are simple in design and easy to use. However, they have several disadvantages such as little or no control over impingement angle and velocity and concentration of particles impinging the sample surface. On the contrary, jet-type rigs provide more control over these operating parameters. The complexity of the design varies with the kind of jet-type rig used. The re-circulating jet-type rigs, wherein erodent particles remain in the cycle and interact with the sample again and again are easier to design and construct. On the other hand, non-recirculating type jet rigs have complex design and require expertise to operate and control the test parameters.

In present chapter, the affect of various factors, especially the microstructure, on the slurry erosion performance of thermal spray coatings have been discussed. Further, a case study on the slurry erosion performance of Ni-Al_2O_3 based thermal spray coatings has also been presented and discussed. Detailed discussion on the slurry erosion mechanism and the influence of microstructural and mechanical properties of the coatings on their slurry erosion behavior has also been discussed. Finally, a parameter for predicting the slurry erosion mechanism of materials and coatings has also been presented and discussed.

BACKGROUND

Slurry erosion is a critical issue and slurry erosion severely degrades and affects the performance of commonly employed structural materials. From the detailed and in-depth review of the literature available on slurry erosion, it has been observed that surface coatings can help to improve performance of these structural materials against the slurry erosion (Barbezat, Nicol, & Sickinger, 1993; Bu Qian & Luer, 1994; Chen & Hutchings, 1998; Davis, Boone, & Levy, 1986; de Souza & Neville, 2003; Erickson, Hawthorne, & Troczynski, 2001; Gang, 2003; Grewal, Bhandari, & Singh, 2012; Halling & Arnell, 1984; Hurricks, 1972; Levy & Buqian, 1988; Shrestha, Hodgkiess, & Neville, 2001, 2005; Souza & Neville, 2007; Stack & Abd El Badia, 2006; Tan, 2003; Wang & Lee, 2000; Wood, Mellor, & Binfield, 1997; Zhao, Goto, Matsumura, Takahashi, & Yamamoto, 1999). Amongst the several coating deposition methods, thermal spraying process has attracted considerable attention during recent times owing to its versatile nature.

Slurry erosion studies have been conducted on several coating compositions belonging to different class of materials such as ceramics, polymers, cermets, composites and metallic. The main focus behind the selection of these coatings has either been the hardness or the toughness. Ceramic-based hard coatings have been observed to provide reasonable protection at low impingement angles. However, their poor slurry erosion performance at normal impingement angles promoted the use of composite and cermet coatings. Tungsten carbide (WC) possessing high hardness and sufficient fracture toughness appears to be an important candidate material for obtaining slurry erosion resistant coatings. As a result, several studies can be found in the literature wherein WC-based coatings have been suggested for protection against slurry erosion.

The WC-based thermal spray coatings can offer sufficient resistance against slurry erosion (Grewal, Bhandari, & Singh, 2012; Mann, 2000), however, studies on cavitation erosion behavior of these coatings are limited. From the slurry erosion studies conducted on flame sprayed WC/Co-FeNiCr and Cr_2O_3 coatings, it was observed that microstructural defects such as porosity, un-melted or partially melted particles play an important role (Santa, Espitia, Blanco, Romo, & Toro, 2009). These coatings showed higher resistance against slurry erosion, however, their performance under cavitation erosion was not satisfactory (Santa,Espitia, Blanco, Romo, & Toro, 2009). Even High Velocity Oxy-Fuel (HVOF) sprayed WC/Co and Cr_2O_3 coatings were severely damaged under cavitating erosion conditions (Santa,Espitia, Blanco, Romo, & Toro, 2009).

Comparison of the slurry erosion behavior of ceramic (Al_2O_3 and Cr_2O_3) and polymeric coatings (Nylon-11) deposited by flame spray technique showed that Nylon-11 coating was more effective in reducing the slurry erosion rates in comparison to its ceramic counterparts (Lathabai, Ottmüller, & Fernandez, 1998). Detailed analysis of the eroded specimens showed the fine erodent debris particles embedded inside the Nylon coating. As a result of this, a composite layer was formed on surface of the Nylon coating and this composite layer reduced the slurry erosion rate. Among the ceramic coatings, Cr_2O_3 outperformed the Al_2O_3 coating, owing to large difference in the porosity, highlighting the importance of the microstructure.

Microstructure plays an important role in determining the slurry erosion performance of the coatings. Comparison of the slurry erosion performance of the coatings deposited by APS (atmosphere plasma spraying) and VPS (vacuum plasma spraying) techniques highlighted this important issue (Kingswell, Rickerby, Bull, & Scott, 1991). The VPS sprayed coatings showed superior performance against slurry erosion in comparison to coatings deposited using APS technique. Detailed investigation of the microstructures showed that in contrast to VPS coatings, the brittle fracture of the splats at the interface was responsible for the slurry erosion of APS coatings. The coatings deposited by VPS technique were free from most of the secondary oxides resulting in well-bonded splat structure. The oxidation of the molten powder particles can take place while moving in the flame. Coatings produced in vacuum or with little exposure to air can help reduce this problem. The presence of secondary oxides can adversely affect the bonding strength of the splats.

Other than the secondary oxides, the porosity content of the coatings is an important microstructural feature which affects the slurry erosion performance of the coatings. Comparison of slurry erosion resistance of several coatings deposited by APS and low pressure plasma spray (LPPS) technique with bulk sintered materials showed that lower porosity content was desirable against slurry erosion (Zhao, Goto, Matsumura, Takahashi, & Yamamoto, 1999). Better slurry erosion resistance of the bulk materials in comparison to coatings was attributed to their lower porosity content. Iit was observed that among different composite coatings of Al_2O_3 formed by adding Titania (3% and 13%) and Zirconia (25% and 40%), one with lowest porosity content (Al_2O_3 + 25 wt.% of Zirconia) has highest slurry erosion resistance (Branco, Gansert, Sampath, Berndt, & Herman, 2004). The heat treatment of the coatings helped in further reducing the porosity and hence the slurry erosion rates of the coatings. The various microstructural parameters such as porosity, presence of secondary oxides, cracks and un-melted particles affects the value of different mechanical properties. These, in turn, further influence the slurry erosion performance of the coatings. Moreover, the influence of different microstructural features on mechanical properties is highly complex in nature and requires further studies. This was pointed out by the result of the slurry erosion studies conducted on series of WC-Co and Cr_3C_2 + 25NiCr coatings. Against the general belief, it was observed that hardness alone should not be considered as sole criteria for the selection of the coating (Factor & Roman, 2002). Various other properties such as fracture toughness and ductility which are affected by the microstructure of the coating do also play an important role.

Slurry erosion testing on series of Co-based coatings developed by HVOF and plasma spray technique indicated Stellite 6 to exhibit superior performance. However, WC-12Co coating evaluated for comparison, showed further improvement in slurry erosion resistance (Kumar, Boy, Zatorski, & Stephenson, 2005). Among the two deposition techniques, coatings developed by HVOF technique showed better slurry erosion resistance mainly due to low porosity level. It is worth mentioning here that among these coatings, none of them was efficient enough in reducing the cavitation erosion rates of the SS 308 steel.

Other than porosity, it was observed that the structure of the coating lamella also significantly influenced the slurry erosion resistance. It was observed that mean bonding ratio defined as the ratio of total bonded interface area to total apparent lamellae contact area exhibited a linear relationship with slurry erosion resistance (Li, Yang, & Ohmori, 2006). With an increase in bonding area, the lamellar bonding increased and defects along the interface decreased and this probably reduced the slurry erosion rates. Fracture toughness was found to be the controlling mechanical property.

The adhesion of the coating with the substrate is also an important parameter, which can influence the slurry erosion response of the coating. In case of plasma spray technique, the adhesion (between the coating and substrate) and cohesion (between the coating lamella) strength is influenced by the input power (Li,Yang, & Ohmori, 2006; Mishra,Sahu, Das, Satapathy, Sen,Ananthapadmanabhan, & Sreekumar 2008). With an increase in power the melting of the powder particles takes place more effectively which enhances the bonding strength. However, it has also been observed that beyond a certain threshold value, with an increase in input power, the bonding strength is affected negatively. This is due to overheating and vaporization of the coating material.

Among different ceramic and metallic coatings, it was reported that WC-based coatings exhibited superior slurry erosion resistance (Hawthorne, Arsenault, Immarigeon, Legoux, & Parameswaran, 1999). Further, the effect of erodent particles on the slurry erosion performance of the coatings was found to be dependent upon the coating microstructure. Coating of WC-12Co gave better results for 200 μm erodent particles whereas, when the size of erodent particles was reduced to 35 μm, WC-10Co-4Cr showed better performance. It has been shown experimentally that specific energy required for the removal of WC-12Co coating is almost 1000 times higher than that of SS 316L (Clark, Hawthorne, & Xie, 1999).

However, when exposed to slurries comprised of corrosive media, the superior performance of the WC-Co coatings diminishes (Toma, Brandl, & Marginean, 2001). Among the batch of 10 different coatings, Cr_3C_2-25NiCr showed 10 times lower slurry erosion rates in comparison to WC-17Co coating. However, in absence of corrosive media, WC-12Co performed better in comparison to Cr_3C_2-25%NiCr coating (Mann, 2000). The substantially higher slurry erosion rate of the WC-Co was due to high corrosion rates of Co. The corrosion of the binder phase would severely affect the adhesion strength resulting in peeling-off of carbide particles by impacting erodent particles. In pure erosive conditions, better results reported for WC-12Co and this were directly related to its higher hardness (1100-1150 HV) as compared to that for Cr_3C_2 – 25%NiCr having hardness of 750-800 HV. Another study on the erosion- corrosion performance of the WC-Co-Cr coating deposited using super D–gun showed almost two fold higher slurry erosion resistance in comparison to SS 316 and SS 327 steels (Desouza & Neville, 2003). A significant interaction between the corrosion and slurry erosion phenomenon was found to exist. The increase in erodent content in the slurry enhanced the corrosion rates substantially.

The surface morphology of the coatings has also been observed to have a significant effect on the slurry erosion resistance (Murthy, Rao, & Venkataraman, 2001).The grinding of WC-10Co-4Cr coating helped in reducing the erosion rates of the coating by almost a factor of two in comparison to as-sprayed coatings. This improvement in slurry erosion resistance was attributed to the high residual stresses developed in the coatings due to grinding. These residual compressive stresses might help in reducing the fracture at carbide –matrix interface. Further, for similar coatings, detonation gun sprayed coatings exhibited higher resistance to erosion in comparison to HVOF coating owing to lower porosity and higher hardness.

Microstructural defects such as pores and cracks reduce the erosion-corrosion resistance of the coatings (Barik, Wharton, Wood, Tan, & Stokes, 2005). These microstructural defects lead to the reduction of the erosion-corrosion resistance of the Ni-Al-Bronze (NAB) coatings in comparison to its bulk counterpart. Negative erosion-corrosion synergy effect was observed with cast NAB giving better performance as compared to coating. Similar results were also reported in case of WC-Co-5Cr (Wood,Mellor, & Binfield, 1997). At low energy level (5×10^{-7} J) the slurry erosion response of both the D-gun sprayed coating and sintered bulk material were almost similar, however, at higher energy levels (6×10^{-6}J) the performance of sintered bulk material was 4 times better than coating. This behaviour was explained on the basis of anisotropic microstructure of coating (Wood, Mellor, & Binfield, 1997). The mechanism responsible for the loss of material was observed to be the interlinking of the cracks parallel to surface causing the removal of coating. Various other defects such as voids, oxides also played an important role. It has also been observed that the influence of impingement angle is preferential only in case of low impact energies (Wood, 2010). Beyond certain threshold impact energy, the slurry erosion rates are not influenced by the impingement angle. This indicates the dominant influence of pores and cracks below the threshold impact energy value wherein low-energy fluctuating stresses leads to the growth of crack.

The lower size distribution of the precursor powder particles used for deposition of the coatings also helps in reducing the slurry erosion rates (Berget, Rogne, & Bardal, 2007). Among the three size distributions (15-45, 25-38 & 36-45 μm), the one with narrower size distribution (36-45 μm) showed highest resistance which could be related to the quality of the coating deposited. The smaller range of powder particles would reduce the differential melting of powder particles and thus promoting the homogenous melting of powder particles. This could help in developing homogeneous defect free coatings. Further, it was also observed that reducing the WC grain size could help in improving the slurry erosion performance of the thermal spraying coatings (Thakur & Arora, 2013). Similar results were also reported for plasma sprayed Cr_2O_3-3%TiO_2 coatings (Singh, Sil,& Jayaganthan, 2011). However, in both of these studies it was observed that in spite of using nano-sized grain particles, the improvement in the slurry erosion performance of these coatings was not remarkably high.

Based on the slurry erosion test on various WC-based thermal spray coatings it was observed that in case of low impingement angles of erodent particles, higher content of harder phase in the coating was preferential for lower slurry erosion rates (Kulu, Hussainova, & Veinthal, 2005). However, for higher impact angles, structure with medium content of high hardness phase, gave better results.

MAIN FOCUS OF THE CHAPTER

Issues, Controversies, Problems

For protection against slurry erosion, most of the attention has been devoted to WC-based coatings owing to its high hardness and fracture toughness. Considerable attention needs to be focused on other compositions such as Ni-Al_2O_3/BC_4 (Grewal, Singh, & Agrawal, 2013c) so as to identify some economical alternatives to WC-based coatings. It was shown that cold sprayed Ni + 40%Al_2O_3 coatings could help in reducing the slurry erosion rates of the Inconel-600 alloy (Hu, Jiang, Tao, Xiong, & Zheng, 2011). However, the cavitation erosion performance of these coatings was not satisfactory. Lower bonding strength was identified as the major factor responsible for the degradation of the coating. The coatings Ni-Al_2O_3 developed using electrodeposition technique showed that the erosion rate increased with the alumina content and hardness (Levin, DuPont, & Marder, 2000). Reason for such an opposite trend shown by hardness was due to fracturing and de-bonding of alumina particles. However, with the addition of Co, erosion-corrosion performance of the Ni-Al_2O_3 coatings was significantly enhanced (Tian & Cheng, 2007; Wu, Li, Zhou, & Mitsuo, 2004). Nano-crystalline Ni-Al_2O_3 coating deposited using HVOF technique with Ni content of 2% and 5% were observed to possess high hardness of around 1300 HV (Turunen Varis, Gustafsson, Keskinen, Fält, & Hannula, 2006). Rubber abrasion test showed that the wear resistance was highly dependent upon coating hardness. When evaluated for toughness, the one having 2% Ni content showed least cracks resistance. The crack resistance of conventional and nano-crystalline alumina (without Ni) coating was almost similar. However, no attempt was made to evaluate the slurry erosion performance of these coatings.

In another study, sealing of the plasma sprayed alumina coating with Ni resulted in a 12% increment in hardness, obviously due to reduction in pores and voids (Westergard, 2004). The toughness was also reported to have enhanced with the reduction in crack propagation rate. As a result of sealing of the coating, erosion rates reduced by two-fold. The sealed coating also showed 40% lower wear rates when tested using micro-abrasion wear tests (Westergard, 2004).

Alumina content in the Ni-Al_2O_3 thermal spray coatings has significant influence on the slurry erosion performance (Grewal, Agrawal, & Singh, 2013d,Grewal, Agrawal, Singh, Shollock, 2013e). The slurry erosion resistance of the coatings showed a non-linear trend with the Al_2O_3 content (Figure 1). Maximum slurry erosion resistance was exhibited by coating containing 40 wt. % of the Al_2O_3.With further increase in Al_2O_3 content, slurry erosion resistance decreased and this could be explained on the basis of microstrutural and mechanical properties of the coatings. Table 1 lists the variation in different mechanical properties with the Al_2O_3 content. It could be observed that with the rise in alumina content, the porosity content also increased and this could be explained using the model shown in Figure 2. According to this model, for coating having high Al_2O_3 content, the probability that during deposition process incoming Al_2O_3 particle will encounter the already deposited Al_2O_3splat will be higher. As Al_2O_3 has lower plasticity in comparison to Ni, it will not be able to deform and fill up the adjacent pores which otherwise would have contributed towards reducing the porosity content in the coatings (Grewal, Singh, et al., 2013c). On the other hand, the higher porosity of 60 wt. % Al_2O_3 coating and brittleness of the Al_2O_3 resulted in lower fracture toughness for this coating composition. In addition to lower fracture toughness, low compressive residual stresses in the 60 wt.% Al_2O_3 content coating resulted in low slurry erosion resistance (Grewal, Agrawal, & Singh, 2013d,Grewal, Agrawal, Singh, et al., 2013e).

It is expected that higher compressive stresses in the coating will reduce the crack propagation rate. Attempt was made to correlate the slurry erosion resistance of these coatings with different microstructural and mechanical properties as illustrated in Figure 3. In these plots, the slurry erosion rates of the individual coatings are normalized using the results of the coating containing 60 wt. % of Al_2O_3. It could be observed that among the different parameters, slurry erosion resistance (inverse of erosion rates) exhibits highest correlation with fracture toughness and least with the surface roughness. Correlation with porosity is also within an acceptable range. As discussed in preceding section, it can be observed that with an increase in porosity, slurry erosion rates also increased. However, the correlation with the hardness is not according to expectations. It is generally believed that with an increase in hardness, slurry erosion rates decrease. However, this was not the case. This might be due to increase in brittleness of the coatings with an increase in hardness. However, the combination of both hardness and fracture toughness into a function of form shown in Figure 4 helped in further increase of the correlation factor (Grewal, Agrawal, & Singh, 2013d). Therefore, it can be deduced that for high slurry erosion resistance both hardness and fracture toughness are important mechanical properties. No doubt, both of these properties will further be the function of porosity content of the coating.

The surface morphology of the eroded Ni-Al_2O_3 thermal spray coatings is shown in Figure 5 and 6 (Grewal, Agrawal, & Singh, 2013d, Grewal, Agrawal, Singh, Shollock, 2013e). It can be observed that for coatings containing 20 wt.% and 40 wt.% of Al_2O_3, primary slurry erosion mechanism responsible for the loss of the coating is the removal of individual splats. The presence of cracks along the splat boundaries can be easily observed in Figure 6. The impact of erodent particles induced sub-surface shear stresses in the coating according to Hertzian contact theory. These shear stresses could aid in generation of cracks. The splat boundaries are more susceptible to cracking owing to low bonding strength. The lamellar structure

Figure 1. Effect of Al_2O_3 content on the normalized slurry erosion rates of high velocity flame sprayed Ni-Al_2O_3 coatings deposited on CA6NM hydroturbine steel(Grewal, Agrawal, & Singh, 2013d)

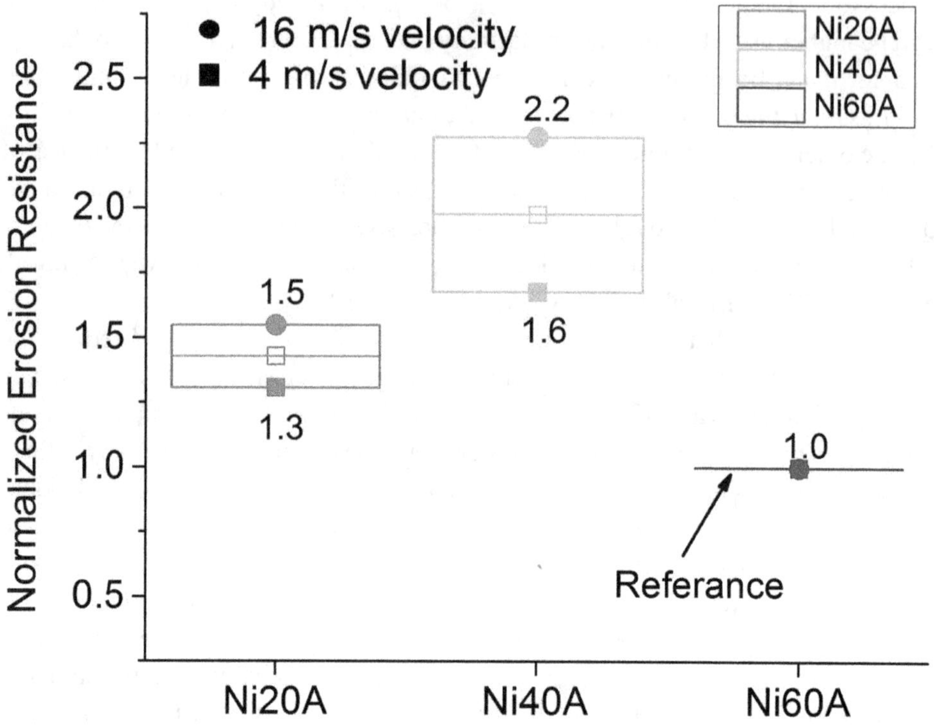

Table 1. Various microstructural and mechanical properties of high velocity flame sprayed Ni-Al_2O_3 coatings deposited on CA6NM hydroturbine steel(Grewal, Singh, &Agrawal, 2013f)

Alumina Content (wt. %)	Coating Thickness (μm) ± SD	Microhardness ($HV_{0.1\ Kgf}$)±SD	Fracture Toughness (MPa$\sqrt{m}$)±SD	Density (Kg/m³)±SD	Porosity (%)±SD	Residual Stress (MPa)±SD
20	684 ± 38	563±90	1.4±0.11	7100±100	1.3±0.20	-122±14
40	628 ± 54	714±78	1.6±0.09	6350±68	1.8±0.16	-93±10
60	583± 24	1151±108	0.9±0.12	5880±83	2.5±0.25	-31±6

of the thermal coatings assisted the propagation of cracks in longitudinal direction (parallel to coating-substrate interface). This finally resulted in the removal of either individual splat or chunk of coating material. A schematic model shown in Figure

Figure 2. Schematic model used to explain the dependency of the porosity of the Ni-Al_2O_3 based thermal spray coatings on the Al_2O_3 content(Grewal, Singh, Agrawal, 2013c)

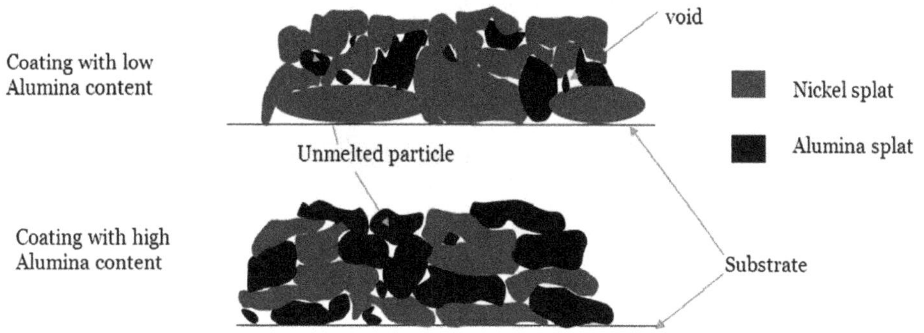

Figure 3. Correlation of the slurry erosion resistance of high velocity flame spray Ni-Al_2O_3 based coatings deposited on to CA6NM hydroturbine steel with different microstrucutral and mechanical properties of the coatings(Grewal, Agrawal, & Singh, 2013d,Grewal, Agrawal, Singh, Shollock, 2013e)

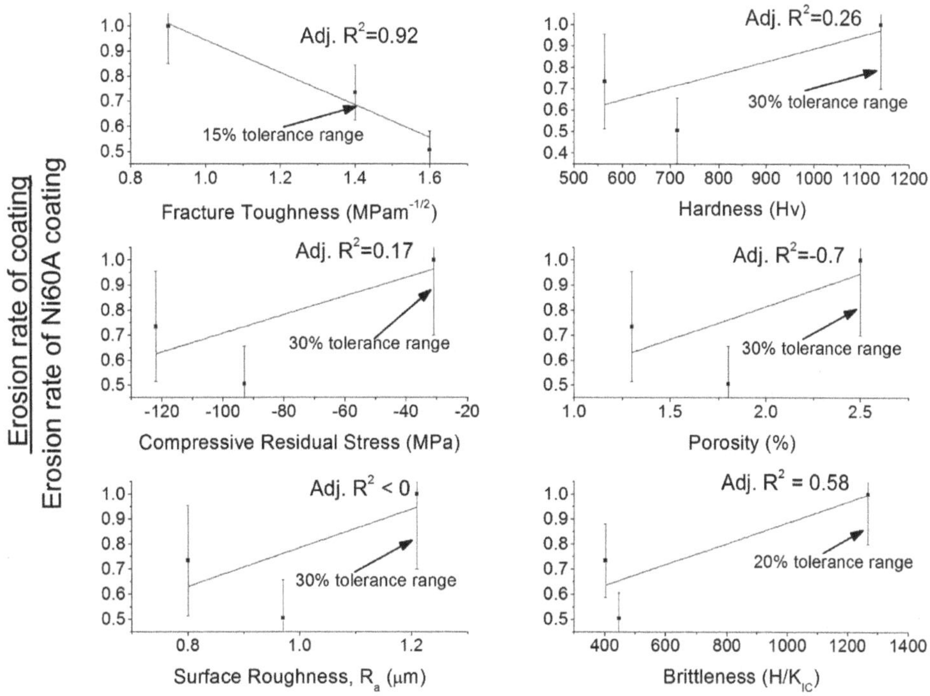

Figure 4. Correlation of the slurry erosion resistance of high velocity flame spray Ni-Al_2O_3 based coatings deposited on to CA6NM hydroturbine steel with a parameter($\sqrt[1/3]{K_{IC}^2 H}$). This parameter is the function of fracture toughness (K_{IC}) and hardness (H) (Grewal, Agrawal, & Singh, 2013d)

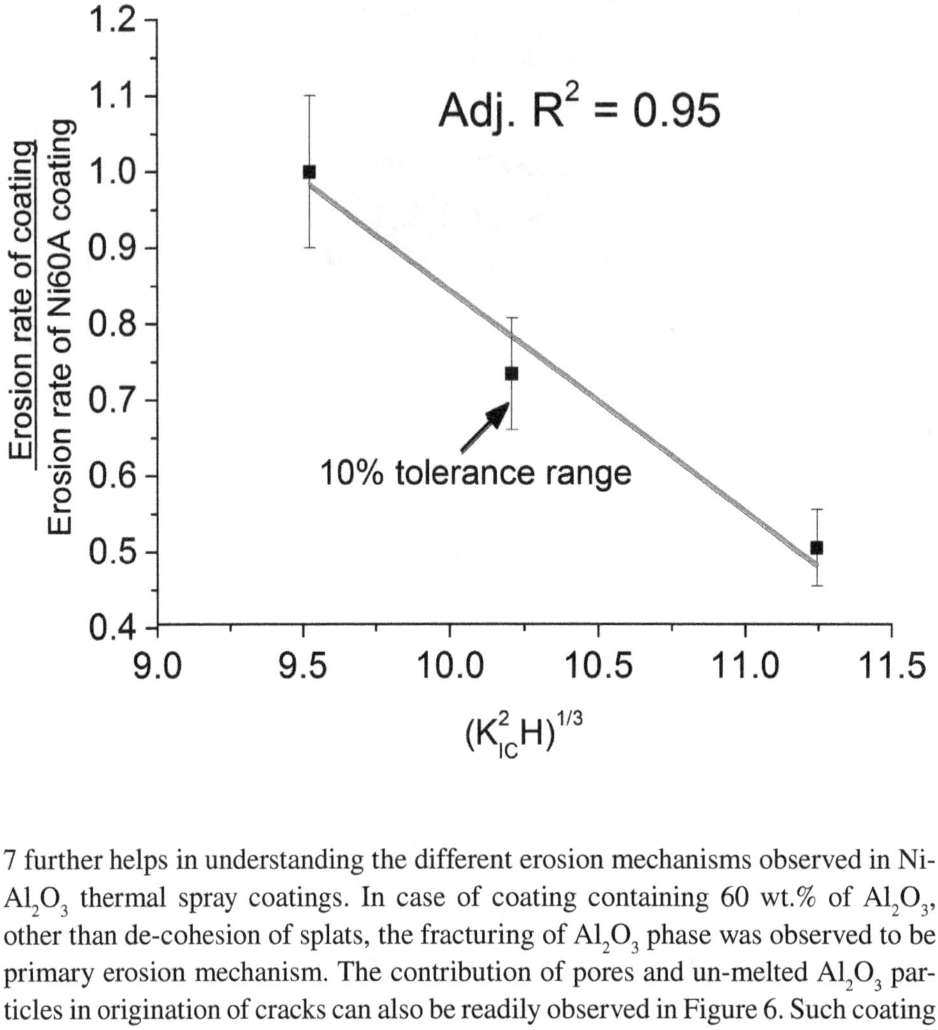

7 further helps in understanding the different erosion mechanisms observed in Ni-Al_2O_3 thermal spray coatings. In case of coating containing 60 wt.% of Al_2O_3, other than de-cohesion of splats, the fracturing of Al_2O_3 phase was observed to be primary erosion mechanism. The contribution of pores and un-melted Al_2O_3 particles in origination of cracks can also be readily observed in Figure 6. Such coating defects gave rise to high stress concentrations in their vicinity. The interaction of these high stress concentration fields with the stresses originating from the impinging erodent particles could cause the initiation of cracks. Presence of such cracks could be observed in Figure 6. Thus, efforts are required to improve the inter-bonding strength of the splats and reducing the coating defects for improving the slurry erosion performance of the coatings.

Figure 5. SEM micrographs showing the surface morphology of the eroded (a) uncoated CA6NM steel (b) Ni-20 wt.% Al_2O_3 coating (c) Ni-40 wt.% Al_2O_3 coating and (d) Ni-60 wt.% Al_2O_3 coating. Ni-Al_2O_3 coatings were deposited using high velocity flame spray technique on CA6NM hydroturbine steel. Test velocity of the erodent particles during slurry erosion experiments was 16 m/s and impingement angle was set at 90°. Concentration of the erodent particles was 0.5 wt. % (5000 ppm) (Grewal, Agrawal, & Singh, 2013d; Grewal, Agrawal, Singh, Shollock, 2013e).

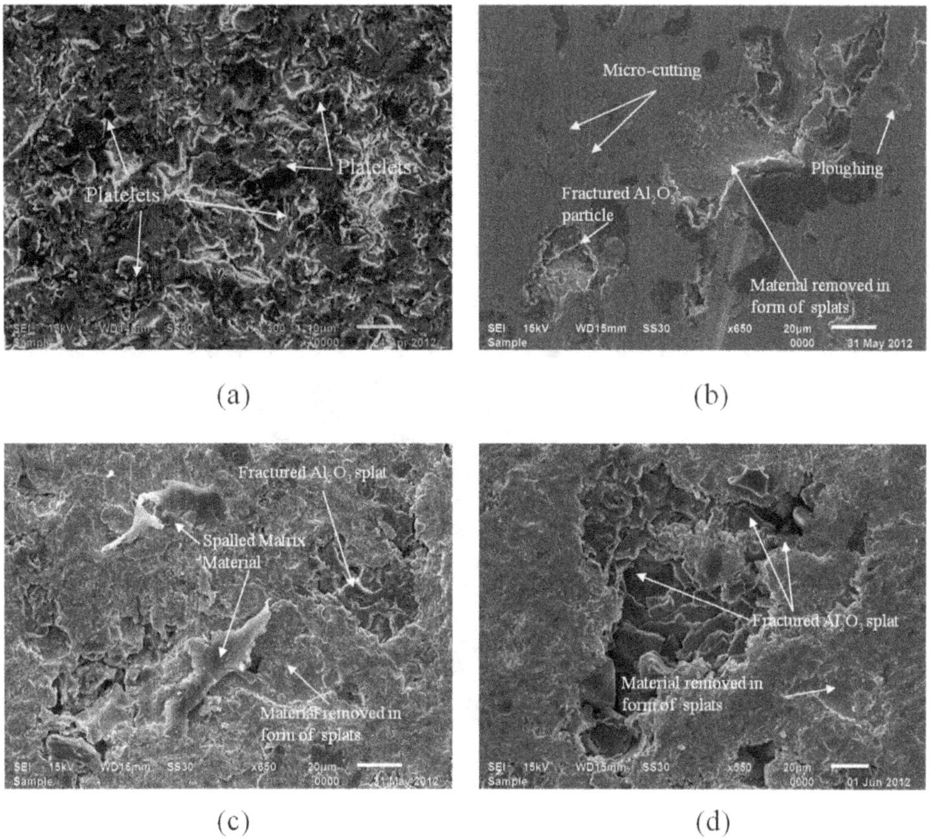

The erosion mechanism of the coatings and bulk materials could also be analyzed using a parameter namely, 'Erosion mechanism identifier, ξ'(Grewal, Agrawal, & Singh, 2013g). In comparison to the earlier model proposed by Sundararajan (1990), the model 'ξ' can be used to predict the erosion mechanism of different categories of materials subjected to various impingement conditions, especially the impingement angle. This parameter is defined as the ratio of the specific energy required to remove the target material (SER) to the specific energy expended in removing the target material (SEE) as shown in Equation (2).

Figure 6. SEM micrographs showing the sub-surface microstructure of the eroded (a) Ni-20 wt.% Al_2O_3 coating (b) Ni-40 wt.% Al_2O_3 coating and (c) Ni-60 wt.% Al_2O_3 coating. Ni-Al_2O_3 coatings were deposited using high velocity flame spray technique on CA6NM hydroturbine steel. Test velocity of the erodent particles during slurry erosion experiments was 16 m/s and impingement angle was set at 90°. Concentration of the erodent particles was 0.5 wt. % (5000 ppm) (Grewal, Agrawal, & Singh, 2013d; Grewal, Agrawal, Singh, Shollock, 2013e).

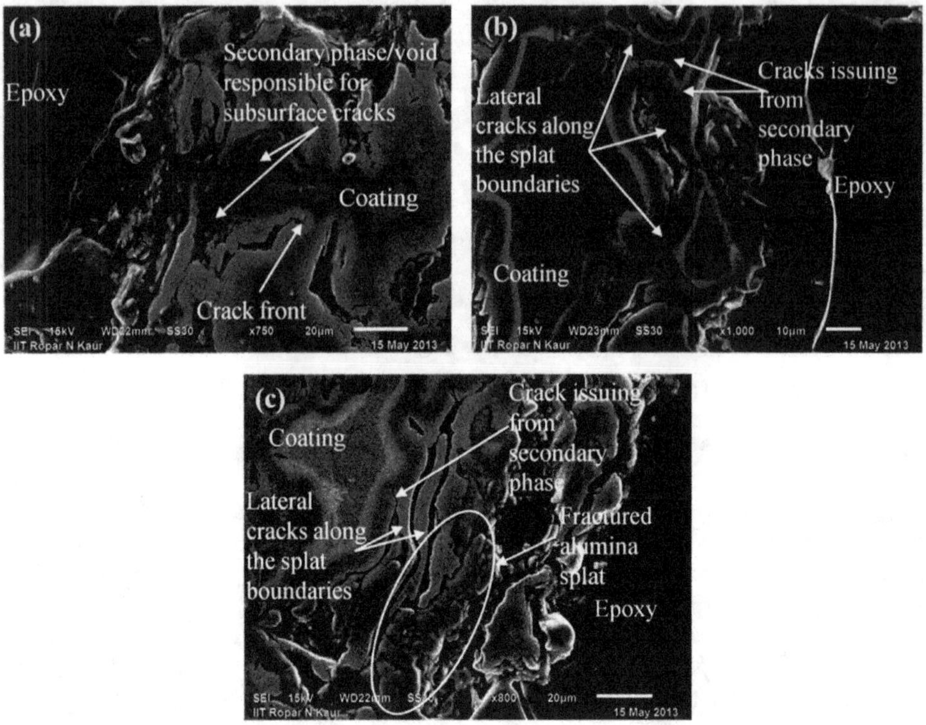

$$\xi = \frac{SER}{SEE} \tag{2}$$

This parameter is quite useful in predicting the erosion mechanism of different categories of materials. Further the substitution of the expressions for SER and SEE result in the form of 'ξ' as shown in Equation (3).

$$\xi = \frac{E\sigma\left(\frac{H}{K}\right)}{0.5mV^2} \tag{3}$$

Figure 7. Schematic model illustrating different slurry erosion mechanisms observed in Ni-Al_2O_3 based high velocity flame sprayed coatings (Grewal, Agrawal, & Singh, 2013d)

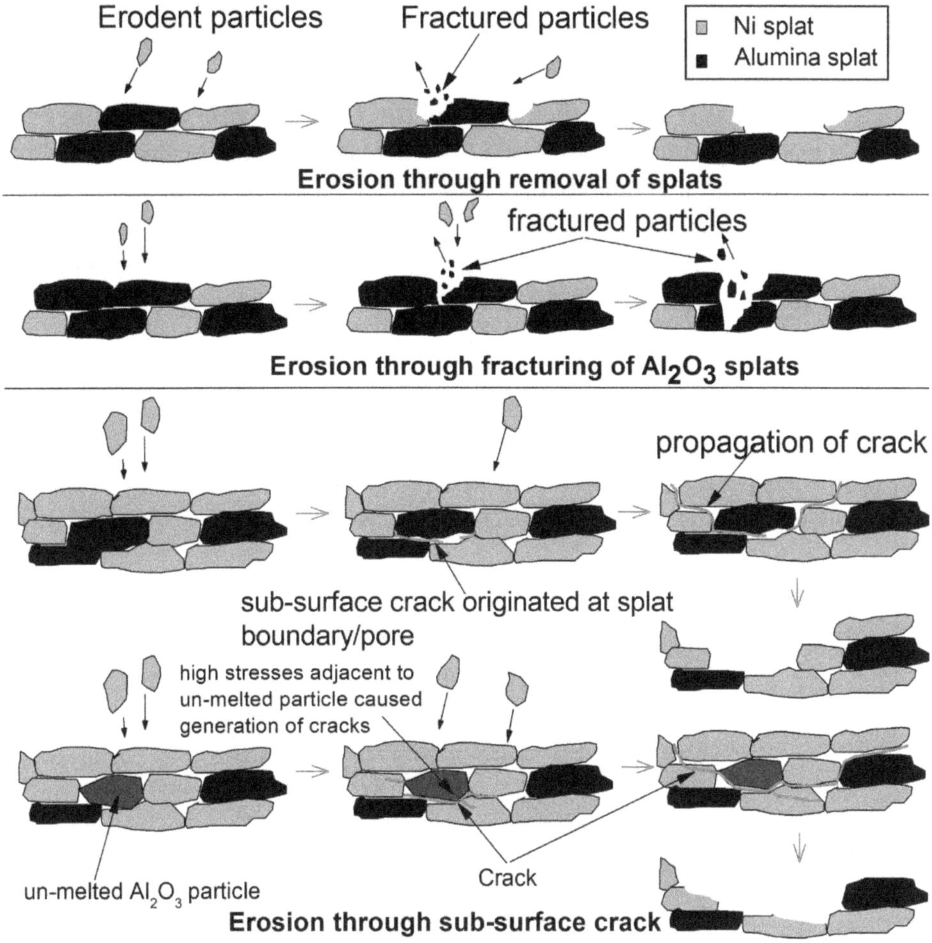

Here, symbols '*H*', '*K*' and 'σ' represent the hardness, toughness and critical strength of the target material. Further, '*m*' and '*V*' stand for the total mass and velocity of the erodent particles and '*E*' the volumetric erosion (m^3). 'ξ' is a dimensionless quantity the value of which can be used for the predicting the erosion mechanism as shown below:

- For 'ξ' < 1, erosion mechanism is primarily of ductile nature where mechanisms such as ploughing and platelet dominate,

- For 'ξ' > 1, erosion mechanism is primarily of brittle nature where mechanisms such as cracking and spalling dominate,
- For 'ξ' = 1, micro-cutting is the primary erosion mechanism

Results for the erosion mechanism predicted using 'ξ' are shown in Figure 8 (Grewal, Agrawal, & Singh, Shollock, 2013e). From the SEM micrographs of the eroded coatings as discussed earlier, the dominant mode for erosion of coatings was primarily brittle in nature (Figure 5 and 6). Whereas, for CA6NM steel, erosion of the material was through ductile mode. From the analysis of Figure 8, it could be observed that 'ξ' was able to successfully predict these erosion modes for respective materials. For CA6NM steel, the value of 'ξ' is nearly equal to 0.1, indicating ploughing to be the dominant erosion mechanism. From the experimental results (Figure 5), it could be observed that primary mechanism responsible for the loss of

Figure 8. Slurry erosion mechanism of the Ni-Al_2O_3 based high velocity flame sprayed coatings and uncoated CA6NM hydroturbine steel predicted using the model "Erosion Mechanism Identifier, ξ" shown as Equation (3)(Grewal, Agrawal, & Singh, et al.,2013e)

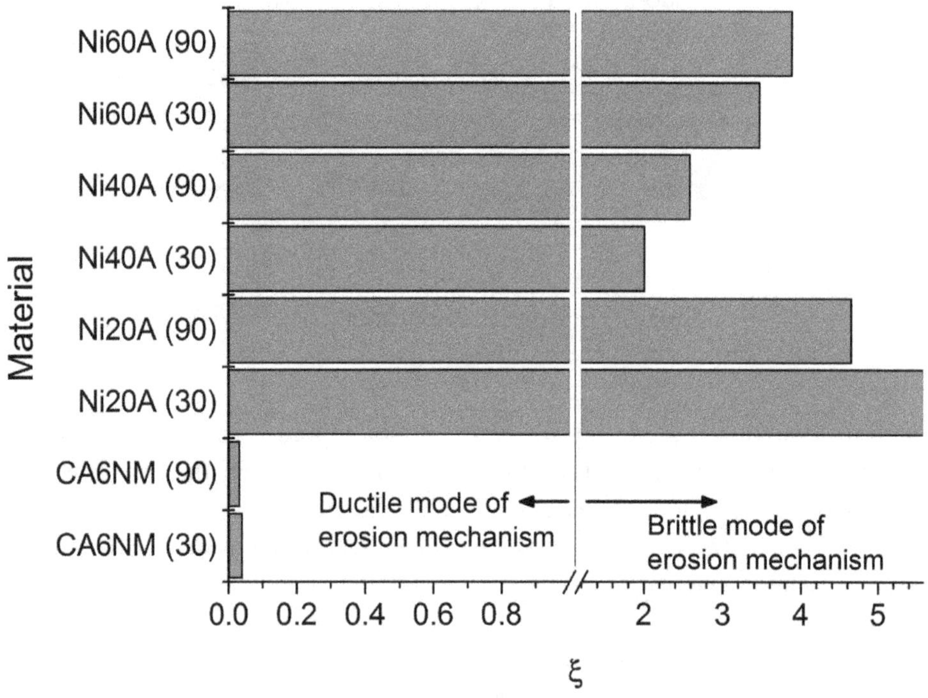

CA6NM steel is ploughing as predicted by ξ. Similar results could also be observed in case of coatings. Therefore, 'ξ' could prove to a useful tool for predicting the erosion mechanism of the materials.

SOLUTIONS AND RECOMMENDATIONS

From a detailed literature review, it has been observed that microstructure, which itself affects the various mechanical properties, plays a crucial role in determining the slurry erosion performance of the materials and coatings. In case of thermal spray coatings, various microstructural features such as content of pores, cracks, un-melted particles and splat boundaries determine their slurry erosion performance. Efforts can be made for reducing the proportionate content of these microstructural defects by either the optimization of the deposition parameters or post-processing of the developed coatings. Use of powder particles having narrow size range is also beneficial for improving the slurry erosion performance of the coatings. The improvement in the microstructure of the coating can help in improving important mechanical properties such as fracture toughness and hardness ultimately leading to improvement in slurry erosion resistance. Composite coatings can readily provide better resistance against slurry erosion, provided the constituent phases possess the required adhesion and has appreciable wettability among each other.

FUTURE RESEARCH DIRECTIONS

Nano-structured thermal spray coatings have shown promising results, however, as discussed, main focus of most of the studies have been the WC-based coatings. Significant efforts can be directed towards evaluating the other nano-structured coating compositions. Further, the effect of different post-processing techniques such as laser and heat treatment in controlled environment on the slurry erosion performance of the thermal spray coating is also required to be evaluated. It is also necessary to have clear understanding of influence of various spraying techniques and spraying parameters of each spraying technique on slurry erosion of potential coatings. This would help in determining the key parameters that have controlling effect on the microstructure and therefore the slurry erosion performance of the coatings. Research also needs to be extended to identify the alternative economical solution to the slurry erosion problems.

CONCLUSION

Thermal spray coatings can significantly improve the slurry erosion resistance of the commonly used structural materials. Microstructure plays an important role in determining the mechanical properties and hence the erosion performance of these coatings. Splat boundaries adversely affect the slurry erosion behavior of coatings. Efforts toward improvement of the microstructure could help in improving the slurry erosion performance of the thermal spray coatings. Among various mechanical properties of the coatings, fracture toughness appears to be the dominant parameter. Slurry erosion mechanism of the materials and coatings could be successfully predicted using the parameter namely, 'Erosion Mechanism Identifier".

REFERENCES

Anand, K., Hovis, S. K., Conrad, H., & Scattergood, R. O. (1987). Flux effects in solid particle erosion. *Wear*, *118*(2), 243–257. doi:10.1016/0043-1648(87)90146-3

Arnold, J. C., & Hutchings, I. M. (1989). Flux rate effects in the erosive wear of elastomers. *Journal of Materials Science*, *24*(3), 833–839. doi:10.1007/BF01148765

Barbezat, G., Nicol, A. R., & Sickinger, A. (1993). Abrasion, erosion and scuffing resistance of carbide and oxide ceramic thermal sprayed coatings for different applications. *Wear*, *162-164*(Part 1), 529–537. doi:10.1016/0043-1648(93)90538-W

Barik, R., Wharton, J., Wood, R., Tan, K., & Stokes, K. (2005). Erosion and erosion–corrosion performance of cast and thermally sprayed nickel–aluminium bronze. *Wear*, *259*(1-6), 230–242. doi:10.1016/j.wear.2005.02.033

Berget, J., Rogne, T., & Bardal, E. (2007). Erosion–corrosion properties of different WC–Co–Cr coatings deposited by the HVOF process—influence of metallic matrix composition and spray powder size distribution. *Surface and Coatings Technology*, *201*(18), 7619–7625. doi:10.1016/j.surfcoat.2007.02.032

Bhandari, S., Singh, H., Kansal, H. K., & Rastogi, V. (2012). Slurry erosion studies of hydroturbine steels under hydroaccelerated conditions. *Proceedings of the Institution of Mechanical Engineers. Part J, Journal of Engineering Tribology*, *226*(3), 239–250. doi:10.1177/1350650111428238

Bhandari, S., Singh, H., Kumar, H., & Rastogi, V. (2012). Slurry Erosion Performance Study of Detonation Gun-Sprayed WC-10Co-4Cr Coatings on CF8M Steel Under Hydro-Accelerated Conditions. *Journal of Thermal Spray Technology*, *21*(5), 1054–1064. doi:10.1007/s11666-012-9799-1

Branco, J. R. T., Gansert, R., Sampath, S., Berndt, C., & Herman, H. (2004). Solid Particle Erosion of Plasma Sprayed Ceramic Coatings. *Materials Research*, *7*(1), 147–153. doi:10.1590/S1516-14392004000100020

Bu Qian, W., & Luer, K. (1994). The erosion-oxidation behavior of HVOF Cr3C2-NiCr cermet coating. *Wear*, *174*(1-2), 177–185. doi:10.1016/0043-1648(94)90100-7

Chen, H., & Hutchings, I. M. (1998). Abrasive wear resistance of plasma-sprayed tungsten carbide cobalt coatings. *Surface and Coatings Technology*, *107*(2-3), 106–114. doi:10.1016/S0257-8972(98)00581-7

Clark, H. M. (1992). The influence of the flow field in slurry erosion. *Wear*, *152*(2), 223–240. doi:10.1016/0043-1648(92)90122-O

Clark, H. M., Hawthorne, H. M., & Xie, Y. (1999). Wear rates and specific energies of some ceramic, cermet and metallic coatings determined in the Coriolis erosion tester. *Wear*, *233-235*, 319–327. doi:10.1016/S0043-1648(99)00213-6

Davis, A. G., Boone, D. H., & Levy, A. V. (1986). Erosion of ceramic thermal barrier coatings. *Wear*, *110*(2), 101–116. doi:10.1016/0043-1648(86)90141-9

de Souza, V. A., & Neville, A. (2003). Corrosion and erosion damage mechanisms during erosion–corrosion of WC–Co–Cr cermet coatings. *Wear*, *255*(1–6), 146–156. doi:10.1016/S0043-1648(03)00210-2

Deng, T., Chaudhry, A. R., Patel, M., Hutchings, I., & Bradley, M. S. A. (2005). Effect of particle concentration on erosion rate of mild steel bends in a pneumatic conveyor. *Wear*, *258*(1-4), 480–487. doi:10.1016/j.wear.2004.08.001

Desouza, V., & Neville, A. (2003). Corrosion and erosion damage mechanisms during erosion-corrosion of WC-CoCr cermet coatings. *Wear*, *255*(1-6), 146–156. doi:10.1016/S0043-1648(03)00210-2

Erickson, L., Hawthorne, H., & Troczynski, T. (2001). Correlations between microstructural parameters, micromechanical properties and wear resistance of plasma sprayed ceramic coatings. *Wear*, *250*(1-12), 569–574. doi:10.1016/S0043-1648(01)00608-1

Factor, M., & Roman, I. (2002). Microhardness as a simple means of estimating relative wear resistance of carbide thermal spray coatings part 1. Characterization of cemented carbide coatings. *Journal of Thermal Spray Technology*, *11*(4), 468–481. doi:10.1361/105996302770348600

Finnie, I. (1995). Some reflections on the past and future of erosion. *Wear*, *186-187*(Part 1), 1–10. doi:10.1016/0043-1648(95)07188-1

Gang, J., Morniroli, J.-P., & Grosdidier, T. (2003). Nanostructures in thermal spray coatings. *Scripta Materialia*, *48*(12), 1599–1604. doi:10.1016/S1359-6462(03)00142-8

Grewal, H. S., Agrawal, A., & Singh, H. (2013a). Slurry erosion mechanism of hydroturbine steel: Effect of operating parameters. *Tribology Letters*, *52*(2), 287–303. doi:10.1007/s11249-013-0213-z

Grewal, H. S., Agrawal, A., & Singh, H. (2013b). Design and development of high-velocity slurry erosion test rig using CFD. *Journal of Materials Engineering and Performance*, *22*(1), 152–161. doi:10.1007/s11665-012-0219-y

Grewal, H. S., Agrawal, A., & Singh, H. (2013d). Slurry erosion performance of Ni–Al2O3 based composite coatings. *Tribology International*, *66*(0), 296–306. doi:10.1016/j.triboint.2013.06.010

Grewal, H. S., Agrawal, A., & Singh, H. (2013g). Identifying erosion mechanism: A novel approach. *Tribology Letters*, *51*(1), 1–7. doi:10.1007/s11249-013-0156-4

Grewal, H. S., Agrawal, A., Singh, H., & Shollock, B. A. (2013e). Slurry erosion performance of Ni-Al_2O_3 based thermal-sprayed coatings: Effect of angle of impingement. *Journal of Thermal Spray Technology*, *23*(3), 389–401. doi:10.1007/s11666-013-0013-x

Grewal, H. S., Bhandari, S., & Singh, H. (2012). Parametric study of slurry-erosion of hydroturbine steels with and without detonation gun spray coatings using taguchi technique. *Metallurgical and Materials Transactions. A, Physical Metallurgy and Materials Science*, *43*(9), 3387–3401. doi:10.1007/s11661-012-1148-y

Grewal, H. S., Singh, H., & Agrawal, A. (2013c). Microstructural and mechanical characterization of thermal sprayed nickel–alumina composite coatings. *Surface and Coatings Technology*, *216*, 78–92. doi:10.1016/j.surfcoat.2012.11.029

Grewal, H. S., Singh, H., & Agrawal, A. (2013f). Understanding liquid impingement erosion behaviour of nickel-alumina based thermal spray coatings. *Wear*, *301*(1-2), 424–433. doi:10.1016/j.wear.2013.01.063

Halling, J., & Arnell, R. D. (1984). Ceramic coatings in the war on wear. *Wear*, *100*(1-3), 367–380. doi:10.1016/0043-1648(84)90022-X

Hawthorne, H. M., Arsenault, B., Immarigeon, J. P., Legoux, J. G., & Parameswaran, V. R. (1999). Comparison of slurry and dry erosion behaviour of some HVOF thermal sprayed coatings. *Wear*, *225-229*(Part 2), 825–834. doi:10.1016/S0043-1648(99)00034-4

Hu, H. X., Jiang, S. L., Tao, Y. S., Xiong, T. Y., & Zheng, Y. G. (2011). Cavitation erosion and jet impingement erosion mechanism of cold sprayed Ni–Al2O3 coating. *Nuclear Engineering and Design*, *241*(12), 4929–4937. doi:10.1016/j.nucengdes.2011.09.038

Hurricks, P. L. (1972). Some aspects of the metallurgy and wear resistance of surface coatings. *Wear*, *22*(3), 291–320. doi:10.1016/0043-1648(72)90391-2

Hussein, M. F., & Tabakoff, W. (1974). Computation and plotting of solid particle flow in rotating cascades. *Computers & Fluids*, *2*(1), 1–15. doi:10.1016/0045-7930(74)90002-4

Hutchings, I. M., & Winter, R. E. (1974). Particle erosion of ductile metals: A mechanism of material removal. *Wear*, *27*(1), 121–128. doi:10.1016/0043-1648(74)90091-X

Kingswell, R., Rickerby, D. S., Bull, S. J., & Scott, K. T. (1991). Erosive wear of thermally sprayed tungsten coatings. *Thin Solid Films*, *198*(1–2), 139–148. doi:10.1016/0040-6090(91)90332-R

Kleis, I., & Kulu, P. (2008). *Solid Particle Erosion: Occurrence Predictio and Control*. Springer Verlag London.

Krishnamoorthy, P. R., Seetharamu, S., & Sampathkumaran, P. (1993). Influence of the mass flux and impact angle of the abrasive on the erosion resistance of materials used in pulverized fuel bends and other components in thermal power stations. *Wear*, *165*(2), 151–157. doi:10.1016/0043-1648(93)90330-O

Kulu, P., Hussainova, I., & Veinthal, R. (2005). Solid particle erosion of thermal sprayed coatings. *Wear*, *258*(1-4), 488–496. doi:10.1016/j.wear.2004.03.021

Kumar, A., Boy, J., Zatorski, R., & Stephenson, L. D. (2005). Thermal spray and weld repair alloys for the repair of cavitation damage in turbines and pumps: A technical note. *Journal of Thermal Spray Technology*, *14*(2), 177–182. doi:10.1361/10599630523737

Kumar Goyal, D., Singh, H., Kumar, H., & Sahni, V. (2012). Slurry erosion behaviour of HVOF sprayed WC–10Co–4Cr and Al2O3+13TiO2 coatings on a turbine steel. *Wear*, *289*(0), 46–57. doi:10.1016/j.wear.2012.04.016

Lathabai, S., Ottmüller, M., & Fernandez, I. (1998). Solid particle erosion behaviour of thermal sprayed ceramic, metallic and polymer coatings. *Wear*, *221*(2), 93–108. doi:10.1016/S0043-1648(98)00267-1

Levin, B. F., DuPont, J. N., & Marder, A. R. (2000). The effect of second phase volume fraction on the erosion resistance of metal-matrix composites. *Wear*, *238*(2), 160–167. doi:10.1016/S0043-1648(99)00363-4

Levy, A. V. (1995). *Solid particle erosion and erosion-corrosion of materials*. ASM International.

Levy, A. V., & Buqian, W. (1988). Erosion of hard material coating systems. *Wear*, *121*(3), 325–346. doi:10.1016/0043-1648(88)90209-8

Li, C., Yang, G., & Ohmori, A. (2006). Relationship between particle erosion and lamellar microstructure for plasma-sprayed alumina coatings. *Wear*, *260*(11-12), 1166–1172. doi:10.1016/j.wear.2005.07.006

Mann, B. S. (2000). High-energy particle impact wear resistance of hard coatings and their application in hydroturbines. *Wear*, *237*(1), 140–146. doi:10.1016/S0043-1648(99)00310-5

Mann, B. S., & Arya, V. (2001). Abrasive and erosive wear characteristics of plasma nitriding and HVOF coatings: Their application in hydro turbines. *Wear*, *249*(5-6), 354–360. doi:10.1016/S0043-1648(01)00537-3

Mann, B. S., & Arya, V. (2002). An experimental study to corelate water jet impingement erosion resistance and properties of metallic materials and coatings. *Wear*, *253*(5-6), 650–661. doi:10.1016/S0043-1648(02)00118-7

Mishra, S. C., Sahu, A., Das, R., Satapathy, A., Sen, S., Ananthapadmanabhan, P. V., & Sreekumar, K. P. (2008). Microstructure, adhesion, and erosion wear of plasma sprayed alumina-titania composite coatings. *Journal of Reinforced Plastics and Composites*. doi:10.1177/0731684407087758

Murthy, J. K. N., Rao, D. S., & Venkataraman, B. (2001). Effect of grinding on the erosion behaviour of a WC-Co-Cr coating deposited by HVOF and detonation gun spray processes. *Wear*, *249*(7), 592–600. doi:10.1016/S0043-1648(01)00682-2

Padhy, M. K., & Saini, R. P. (2009). Effect of size and concentration of silt particles on erosion of Pelton turbine buckets. *Energy*, *34*(10), 1477–1483. doi:10.1016/j.energy.2009.06.015

Papini, M., Ciampini, D., Krajac, T., & Spelt, J. K. (2003). Computer modelling of interference effects in erosion testing: Effect of plume shape. *Wear*, *255*(1-6), 85–97. doi:10.1016/S0043-1648(03)00125-X

Prasad, B., Jha, A., Modi, O., & Yegneswaran, A. H. (2004). Effect of sand concentration in the medium and travel distance and speed on the slurry wear response of a zinc-based alloy alumina particle composite. *Tribology Letters*, *17*(2), 301–309. doi:10.1023/B:TRIL.0000032468.27468.2c

Santa, J. F., Espitia, L. A., Blanco, J. A., Romo, S. A., & Toro, A. (2009). Slurry and cavitation erosion resistance of thermal spray coatings. *Wear, 267*(1-4), 160–167. doi:10.1016/j.wear.2009.01.018

Shivamurthy, R. C., Kamaraj, M., Nagarajan, R., Shariff, S. M., & Padmanabham, G. (2009). Influence of microstructure on slurry erosive wear characteristics of laser surface alloyed 13Cr–4Ni steel. *Wear, 267*(1-4), 204–212. doi:10.1016/j.wear.2008.12.027

Shrestha, S., Hodgkiess, T., & Neville, A. (2001). The effect of post-treatment of a high-velocity oxy-fuel Ni-Cr-Mo-Si-B coating part 2: Erosion-corrosion behavior. *Journal of Thermal Spray Technology, 10*(4), 656–665. doi:10.1361/105996301770349187

Shrestha, S., Hodgkiess, T., & Neville, A. (2005). Erosion–corrosion behaviour of high-velocity oxy-fuel Ni–Cr–Mo–Si–B coatings under high-velocity seawater jet impingement. *Wear, 259*(1–6), 208–218. doi:10.1016/j.wear.2005.01.038

Singh, V. P., Sil, A., & Jayaganthan, R. (2011). Tribological behavior of plasma sprayed Cr_2O_3–3%TiO_2 coatings. *Wear, 272*(1), 149–158. doi:10.1016/j.wear.2011.08.004

Souza, V. A. D., & Neville, A. (2007). Aspects of microstructure on the synergy and overall material loss of thermal spray coatings in erosion–corrosion environments. *Wear, 263*(1–6), 339–346. doi:10.1016/j.wear.2007.01.071

Stachowiak, G. W., & Batchelor, A. W. (2005). *Engineering tribology*. Butterworth-Heinemann.

Stack, M. M., & Abd El Badia, T. M. (2006). On the construction of erosion-corrosion maps for WC/Co-Cr-based coatings in aqueous conditions. *Wear, 261*(11-12), 1181–1190. doi:10.1016/j.wear.2006.03.038

Stack, M. M., & Abd El-Badia, T. M. (2008). Some comments on mapping the combined effects of slurry concentration, impact velocity and electrochemical potential on the erosion-corrosion of WC/Co-Cr coatings. *Wear, 264*(9-10), 826–837. doi:10.1016/j.wear.2007.02.025

Sun, D., Wharton, J. A., & Wood, R. J. K. (2009). Abrasive size and concentration effects on the tribo-corrosion of cast CoCrMo alloy in simulated body fluids. *Tribology International, 42*(11-12), 1595–1604. doi:10.1016/j.triboint.2009.03.018

Sundararajan, G., Roy, M., & Venkataraman, B. (1990). Erosion efficiency-a new parameter to characterize the dominant erosion micromechanism. *Wear, 140*(2), 369–381. doi:10.1016/0043-1648(90)90096-S

Tan, K., Wood, R. J. K., & Stokes, K. R. (2003). The slurry erosion behaviour of high velocity oxy-fuel (HVOF) sprayed aluminium bronze coatings. *Wear*, *255*(1-6), 195–205. doi:10.1016/S0043-1648(03)00088-7

Telfer, C. G., Stack, M. M., & Jana, B. D. (2012). Particle concentration and size effects on the erosion-corrosion of pure metals in aqueous slurries. *Tribology International*, *53*(0), 35–44. doi:10.1016/j.triboint.2012.04.010

Thakur, L., & Arora, N. (2013). A comparative study on slurry and dry erosion behaviour of HVOF sprayed WC–CoCr coatings. *Wear*, *303*(1–2), 405–411. doi:10.1016/j.wear.2013.03.028

Tian, B. R., & Cheng, Y. F. (2007). Electrolytic deposition of Ni–Co–Al2O3 composite coating on pipe steel for corrosion/erosion resistance in oil sand slurry. *Electrochimica Acta*, *53*(2), 511–517. doi:10.1016/j.electacta.2007.07.013

Toma, D., Brandl, W., & Marginean, G. (2001). Wear and corrosion behaviour of thermally sprayed cermet coatings. *Surface and Coatings Technology*, *138*(2-3), 149–158. doi:10.1016/S0257-8972(00)01141-5

Turenne, S., Fiset, M., & Masounave, J. (1989). The effect of sand concentration on the erosion of materials by a slurry jet. *Wear*, *133*(1), 95–106. doi:10.1016/0043-1648(89)90116-6

Turunen, E., Varis, T., Gustafsson, T. E., Keskinen, J., Fält, T., & Hannula, S.-P. (2006). Parameter optimization of HVOF sprayed nanostructured alumina and alumina–nickel composite coatings. *Surface and Coatings Technology*, *200*(16–17), 4987–4994. doi:10.1016/j.surfcoat.2005.05.018

Uuemo, H., & Kleis, I. (1975). A critical analysis of erosion problems which have been little studied. *Wear*, *31*(2), 359–371. doi:10.1016/0043-1648(75)90169-6

Wang, B., & Lee, S. W. (2000). Erosion-corrosion behaviour of HVOF NiAl-Al2O3 intermetallic-ceramic coating. *Wear*, *239*(1), 83–90. doi:10.1016/S0043-1648(00)00309-4

Westergard, R., & Hogmark, S. (2004). Sealing to improve the wear properties of plasma sprayed alumina by electro-deposited Ni. *Wear*, *256*(11-12), 1153–1162. doi:10.1016/j.wear.2003.07.005

Wong, K. K., & Clark, H. M. (1993). A model of particle velocities and trajectories in a slurry pot erosion tester. *Wear*, *160*(1), 95–104. doi:10.1016/0043-1648(93)90409-F

Wood, R. J. K. (2010). Tribology of thermal sprayed WC–Co coatings. *International Journal of Refractory Metals & Hard Materials*, *28*(1), 82–94. doi:10.1016/j.ijrmhm.2009.07.011

Wood, R. J. K., Mellor, B. G., & Binfield, M. L. (1997). Sand erosion performance of detonation gun applied tungsten carbide/cobalt-chromium coatings. *Wear*, *211*(1), 70–83. doi:10.1016/S0043-1648(97)00071-9

Wu, G., Li, N., Zhou, D., & Mitsuo, K. (2004). Electrodeposited Co–Ni–Al2O3 composite coatings. *Surface and Coatings Technology*, *176*(2), 157–164. doi:10.1016/S0257-8972(03)00739-4

Zhao, H. X., Goto, H., Matsumura, M., Takahashi, T., & Yamamoto, M. (1999). Slurry erosion of plasma-sprayed ceramic coatings. *Surface and Coatings Technology*, *115*(2-3), 123–131. doi:10.1016/S0257-8972(99)00164-4

KEY TERMS AND DEFINITIONS

Cavitation Erosion: The degradation of the solid surface due to imploding of vapor bubbles in its proximity.

Composite: Microstructure composed of two or more different materials/phases.

Erosion Mechanism Identifier: A mathematical model used to predict erosion mechanism from the measured erosion rates.

Erosion Mechanism: The phenomenon responsible for the loss or detachment of the material from the surface.

Microstructure: The structure of the material at the micron scale.

Porosity: The void space, either open or closed, in the material.

Slurry Erosion: The degradation of solid surface due to impact of solid particles entrained in a liquid medium.

Chapter 10
Detonation Sprayed Coatings and their Tribological Performances

D. Srinivasa Rao
International Advanced Research Centre for Powder Metallurgy and New Materials (ARCI), India

G. Sivakumar
International Advanced Research Centre for Powder Metallurgy and New Materials (ARCI), India

D. Sen
International Advanced Research Centre for Powder Metallurgy and New Materials (ARCI), India

S.V. Joshi
International Advanced Research Centre for Powder Metallurgy and New Materials (ARCI), India

ABSTRACT

The Detonation Spray Coating (DSC) process is a unique variant among the wide choice of thermal spray processes. The typical functionalities of DSC coatings include wear and corrosion resistance, elevated temperature oxidation resistance, thermal barrier, insulative/conductive, abradable, lubricious surface, etc. Among the coatings for wear resistance, the cermet coatings based on WC–Co and Cr_3C_2*–NiCr are the most popular materials of choice and contribute to bulk of the utilization*

DOI: 10.4018/978-1-4666-7489-9.ch010

by the industry towards wear resistance. Notwithstanding the above materials, alternative materials involving modifications in both hard and binder phases like TiMo (CN)–NiCo, WC-CrC-Ni, WC-Co-Cr, WC-Ni, Cr_3C_2-Ni, Cr_3C_2-Inconel, etc. exhibit great promise towards tribological applications under diverse wear modes. This chapter on the tribological characteristics of the detonation sprayed coatings provides a comprehensive overview on the characteristics of various cermet coatings generated at varied process conditions and its influence on the tribological properties under abrasive, sliding, and erosive wear modes.

INTRODUCTION

Surface modification approach has been attracting a great deal of attention as it presents a cost-effective way to combat degradation resulting from mechanisms such as wear, oxidation, corrosion, or failure under an excessive heat load without sacrificing the bulk properties of the component material. Various surfacing techniques are now available which offer a wide range of quality and cost. Since a vast majority of industrial components deteriorate, and eventually fail, due to one of several wear phenomena that may be experienced during normal operation, considerable attention has been devoted to the development of coating materials and processes specifically to combat the routinely confronted wear modes, viz. erosion, abrasion and sliding. Amongst all the currently available coating processes, the thermal spray technique has gradually emerged as the most useful method of developing a wide variety of coatings to enhance the performance and durability of engineering components exposed to the above forms of wear.

Of all the thermal spray variants, the detonation spray coating (DSC) technology is widely acknowledged to be the most superior by virtue of the very high particle velocities that it imparts, thereby leading to dense and well-bonded coatings which exhibit excellent tribological properties such as high abrasion and erosion resistance. Initially developed specifically for depositing wear-resistant coatings of tungsten carbide, which undergoes substantial decarburization during plasma spraying, the use of the DSC technology currently extends to all industry segments ranging from gas turbines to textile industries and from nuclear energy to data processing. The technique has grown substantially with a range of materials that can be sprayed, which includes metals, metal alloys, ceramics, cermets, composites and combinations of these.

Detonation spray coating (DSC) process is a unique thermal spray variant which relies upon the controlled explosive combustion of a measured mix of oxy-acetylene gases. The powder feedstock of the desired coating material is heated and accelerated by high-pressure, high-temperature gaseous products resulting from the detonation, to impact a substrate at very high velocities to yield a well-adherent layer (Sundararajan *et al.*, 2013). The DSC process is cyclic in nature and operates at a firing frequency ranging between 2 to 10 shots per sec. The interesting aspect of DSC process is the versatility in depositing a wide variety of materials ranging from metals to ceramics, expect for few materials with very high melting points, which makes it worthwhile for many industrial applications. The schematic of the DSC process is shown in Figure 1, which is already described elsewhere in detail (Sundararajan *et al.*, 2013).

Like any other thermal spray variant, the velocity and the physical state of the powder particles at the time of impact are the most critical factors that influence the quality of detonation sprayed coatings (Sundararajan *et al.*, 2001; Kharlamov, 1986). These factors are directly governed by the process parameters employed as well as the starting powder characteristics which eventually manipulate the resultant coating properties.

WEAR RESISTANT COATINGS

Over the years, the hardness of the material is considered as the important criterion for selection towards wear resistant applications which is found to influence the material wear rates (Fauchais *et al.*, 2014). Accordingly, materials like WC-Co, TiC, Al_2O_3, Cr_2O_3 etc are the materials of choice for wear resistant applications under diverse wear modes. In terms of spray variants, DSC and high-velocity oxy-fuel (HVOF) processes are the ideal techniques to achieve dense, phase pure and therefore, better wear resistant coatings (Wang, 1993; Barbezat *et al.*, 1993). Cermets or cemented carbides consisting one or more carbides of tungsten/titanium/tantalum/vanadium

Figure 1. Schematic of detonation spray coating process (Sundararajan et al., 2013)

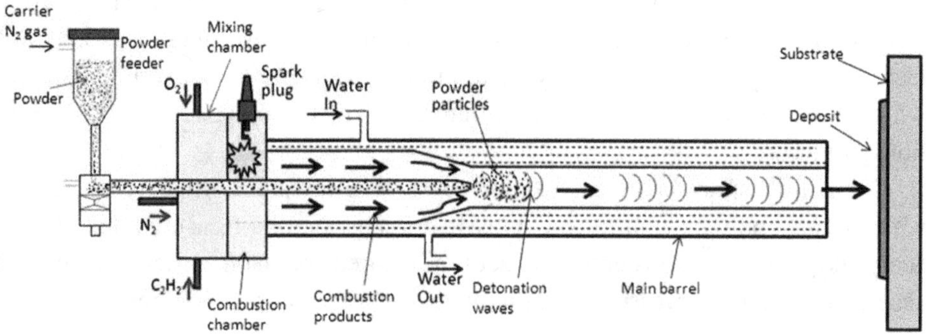

embedded in a matrix of softer metal and/or alloys like Ni, Co, NiCr etc forms the major category of materials sprayed for wear resistant applications. Ceramics like Al_2O_3, Cr_2O_3, with and without additives like ZrO_2, TiO_2, SiO_2, etc also constitute the wear resistant coatings deposited using DSC system.

Although the choice of materials are huge, experimental studies have shown that the hardness and elastic modulus of coatings deposited using the above powders are substantially lower than that of bulk material having an identical composition (Rybicki *et al.,* 1995; Zhang *et al.,* 1998). Coating characteristics like phases formed, coating microstructure, porosity, the nature of residual stress and its magnitude within the coating, inter-splat cohesion and finally the coating-substrate adhesion, govern the mechanical properties of the coatings (Nakamura *et al.,* 2000; Fauchais & Vardelle, 2000). It is pertinent to note that the experimental values of coating hardness and elastic modulus are dependent not only on the coating technique, but also on the choice of process parameters for a given coating technique. For example, a minor variation in the oxygen to acetylene ratio during detonation spraying of WC-Co or Al_2O_3 coatings can substantially influence the hardness, porosity and its wear properties (Sundararajan *et al.,* 2001; Sivakumar *et al.,* 2001). The variation in the oxy-fuel ratio is known to influence the detonation dynamics, thereby, altering the in-flight particle characteristics like velocity and temperature. This eventually will have a bearing on the splatting & melting behavior of the impacting particles, which greatly influence the coating microstructure. Therefore, there is a need for continuous quest in developing better coatings that can exhibit properties closer to that of its equivalent bulk material. In view of the above, the following sections will give an overview of the tribological characteristics of detonation sprayed coatings through various case studies which are specifically focused on achieving improved performance.

Effect of Particle Size

The commercially available spray grade powders differ significantly in terms of its morphology, particle size distribution, the manufacturing route, etc (www.hcstarck.com, www.oerlikon.com, www.praxairsurfacetechnologies.com). Though the obvious reason of increased production cost with narrow size ranged powders resists the use of it, the ultimate indicator is wear performance. A detailed study undertaken by (Suresh Babu *et al.*, 2007) involved three different chemistries of NiCr, WC-Co and Al_2O_3 powders, which were classified into three categories as fine, coarse and as-received (shown in Figure 2). The powders were classified using a centrifugal air classifier (Suresh Babu *et al.*, 2007). In terms of microstructure shown in Figure 3, it was clear that the porosity levels were low with fine sized particles and continues to increase with increase in particle size. However, it is interesting to note that the retention of preferred phases like α-Al_2O_3 and WC in Al_2O_3 and WC-12Co coatings, respectively, were higher at higher mean particle sizes.

More significantly, the wear performance of the above coatings under different wear modes of abrasion, sliding and erosion are illustrated in Figure 4. The abrasion tests were performed using a dry rubber when three body abrasion test rig designed as per ASTM G65 standards using SiO_2 particles at a load of 5 kgf. The sliding wear was assessed using a pin-on-disc tribometer, tested against sintered WC-6Co disc under a load of 5 kgf. Dry sand erosion test designed as per ASTM standard G79 was used to assess the erosion resistance against SiO_2 particles impacting at about 60 m/s impact velocities. As shown in Figure 4, Al_2O_3 coatings exhibited better abrasion and sliding wear resistance at higher particle size, which can attributed to the higher hardness associated with higher retention ofα-Al_2O_3 phases. As proposed by McPherson (McPherson, 1980), the higher particle size allows the higher retention of α-Al_2O_3 phase due to the partial melting and pre-existing nuclei. Hence, higher retention of α-Al_2O_3 phase ensures that the hardness exhibited by the Al_2O_3 coatings is better as well as provide better wear resistance. The erosion resistance was highest at intermediate particle size level due to the combined requirement of hardness as well as toughness. In case of WC-12Co coatings, wear rates under abra-

Figure 2. SEM images of Al_2O_3, Ni-20Cr and WC-12Co powders. (a-c)—as received; (d-f)—coarse; (g-i)—fine fractions of respective powders (Suresh Babu et al., 2007)

Figure 3. SEM images of detonation sprayed Al_2O_3, Ni-20Cr and WC-12Co coatings deposited using (a-c)—as received; (d-f)—coarse;(g-i)—fine fractions of respective powders (Suresh Babu et al., 2007)

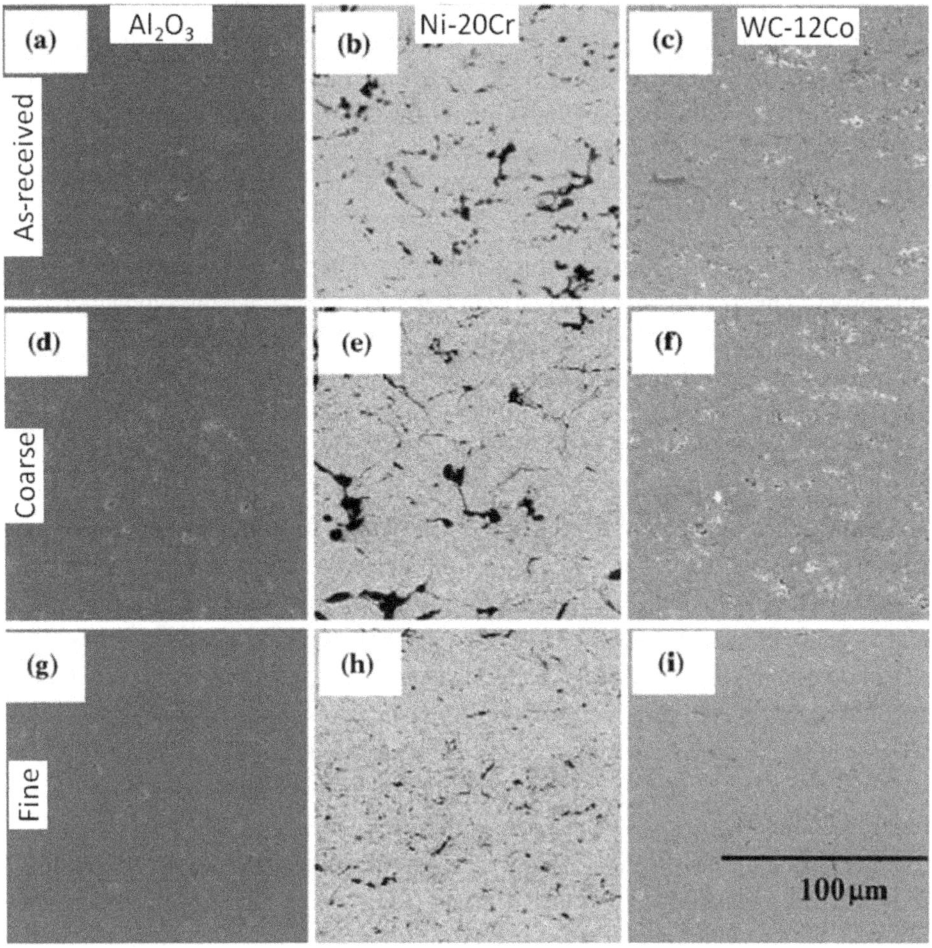

sion, sliding and erosion modes shown an increasing trend with increasing particle size, which was against the expected trend. This could be attributed to a complex influence of the phases formed, hardness and the fracture toughness of the WC-12Co coatings. Though the possibility of grain size effect was not reported, this may not be ruled out as the observed WC cuboids undergo higher carbide dissolution at higher oxy-fuel ratios (Sundararajan and Suresh Babu, 2009). Similarly, the WC cuboids grain sizes of finer particles might undergo higher carbide dissolution leading to difference in the tribological trend shown n Figure 4. The lower hardness values resulted in poor abrasion and sliding wear performance for Ni-20Cr coatings, however, exhibited better erosion resistance almost matching with that of Al_2O_3 coatings. Thus, the observed trend implies that the hardness play a significant role in determining the abrasive and sliding wear rate, whereas, an optimum level of hardness and toughness is essential for better erosion resistance. The major conclusion derived from the above study is that the standard particle size distribution for high velocity processes like HVOF/DSC is sufficient to provide the necessary wear resistance and further modification at an additional cost is not worthwhile.

Effect of Grain Size

The driving force for the assessing the effect of grain size can be attributed to the gap observed in arriving at a conclusive understanding on the tribological behavior of WC-12Co coatings. In addition to the powder processing characteristics, WC grain sizes are found to affect the coating behavior. Usually, the conventional WC-Co powder contains 0.5-6 μm of angular WC grains. Different studies on the grain size effect of WC cuboids in WC-Co coatings revealed contrasting results against that of sintered bulk materials. Sintered fine grained WC-Co cermets were found to exhibit better mechanical properties like hardness, fracture toughness and wear resistance (Suresh Babu *et al.*, 2008). Encouraged by the results obtained with sintered nanostructured carbides, there is always a growing interest in thermal spraying of nanostructured WC-based materials, since it is a convenient and economical consolidation method, with low particle residence times disallowing prolonged high temperature exposure and, thereby, favoring retention of starting nanograins in the coating. However, due to extensive decomposition of WC grains to W_2C and amorphous $Co_xW_yC_z$ phases, it was found that fine grained WC-Co coatings exhibit lesser fracture toughness and, thereby, poor abrasion resistance values (Qiao Yunfei *et al.,* 2003). But the sliding wear performance exhibited mixed results with the possibility of metastable W_2C phases aiding better frictional characteristics. Various other results with nano-grained and ultrafine grained WC-Co coatings through different HVOF, plasma spraying systems also confirmed conflicting performance as against the conventional micron sized grains (Qiao Yunfei *et al.,* 2003; Qiaoqin Yang *et al.,* 2003). Though, many such results are available with HVOF, no such detailed reports are available with DSC. In view of the above, a detailed study on the effect of WC grain size was undertaken at the author's laboratory using three commercially available powder feedstock materials and the major findings are given below. Three WC-12Co powders used for the study were Bimodal (mixture of micron and sub-micron sized WC grains) type Nanomyte M1 from NEI Corporation, USA, coarse grained (Diamalloy, 2004) WC-12Co from Sulzer Metco, USA and fine sized (Infralloy S7412) WC-Co from Inframat Corporation, USA.

Figure 4. Wear performance of Al_2O_3, Ni-Cr, and WC-Co coating as a function of mean particle size (Suresh Babu et al., 2007)

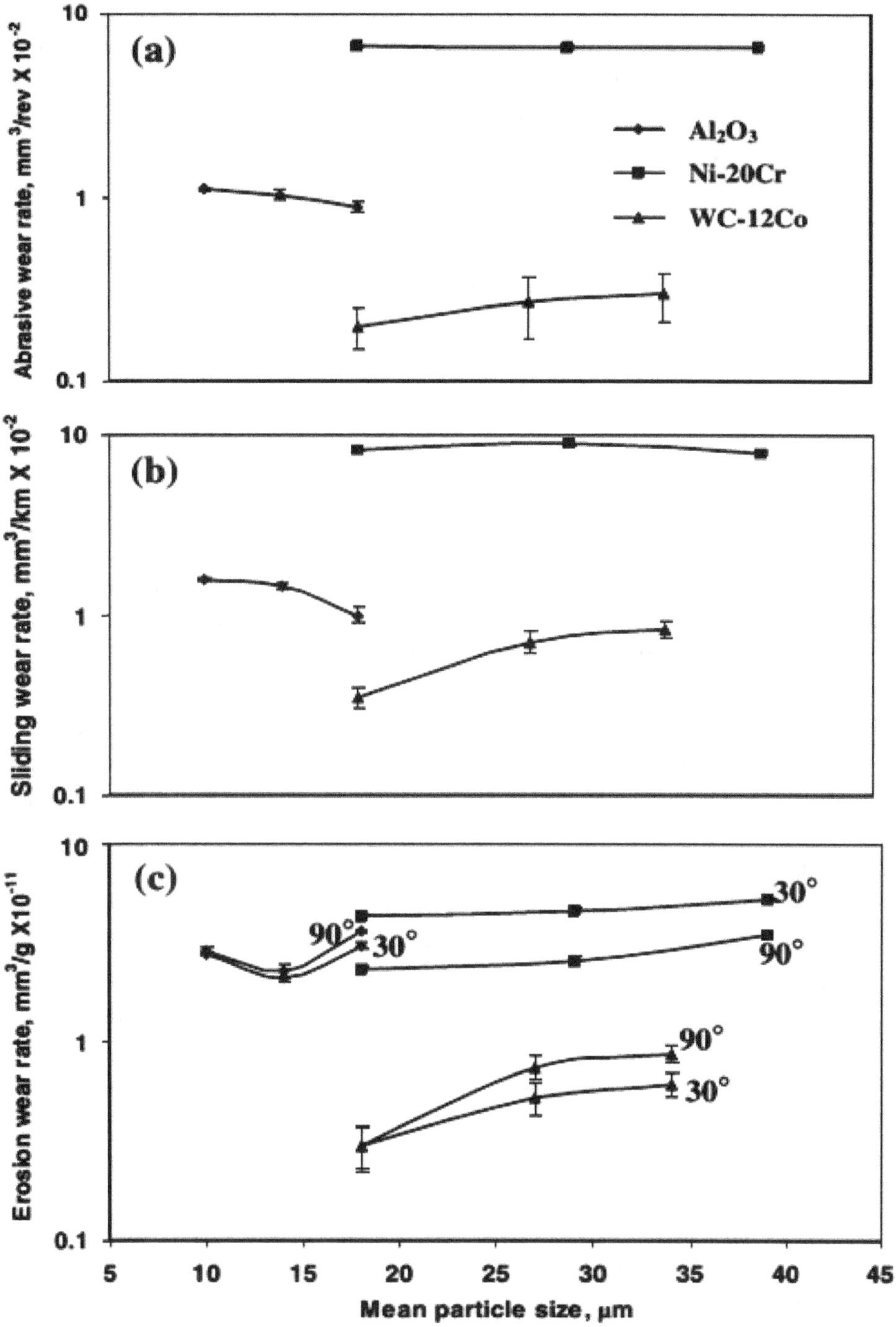

Typical cross-sectional micrographs of DSC sprayed WC-12Co generated at similar conditions are shown in Figure 5. As evident from Figure 5, there is a clear distinction between the grain sizes observed for the three coatings, wherein, the fine sized WC-12Co coatings clearly exhibiting fine WC cuboid, closely followed by the Bimodal type of WC-12Co coatings. The grain size ranges were measured through the high magnification micrographs, which were in the range of 0.05 μm to 1.3 μm for coarse WC-12Co, 0.02 μm to 1 μm for bimodal WC-12Co, 0.01 μm to 0.65 μm for fine–sized WC-12Co coatings. The hardness and modulus data measured for the above coatings are shown in Figure 6, which indicates that fine sized WC-12Co coatings exhibited the highest hardness and better modulus values, similar to various observations made for nanostructured WC-12Co coatings (Guilemany *et al.,* 2006). The higher hardness was attributed to the formation of fine W_2C rims over the WC grains. This was also noted in previous reports of the possibilities of decarburization of fine sized WC grains (Suresh Babu *et al*., 2008). Incidentally, the fine sized coatings exhibited highest fracture toughness amongst the WC-12Co coatings.

The wear performance of these coatings was tested under abrasion (same conditions with additional tests against SiC particles), sliding (at varied stress levels) and erosion wear modes (using the conditions previously mentioned), as illustrated in Figure 7. It is clear that the wear loss for WC-12Co tested against SiC particles was significantly higher than against SiO_2 particles as shown in Figure 7(a). The three times difference in hardness among the abrasive particles yielded nearly 10 times increase in abrasive wear for the same coatings. Amongst the WC-12Co grades, bimodal type coatings exhibited relatively poorer performance than the other grades of WC-12Co under abrasion testing. This was contradictory to the reported results for bimodal type WC-12Co coatings (Guilemany *et al.,* 2006; Skandan *et al.,* 2000). The general wear mechanism noted for various WC-Co coatings under abrasive wear is the weakening of splats under removal of the binder materials and the subsequent removal of WC grains. The extent of decarburization leads to the formation of W_2C over the WC grains (Suresh Babu *et al*., 2008; Guilemany *et al.,* 2006), which are reportedly reduces the adhesion to the soft metallic matrix and results in higher abrasive wear loss. In case of less decarburized WC grains and higher sized grains, the adhesion with the binder is good, which ensures better resistance to abrasion. Therefore, the higher loss noted for the bimodal WC-12Co coatings can be partially attributed to higher decarburization and also to the selection of process parameters. For fine grained WC coatings, better abrasive wear performance can be attributed to better homogeneity of dispersion of WC grains within the microstructure and partially to the adhesion of WC grains. In case of coarse WC grains, the better adhesion of the grains ensured better abrasion resistance.

Under sliding wear mode, the performance of WC-Co coatings is of great interest due to the repeated fatigue of the sliding pair usually stresses the sub-surface

Figure 5. Cross-sectional microstructure of detonation sprayed WC-Co coatings using (a-b)-bimodal type; (c-d)-coarse; (e-f)-fine grained WC-12Co powders

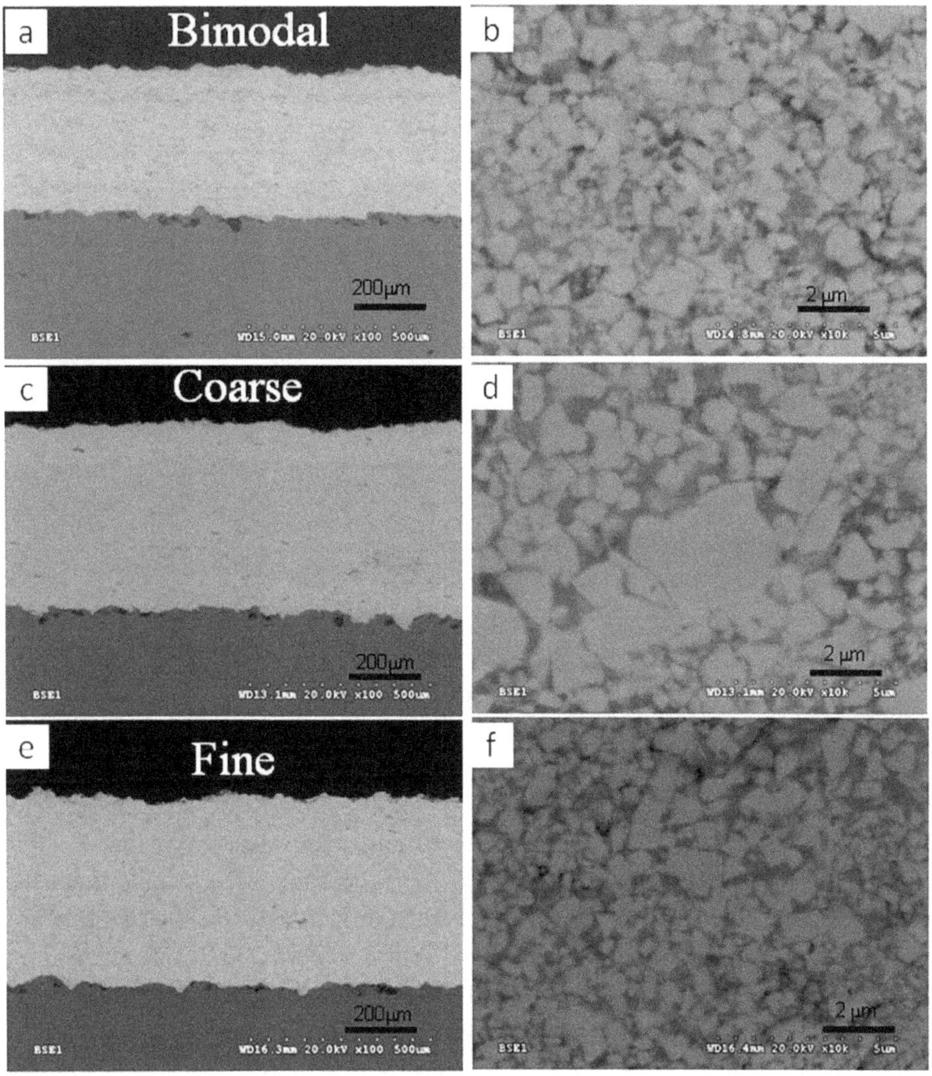

and causes greater splat delamination and binder/carbide interface fracture (Guilemany *et al.*, 2006). Upon testing at varied stress levels, the bimodal type coatings performed better amongst the WC-12Co coatings as shown in Figure 7(b). Similar trend was also observed for the reported HVOF WC-12Co coatings (Qiao Yunfei *et al.*, 2003), though no significant damage could be seen on the counterpart sintered

Figure 6. Elastic modulus and hardness of detonation sprayed WC-Co coatings using bimodal type, coarse and fine grained WC-12Co powders

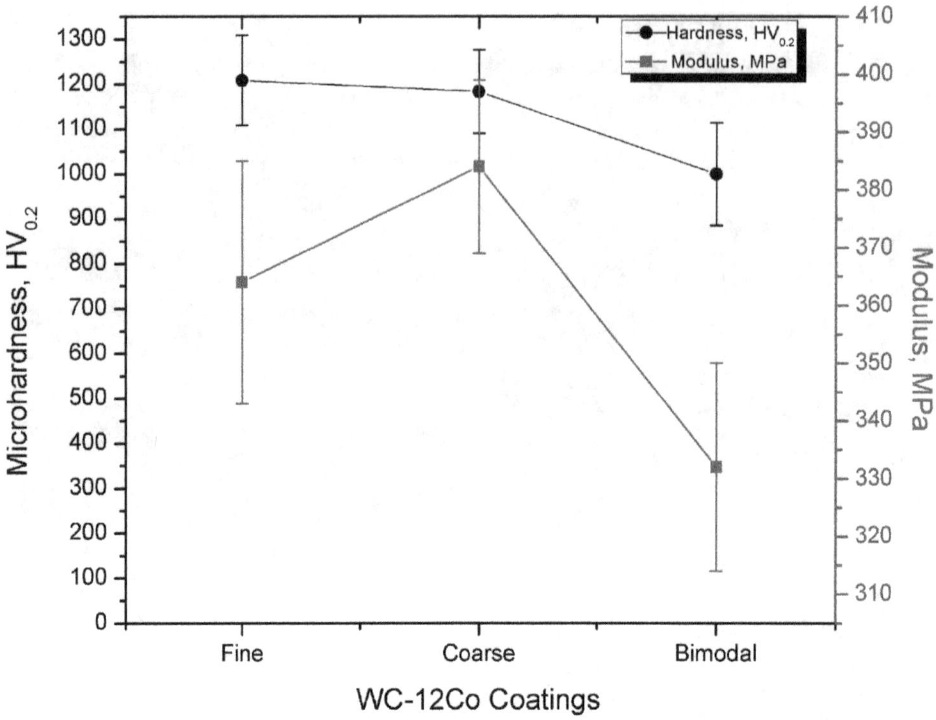

WC-6Co disc. This could be attributed to the presence of brittle W_2C phase yielding better sliding wear resistance. The increase in wear rate with increased stress levels was substantial for fine grained WC-12Co coatings, which could be due to the loss of ductility of the binder phase resulted in increased sub-surface fracture, as observed for HVOF WC-12Co coatings (Shipway *et al.*, 2005). Under such high speed of testing, the fracture leads to generation of wear debris which also could have acted as abrasive to accelerate the wear rate.

The performance of the WC-12Co coatings clearly correlates well with the grain size under erosive wear mode, as the smaller grain sized coatings exhibited better wear resistance compared to coarse grained coating as observed in Figure 7(c). This followed the general trend of higher fracture toughness yielding lower wear rate. The ratio of erosion rate of a material at impact angles of 90° and 30° indicate the mechanism of erosion behavior (Suresh Babu *et al.*, 2003). For values of $E90^{o}/E30^{o} < 1$, the behavior is observed to be ductile and for results yielding >1, it corresponds t brittle behavior. The $E90^{o}/E30^{o}$ values also imply that the bimodal

Figure 7. Wear performance of detonation sprayed WC-12Co coatings with varied grain sizes tested under (a) abrasive, (b) sliding, and (c) erosive wear modes

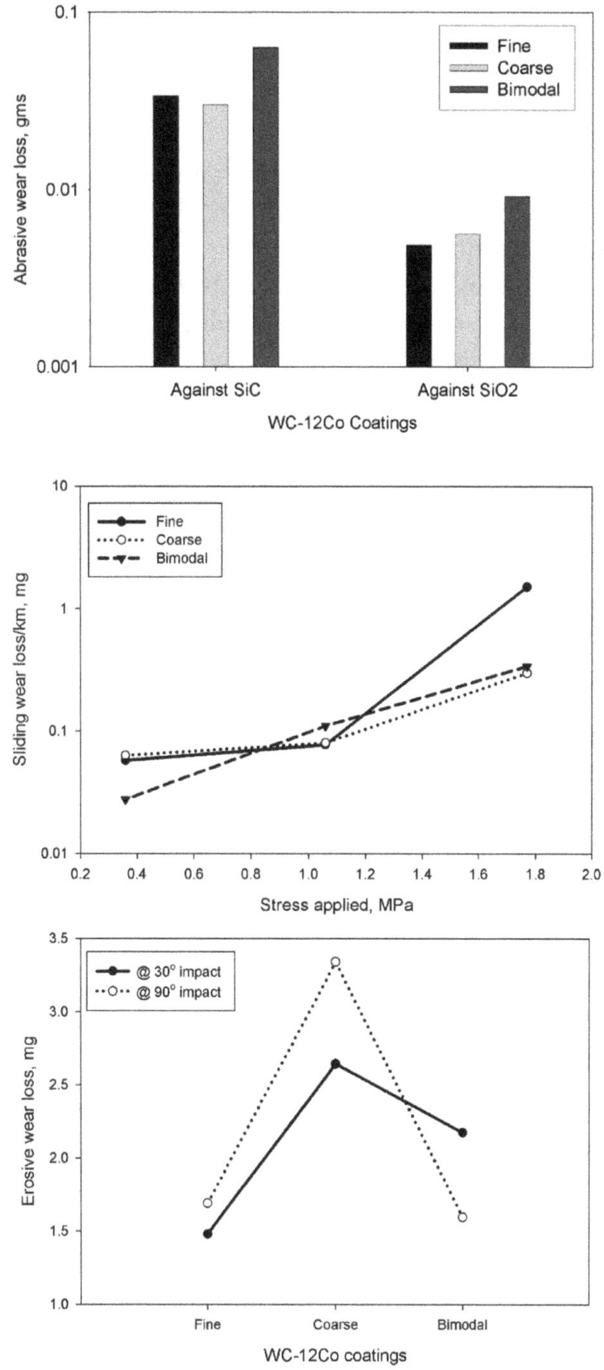

WC-12Co coatings exhibited intermediate ductile-brittle behavior, whereas the fine and coarse grained coatings exhibited clear brittle behavior. Though the fine grained WC-Co coatings exhibited brittle behavior, the higher toughness values influenced the reduction in the wear rates.

Effect of Process Parameters

For many years, the detonation spray coatings is an established technique for aerospace applications (Budinski, 1998; Pejryd *et al.,* 1995), and many other application areas especially in the civilian sector have emerged in the last few years (Guest, 1986, Pawlowski, 1995). Notwithstanding the advantages of the DSC technology over other thermal spray variants, the understanding of the fundamental aspects of this technology is extremely limited due to less number of publications. Specifically, details regarding splat morphology - microstructure - properties correlation provides valuable insights towards the coating formation mechanisms. In view of the above, a detailed assessment of parametric impact with the key DSC process variables (oxy-fuel ratio, spray distance, powder feed rate and shot frequency) on the mechanism of coating formation and the properties of the resulting coating is essential. The above study is also important to assess the repeatability and robustness of the process in order to successfully commercialize the technology. In view of the above, the present section deals with the evaluation of coating properties under systematic variation of DSC process variables.

Design of experiments using the Taguchi method is considered to be an effective method to identify the optimum process parameters and also can be utilized to have an enhanced understanding of a spray variant (Taguchi & Konishi, 1987; Saravanan *et al.,* 2000). The process optimization studies involving oxy-fuel ratio, powder feed rate, spray distance and shot frequency was assessed for various coatings like Al_2O_3, WC-12Co, Al_2O_3-TiO_2, Cr_3C_2-25NiCr. The detailed analysis indicated that the variation in process parameters yielded significant variation in the abrasive wear loss, as noted for WC-12Co coatings shown in Figure 8 (Sivakumar *et al.,* 2001). Therefore, the coating performance is substantially influenced by the choice of process parametric combinations. Upon further statistical analysis, the relative variation in the response is estimated through the averaging roué. Figure 9 indicates that the abrasive wear trend (response) with respect to the variation between the levels of interest. It is very informative about the preferred selection and is required to exceed the inherent experimental scatter. Most of the material characteristics indicated that the DSC process is highly robust within the range of process parameters studied (Sivakumar *et al.,* 2001). Further correlation between the intrinsic properties and its performance may also be required to have better understanding of the process.

Figure 8. Abrasive wear performance of WC-12%Co coatings at varied process parameters (Sivakumar et al., 2001)

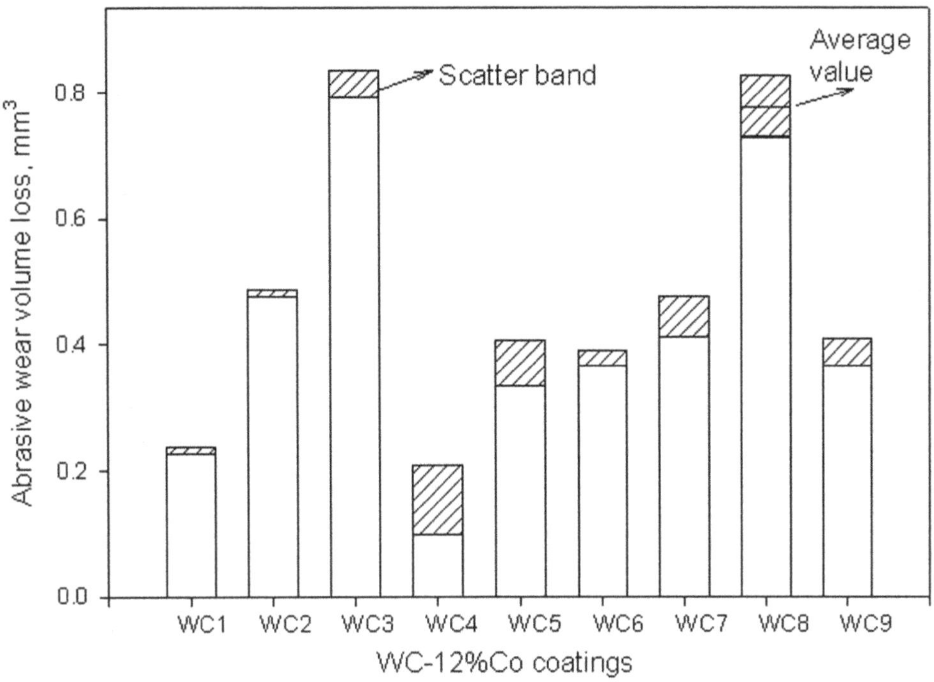

The second approach is to understand the correlation between the splat morphology - microstructure – properties under varied processing conditions. This is very relevant due to the limited data available for detonation sprayed coatings, in-spite of the fact that the technique being available commercially for a long time (Sundararajan *et al.,* 2001). Specifically, the influence of particle velocity and temperature on the splat morphology and also area coverage of the splat and its correlation with tribological performance was reported for detonation sprayed Al_2O_3 particles on a mild steel substrate (Sundararajan *et al.,* 2001). The in flight particle temperature and velocity were measured using the DPV 2000 system supplied by Technar, Canada. Based on the particle diagnostic values, the approximate thermal and kinetic energy values were estimated and compared as a function of DSC process variables shown in Figure 10. The contribution of kinetic energy is observed to be about 8 to 15%, which implies that the deposition of ceramic Al_2O_3 particles is completely dominated by the thermal energy only, even in case of high velocity processes like DSC. This is also the same reason that many of HVOF systems although having high velocities, but with limited temperature component may not be

Figure 9. Typical influence of process parameters on the relative variation of abrasive wear loss of WC-12Co coatings (Sivakumar et al., 2001)

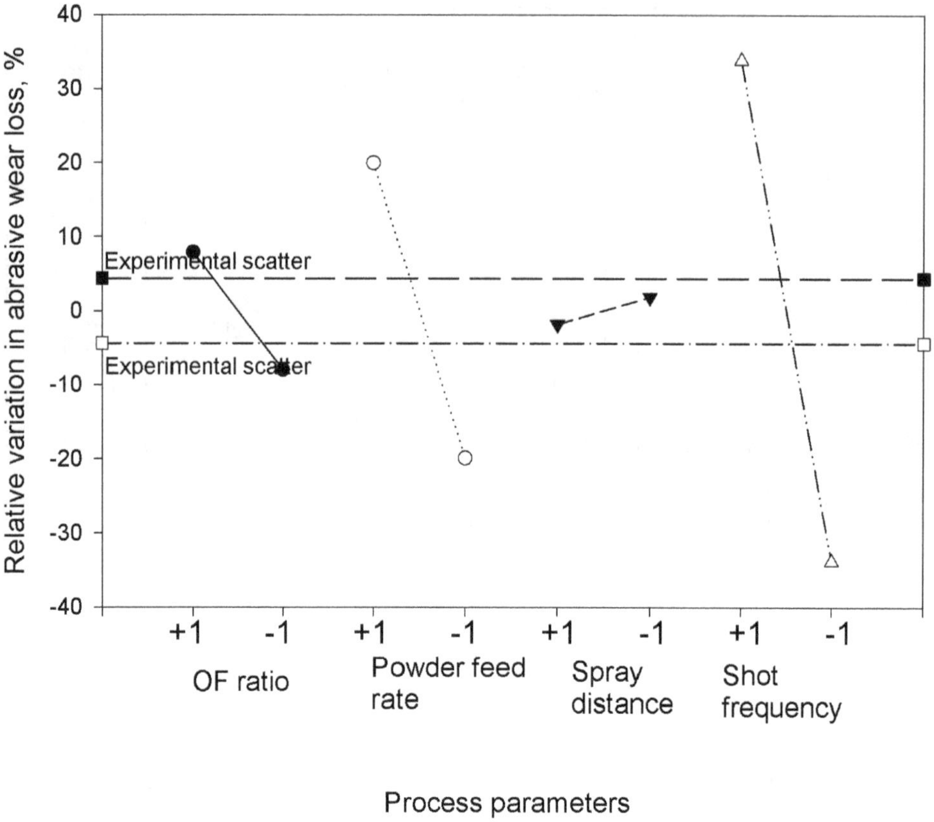

able to deposit ceramics. Whereas, DSC process is more versatile in terms of its velocity and temperature profiles, which can be conveniently used for depositing wide choice of materials including ceramics.

During deposition of ceramics, a well-flattened splat ensures better bonding with the substrate and also with the underlying splats. The particle deformation at the substrate is governed by particle velocity and temperature present in two factors namely, $\rho_p V_p^2/H$ (ρ_p= particle density, V_p= particle velocity; H= particle material hardness) which is often referred to as the Best number (B) and T_p/T_{mp} (T_p= temperature of particle; T_{mp}= melting point of the particle material) defined usually as the homologus temperature (T_H), respectively (Sundararajan *et al.*, 2001). For conditions exceeding the melting point of Al_2O_3, splash type splats were observed which tend to improve deposition efficiency and at lower temperatures, rough splats were achieved which are known for imparting unmelted particles. The correlation between

Figure 10. The influence of process parameters on the total energy and its components for the Al_2O_3 particle (The numbers indicated by arrows represent the proportion of kinetic energy in % of the total energy) (Sundararajan et al., 2013)

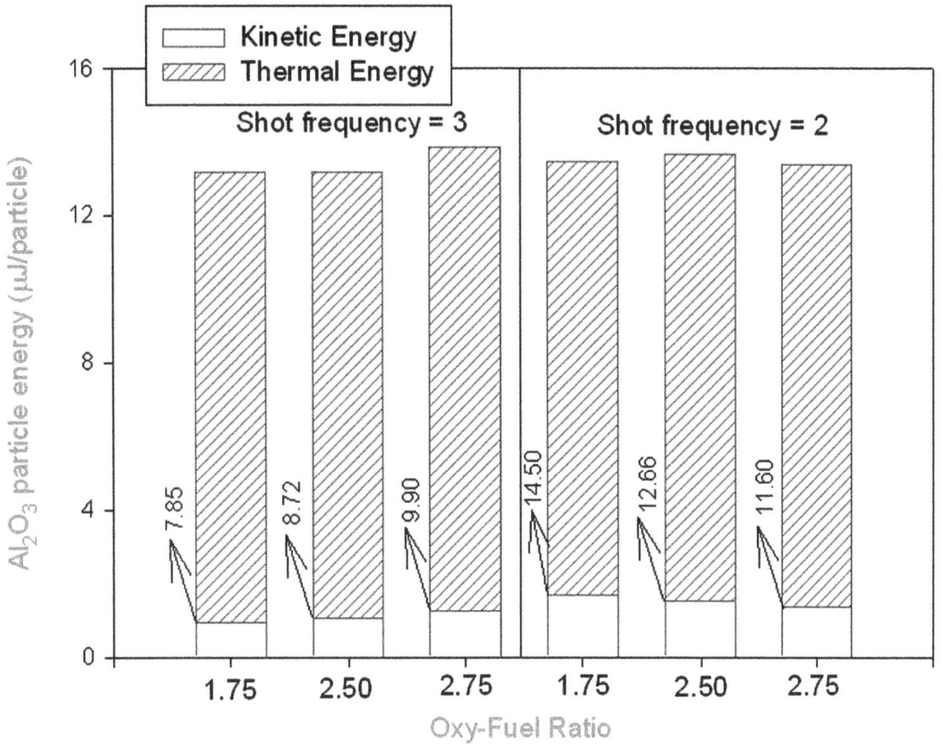

the particle diagnostic values and the abrasive wear rate shown in Figure 11 clearly indicates that the melting point is an important criterion which is also influenced the deposition efficiency. In order to achieve enhanced wear performance, particle velocity may not be so critical and the increase in particle velocity may assist in roughening the substrate/splat surface to improve the bonding of splats and thereby, better deposition efficiency. The low range of deposition efficiency achieved in DSC process (shown in Figure 11c) can be attributed to the delamination of few splats upon cooling from molten state experiencing thermal expansion mismatch as well as the quenching residual stresses. The above correlation studies provide greater understanding of coating characteristics and the resultant performance.

Figure 11. Influence of Al_2O_3 particle (a) temperature, (b) velocity, and (c) deposition efficiency on the abrasive wear performance of detonation sprayed Al_2O_3 coatings

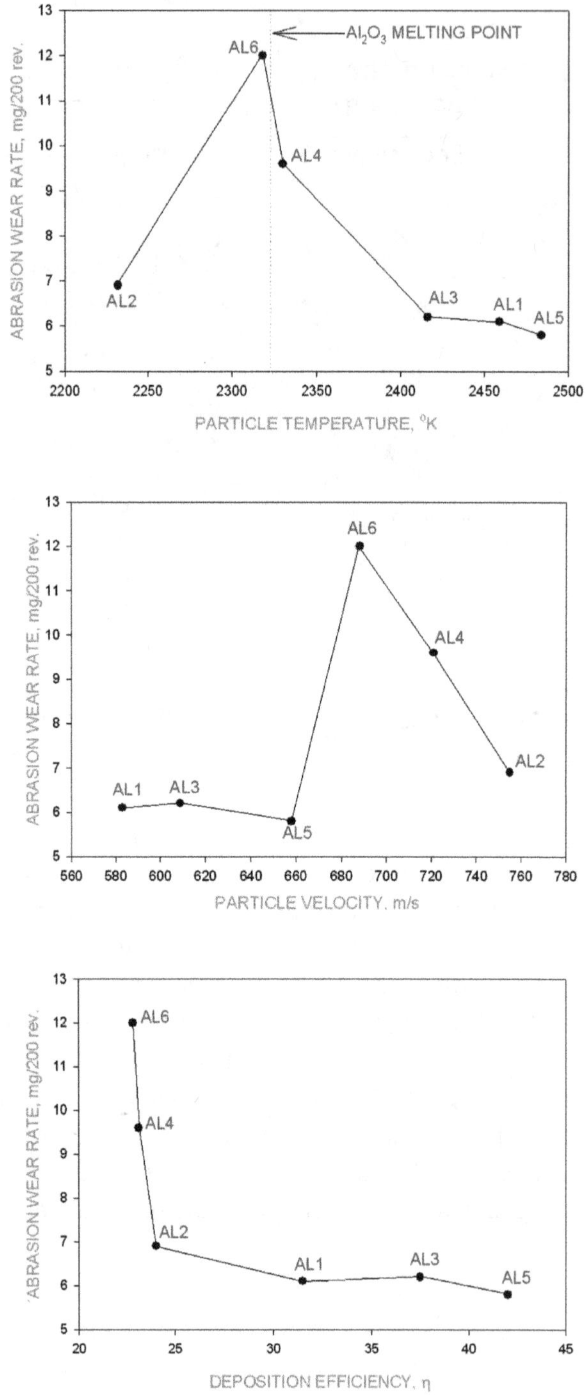

Effect of Binder Content

Though the detonation sprayed WC-Co based coatings are well–proven in imparting enhanced tribological performance, it is mostly decided by the phase changes that occur during spraying. Apart from the role of process parameters, the binder type (Co, Ni, CoCr, etc) and its content also influence the overall outcome and also can be directed towards diverse functionalities. For example, the addition of chromium in WC-Co is envisaged towards improving the corrosion resistance in addition to improvement of erosion resistance (Berget John *et al.*, 1998). The previously published work with detonation spraying of WC with different binders shown that the coatings exhibited brittle phases W_2C, Co_3W_3C, Co_6W_6C, Co_2W_4C, $Co_3W_9C_4$, and even tungsten. In view of the conflicting reports, a detailed study on the role of process parameters for WC based coatings with different binders was studied and reported in this section (Sundararajan *et al.*, 1998).

WC-11Co, WC-10.5Ni and WC-10Co-4Cr powders were selected to investigate the influence of binder type. The key process variables like oxy-fuel (O_2/C_2H_2) gas ratio, volume of the $O_2+C_2H_2$ gas mix and spray distance were varied over a wide range with a two-fold objective viz., to investigate the influence of the above process variables on coating properties and to identify the optimum process parameters for coating each powder. The best characteristics achieved in case of each powder alone are being presented and the detailed process parametric effect is detailed in the subsequent section. As shown in Figure 12, the microhardness values of WC-11Co was the highest among the coatings and the least being in case of WC-10Co-4Cr coatings. This is also expected due to the relative higher content of soft binder phase in case of WC-10Co-4Cr coatings. It is pertinent to note that the coatings exhibited very less porosity values less than 0.5%, a typical nature of the detonation sprayed WC-based coatings. Results have revealed that the WC-10Co-4Cr powder yielded by far the highest WC phase content and this presumably accounts for the superior performance of this coating under abrasive and erosive wear conditions as illustrated in Figure13. However, its relatively poor performance under sliding wear conditions remains yet to be explained and possibly it could be accounted to segregation of Cr, as observed in the microstructures. The WC-11Co coating yielded a overall better performance under all wear modes, while its erosive and abrasive wear rate was marginally less than that of WC-10C0-4Cr and its sliding wear resistance was exceptionally high. This may be attributed to better retention of carbides with higher WC to W_2C ratio and also, due to the presence of tough Co-containing phases in the coating.

Figure 12. Microhardness and porosity characteristics of detonation sprayed WC-based coatings with different binder content

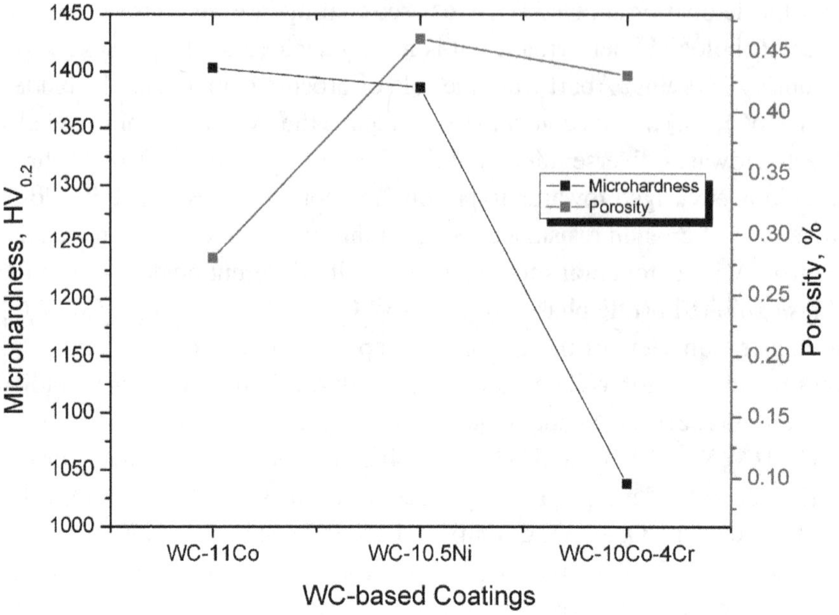

Figure 13. Tribological characteristics under diverse wear modes for detonation sprayed WC-based coatings with different binder content

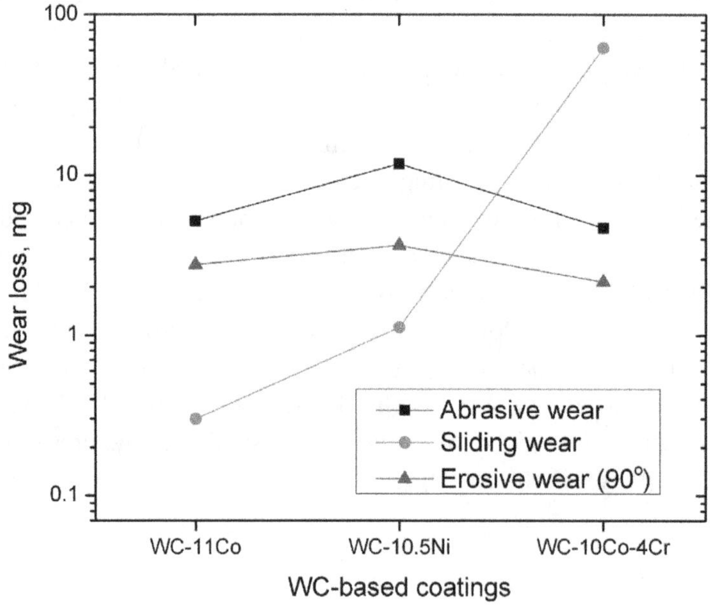

Effect of Hard Phase Characteristics

Cermet coatings based on WC-Co and Cr_3C_2-NiCr are the most popular among the thermal spray feedstock for wear resistant applications and is being utilized in bulk by various industries. There were attempts been made over the years to find alternatives for the above materials, specifically using TiC based compositions due to its low density, high hardness and better alloyability (Bolognini *et al.*, 2003; Mari *et al.*, 2003; Qi *et al.*, 2006). One such powder with considerable promise of performance matching with that of conventional WC and Cr_3C_2 based cermets is (Ti, Mo) (CN) based cermets with either Co or NiCo as the binder phase (Berger *et al.*, 2004).

A second approach involves a development of wear resistant composite coatings with varied hard phases like WC, Cr_3C_2, TiN, B_4C etc. Though, the possibility of achieving wear performance as close to the bulk characteristics is highly unattainable, the proper choice is hard phase is more relevant to the perceived applications. In view of the above, a series of cermet powders having varied hard phase with Ni as the binder phase was attempted for better wear resistance. The powders include, TiMo (CN)–36NiCo, TiMo (CN)-28Co, Cr_3C_2-17Ni, WC-17Ni, WC-12Ni, B_4C-20Ni and TiN-20Ni all procured from different sources. The cross-sectional microstructures under optimized spray conditions are shown in Figure 14 for the above powders, which appear dense and well-adherent to the substrate. Microhardness values of various coatings shown in Figure 15 clearly indicate that WC-based coatings exhibit the highest hardness. TiMo (CN) based coatings exhibited the lowest due to its higher binder content.

The relative wear resistance with respect to the mild steel characteristics is presented in Table 1 comparing with WC-Co based data (Sundararajan *et al.*, 2010). The table clearly reveals that the hardness was significantly lower than WC-Co coatings, though the porosity values were comparable. Though the tribological performances under sliding and erosive wear modes were not encouraging, the abrasive wear resistance was found to be comparable with that of WC-17Co coatings. This is understandable due to the reduced hard phases and higher binder content for TiMo (CN) based coatings. These materials possess additional interesting characteristics in terms of its ability to operate under elevated temperature makes it another interesting candidate for wear resistance.

In case of composite coatings, the tribological performance was evaluated under erosion mode at two impact angles (30°, 90°) and two impact velocities (60, 108 m/s). Among the coatings, B_4C-Ni and WC-12Ni exhibited the lowest erosion volume and WC-17Ni suffered the highest wear rate. The $E90^o/E30^o$ ratio shown in Figure 16 implies that the WC-Ni based coatings exhibited brittle behavior while the other showing an intermediate behavior. The erosion efficiency (η) is another important parameter which is used for describing the erosion mechanism occurring

Figure 14. Typical microstructures of detonation sprayed coatings: (a) Cr_3C_2-17Ni, (b) WC-12Ni, (c) B_4C-20Ni, (d) TiN-20Ni, (e) TiMo(CN)–36NiCo, and (f) TiMo(CN)–28.4Co (Sundararajan et al., 2010, Murthy et al., 2006)

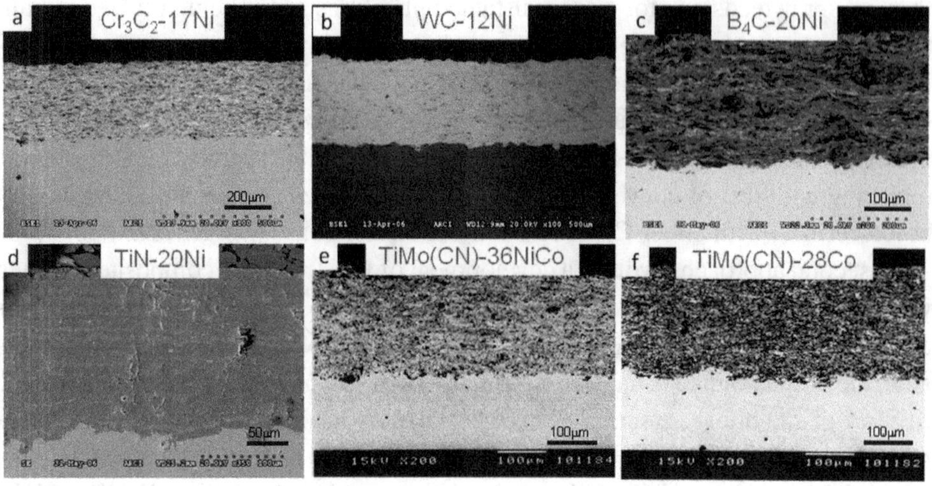

Figure 15. Microhardness values of various detonation sprayed coatings (Sundararajan et al., 2010, Murthy et al., 2006)

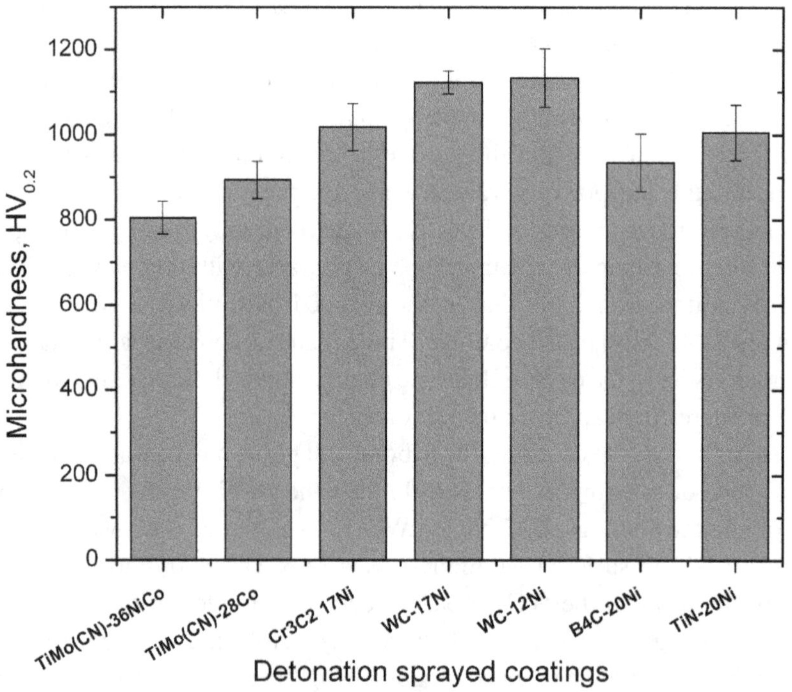

Table 1. Comparison of the tribological properties of TiMo (CN) type coatings with WC-Co type coatings (Sundararajan et al., 2010)

S.No.		HV	Porosity	RAR	RSWR	RER	
						30°	90°
1	Mild steel	123	0	1.0	1.0	1.000	1.000
2	TiMo(CN)-28Co	740	0.83	37.7	17.5	0.021	0.012
3	TiMo(CN)-36NiCo	789	0.69	50.6	7.6	0.037	0.020
4	WC-12Co	1252	0.76	87.9	200.3	0.490	0.430
5	WC-17Co	1240	0.52	42.3	87.4	0.520	0.310
6	WC-25Co	1167	0.61	11.2	64.2	0.480	0.330

in various materials and coatings (Sundararajan *et al.,* 1990, Sundararajan *et al.,* 2006). The observed increase in η with increase in hardness shown in Figure 17 clearly indicates that coating gets more brittle. This is also consistent with the earlier observation of E90°/E30° ratio indicating the harder coatings being brittle. Higher brittleness tends to increase the erosion rate and hence, the desired coating needs to be designed with a compromise between the hardness as well as its hard phase content.

Comparative Tribological Studies

Detonation sprayed coatings finds applications in various segments including aeronautical, agriculture, textile, mining and industrial sectors and being the process of choice for wear resistant applications. The fact that the DSC sprayed coatings is superior to other techniques is because of its versatility, sprayability of dense, well-bonded coatings with less decarburization /oxidation /elemental removal etc. The choice of DSC is also dictated by the superior wear performance compared to plasma sprayed equivalents as shown in Figure 18 (Sundararajan *et al.,* 1998), discussed in terms of its relative wear intensity. Relative wear intensity is described as the efficiency with which the material is removed under a particular wear mechanism or wear constant (Sundararajan *et al.,* 1998). Usually, the wear intensity is too low for sliding, followed by abrasion and the highest under erosion wear modes. Additionally, few coatings may perform well under sliding and may not be under erosive wear conditions. Therefore, combining the wear intensity with the relative wear resistance as shown in Figure 18, the comparison provides significant information on the methodology to be adopted for designing coatings for wear resistance which also considers the operational conditions as well. A comparative study involving

Figure 16. Erosive wear behaviour characteristics of Ni based cermet coatings (Murthy et al., 2006)

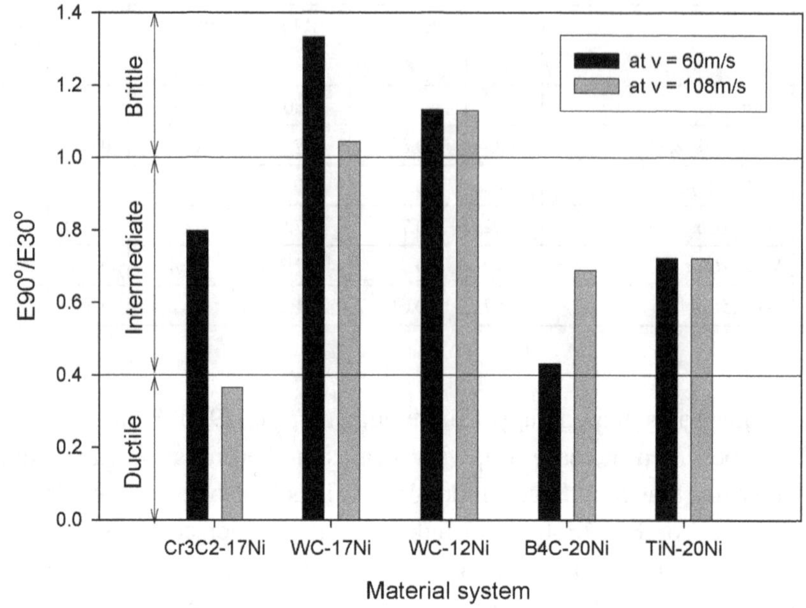

Figure 17. Effect of microhardness on erosion efficiency of the Ni based cermet coatings (Murthy et al., 2006)

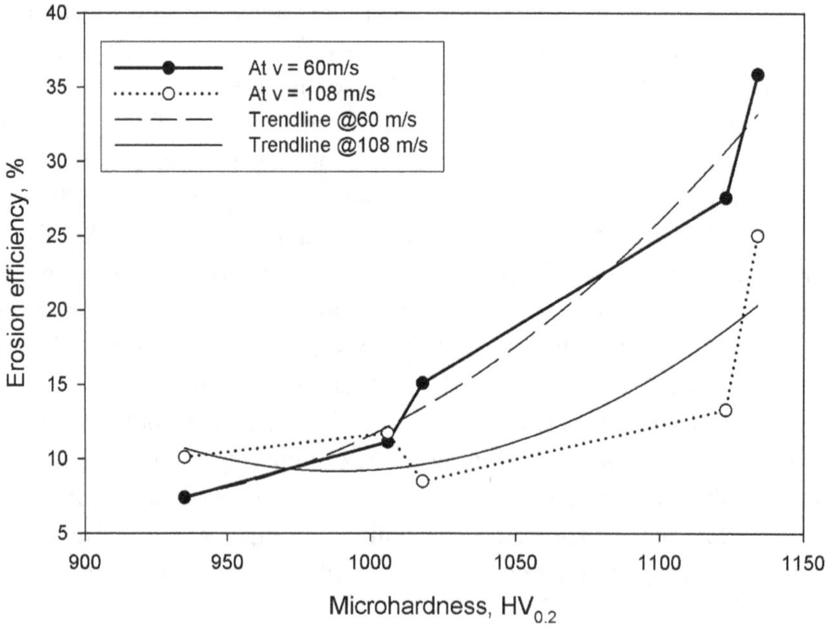

abrasion wear performance of detonation and plasma sprayed WC-Co coatings with carburized, nitrided surfaces implies that the DSC coatings exhibited excellent resistance to wear as shown in Figure 19 (Roy *at al.,* 1999).

In general, all thermal sprayed coatings in the as-deposited state are usually rough to the extent of about R_a: 3-8 μm. Therefore, before the components are put in use, they are subjected to post-spray grinding process to finish the surfaces acceptable for the component assembly. In this regard, a comparative study was performed with HVOF technique, on the role of grinding process on erosive wear behavior of WC-CoCr coatings (Murthy *et al.,* 2001). The erosive wear performance of the as-sprayed and ground coatings is shown in Figure 20. It is clearly evident that the erosion resistance of detonation sprayed was significantly higher in the as-sprayed state and upon grinding, the performance levels of both HVOF and DSC coatings were almost similar at 90° impact. At 30° impact angle, DSC coatings retain their better performance mark over HVOF even after the surface grinding treatment. It is also interesting to note the level of performance improvement offered by the post-spray grinding treatment, which is known to introduce higher compressive residual stresses in the coatings. The augmentation of the favourable residual stresses is found to influence the erosive wear performance. In terms of abrasive wear performance, DSC and HVOF sprayed WC-CoCr and Cr_3C_2-20NiCr coatings were compared as shown in Figure 21 (Murthy *et al.,* 2006). WC-CoCr coatings exhibited higher wear resistance on account of higher fracture strength, high fracture toughness and better binding strength offered by the CoCr matrix (Murthy *et al.,* 2006). The repeated fatigue caused by the abradant movement over the surface caused the damage to the coated surface. Since, the DSC coatings had higher compressive residual stress levels, the damages were less compared to the HVOF equivalents. On a whole, detonation sprayed coatings still holds the position of being premium wear resistant coatings.

New Directions in Detonation Spray Coatings

Detonation sprayed coatings are known for providing improved tribological properties or corrosion or oxidation resistance at elevated temperatures. An additional improvement in coating properties can be realized through the development of graded or layered coating architecture across the thickness. Thereby, the coatings can be tailored to address specific industrial problems through the microstructural changes. In view of the above, a novel approach was adopted to develop multiple powder feeding system (Nirmala et al., 2013), wherein the quantity of powder can be accurately fed and also, parallel operation of multiple powder feeders also can be enabled through electronic controls. Using the above configuration, the quantity

Figure 18. Relative wear resistance characteristics of detonation sprayed (a) WC-12Co, (b) Cr_3C_2-NiCr and (C) Al_2O_3 coatings under diverse wear modes

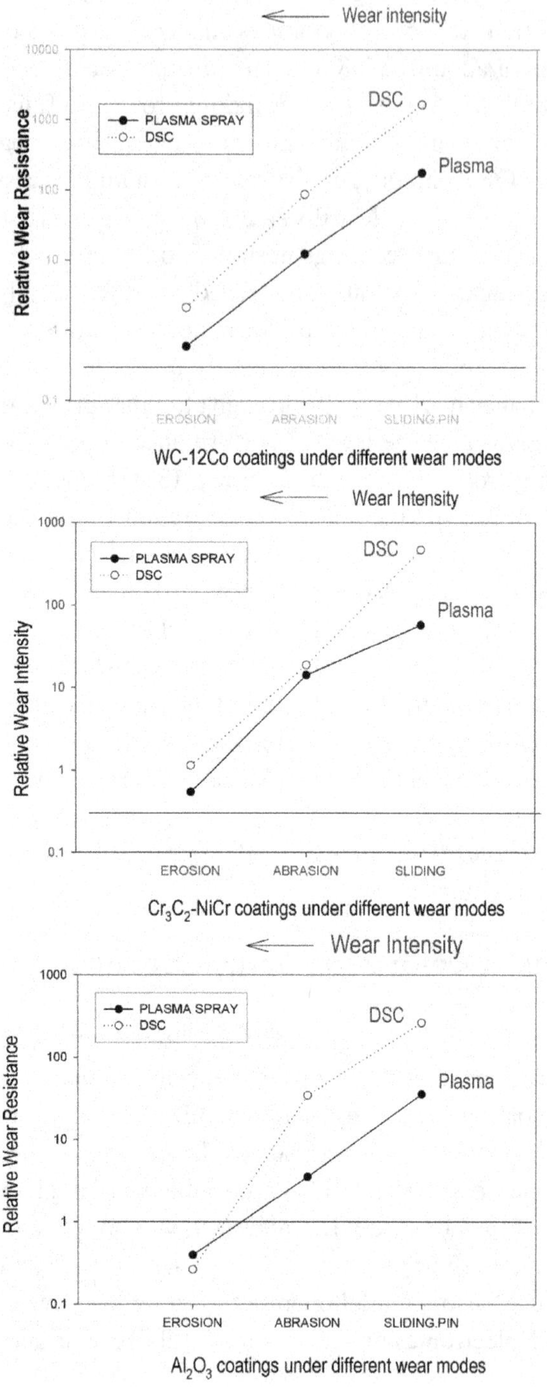

Figure 19. Comparative abrasive wear resistance of detonation sprayed WC-Co coatings with diverse surface engineered coatings and substrates (Sundararajan et al., 2006)

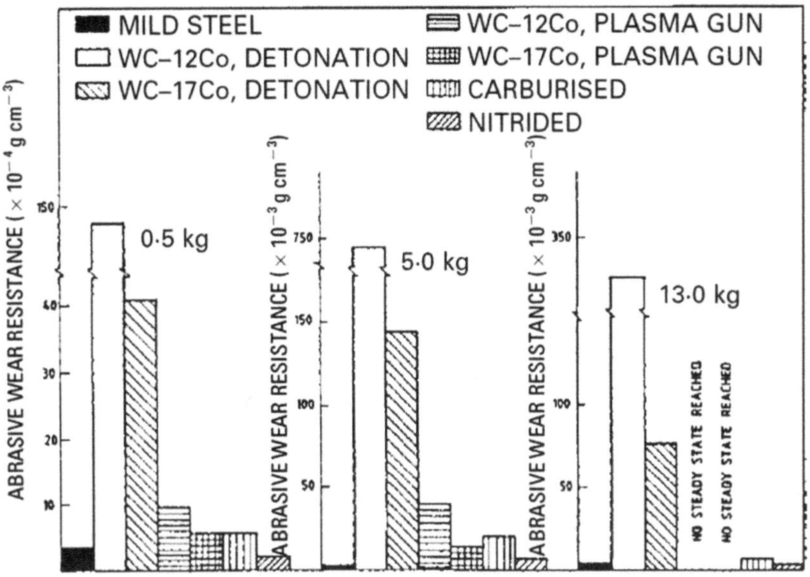

Figure 20. Steady state erosive wear rate for as-sprayed and ground WC-12Co coatings deposited by detonation and HVOF spraying, in comparison with the substrate (Roy et al., 1999)

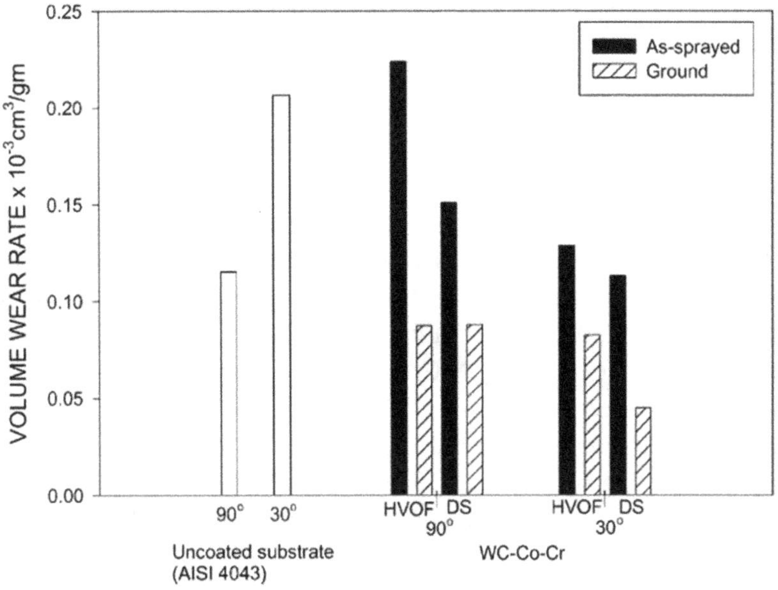

Figure 21. Cumulative volume wear loss of detonation and HVOF sprayed coatings in comparison with the hard chrome plating (Murthy et al., 2001)

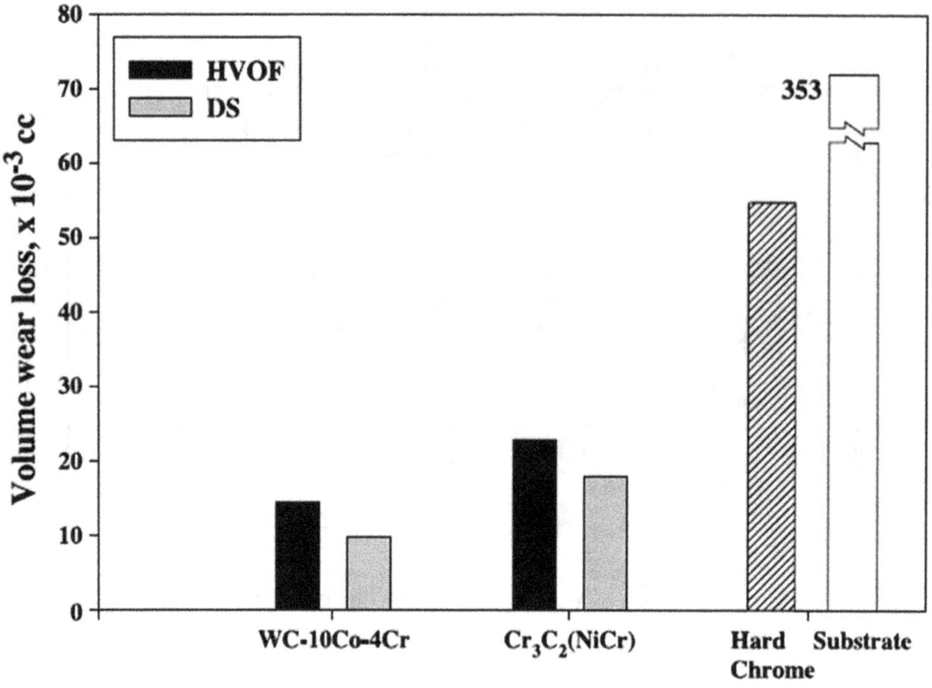

and composition of the injected powder can be controlled, either in a sequential manner or simultaneous manner to that layered or gradient coatings, respectively, can be achieved. By proper selection of the coating feedstock and proper tailoring of coating materials, several desired functional properties like wear resistance, thermal barrier, electrical insulation, corrosion etc can be augmented over the conventional mono-layer systems.

In the present study, the detonation spray coating systems with multiple powder feeding mechanism setup was utilized to develop multilayered and graded coatings of WC-12Co and NiCrBSiFe powders to examine the microstructural features. Some preliminary tribological studies were also carried out under diverse wear modes. It is pertinent to note that WC-12Co coatings are very effective in providing enhanced wear resistance coatings whereas, NiCrBSiFe is a self-fluxing alloy with enhanced corrosion resistance.

Typical microstructures of layered and graded WC-12Co + NiCrBSiFe coating is shown in Figure 22a-b, which exhibits a clear difference in contrast, between the two different material systems. The dark grey feature corresponds to NiCrBSiFe

Figure 22. Cross-sectional microstructure of (a) graded WC-12Co+NiCrBSiFe coating, (b) multilayer WC-12Co+NiCr coating and microhardness profile of (c) graded WC-12Co+NiCrBSiFe coating (Nirmala et al., 2013)

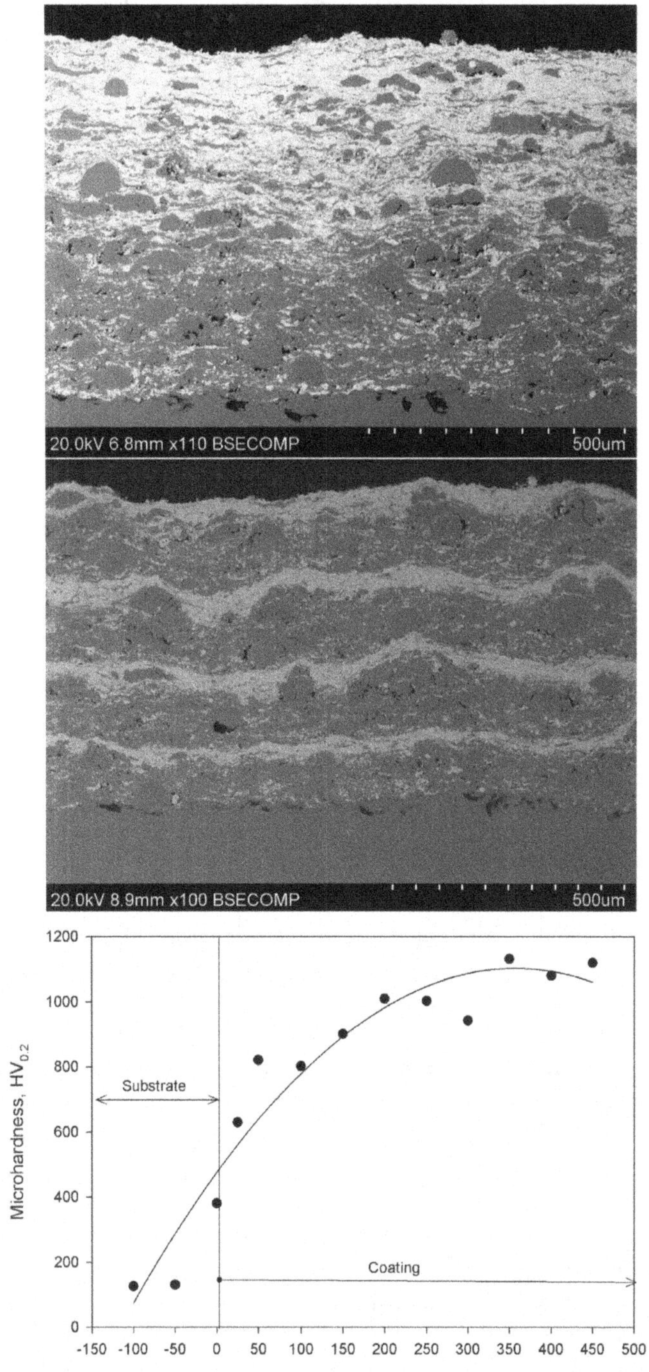

coating and the light colored to WC-12Co. Additional proof of the gradient structure is provided by microhardness measurements taken at periodic intervals across the coating thickness. The microhardness values recorded at 50 μm intervals, exhibited a gradual increase in the values as shown in Figure 22c. Preliminary wear studies with multilayered coatings shown better abrasion wear rate in comparison with conventional WC-12Co coating at 50N load which can be attributed to the superior toughness of well modulated multilayered coatings compared to that of conventional WC-12Co coatings. The multilayered coatings also exhibited improvement in erosion resistance on account of homogenously distributed metallic phase within the hard carbide phases.

SUMMARY

The detonation spray coating (DSC) facility available at author's institute is being used extensively for developing protective coatings to employ on wide range of engineering components for enhancing the service life. Amongst list of the coatings development, all the commercially available carbides and oxides based powders have been successfully accomplished including the process parameters optimization, studies on influence of process parameters on particle velocity and temperature and their effect quality of the coating properties. Studies also carried out effect of particle size and binder content on tribological characteristics of wear resistance coating materials such as WC–Co, Cr_3C_2–NiCr, TiMo (CN)–NiCo, WC-CrC-Ni, WC-Co-Cr, WC-Ni, Cr_3C_2-Ni, Cr_3C_2-Inconel, etc.

REFERENCES

Barbezat, G., Nicoll, A., & Sickinger, A. (1993). Abrasion, erosion scuffing resistance of carbide and oxide ceramic thermal sprayed coatings. *Wear*, *162–164*, 529–537. doi:10.1016/0043-1648(93)90538-W

Berger, L. M., Woydt, M., Zimmermann, S., Keller, H., Schwier, G., Enžl, R., & Thiele, S. (2004). Tribological behavior of HVOF-sprayed Cr3C2-NiCr and TiC-based coatings under high temperature dry sliding conditions. In *Proceedings of International Thermal Spray Conference*. DVS-Verlag.

Berget, J., Einar, B., & Trond, R. (1998). *Effects of matrix composition and particle size on structure and erosion –corrosion resistance of HVOF sprayed WC-Co-Cr coatings*. London: Institute of Materials.

Bolognini, S., Feusier, G., Mari, D., Viatte, T., & Benoit, W. (2003). TiMoCN based cermets: High temperature deformation. *International Journal of Refractory Metals & Hard Materials*, *21*(1-2), 19–29. doi:10.1016/S0263-4368(02)00091-4

Budinski, K. G. (1988). *Surface engineering for wear resistance*. Prentice Hall.

Fauchais, P. L., Heberlein, J., Boulos, V. R., & Maher, I. (2014). Thermal spray fundamentals. Springer Science + Business Media.

Fauchais, P. L., & Vardelle, A. (2000). Heat, mass and momentum transfer in coating formation by plasma spraying. *International Journal of Thermal Sciences*, *39*(9-11), 852–870. doi:10.1016/S1290-0729(00)01195-9

Guest, C. J. S. (1986). Plasma and detonation gun coatings. *Trans IMF*, *64*, 33–38.

Guilemany, J. M., Dosta, S., & Miguel, J. R. (2006). The enhancement of the properties of WC-Co HVOF coatings through the use of nanostructured and microstructured feedstock powders. *Surface and Coatings Technology*, *201*(3-4), 1180–1190. doi:10.1016/j.surfcoat.2006.01.041

Kharlamov, Y. A. (1986). Detonation spraying of protective coatings. *Materials Science and Engineering*, *93*, 1–37. doi:10.1016/0025-5416(87)90409-5

Li, Y., Huang, J., & Ding, C. (1998). Mechanical and tribological properties of plasma-sprayed Cr_3C_2-NiCr, WC-Co and Cr_2O_3 coatings. *Journal of Thermal Spray Technology*, *7*, 242-246.

Mari, D., Bolognini, S., Feusier, G., Cutard, T., Verdon, C., Viatte, T., & Benoit, W. (2003). TiMoCN based cermets part I: Morphology and phase composition. *International Journal of Refractory Metals & Hard Materials*, *21*(1-2), 37–46. doi:10.1016/S0263-4368(03)00010-6

McPherson, R. (1980). On the formation of thermally sprayed alumina coatings. *Journal of Materials Science*, *15*(12), 3141–3149. doi:10.1007/BF00550387

Murthy, J. K. N., Rao, S., & Venkataraman, B. (2001). Effect of grinding behaviour of a WC-Co-Cr coating deposited by HVOF and detonation gun spray processes. *Wear*, *249*(7), 592–600. doi:10.1016/S0043-1648(01)00682-2

Murthy, J. K. N., & Venkataraman, B. (2006). Abrasive wear behaviour of WC–CoCr and Cr3C2–20(NiCr) deposited by HVOF and detonation spray processes. *Surface and Coatings Technology*, *200*(8), 2642–2652. doi:10.1016/j.surfcoat.2004.10.136

Nakamura, T., Qian, G., & Berndt, C. C. (2000). Effects of pores on mechanical properties of plasma-sprayed ceramic coating. *Journal of the American Ceramic Society*, *83*(3), 578–584. doi:10.1111/j.1151-2916.2000.tb01236.x

Pawlowski, L. (1995). *The science and engineering of thermal spray coatings*. John Wiley & Sons.

Pejryd, L., Wigren, J., Greving, D. J., Shadley, J. R., & Rybicki, E. F. (1995). Residual stresses as a factor in the selection of tungsten carbide coatings for a jet engine application. *Journal of Thermal Spray Technology*, *4*(3), 268–274. doi:10.1007/BF02646970

Praxair. (n.d.). *Powders*. Retrieved from http://www.praxairsurfacetechnologies.com/na/us/pst/pst.nsf/0/085C6CF7641FE446852576A50058300A?OpenDocument

Qi, X., Eigen, N., Aust, E., Gartner, F., Klassen, T., & Bormann, R. (2006). Two-body abrasive wear of nano- and microcrystalline TiC–Ni based thermal spray coatings. *Surface and Coatings Technology*, *200*(16-17), 5037–5047. doi:10.1016/j.surfcoat.2005.05.007

Roy, M., Rao, C. V. S., Rao, S., & Sundararajan, G. (1999). Abrasive wear behaviour of detonation sprayed WC-Co coatings on mild steel. *Surface Engineering*, *15*(2), 129–136. doi:10.1179/026708499101516470

Rybicki, E. F., Shadley, J. R., Xiong, Y., & Greving, D. J. (1995). A cantilever beam method of evaluating Young's modulus and Poisson's ratio of thermal spray coatings. *Journal of Thermal Spray Technology*, *4*(4), 377–384. doi:10.1007/BF02648639

Sanikommu, N., Sivakumar, G., Joshi, A. S., Aruna, N. S., Rao, D., & Sundararajan, G. (2013). A computer-based approach for developing functionally graded and layered coatings with detonation spray coating process. *Journal of Scientific &Industrial Research*, *72*, 477-480. Retrieved from http://www.oerlikon.com/metco/en/products-services/coating-materials/coating-materials-thermal-spray/

Saravanan, P., Selvarajan, V., Srinivasa Rao, D., Joshi, S. V., & Sundararajan, G. (2000). Influence of process variables on the quality of detonation gun sprayed alumina coatings. *Surface and Coatings Technology*, *123*(1), 44–54. doi:10.1016/S0257-8972(99)00252-2

Shipway, P. H., McCartney, D. G., & Sudaprasert, T. (2005). Sliding wear behaviour of conventional and nanostructured HVOF sprayed WC–Co coatings. *Wear*, *259*(7-12), 820–827. doi:10.1016/j.wear.2005.02.059

Sivakumar, G., Ramakrishna, L., Jain, V., Srinivasa Rao, D., & Sundararajan, G. (2001) The influence of the process parameters on the properties of detonation sprayed WC-12Co coatings. In *Proceedings of International Thermal Spray Conference.* ASM Materials.

Skandan, G., Jain, M., Fischer, T. E., Kear, B. H., Rigney, W. R., Shropshire, R., & Brunhouse, S. (2000). *Thermal spray: Surface engineering via applied research.* Materials Park, OH: ASM International.

Stark, H. C. (n.d.) *AMPERIT® thermal spray powders.* Retrieved from http://www.hcstarck.com/en/products/amperitreg_thermal_spray_powders.html

Sundararajan, G., Roy, M. B., & Venkataraman, B. (1990). Erosion efficiency-a new erosion micromechanism parameter to characterize the dominant erosion micromechanism. *Wear, 140*, 369–381. doi:10.1016/0043-1648(90)90096-S

Sundararajan, G., Sivakumar, G., Kavitha, R., & Phanikrishna, A. R. (2006), Erosion behaviour of composite cermet coatings. In *Proc. ACUN-5 "Developments in Composites: Advanced, Infrastructural, Natural, and Nanocomposites.* UNSW.

Sundararajan, G., Sivakumar, G., Sen, D., Srinivasa Rao, D., & Ravichandra, G. (2010). The tribological behaviour of detonation sprayed TiMo(CN) based cermet coatings. *International Journal of Refractory Metals & Hard Materials, 28*(1), 71–81. doi:10.1016/j.ijrmhm.2009.07.007

Sundararajan, G., Sivakumar, G., & Srinivasa Rao, D. (2001), The interrelationship between particle velocity and temperature, splat formation and deposition efficiency in detonation sprayed alumina coatings. In *Proceedings of International Thermal Spray Conference.* Academic Press.

Sundararajan, G., Srinivasa Rao, D., Sen, D., & Somaraju, K. R. C. (1998). *Tribological behavior of thermal sprayed coatings.* London: Institute of Materials.

Sundararajan, G., Srinivasa Rao, D., Sivakumar, G., & Joshi, S. V. (1998). A comparative study of tribological behavior of plasma and D-gun sprayed coatings under different wear modes. *Journal of Materials Engineering and Performance, 7*(3), 343–351. doi:10.1361/105994998770347783

Sundararajan, G., Srinivasa Rao, D., Sivakumar, G., & Joshi, S. V. (2013). Detonation spray coatings. In *Encyclopedia of tribology.* Springer.

Sundararajan, G., & Suresh Babu, P. (2009). Detonation sprayed WC-Co coatings: Unique aspects of their structure and mechanical behavior. *Transactions of the Indian Institute of Metals, 62*(2), 1–9. doi:10.1007/s12666-009-0013-1

Suresh Babu, P. (2011). The influence of erodent hardness on the erosion behavior of detonation sprayed WC-12Co coatings. *Wear, 270*(11-12), 903–913. doi:10.1016/j.wear.2011.02.019

Suresh Babu, P., Basu, B., & Sundararajan, G. (2008). Processing–structure–property correlation and decarburization phenomenon in detonation sprayed WC–12Co coatings. *Acta Materialia, 56*(18), 5012–5026. doi:10.1016/j.actamat.2008.06.023

Suresh Babu, P., Srinivasa Rao, D., Rao, G.V.N., & Sundararajan, G. (2007). Effect of feedstock size and its distribution on the properties of detonation sprayed coatings. *Journal of Thermal Spray Technology, 16*(2), 281-290.

Taguchi, G., & Konishi, S. (1987). *Taguchi methods: Orthogonal arrays and linear graphs*. Dearborn, MI: ASI Press.

Wang, Y. (1993). Friction, wear performances of detonation-gun and plasma sprayed ceramic and cermet hard coatings. *Wear, 161*(1-2), 69–78. doi:10.1016/0043-1648(93)90454-T

Yang, Q., Senda, T., & Ohmori, A. (2003). Effect of carbide grain size on microstructure and sliding wear behavior of HVOF-sprayed WC–12% Co coatings. *Wear, 254*(1-2), 23–34. doi:10.1016/S0043-1648(02)00294-6

Yunfei, Q. (2003). The effects of fuel chemistry and feedstock powder structure on the mechanical and tribological properties of HVOF thermal-sprayed WC–Co coatings with very fine structures. *Surface and Coatings Technology, 172*(1), 24–41. doi:10.1016/S0257-8972(03)00242-1

KEY TERMS AND DEFINITIONS

Abrasive Wear: Wear due to hard particles or hard protuberances forced against and moving along a solid surface.

Binder: A cementing medium used in producing composite or agglomerate powders.

Bond Strength: The strength of the adhesion between the coating and the substrate. A number of test methods are in use to measure the bond strength of coatings.

Deposition Efficiency: The ratio, usually expressed as a percentage, of the weight of the thermal spray deposit to the weight of the thermal spray powder processed using any of the thermal spray processes.

Detonation Spray Coating or D-Gun: A thermal spray process in which the coating material is heated and accelerated to the work piece by a series of detonations or explosions from oxy-fuel gas mixtures.

Erosion Wear: Removal of material from a surface due to mechanical interaction between that surface and a fluid, a multicomponent fluid, or impinging liquid or solid particles.

High Velocity Oxy Fuel (HVOF) Spraying: A thermal spray process in which a fuel gas is mixed with oxygen and delivered at high pressure to the HVOF gun and ignited to form a high velocity oxygen/fuel gas stream in to which thermal spray powders are introduced and propelled on to the substrate. A Thermal spray process. The spray powder particles are injected into a high velocity jet formed by the combustion of oxygen and fuel, heated and accelerated to the work piece.

Plasma Spraying: A thermal spray process in which a non-transferred arc is produced by ionizing an inert gas that then forms the heat source in to which thermal spray materials are injected which are then subsequently propelled to the substrate to form a thermal spray coating.

Porosity: Cavity type discontinuities within a thermal sprayed coating.

Thermal Spraying: A group of processes in which finely divided metallic or non-metallic thermal spray materials are deposited in a semi molten state to a substrate to form a thermal spray coating.

Tribology: The science and technology concerned with interacting surfaces in relative motion.

Wear: Loss of material from a surface by means of some mechanical action.

Chapter 11
Characterization of Mechanical Properties and the Abrasive Wear of Thermal Spray Coatings

Salim Barbhuiya
Curtin University of Technology, Australia

Ikbal Choudhury
National Institute of Technology Silchar, India

ABSTRACT

Thermal spray is a generic term used to define a group of coating processes used to apply both metallic and non-metallic coatings. These coatings are usually defined by their hardness, strength, porosity, roughness, and wear resistance. In this chapter, the authors turn their attention to the mechanical and tribological properties of thermal spray coatings. The individual phase plays a very important role in determining the performance of the coating. However, evaluating the mechanical and tribological properties at a nano-level requires new test methods and their validation. In this chapter, elaborate discussion of some techniques to evaluate and analyze the mechanical and tribological properties of different thermal spray coatings is done. This chapter is intended to help the reader to firstly understand the basic principle and methods of characterization of thermal spray coatings using instrumented nanoindentation, nanoscratch, abrasive wear testing techniques, and secondly to get an idea of the recent techniques and review the research and development in the same field.

DOI: 10.4018/978-1-4666-7489-9.ch011

INTRODUCTION

The most important advantage of thermal spray process, compared to the other coating processes, is that it has a great range of coating materials, coating thickness and possible coating characteristics. To predict their behavior, lifetime and application area, it is necessary to completely understand the relationships between technology, process parameters, microstructure and properties of the coatings. In the characterization of thermally sprayed coatings it is necessary to take into consideration their unique lamellar microstructure as shown in Figure 1.

The purpose of this chapter in tribology is understandably the minimization and elimination of losses resulting from friction and wear at all levels of technology where the rubbing of surfaces is involved. Now a days, cermets (i.e ceramics and metals) are used for coating purposes. The hardness of ceramics and the toughness of metals when combined gives unconventional strength and wear resistance to these thermally sprayed coatings. In order to achieve the best performance of a thermal spray coating we need to know the relationships between the mechanisms of coating formation and its mechanical properties. While the mechanism of formation of coating has rather been well discussed and understood, the detailed information on its mechanical properties still remains a difficult job. The primary reason is the heterogeneity of the coating as shown in Figure 2.

Now a days, analytical and numerical modeling of thermal spray coatings is possible using various software. These models are capable of demonstrating the

Figure 1. Diagrammatic representation of the cross-sectional view of thermal spray coatings

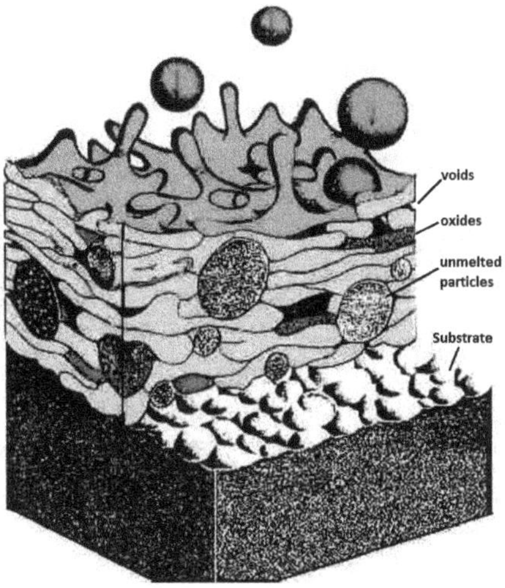

Figure 2. Typical microstructure of a wc-17co HVOF-sprayed coating showing WC grains and binding metal matrix

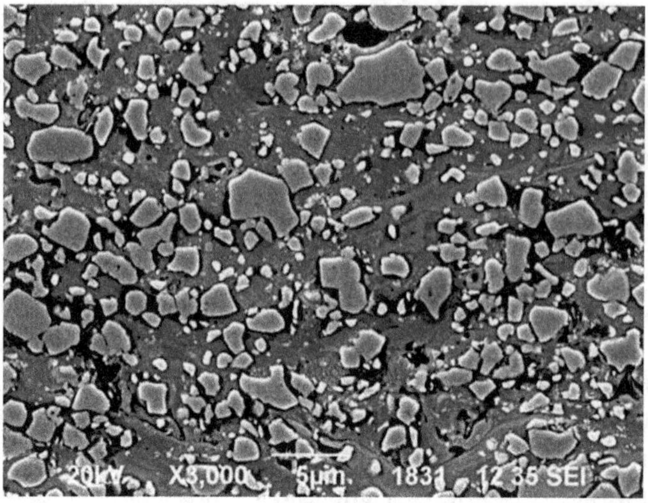

complexity of gas-particle interactions like distribution of non-molten particles in a spray. Modeling and software simulation is also done to validate new spray techniques. Due to splattering up of molten coating particles at supersonic speed, pores are generated in thermally sprayed coatings. As a result of these pores, residual stresses are developed. Various research works are aimed at determining the mechanical and tribological properties of these coatings in order to develop coatings with superior performance.

In view of the previous discussion the objective of this chapter is to help the reader to firstly understand the basic principles and methods of characterization of thermal spray coatings using instrumented nanoindentation, nanoscratch techniques and abrasion wear testing and secondly to get an idea of the recent techniques, related sources of error and review the research and development in the same field. This chapter is not intended to describe the tribological properties and aspects of thermal spray coatings, rather it elaborately demonstrates the working principle of some of the most important characterization techniques to evaluate the mechanical and tribological properties.

MECHANICAL AND TRIBOLOGICAL CHARACTERIZATION

Need to Evaluate Mechanical and Tribological Properties

A coating is usually defined by its intrinsic and extrinsic properties. Both these properties are equally important in determining the performance of the coating. When a coating is in its research and development stage, understanding the mechanical and tribological behavior of the coating is of utmost importance in order to decide it's performance. In the case of thermally sprayed coatings, it is necessary to take into consideration their unique lamellar microstructure. Together with materials characteristics, such as hardness, modulus of elasticity or fracture toughness, the coatings porosity, cohesive strength, content of oxides and other microstructure defect also plays its role.

The tribological properties of parts surface, namely their wear resistance and friction properties, are in many cases determining for their proper function. To improve surface properties, it is possible to create hard, wear resistant coatings by thermal spray technologies. Using these versatile coatings it is possible to increase parts lifetime, reliability and safety. To predict behavior, lifetime and application area of thermally sprayed coatings it is necessary to completely understand the relationships between technology, process parameters, microstructure and properties of the coatings.

Important Mechanical and Tribological Properties

The relative importance given to different properties depends on a number of factors, such as coating's level of development, e.g the need to further optimization and the intended application. A coating in full production for many years may be subjected to less rigorous testing than one which is newly introduced. When the coating is in production stage, the evaluation of its mechanical and tribological properties is a measure of quality control and repeatability. However, if the coating is still in its research and development stage, then characterization is needed to improve specific properties of the coating and the way in which they are used for process variables. Thus, adhesion and thickness are the key properties that are important when a coating has already been optimized and is in production, whereas properties like friction, wear, hardness, Young's modulus and composition are usually of most importance when a tribological coating is under development. Thermal spray coatings are porous, which results in decrease in toughness and higher permeability to water and corrosive ions.

Methods Applied to Evaluate the Properties

Assessing the usefulness of thermal spray coatings requires understanding, developing and utilizing appropriate testing and characterization methods. The goal to mechanical and tribological characterization can be addressed by a wide range of techniques. As described in the previous section, hardness and modulus of elasticity are the important properties of coatings. There are a number of ways of testing the mechanical and tribological properties of thermal spray coatings. The modulus of elasticity can be calculated by methods like tensile tests, bending tests and indentation measurements. Micro and nanoindentation are used now a days to evaluate Young's modulus, hardness, creep, stiffness, residual stress of the sample. The tribological properties like wear rate, coefficient of friction, can be measured using ball-on-disc tests. This test can also be used to evaluate adhesion strength of the coating. With the help of Pull off tests the nature and numerical value of bond strength can be measured. However, the most advanced method to evaluate the adhesion, cohesion and coefficient of friction of a coating is scratch test. Scratches can be run at micro or nano level and material properties can be accurately measured by analyzing the data obtained.

INDENTATION TESTS

Principle

Young's modulus and hardness are key parameters in the study of the adhesion of thick or thin films to various substrates and their response to mechanical loads. One of the techniques most recently used to study the mechanical properties of thin films is the instrumented indentation, known as nano-indentation. Indentation is perhaps the most commonly known means to study the mechanical properties of materials. This is a superficial technique to quasi-statically measure the penetration generated by increasing loads applied to a material. This section briefly describes the theoretical aspects of an instrumented nanoindentation technique and then the experimental methodology. Standard nanoindentation equipment as shown in Figure 3 consists mainly of an actuator to apply the force, an indenter mounted on a rigid column through which the force is applied on the sample and a sensor that measures the displacement of the indenter (Bull, 2005).

The principal aim of majority nanoindentation test is to determine the Young's modulus and hardness of the specimen material from load-displacement measurements. In nanoindentation testing the depth of penetration beneath the coating layer is measured by applying load on the indenter. Usually the properties of the

Figure 3. Schematic diagram of a depth-sensing nanoindentation tester

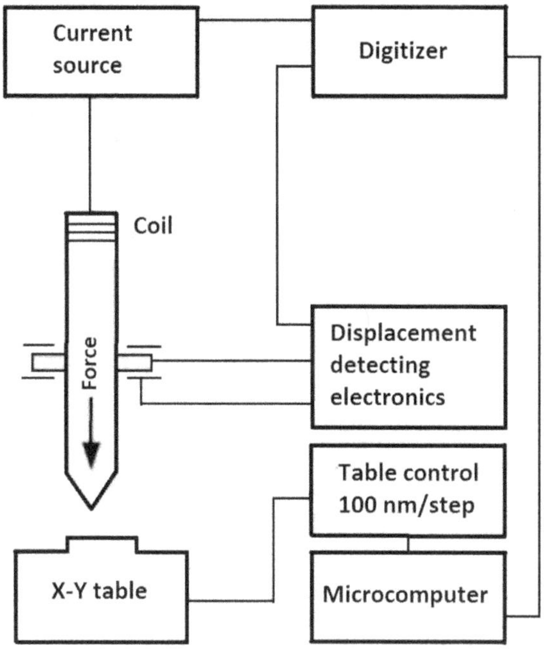

coating are significantly affected by the substrate. In order to avoid any changes in property of coating the depth of penetration in this case is limited to 10% of the dry film thickness (DFT) (Oliver and Pharr, 1992; Rabei et al., 1999; Hay and Pharr, 2000). The area of contact can be measured easily as the geometry of the indenter is known. Generally a wide variety of indenters (Vickers, Berkovich, Knoop, spherical etc.) are used to get an imprint on the sample. Figure 4 is a nanoindentation curve plotted between load and displacement. It clearly shows the loading and unloading pattern of the sample under consideration. This curve is used to determine the hardness and elastic modulus of the coating material. The hardness of the coating is evaluated as the maximum indenter load P_{max} divided by the projected area A of the indenter at maximum load as shown in equation (1).

$$H = \frac{P_{\max}}{A} \tag{1}$$

The projected area can be calculated from the indenter geometry as shown in Figure 4. The relation between the projected area A and the contact depth h_c is shown by equation (2).

Figure 4. Single cycle load on sample-displacement of indenter curve of an indentation test

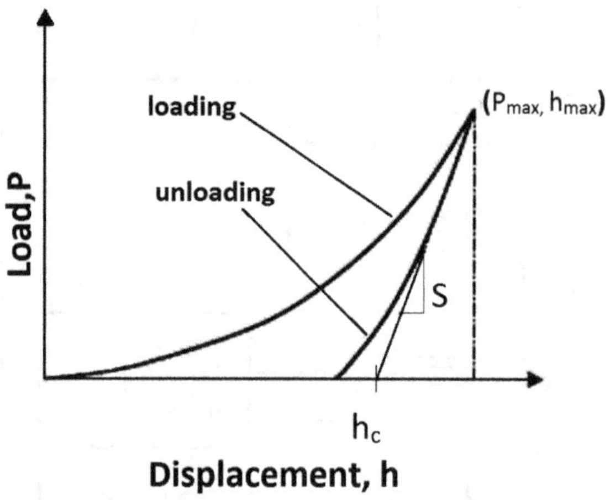

$$A = c_0 h_c^2 \tag{2}$$

The above equation is derived from the polynomial equation (3). The area function $A(h_c)$ is a fifth order polynomial as shown below:

$$A = c_0 h_c^2 + c_1 h_c + c_2 h_c^{1/2} + c_3 h_c^{1/4} + c_4 h_c^{1/6} + c_5 h_c^{1/8} \tag{3}$$

The value of C_0 depends on the indenter tip and for Berkovich indenter with a pyramidal diamond tip the value is 24.5. In Figure 5, we can observe that the contact depth h_c is less than the maximum indentation depth h_{max} as because h_{max} is as a result of elastic deformation of the sample area around the indenter tip. The contact depth is given by equation (4) as below.

$$h_c = h_{max} - \varepsilon \frac{P_{max}}{S} \tag{4}$$

where S denotes the stiffness of the coating that can be calculated from the slope of the unloading curve in Figure 3. The value of ε for the pyramidal indenter is considered as 0.75. Usually fused quartz silica is used to calibrate the indenter shape, which is very critical in the nanoindentation technique. The stiffness S_{max} is obtained from the nanoindentation curve and the projected area can be evaluated as a function of S_{max} as per equation (5)

Figure 5. Schematic diagram of an ideal conical indenter at maximum load

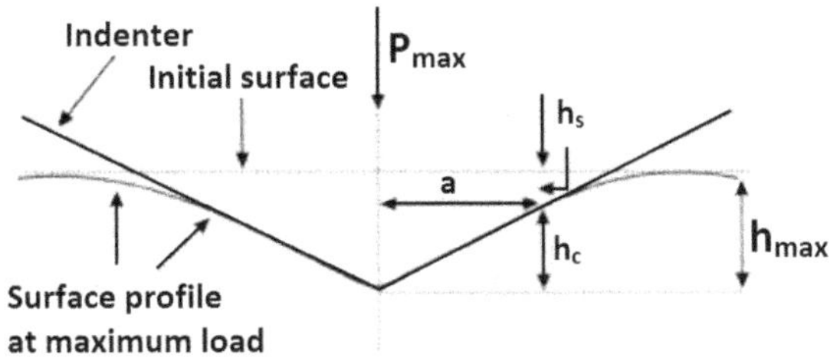

$$A = \frac{\pi}{4}\left(\frac{S_{max}}{E_r}\right)^2 \quad (5)$$

where E_r is the reduced modulus, which is as a result of the elastic deformation occurring in both the indenter and the coating. The value of E_r can be calculated as per equation (6).

$$\frac{1}{E_r} = \frac{1-v_s^2}{E_s} + \frac{1-v_i^2}{E_i} \quad (6)$$

where E_s is the elastic modulus of the test specimen, E_r is the elastic modulus of the indenter, ϑ_s is the poison's ratio of the test specimen and ϑ_i is the poison's ratio of the indenter. The elastic modulus E_r is given by equation (7) as shown below

$$E_r = \frac{dp}{dh} \cdot \frac{1}{2} \cdot \frac{\sqrt{\pi}}{\sqrt{A}} = S \cdot \frac{1}{2} \cdot \frac{\sqrt{\pi}}{\sqrt{A}} \quad (7)$$

and the E_s of the coating is determined by using equation (6) and (7).

Indentation experiments can be performed using both load control and depth control methods (CSM). In load control measurements the values of Young's modulus and hardness are obtained at maximum penetration depth. However in case of continuous stiffness option (CSM), the Young's modulus and hardness are measured as a continuous function of indenter penetration.

Mechanical Characterization Using Nanoindentation Tests

With the advent and development of microsystems and nanotechnology, researchers have developed better thermal spray coatings with superior tribological properties. The new generation thermal spray coatings have nanoparticles embedded in it to enhance their properties. During application the coatings gets deposited on the substrate in the form of splats at high speed. Knowing the material properties of individual splats becomes imperative to characterize the coating and to further develop it. Thermal spray coatings are now a days being used in various micromechanical systems and nanotechnological applications like MEMS, biomedical devices, electronic items etc. These coatings are used to coat mechanical miniature panels (micro-machines), such as membranes, cantilever, gears, engines, and valves, which are subjected to high temperature. Plasma-deposited silicon nitride (a-SiN_xH_y) thin films and similar materials are increasingly being used as flexible membranes in microsystems. Thermal spraying of nanoscale materials has generated coatings with properties reportedly superior to those of micron-sized counterparts. Recently, some researchers produced super-nano-crystalline coatings of diamond with CVD method, by short waves using unique chemical such as C–Ar or methane-Ar. The traditional tests like bulge or tensile tests are impractical and/or unsuitable, as they do not scale into the micro or nano-scale. Microindentation (Knoop or Vickers indentation) promises to be rather effective in the measurement of thermal spray coatings, but for thin films they cannot provide accurate results at a nano-scale. Moreover, the area of investigation is larger and the properties at nano-scale remain hidden and only the composite value of mechanical properties is revealed. To date, nanoindentation has not been a popular method of determining mechanical properties of thermal spray coatings due to their relatively higher thickness (> 30 um), which relies on a shallower penetration depth (submicron scale). However in order to clearly understand the mechanical behavior of thermal spray coatings, the submicron scale mechanical properties of individual splats need to be evaluated. This is because splats are the very basic structural unit of thermal spray deposit. Thus, nanoindentation is particularly important for the investigation of nanocrystalline coatings where, despite the fact that the basic lamellar structure is similar to that of conventional coatings, the intrinsic properties and the microstructure are quite different.

Nanoindentation and microindentation (Vickers microindentation) differ in a couple of ways, but researchers can expect to be able to compare the results if tip geometries are used that provide similar strains in the material when testing-such as Vickers and Berkovich shaped tips. Both tips exhibit the same projected contact area with respect to the penetration depth into the sample. The difference between the tip shapes is that the Vickers tip is a 4-sided pyramid while the Berkovich tip is a 3-sided pyramid. The Berkovich tip is much sharper than the Vickers tip (usual tip radii for a Berkovich tip is < 20nm). As both tip shapes – the Berkovich and Vickers tips - produce approximately 8% strain in the material during indentation, the hardness results obtained from test using the different tip geometries can be directly compared. However, care must be taken to correctly perform the conversion of the results for the Vickers hardness tests using micro-indentation. There is a definition change between the hardness measured using micro-indentation to the hardness used in nano-indentation. While both hardness values are calculated as the peak force divided by the area of contact (as shown in Equation 1), the definition of the contact area differs between the test techniques. The detailed information pertaining to the difference between the various types of tips available in market is provided elsewhere (Fischer-Cripps, 2004).

Nanoindentation is a versatile technique wherein the properties of thermal spray coatings can be mapped accurately. The grain size is a very important factor for the determination of indentation parameters in thermal spray coatings. In case of nanoindentation (with a Diamond Berkovich tip), very small loads of a few millinewtons (mN) can be applied to calculate the properties of individual grains or splats. The young's modulus and hardness values of individual splats can be calculated by continuous measurement of force and indentation depth. This allows a better understanding of elastic-plastic behavior of thermal spray coating, which can be closely related to its wear and failure resistance. For instance, Saeed and Gross (2009) evaluated the mechanical properties of thermally sprayed Hydroxyapatite (HAp) coating on Ti using nanoindentation technique. The samples were mounted on an epoxy resin and the indentations were performed on the cross-section of the coating. The variation of the hardness and elastic modulus with the number of indents is shown in Figure 6. Figure 7 indicates the variation of hardness with the distance of three different particle size of Hap at low loads. The Authors concluded that at low load and closely spaced indents the hardness value at the interface can be evaluated more precisely.

The hardness curves can also be used as an indicator of the stress state within a material. Is a fact that Hardness values decreases with tensile stress and increases with compressive stress. In Figure 8, the hardness values increase abruptly from 0 to 20 um, followed by a slow increase from 20 to 60 um and a plateau in 40 to 60 um range. It can be concluded that 40 um is the critical thickness where, there is a balance between the tensile and compressive residual stresses.

Nohava et al., (2010) performed instrumented indentation on a wide range of thermal sprayed ceramic coating. A scale-dependent behavior of hardness was observed as a function of indentation depth for all coatings: at low penetration depths, the hardness value depends on the intralamellar material properties, where-

Figure 6. Variation of Hardness and Elastic modulus with the number of indents in a HAp coating (Saeed and Gross, 2009)

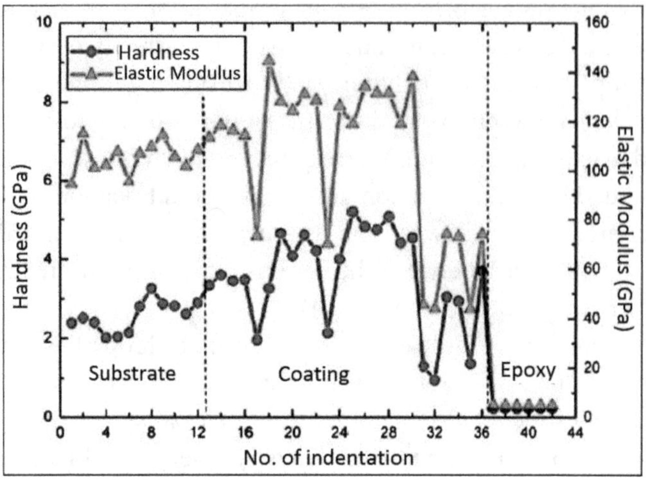

Figure 7. Hardness vs distance of Hap by applying a low load of 10 mN (Saeed and Gross, 2009)

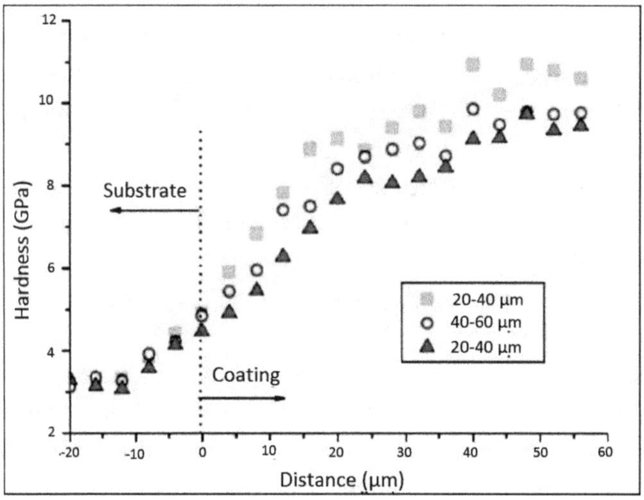

as at larger depths it reflects the long-range cohesive strength of the coating. In all cases, hardness becomes independent of the indentation depth above a threshold value of ~2000 nm.

Figure 8. Hardness as a function of indentation depth of thermal sprayed ceramic coatings (Saeed and Gross, 2009)

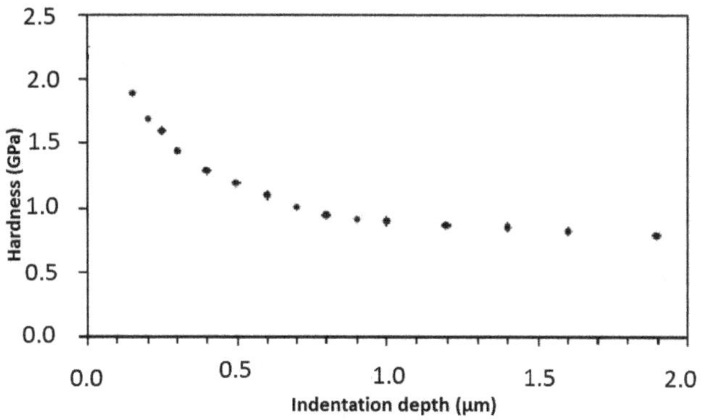

A Typical nanoindentation curve having dual cycles of loading up to a maximum load of 1 mN is shown in Figure 9. This figure is obtained from a nanoindentation experiment done on a glass-flake embedded vinyl ester coating (VE/GF). The indentations were carried out in grids of size 10X4 with each indent 20 um apart, so that a large area gets covered and an ample amount of data is generated for analysis. The experiment was done using a diamond Berkovich tip indenter using load control mode on the top surface of the coating. The indentation instrument may be set into either load or depth control. In load control, the user specifies the maximum test force (usually in mN) and the number of load increments or steps to use. Such a type of indentation gives us a single value of the elastic modulus or hardness at the point of indentation. It doesn't matter how much depth the indenter penetrates to achieve the predetermined load. This type of indentations can be done on both the coating surface and the coating-substrate interface. In case of depth control, the user specifies a maximum depth of penetration. This type of tests gives us a variation of elastic modulus or hardness values with the depth of the coating. It helps to understand the critical depth from the surface where the hardness values stabilize. Such tests are mostly carried out on the topcoat of the sample. It should be noted that most nanoindentation instruments are inherently load-controlled devices. It is customary for a nanoindentation instrument to allow for a dwell of hold period at each load increment and at maximum load. The load period data at maximum load can be used to measure creep within the specimen or thermal drift of the apparatus during test. Better results are obtained when multiple loading are done for an indentation test. This is because before taking the hardness or modulus reading from the final unloading curve the thermal drift of the instrument and other errors needs to be minimized.

Figure 9. Load vs displacement curve of indentations performed on a glass flake filled vinyl ester coating. This diagram represents double loading cycle of a nanoindentation test

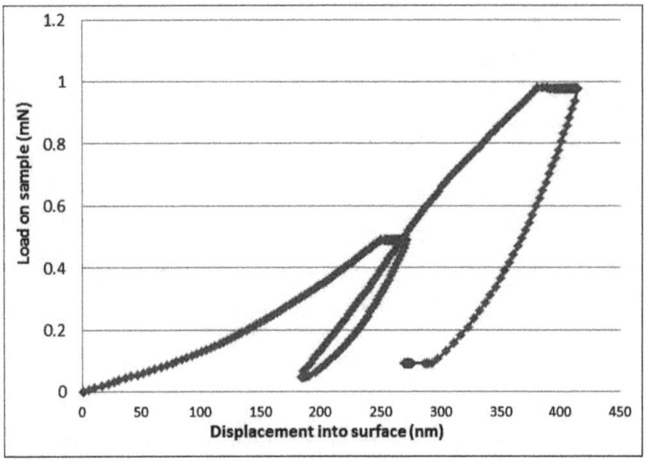

In general a single cycle (as in Figure 4) or multi cycle indentations are done in a grid like fashion (e.g. 10x4 = 40 indents). This is done to cover areas with all the phases in the coating in order to obtain accurate results. Indentations can be load dependent and depth of penetration dependent as well. The statistical analysis of the obtained data is done to develop variation of Young's modulus, hardness with depth and also their probability functions. These data can also be used to define the percentage of various phases in the composite structure of the coating. Indentations can also be carried out along the cross-section of the coating-substrate interface. A Scanning electron micrograph of a grid of nanoindentation made with a Berkovich indenter in a cross-section of HAp coating on Ti is illustrated in Figure 10.

Nanoindentation provides information on mechanical properties by calculation from indenter load and displacement. The basic formulae for these computations are based on certain assumptions on the material response, which are valid only under some conditions. If a factor is neglected that is important at these conditions, the results can be wrong. The user of nanoindentation techniques thus should be aware of possible influences and errors, in order to prepare the tests and process the data so as to avoid or reduce them. The errors might be due to properties of specimen and indenter, the models followed for evaluation etc. Various works have been done to investigate the errors involved in the mechanical characterization of thermal spray coatings (Menčík and Swain, 1995; Fischer-Cripps, 2004). It should also be

Figure 10. Scanning electron micrograph of a grid of nanoindentation made with a Berkovich indenter in a cross-section of HAp coating on Ti

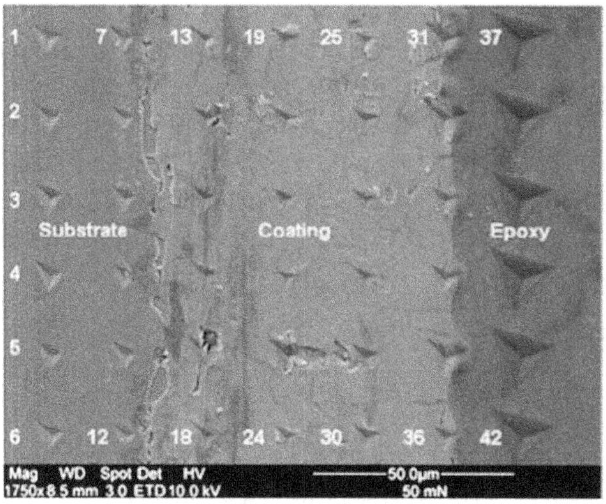

noted that no two indentations should be very close to each other, as one indent affects the results of the other. A minimum of 25 μm is to be maintained between two indents.

Review of Research and Development

Advanced nanoindentation techniques offer the possibility to investigate even extremely thin films attached to substrates down to a film thickness of 1 to 3 nm (Soh et al., 2006; Wang and Shao, 2012). The interpretation of indentation results is supported by theoretical analysis such as analytical modelling, finite element modelling and even atomic-scale FEM modeling (Chen and Lu, 2012). The testing of very thin films by nanoindentation is a very challenging task and needs extremely careful procedures. The commonly occurring sources of errors in using this technique have been reviewed by Fischer-Cripps (2004). Soh et al., (2006) performed nanoindentation of plasma-deposited nitrogen-rich silicon nitride thin films. They elaborated the contribution of elastic and plastic deformations from nanoindentation to clarify the physical meaning of hardness and reduced modulus. Wang and Shao, (2012) performed indentation tests on polysilicon thin films and evaluated the effect of surface roughness and internal stresses.

Saeed and Gross (2009) performed nanoindentation tests on the cross-sections of hydroxyapatite (HAp) thermal spray coatings having three different particle size. It was concluded that use of with the use of smaller particles, the hardness and elastic modulus increased as compared to that of larger particles. Further qualitative SEM analysis proved the uniformity of oxygen content throughout the coating. Thermal barrier coating (TBC) is another type of coatings, which consists of a super alloy substrate, an oxidation-resistant metallic bond coat, and a zirconia-based topcoat that is thermally sprayed onto the bond coat. Rayon et al., (2011) performed statistical indentation analysis of WC-Co coatings obtained by atmospheric plasma spray (APS) with two secondary plasma gases, H and He. The nano-hardness, and Young's modulus resulting from a grid of 300 indentations identified the characteristics of each phase in the heterogeneous structure produced by decarburization.

SCRATCH TESTS

Principle

The principal aim of majority scratch tests is to evaluate the nature of bond strength of the coating with the substrate. Basically there are two types of bond strengths-adhesive and cohesive bond strength. The scratch test is also used to evaluate the coefficient of friction and for surface profiling as well. Coefficient of friction and wear resistance can be measured by ball-on-disc type tribological tests. The nature and numerical value of bond strengths can also be evaluated using pull-off tests, where the coated sample is glued to an uncoated sample and both the pieces are pulled apart using an instrument. The samples are loaded with a constant loading rate until fracture takes place and then the corresponding fracture stress is recorded. If failure occurs at the coating-substrate interface then it is termed as adhesive failure and if the failure occurs in the layers of coating itself then it is termed as cohesive failure.

The adhesion is a fundamental property of hard coatings in view of tribological-mechanical demands. Scratch test is one of the most discussed and well-established test methods to evaluate this property. Though the scratch test technique is often used for characterization of adhesion of surface coatings, it can also be used for characterization of bulk materials. In such a cases, scratch resistance is usually obtained rather than adhesion. The scratch tests are often used to study the mechanisms that take place during abrasion or wear. Further, the scratch tests offer a unique possibility of studying the effect of single particle abrasion, which is difficult or impossible to simulate by other abrasion or wear tests. Various information such as elastic-plastic deformation, formation of median, radial and lateral cracks and material failure mechanisms occurring during scratch testing can be obtained. This information can substantially contribute to the understanding of the mechanisms involved in sliding wear and abrasion of brittle materials. Scratch testing can also be used to evaluate the nature of bond strength of the coating by running constant load scratches on the

Figure 11. A schematic diagram of a scratch tester

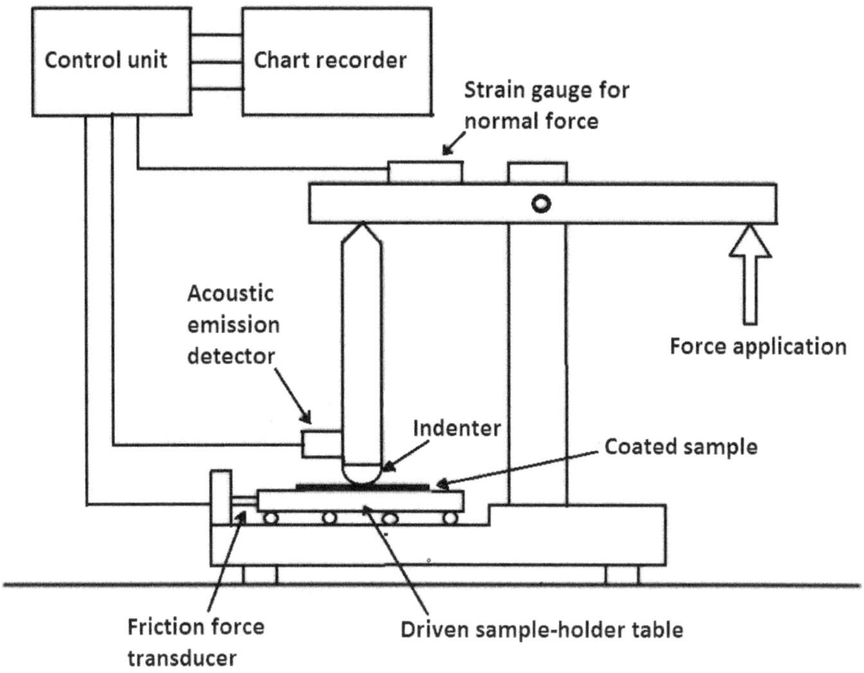

coating-substrate interface. The technique involves generating a controlled scratch with a sharp tip on a specified area. The tip (usually diamond) is drawn against the surface at a constant, incremental, progressive load. At a critical load the coating will start to fail. The critical point is noted by means of an acoustic sensor attached to the load arm. It can also be noted through a built in optical microscope in the instrument. This critical load is used to quantify the adhesive strength of the coating substrate combination. A schematic diagram of a scratch tester is shown in Figure 11.

Constant load scratch tests provide better differentiation of damage levels. Such test requires more specimens' area and test time, it is used for research and process development of coatings. This type of scratches is usually run at the coating-substrate interface. The resistance to propagation of the scratch is used as a measure to evaluate the adhesion and cohesion of the coating. Whereas the progressive load scratch test covers full load range with a single test without any gap, therefore, it is faster than constant load test. This type of scratch is run on the surface of coating and the load at which the coating fails is noted. It is mostly popular in research and development and quality control of coatings.

Mechanical Characterization Using Nanoscratch Tests

The scratch test gives very reproducible quantitative data that can be used to compare the behavior of various coatings. A series of scratch test are performed to obtain the value of the critical load at which the coating exhibits failure. The critical loads depend on the mechanical strength (adhesion, cohesion) of a coating-substrate composite but also on several other parameters: some of them are directly related to the test itself, while others are related to the coating-substrate system. The constant load scratch test require more time but give more statistical data, whereas progressive load (ramp load) scratch tests are used for rapid assurance and quality assurance of coatings. It is more popular for research and development of thermal spray coatings. The variation of the displacement of the indenter into surface of the coating and the scratch distance for constant load and ramp load nanoscratch is shown in Figures 12 and 13. A little consideration will show that during constant loads scratch test the indenter goes inside the coating surface up to a certain depth and then starts the scratch at a constant vertical load. However in case of ramp load scratch test, the indenter starts to run the scratch as soon as it reaches the surface. Simply knowing the value of critical load is not sufficient to predict nature of bond strength or failure mechanism of the coating. To identify the nature of failure, a clear understanding of the types of cracks is imperative. When the coating is sufficiently stressed, cracks generate and start propagating. Other phenomena like chipping, delamination also takes place. There are two basic types of coating failure cohesive failure and adhesive failure.

Cohesive Failure: This type of failure occurs as a result of generation of tensile stress behind the tip of the indenter. This results in the formation of through-thickness cracks like brittle tensile cracking, hertz cracking, conformal cracking as shown in Figure 14. In brittle tensile cracking, patterns of micro-cracks are formed behind the indenter. The shape of the micro-cracks is either straight or curved and directed towards the direction of scratch. In case of hertz cracking, a series of nested micro-cracks within the groove is formed. They are circular in shape. Conformal cracks are formed when the coating tries to conform to the groove. These are open away from the direction of scratch.

Adhesive Failure: This type of failure is as a result of compressive stress. Here the coating either separates from the surface either by cracking and lifting (Buckling) or by full separation (Spalling, Chipping) as shown in Figure 15. The scratch adhesion value can be defined as the minimum value of the critical load and it is a very useful tool in the comparative study in characterizing thermal spray coatings. Chipping means removal of rounded regions of coatings laterally at the end of the scratch groove. Spallation may be subdivided into Buckling, Wedging, recovery, gross spallation. In Buckling, the coating buckles ahead of the stylus tip and irregular

Figure 12. Displacement into surface vs scratch distance in case of constant load

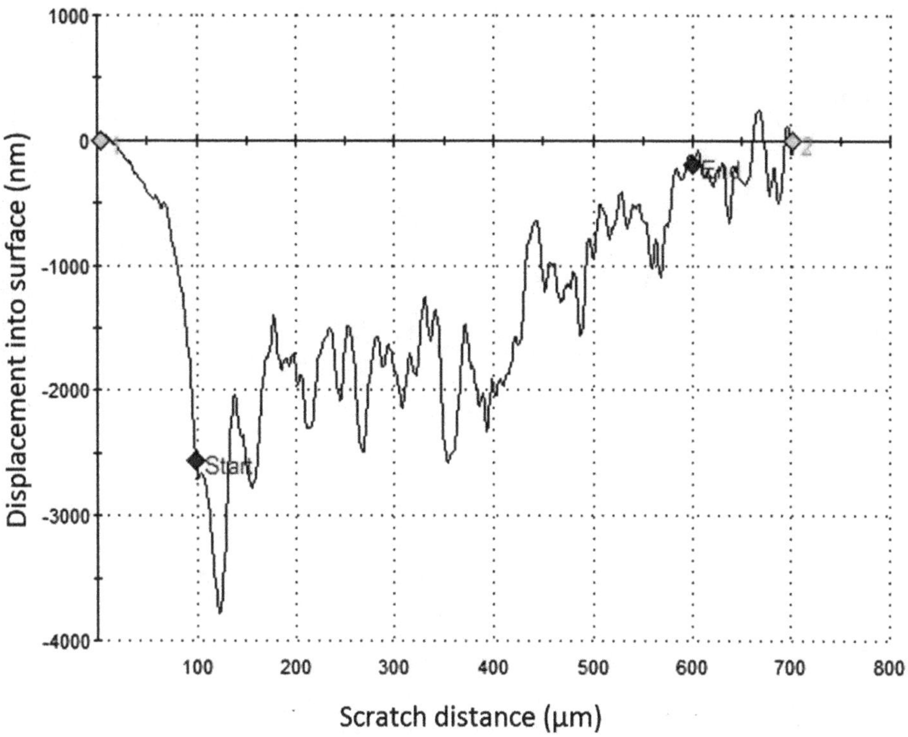

Figure 13. Displacement into surface vs scratch distance in case of ramp load

Figure 14. Schematic diagram of various types of cohesive failures

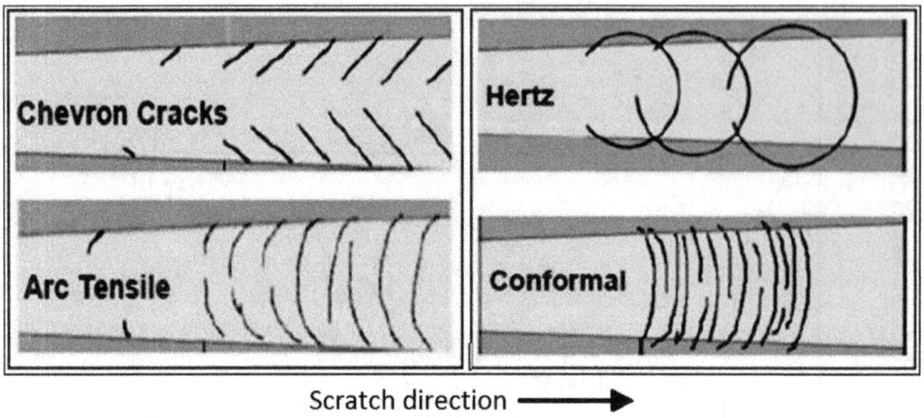

wide-arc shaped patches are formed and they open away from the direction of scratch. Wedging is caused due to delamination where the coating tries to come out of the substrate. Optically, they are characterized by regular annular circles that move away from the edges of the scratch groove. Recovery means regions of detached coating on both sides of the groove. This is formed as a result of elastic recovery behind stylus and plastic deformation in the substrate. Gross spallation is large detached coatings along the edges of the scratch groove. These are indicative of the weak adhesive strength of the coating-substrate combination. Therefore, the coating can be characterized by observing the scratch grooves under microscope and by calculating minimum value of the critical load.

Recently, there have been attempts to determine the adhesion of thermal spray coatings using scratch tests in the same manner as in the thin film domain, i.e. scratching on the top surface of the coating with increasing load. However, due to much higher thickness and surface roughness of thermal spray coatings the method has been found to be unsatisfactory. However Lopez et al., (1989) proposed a method where a constant load scratch is run on the cross-section of the coating. In this technique, the sample is cross-sectioned, embedded in resin and then polished. The polishing has to be done very carefully to achieve a mirror finish on the surface. The results of scratch tests change considerably due to surface roughness of the sample. According to this method the scratch runs from the substrate through coating and into then to the resin where the sample is embedded as shown in Figure 16. The projected area of the cone at the interface is in indirect correlation with the bond strength of the coating. If the cone formed at the coating-substrate interface then adhesive failure is said to occur and if the cone if formed at the coating-resin

Figure 15. Schematic diagram of various types of adhesive failures

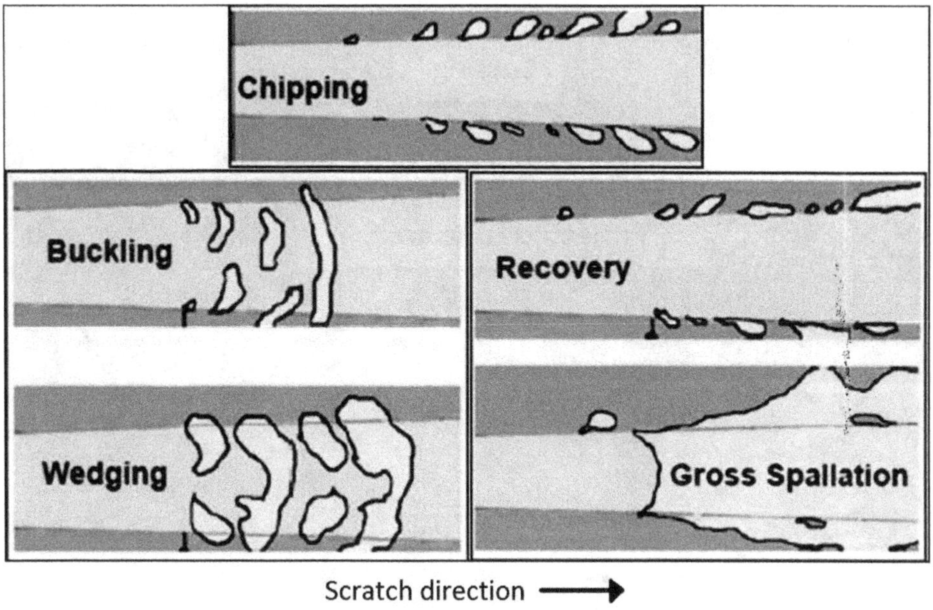

interface, cohesive failure dominates. Larger the projected area of the cone, lower the bond strength of the coating-substrate combination. This method is quite simple and it can be used to draw comparison between two or more coatings.

The main source of error in the mechanical characterization of thermally sprayed coatings and the evaluation of test results is the damage that could occur during sample preparation for the analysis. For the majority of sprayed materials, the damage is mainly due to localized material removal (void formation) during sectioning, grinding and polishing stages. For soft, ductile coatings, smearing of the material over the voids, mainly during sectioning and planar grinding stages, may have an additional effect as well. Microhardness of the coating is identified as a key factor for polishing procedure development. It's a common trend that higher the microhardness, more severe the polishing parameters (higher polishing force and time, increased surface speed) for proper sample preparation. This process holds good for indentation tests and their analysis as well. The orientation of the stylus should be in such that the scratch is run along the edge of the stylus not the side (in case of a triangular pyramid diamond Berkovich tip). This is done to get a clean scratch and avoid damage to the stylus. Another way to avoid errors in results is that the scratches should not be very close to each other. This is because the lateral pileups caused by one scratch affect the results of other scratch if they are close to each other.

Figure 16. Micrograph of a scratch and the measurement of projected area (Lopez et al., 1989)

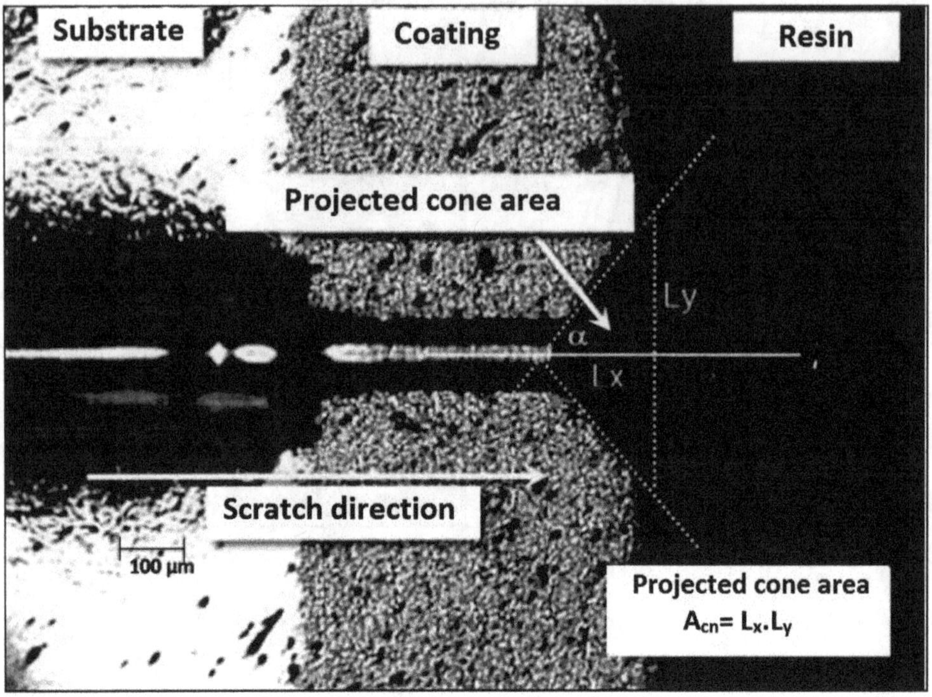

Review of Research and Developments

Evaluating the tribological properties using nanoscratch is a very new concept and requires further validation. A scratch test is a very versatile method, which along with nanoindentation can evaluate different mechanical and tribological properties. The problem with nanoindentation technique is that the interaction on the indented material and the indenter tip in lateral direction is not characterized as the indenter applies load on the sample in perpendicular direction of the surface. However, scratch test takes into consideration both the normal and lateral interaction of the surface and the indenter. Arash Ghabchi (2013) performed scratch tests on HVOF sprayed WC-CoCr coatings. The author observed the presence of semi-circular cracks on the scratch groove. Grewal et al., (2013) studied the scratch resistance, fracture toughness, microhardness of Ni-Al_2O_3 bases composite coating using scratch tests. It was found that cracking and spallation were found to be the mechanisms responsible for damage of coatings during scratch testing. Nordin et al., (1998) performed scratch tests on thermally sprayed TiN/CrN, Ti/MoN, TiN/NbN, TiN/TaN coatings on cemented carbide to evaluate their mechanical and tribological properties. The coating morphology was observed along with properties like residual stress, hardness, cohesion and adhesion were evaluated. SEM images of scratch tests reveal small scale and large-scale adhesive failure as shown in Figure 17. The authors concluded that TiN/CrN coating revealed the highest potential as a tribological coating material. Thus scratch test along with other tests like indentation, abrasive wear etc. can be used to compare the mechanical and tribological properties of thermal spray coatings.

Abrasive Wear Testing

One of the primary functions for thermal spray coatings is to improve the wear resistance and thus the durability of components and products. To determine these aspects related to the performance of the coatings, robust wear testing techniques are required. Of the various wear testing methods abrasive wear testing is the most promising in terms of standardization and reproducibility. This type of testing generally makes use of fixed or free flowing abrasives to wear the specimen.

Classification and Principles

Abrasive wear tests can be primarily classified according to the type of contact and contact environment. So, depending on the type of contact the two modes of abrasive wear are known as two-body and three body abrasive wear. Two-body wear is a well-defined and discussed process. Here the grits or abrasive particles

Figure 17. SEM image of adhesive failure of TIN/CrN coatings during scratch tests (Nordin et al., 1998)

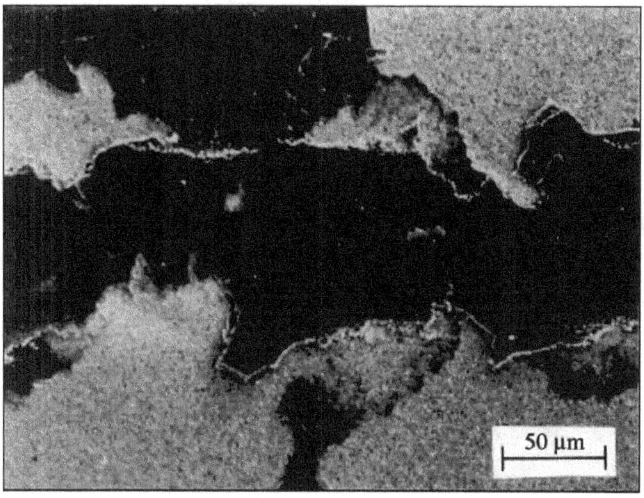

remove material from the opposite direction by a plowing operation. In literature three-body abrasive wear is often associated with loose abrasive particles trapped between two surfaces. Thus the two surfaces and the abrasive particle form the three bodies. However, the two surfaces are not necessarily required. In many practical situations only one surface contacts the abrading particles. In other cases when two surfaces are involved they may be so far apart that the mechanical properties of one surface have no influence on the wear of the other surface. This leads us to define "closed" and "open" three-body abrasive wear.

Closed three-body abrasive wear occurs when loose abrasive particles are trapped between two sliding or rolling surface which are close to one another. In this type of wear the particles often indent and settle into the softer surface with subsequent abrasive wear of the other surface which resembles the two-body situation. Here a change in properties of one surface may influence the wear of the other surface. Open three-body abrasive wear occurs when two surfaces are far apart or when only one surface is only involved in the wear process. In contrast to closed three-body wear, which can be often eliminated by filters, seals or flushing, open three-body wear is often an inherent feature of a process and has to be controlled by a suitable choice of material (Misra and Finnie, 1980). It is convenient to classify it further to high-stress and low stress abrasive wear.

Open abrasive wear testing as currently practiced can be categorized into (a) high-stress abrasive wear and (b) low stress abrasive wear. A schematic diagram of the working principle of the high-stress abrasive wear testing is shown in Figure 18. This method corresponds to the standard ASTM B611-85 test method. In this method a steel wheel having vanes at both sides is rotated in a abrasive slurry. The vanes are used to continuously agitate and mix the abrasive slurry and to propel it towards the specimen during the test. The specimen is placed in contact with the wheel by maintaining a normal load on it. The main feature of this test that discriminates it from other abrasion tests is that the abrasive is forced against the test specimen with a steel wheel with sufficient force to cause fracture of the abrasive particles.

In low-stress abrasive wear test, rubber wheel abrasion tester is normally used as shown in Figure 19. The rubber wheel is used to force the abrasive particles into the samples. In this method a rubber-rimmed wheel slides against the surface of a plane test sample in the presence of abrasive particles at a constant applied load. The test method is standardized as per ASTM G 65-85. Rubber wheel produces low stress whereas steel wheel produces high stress. In both cases the samples are polished using diamond abrasives (9μm, 6μm, 3μm, 1μm) before wear tests. Clean and weighted samples are exposed to abrasion for a predetermined time and then cleaned ultrasonically and weighed to determine wear rate after taking the samples out of

Figure 18. Schematic diagram of high stress abrasive wear testing (Liao et al., 2000)

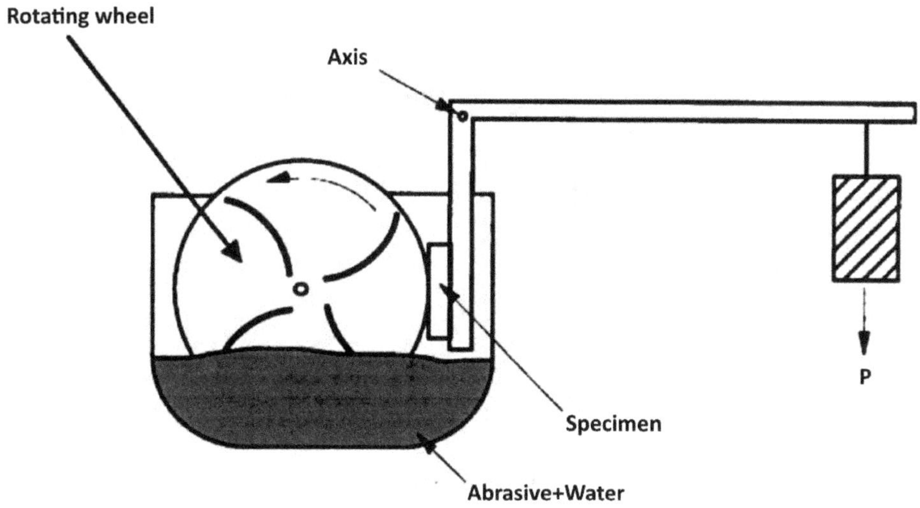

the test rig. Silica sand, silicon carbide, alumina abrasives and diamond polishing grits with wide range of sizes are used as abrasives even though rubber wheel abrasive tests is standardised by ASTM with dry silica sand as abrasive.

The experimental set for high-stress abrasive wear can also be occasionally used for low-stress abrasive wear at low applied load. The main difference in the above two configuration is that in the first case the area exposed to wear remains constant with change of time, whereas, in rubber wheel abrasion test wearing area changes continuously. Taber Abraser, shown in Figure 20 is one such technique used for low-stress abrasion testing. Here the specimen is mounted on a rotating turntable, which is subjected to the wearing action of two abrasive wheels applied at a specific pressure. The wheels are driven by the rotating specimen about a horizontal axis displaced tangentially from the vertical axis of the specimen. Various types of wheels are available containing abrasives dispersed in an elastomeric binder. The test is standardized by ASTM for organic paint coating. This test has less reproducibility than what is normally obtained for sliding wear. The reason for this is because of the variability in the properties of abrasives, frictional heating of the wheels and specimen during the test, and the influence of wear debris, which is transferred to some extent to the wheel surface.

To overcome this problem, and to ensure constant wear surface during testing Axen et al., (1997) proposed another technique. The schematic diagram of the test set-up is shown in Figure 21. In this method a hollow cylinder is rotated with its

Figure 19. Schematic diagram of low stress abrasive wear testing

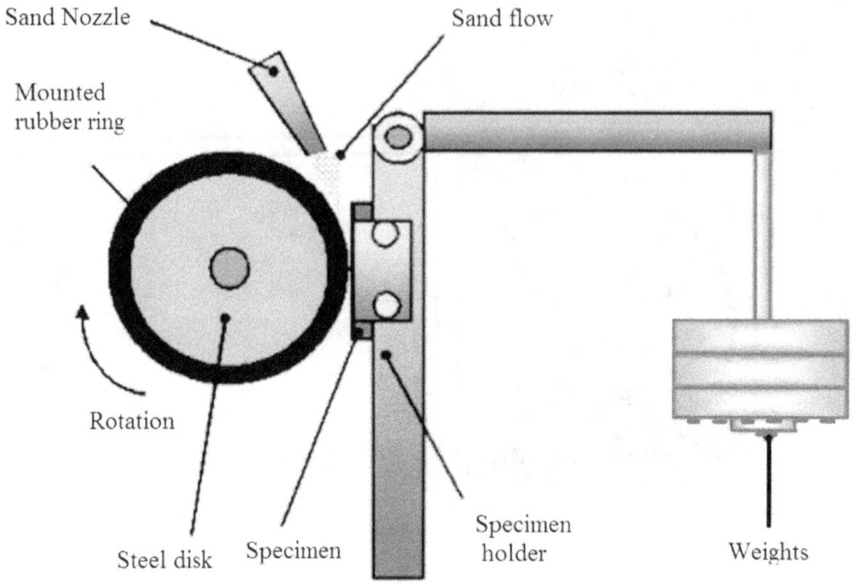

axis normal to the specimen surface, in the presence of a slurry of abrasive particles. Similar to Taber test, an annular scar results, but constant supply of fresh abrasive particles maintains constant wearing area and avoids many of the problems described above. It is possible to have a cylinder of very small diameter so that the wear test can be performed on a specimen area only a few millimeters. Only preliminary results have so far been published though the method appears to offer significant potential for future development.

The series of test described above are known as two-body abrasive wear test, except the rubber wheel abrasion tests. Rubber wheel abrasion test is a three-body abrasive wear test method. However it is possible to simulate two-body abrasive wear test without altering geometry. In such case, a wheel with embedded abrasive is used. As there are no loose abrasives, this acts as two-body abrasive wear test equipment. Thus, the basic difference in the two techniques lies in the fact that sliding wear test can be carried out either in interrupted or continuous manner, but, abrasive wear test can be carried out only in interrupted manner. The abrasive wear test methods discussed so far have many problems when it comes to practical application. Kassman et al., (1991) introduced a new method in order to have a quick, simple and reproducible method for the characterization of the wear resistance of thin coatings. Over the years this researchers modified this method and today it is called Crater Grinding Test. A schematic diagram of the test is shown in Figure 22.

Figure 20. Schematic representation of the working of a Taber abraser

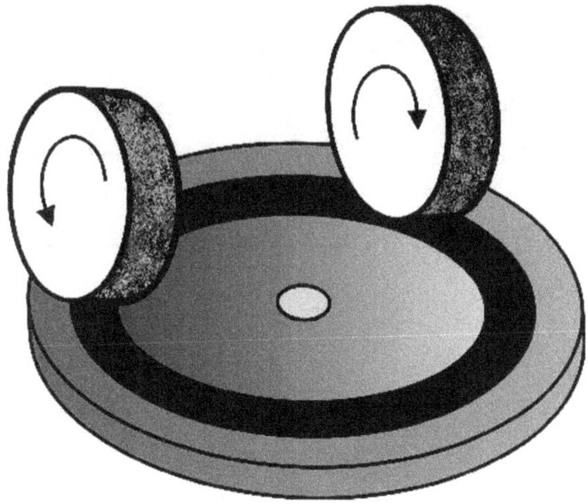

Figure 21. Schematic diagram of an abrasive wear tester based on a tube shaped drill with abrasive slurry (Axen et al., 1997)

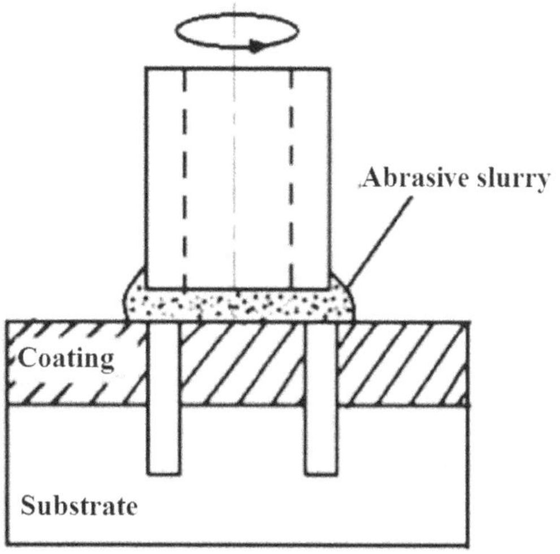

With the ball-cratering test the wear resistance of coatings and surfaces can be measured precisely. Abrasive slurry is drip-fed onto a rotating steel ball, which presses against the sample to grind a crater (hemisphere) into the coating. The wear coefficient can be calculated from the volume of the crater. This method is suitable

Figure 22. A schematic representation of the ball-cratering test (Gee et al., 2003)

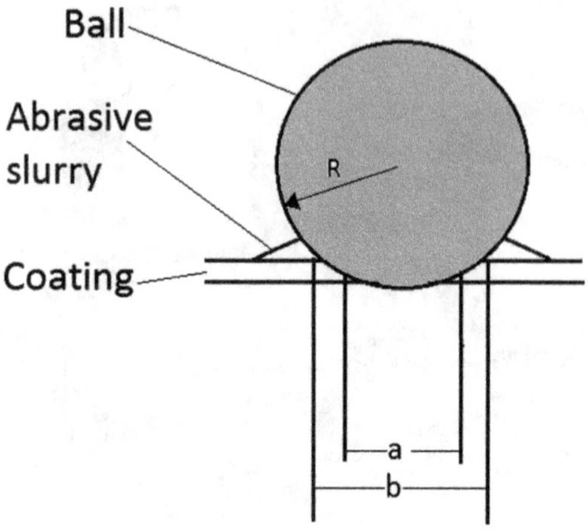

for testing coatings with coating thickness ³ 1 μm. The grinding is interrupted at regular intervals and the crater diameter is measured using an optical microscope or a SEM. The used slurry is also replaced with a new one during such measurements. Tests can be carried out limiting the duration of the test so that no perforation of the coating occurs. When perforation does occur, analysis of measurements of overall crater diameter (b in Figure 22) and the diameter of the crater in the substrate (a in Figure 22) for sets of tests performed too different test durations can yield wear rates for the substrate and the coating. A ball cratering (micro-abrasion) test instrument is shown in Figure 23.

Review of Research and Development

At the initial stages the ball-crater abrasion test suffered from many serious setbacks. There was no standardized test procedure to compare the results from different tests performed by different users. The calculation of the volume is also a problem, as the crater shape in practice is different from ideal sphere. However, this test method, which is also known as micro-abrasion test, is becoming popular now a days as a method for the abrasion testing of surface engineered materials. It possesses many advantages over conventional tests including the ability to make small volumes of materials and thin coatings. The test procedure is easy to use, low cost and versatile. Moreover, the examination of the crater in SEM yields information about the adhesion between the coating and the substrate. This test also helps in throwing light on the coating cohesion, porosity and in depth profiling of the composition of the

Figure 23. Micro abrasion test instrument (Schiffmann, 2005)

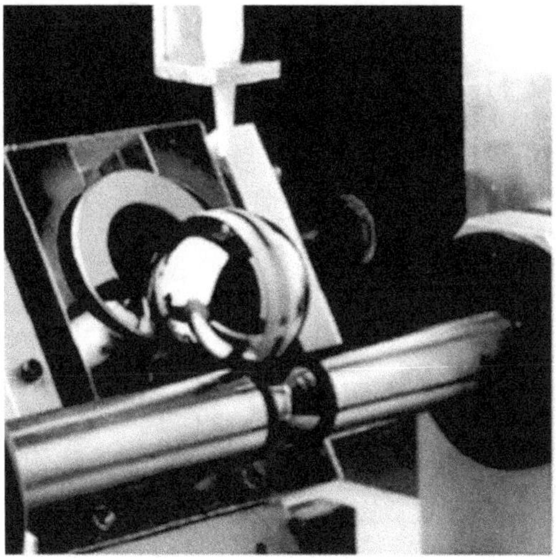

coating. Various efforts to standardize these tests are being made by researchers both in the USA and Europe (Gee et al., 2003; Leroy et al., 2005).

Leroy et al., (2005) used ball-cratering tests on thermally sprayed DLC, TiN and Al coatings as part of a completed standard measurement and testing programme "crater" addressing the subject of using this type of instrument for determination of the intrinsic wear rates of thin coatings. Effective crater diameter profiling was done (Figure 24) on both the coated and uncoated Ti-alloy samples. Gee et al., (2003) performed micro-abrasion tests using three different setups of ball-crater instrument and standardized the test. They also explained in detail the parameters affecting results like abrasive material, size, shape, loading and speed. The variation of the wear volume with the speed of the ball is shown in Figure24.

Kennedy and Hashmi (1998) examined the various methods of assessing the wear resistance of coated and uncoated materials. They explained in detail the test equipment for sliding, abrasion wear, erosion, impact and dynamic wear tests. Bromark et al., (1997) studied the wear characteristics of PVD Ti/TiN multilayer coatings subjected to two-body abrasion using diamond slurry and silicon carbide particles as abrasive medium and erodent. The abrasive wear rate of the coating was found to increase with the relative amount of metallic Ti in the coating. The authors concluded that the concept of multilayered coatings offer a potential means to tailor the properties of tribological coatings. Wang et al., (2000) calculated the abrasive wear resistance of thermally sprayed coatings with reconstituted nanostructured Al_2O_3/TiO_2 powders. The result showed that the abrasive wear resistance of the

Figure 24. Variations of wear volume with the cratering ball speed (Gee et al., 2003)

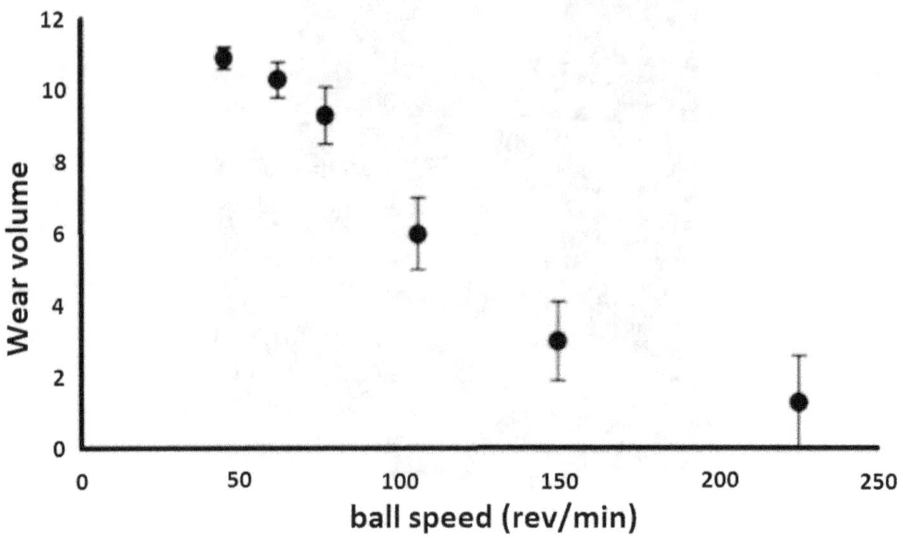

coatings produced using the nanostructured Al_2O_3/TiO_2 powders is greatly improved compared with the conventional coating. They also evaluated the relationship between the wear volume with the hardness of the coating.

CONCLUDING REMARKS

Advanced design of decorative and wear resistant coatings hinges on the characterization and optimization of the mechanical properties like elastic modulus, hardness, yield strength, internal stress, fracture and wear resistance of the coating-substrate system. The goal is to find material and structural solutions, which keep the resulting stress-strain field under typical; application conditions below the stability limits of the system. Based on nanoindentation measurements obtained from the coating-substrate system, which should be optimized, a scratch is dimensioned as a function of the load range and indenter geometry. The obtained result from both indentation tests and scratch measurements has to be correlated to characterize the coating-substrate system. In the case of thermal spray coatings, the calculation of coating hardness and abrasive wear rate is of utmost priority. The abrasion wear tests discussed so far shall solve the problems as of now, but there is a huge scope of research work still pending to design and analyze better test methods and their standardization procedure.

REFERENCES

American Society for Testing and Materials. (1984). *Standard test method for abrasion resistance of organic coatings by the tabor abraser*. Author.

Annual Book of ASTM Standards. (1990). *Metal test methods and analytical procedures, vol.03, 02, wear and erosion, metal corrosion*. Author.

Annual Book of ASTM Standards. (2005). *Standard test method for determining the high stress abrasion resistance of hard materials*. Author.

Axen, N., Jacobson, S., & Hogmark, S. (1997). Principles for tribology evaluation of intrinsic coating properties. *Wear*, *203-204*, 637–645. doi:10.1016/S0043-1648(96)07429-7

Bromark, M., Larsson, M., Hedenqvist, P., & Hogmark, S. (1997). Wear of PVD Ti/TiN multilayer coatings. *Surface and Coatings Technology*, *90*(3), 217–223. doi:10.1016/S0257-8972(96)03141-6

Bull, S. J. (2005). Nanoindentation of coatings. *Journal of Physics. D, Applied Physics*, *38*(24), 393–413. doi:10.1088/0022-3727/38/24/R01

Chen, J., & Lu, G. (2012). Finite element modelling of nanoindentation based methods for mechanical properties of cells. *Journal of Biomechanics*, *45*(16), 2810–2816. doi:10.1016/j.jbiomech.2012.08.037 PMID:23017378

Fischer-Cripps, A. C. (2004). *Nanoindentation* (2nd ed.). Springer-Verlag.

Gee, M. G., Gant, A., Hutchings, I., Bethke, R., Schiffman, K., Van Acker, K., & von Stebut, J. et al. (2003). Progress towards standardization of ball cratering. *Wear*, *255*(1-6), 1–13. doi:10.1016/S0043-1648(03)00091-7

Ghabchi, A. (2013). Wear resistant carbide-based thermal sprayed coatings: Process, properties, mechanical degradation and wear. VTT Science, 29.

Grewal, H. S., Singh, H., & Agrawal, A. (2013). Microstructural and mechanical characterization of thermal sprayed nickel-aluminium composite coatings. *Surface and Coatings Technology*, *216*, 78–92. doi:10.1016/j.surfcoat.2012.11.029

Hay, J. L., & Pharr, G. M. (2000). Instrumented indentation testing. In ASM handbook, mechanical testing and evaluation. ASM International.

Kassman, A., Jacobson, S., Erickson, L., Hedengvist, P., & Olson, M. (1991). A new test method for the intrinsic abrasion resistance of thin coatings. *Surface and Coatings Technology*, *50*(1), 75–89. doi:10.1016/0257-8972(91)90196-4

Kennedy, D. M., & Hashmi, M. S. (1998). Methods of wear testing for advanced surface coatings and bulk materials. *Journal of Materials Processing Technology*, *77*(1-3), 246–253. doi:10.1016/S0924-0136(97)00424-X

Leroy, C., Schiffmann, K. I., van Acker, K., & von Stebut, J. (2005). Ball cratering an efficient tool for 3 body microabrasion of coated samples. *Surface and Coatings Technology*, *200*(1-4), 153–156. doi:10.1016/j.surfcoat.2005.02.076

Liao, H., Normand, B., & Coddet, C. (2000). Influence of coating microstructure on the abrasive wear resistance of WC/Co cermet coatings. *Surface and Coatings Technology*, *124*(2-3), 235–242. doi:10.1016/S0257-8972(99)00653-2

Lopez, E., Beltzung, F., & Zambelli, G. (1989). Measurement of cohesion and adhesion strengths in alumina coatings produced by plasma spraying. *Journal of Materials Science Letters*, *8*(3), 346–348. doi:10.1007/BF00725519

Menčík, J., & Swain, M. V. (1995). Errors associated with depth-sensing micro-indentation Tests. *Journal of Materials Research*, *10*(6), 1491–1501. doi:10.1557/JMR.1995.1491

Misra, A., & Finnie, I. (1980). A classification of three-body abrasive wear and design of a new tester. *Wear*, *60*(1), 111–118. doi:10.1016/0043-1648(80)90252-5

Nohava, J., Bonferroni, B., Bolelli, G., & Lusvarghi, L. (2010). Interesting aspects of indentation and scratch methods for characterization of thermally sprayed coatings. *Surface and Coatings Technology*, *205*(4), 1127–1131. doi:10.1016/j.surfcoat.2010.08.086

Nordin, M., Larsson, M., & Hogmark, S. (1998). Mechanical and tribological properties of multilayered PVD TiN/CrN, TiN/MoN, TiN/NbN, TiN/TaN coatings on cemented carbide. *Surface and Coatings Technology*, *106*(2-3), 234–241. doi:10.1016/S0257-8972(98)00544-1

Oliver, W. C., & Pharr, G. M. (1992). An improved technique for determining hardness and elastic modulus using load and depth sensing indentation experiments. *Journal of Materials Research*, *7*(6), 1564–1583. doi:10.1557/JMR.1992.1564

Rabei, A., Mumm, D. R., Hutchinson, J. W., Schweinfest, R., Ruhle, M., & Evans, A. G. (1999). Microstructure, deformation and cracking characteristics of thermal spray ferrous coatings. *Materials Science and Engineering A*, *269*(1-2), 152–165. doi:10.1016/S0921-5093(99)00132-X

Rayon, E., Bonache, V., Salvador, M. D., Roa, J. J., & Sanchez, E. (2011). Hardness and Young's modulus distributions in atmospheric plasma sprayed WC-Co coatings using nanoindentation. *Surface and Coatings Technology*, *205*(17-18), 4192–4197. doi:10.1016/j.surfcoat.2011.03.012

Saeed, S. S., & Gross, K. A. (2009). Nanoindentation reveals mechanical properties within thermally sprayed hydroxyapatite coatings. *Surface and Coatings Technology*, *203*(12), 1660–1664. doi:10.1016/j.surfcoat.2008.12.025

Schiffmann, K. I., Bethke, R., & Kristen, N. (2005). Analysis of perforating and non-perforating micro-scale abrasion tests on coated substrates. *Surface and Coatings Technology*, *200*(7), 2348–2357. doi:10.1016/j.surfcoat.2005.01.015

Soh, M. T. K., Fischer-Cripps, A. C., Musca, C. A., & Faraone, L. (2006). Nanoindentation of plasma-deposited nitrogen-rich silicon nitride thin films. *Journal of Applied Physics*, *100*(2), 1–8. doi:10.1063/1.2217105

Wang, Q., Hu, R., & Shao, Y. (2012). The combined effect of surface roughness and internal stresses on nanoindentation tests of polysilicon thin films. *Journal of Applied Physics*, *112*(4), 1–8. doi:10.1063/1.4748176

Wang, Y., Jiang, S., Wang, M., Wang, S., Xiao, T. D., & Strutt, P. R. (2000). Abrasive wear characteristics of plasma sprayed nanostructured alumina/titania coatings. *Wear*, *237*(2), 176–185. doi:10.1016/S0043-1648(99)00323-3

Compilation of References

Barik, R.C., Wharton, J., Wood, R. J. K., & Strokes, K. (n.d.). *Erosion and erosion-corrosion performance of cast and thermally sprayed nickel-aluminium bronze (NAB).* Surface Engineering and Tribology Group, University of Southampton.

Abdel-Samad, A. A., El-Bahloul, A. M. M., Lugscheider, E., & Rassoul, S. A. (2000). A comparative study on thermally sprayed alumina based ceramic coatings. *Journal of Materials Science*, *35*(12), 3127–3130. doi:10.1023/A:1004824104162

Ageorges, H., & Ctibor, P. (2008). Comparison of the structure and wear resistance of Al_2O_3 – 13 wt.% TiO_2 coatings made by GSP and WSP plasma process with two different powders. *Surface and Coatings Technology*, *202*(18), 4362–4368. doi:10.1016/j.surfcoat.2008.04.010

Ahmaniemi, S., Vippola, M., Vuoristo, P., Mäntylä, T., Buchmann, M., & Gadow, R. (2002). Residual stresses in aluminium phosphate sealed plasma sprayed oxide coatings and their effect on abrasive wear. *Wear*, *252*(7-8), 614–623. doi:10.1016/S0043-1648(02)00020-0

Ahn, Y., Cho, N.-G., Lee, S.-H., & Lee, D. (2003). Lateral crack in abrasive wear of brittle materials. *JSME International Journal, Series A*, *46*(2), 140-144.

Ahn, H. S., & Kwon, O. K. (1999). Tribological behaviour of plasma-sprayed chromium oxide coating. *Wear*, *225–229*, 814–824. doi:10.1016/S0043-1648(98)00390-1

Ahn, H., Kim, J., & Lim, D. (1997). Tribological behaviour of plasma-sprayed zirconia coatings. *Wear*, *203–204*, 77–87. doi:10.1016/S0043-1648(96)07395-4

Al-Fadhli, H. Y., Stokes, J., Hashmi, M. S. J., & Yilbas, B. S. (2006). The erosion–corrosion behaviour of high velocity oxy-fuel (HVOF) thermally sprayed inconel-625 coatings on different metallic surfaces. *Surface and Coatings Technology*, *200*(20-21), 5782–5788. doi:10.1016/j.surfcoat.2005.08.143

American Society for Testing and Materials. (1984). *Standard test method for abrasion resistance of organic coatings by the tabor abraser*. Author.

Anand, K. (1985). Erosion of multiphase materials. In *Proceedings of International-Conference on Science of hard Materials*. Rhodes, Greece: Academic Press.

Anand, K., & Conard, H. (1989). Microstructural effect sin the erosion of cemented carbides. In *Proceedings of International Conference on Wear of Materials* (pp. 135-142). Academic Press.

Anand, K., & Conard, H. (1988). Microstructure and scaling effects in the damage of WC-Co alloys by single impacts of hard particles. *Materials Science and Engineering*, *23*, 2931–2942.

Anand, K., Hovis, S. K., Conrad, H., & Scattergood, R. O. (1987). Flux effects in solid particle erosion. *Wear*, *118*(2), 243–257. doi:10.1016/0043-1648(87)90146-3

Ananthapadmanabhan, P. V., Thiyagarajan, T. K., Sreekumar, K. P., Satpute, R. U., Venkatramani, N., & Ramachandran, K. (2003). Co-spraying of alumina-titania: Correlation of coating composition and properties with particle behaviour in the plasma jet. *Surface and Coatings Technology*, *168*(2-3), 231–240. doi:10.1016/S0257-8972(03)00006-9

Ang, A. S. M., & Berndt, C. C. (2014). A review of testing methods for thermal spray coatings. *International Materials Reviews*, *59*(4), 179–223. doi:10.1179/174328041 4Y.0000000029

Annual Book of ASTM Standards. (1990). *Metal test methods and analytical procedures, vol.03, 02, wear and erosion, metal corrosion*. Author.

Annual Book of ASTM Standards. (2005). *Standard test method for determining the high stress abrasion resistance of hard materials*. Author.

Anstis, G. R., Chantikul, P., Lawn, B. R., & Marshall, D. B. (1981, September). A critical evaluation of indentation techniques for measuring fracture toughness: I, direct crack measurements. *Journal of the American Ceramic Society*, *64*(9), 533–538. doi:10.1111/j.1151-2916.1981.tb10320.x

Archard, J. F., & Hirst, W. (1956). The wear of metals under unlubricated conditions. *Proceedings of the Royal Society of London. Series A, Mathematical and Physical Sciences*, *236*(1206), 397–410. doi:10.1098/rspa.1956.0144

Arnold, J. C., & Hutchings, I. M. (1989). Flux rate effects in the erosive wear of elastomers. *Journal of Materials Science*, *24*(3), 833–839. doi:10.1007/BF01148765

Ashary, A. A., & Tucker, R. C. Jr. (1989). Electrochemical corrosion studies of alloys plasma sprayed with Cr_2O_3. *Surface and Coatings Technology*, *39-40*(2), 701–709. doi:10.1016/S0257-8972(89)80032-5

Ashary, A. A., & Tucker, R. C. Jr. (1991). Corrosion characteristics of several thermal spray cermet-coating/alloy systems. *Surface and Coatings Technology*, *49*(1-3), 78–82. doi:10.1016/0257-8972(91)90035-U

Assadi, H., Gärtner, F., Stoltenhoff, T., & Kreye, H. (2003). Bonding mechanism in cold gas spraying. *Acta Materialia*, *51*(15), 4379–4394. doi:10.1016/S1359-6454(03)00274-X

Assi, F., & Bohni, H. (1999). Study of wear–corrosion synergy with a new microelectrochemical technique. *Wear*, *233–235*, 505–514. doi:10.1016/S0043-1648(99)00237-9

ASTM G119-93. (1998). Standard guide for determining synergism between wear and corrosion. In *ASTM handbook* (pp. 529-534). West Conshohocken: ASTM.

ASTM Standard: G3. (n.d.). *Annual book of ASTM standards* (vol. 03.02). Author.

Aw, P. K., & Tan, B. H. (2006). Study of microstructure, phase and microhardness distribution of HVOF sprayed multi-modal structured and conventional WC–17Co coatings. *Journal of Materials Processing Technology, 174*(1-3), 305–311. doi:10.1016/j.jmatprotec.2006.02.006

Axen, N., & Jacobson, S. (1994). A model for abrasive wear resistance for multiphase materials. *Wear, 174*(1-2), 187–199. doi:10.1016/0043-1648(94)90101-5

Axen, N., Jacobson, S., & Hogmark, S. (1997). Principles for tribology evaluation of intrinsic coating properties. *Wear, 203-204*, 637–645. doi:10.1016/S0043-1648(96)07429-7

Babu, P., Basu, B., & Sundararajan, G. (2011). The influence of erodent hardness on the erosion behavior of detonation sprayed WC-12Co coatings. *Wear, 270*(11–12), 903–913. doi:10.1016/j.wear.2011.02.019

Bacon, J. L., Davis, D. G., Polizzi, R. J., Sledge, R. L., Uglum, J. R., & Zowarka, R. C. (1997). A new electromagnetic power deposition system. In *Thermal spray: A united forum for scientific and technological advances* (pp. 393–397). Materials Park, OH: ASM International.

Ball, A., & Patterson, A. W. (1985, May). Microstructural design of erosion resistant hard materials. In *Proc. 11th Int. Plausee Seminar on New Applications, Recycling and Technology of Refractory Metals and Hard Metals*. Reutte Tirol, Austria: Academic Press.

Barber, J., Mellor, B. G., & Wood, R. J. K. (2005). The development of sub-surface damage during high energy solid particle erosion of a thermally sprayed WC-Co-Cr coating. *Wear, 259*(1-6), 125–134. doi:10.1016/j.wear.2005.02.008

Barbezat, G., & Nicoll, A. R. (1993). Abrasion, erosion and scuffing resistance of carbide and oxide ceramic thermal sprayed coatings for different applications. *Wear, 162*, 529–537. doi:10.1016/0043-1648(93)90538-W

Barik, R., Wharton, J., Wood, R., Tan, K., & Stokes, K. (2005). Erosion and erosion–corrosion performance of cast and thermally sprayed nickel–aluminium bronze. *Wear, 259*(1-6), 230–242. doi:10.1016/j.wear.2005.02.033

Barradja, A., Bratu, F., Benea, L., Willems, G., & Celis, J.-P. (2006). Effect of sliding wear on tribocorrosion behaviour of stainless steels in a ringer's solution. *Wear, 261*(9), 987–993. doi:10.1016/j.wear.2006.03.003

Barradja, A., De'forge, D., Nogueira, R. P., Ponthiaux, P., & Celis, J.-P. (2006). An electrochemical noise study of tribocorrosion processes of AISI 304 L in Cl^-□and media. *Journal of Physics. D, Applied Physics, 39*(15), 3184–3192. doi:10.1088/0022-3727/39/15/S08

Barril, S., Mischler, S., & Landolt, D. (2001). Triboelectrochemical investigation of the friction and wear behaviour of TiN coatings in a neutral solution. *Tribology International, 34*(9), 599–608. doi:10.1016/S0301-679X(01)00052-4

Barril, S., Mischler, S., & Landolt, D. (2005). Electrochemical effects on the fretting corrosion behaviour of Ti6Al4V in 0.9% sodium chloride solution. *Wear*, *259*(1-6), 282–291. doi:10.1016/j.wear.2004.12.012

Basak, A. K., Matteazzi, P., Vardavoulias, M., & Celis, J.-P. (2006). Corrosion–wear behaviour of thermal sprayed nanostructured FeCu/WC–Co coatings. *Wear*, *261*(9), 1042–1050. doi:10.1016/j.wear.2006.03.026

Batchelor, A. W., & Stachowiak, G. W. (1988). Predicting synergism between corrosion and abrasive wear. *Wear*, *123*(3), 281–291. doi:10.1016/0043-1648(88)90144-5

Beczkowiak, J., Keller, H., & Schwier, G. (1996). Al_2O_3-TiO_2-Schichten - eine Alternative zu Cr_2O_3. In E. Lugscheider (Ed.), *Thermische Spritzkonferenz 1996, DVS-Berichte 175* (pp. 68–75). Düsseldorf: DVS-Verlag.

Bellman, R. Jr, & Levy, A. V. (1981). Erosion mechanism in ductile metals. *Wear*, *70*(1), 1–27. doi:10.1016/0043-1648(81)90268-4

Benea, L., Ponthiaux, P., Wenger, F., Galland, J., Hertz, D., & Malo, J. Y. (2004). Tribocorrosion of stellite 6 in sulphuric acid medium: Electrochemical behaviour and wear. *Wear*, *256*(9-10), 948–953. doi:10.1016/j.wear.2003.06.003

Benish, M. L., Sikora, J., Shaw, B., Sikora, E., Yaffe, M., Krebs, A., & Martinchek, G. (1998), A new electrochemical noise technique for monitoring the localised corrosion of 304 stainless steel in chloride containing solutions. In *Proceedings of the CORROSION '98* (Paper No. 370). National Association of Corrosion Congress.

Berger, L. M., Woydt, M., Zimmermann, S., Keller, H., Schwier, G., Enžl, R., & Thiele, S. (2004). Tribological behavior of HVOF-sprayed Cr3C2-NiCr and TiC-based coatings under high temperature dry sliding conditions. In *Proceedings of International Thermal Spray Conference*. DVS-Verlag.

Berger, L.-M. (2004). Titanium oxide – New opportunities for an established coating material, In *Thermal Spray 2004, Advances in Technology and Application, Proc. Int. Thermal Spray Conf. ITSC 2004*. Düsseldorf: DVS-Verlag.

Berger, L.-M. (2014). Coatings by thermal spray. In Comprehensive hard materials (Vol. 1, pp. 471-506). Elsevier.

Berger, L.-M., & Stahr, C. C. (2008). State and perspectives of thermally sprayed ceramic coatings in the Al2O3-Cr2O3-TiO2 system, In T.S. Sudarshan & M. Jeandin (Eds.), *Proc. 21st Int. Conf. Surf. Modification Technol.* (pp. 469–478). Valar Docs.

Berger, L.-M., Saaro, S., & Woydt, M. (2007). Reib-/Gleitverschleiß von thermisch gespritzten Hartmetallschichten, In R. Suchentrunk (Ed.), Jahrbuch Oberflächentechnik 2007 (vol. 63, pp. 242–267). Bad Saulgau: Eugen G. Leuze Verlag.

Berger, L.-M., Stahr, C. C., Toma, F.-L., Stehr, G. C., & Beyer, E. (2007). Ausgewählte Entwicklungstendenzen bei der Herstellung thermisch gespritzter keramischer Schichten. In R. Suchentrunk (Ed.), Jahrbuch Oberflächentechnik 2007 (vol. 63, pp. 71–84). Bad Saulgau: Eugen G. Leuze Verlag.

Berger, L.-M., Schneider, D., Barbosa, M., & Puschmann, R. (2012). Laser acoustic surface waves for the non-destructive characterisation of thermally sprayed coatings. *Thermal Spray Bulletin*, *5*(1), 56–64.

Berger, L.-M., Stahr, C. C., Saaro, S., Thiele, S., & Woydt, M. (2009a). Development of ceramic coatings in the system Cr_2O_3-TiO_2. *Thermal Spray Bulletin*, *2*(1), 64–77.

Berger, L.-M., Stahr, C. C., Saaro, S., Thiele, S., Woydt, M., & Kelling, N. (2009b). Dry sliding up to 7.5 m/s and 800°C of thermally sprayed coatings of the TiO_2-Cr_2O_3 system and (Ti,Mo)(C,N)-Ni(Co). *Wear*, *267*(5-8), 954–964. doi:10.1016/j.wear.2008.12.105

Berger, L.-M., Stahr, C. C., Toma, F.-L., Saaro, S., Herrmann, M., Deska, D., & Thiele, S. et al. (2009c). Corrosion of thermally sprayed oxide ceramic coatings. *Thermal Spray Bulletin*, *2*(1), 40–56.

Berger, L.-M., Toma, F.-L., & Potthoff, A. (2013). Thermal spraying with suspensions – An economic spray process. *Thermal Spray Bulletin*, *6*(2), 98–101.

Berger, L.-M., Toma, F.-L., Scheitz, S., Trache, R., & Börner, T. (2014). Thermisch gespritzte Schichten im System Al_2O_3-Cr_2O_3-TiO_2 – ein Update. *Materialwissenschaft und Werkstofftechnik*, *45*(6), 465–475. doi:10.1002/mawe.201400260

Berger, L.-M., Woydt, M., Saaro, S., Stahr, C. C., & Thiele, S. (2010). High-temperature sliding wear of APS-sprayed coatings of the Cr2O3-TiO2 system. In *Proceedings 14th Nordic Symposium on Tribology*. Luleå, Sweden: Luleå University of Technology.

Berget, J. E. B. (1998). *Effects of powder composition on the erosion, corrosion and erosion–Corrosion properties of HVOF sprayed WC based coatings.* Paper presented at the 15th Int. Thermal Spray Conference - Meeting the Challenges of 21st Century, Nice, France.

Berget, J., Bardal, E., & Ronge, T. (1998). Article. *Surface Modification Technology*, *11*, 501.

Berget, J., Einar, B., & Trond, R. (1998). *Effects of matrix composition and particle size on structure and erosion–corrosion resistance of HVOF sprayed WC-Co-Cr coatings.* London: Institute of Materials.

Berget, J., Rogne, T., & Bardal, E. (2007). Erosion–corrosion properties of different WC–Co–Cr coatings deposited by the HVOF process—Influence of metallic matrix composition and spray powder size distribution. *Surface and Coatings Technology*, *201*(18), 7619–7625. doi:10.1016/j.surfcoat.2007.02.032

Bever, M. B., & Duwez, P. E. (1972). Gradients in composite materials. *Materials Science and Engineering*, *10*, 1–8. doi:10.1016/0025-5416(72)90059-6

Bhagat, R. B., Amateau, M. F., Papyrin, A., Conway, J. C. Jr, Stutzman, B., & Jones, B. (1997). Deposition of nickel-aluminium bronze powder by cold gas-dynami spray method on 2618 Al for developing wear resistant coatings. In *Thermal spray: A united forum for scientific and technological advances* (pp. 361–367). Materials Park, OH: ASM International.

Bhandari, S., Singh, H., Kansal, H. K., & Rastogi, V. (2012). Slurry erosion studies of hydroturbine steels under hydroaccelerated conditions. *Proceedings of the Institution of Mechanical Engineers. Part J, Journal of Engineering Tribology*, *226*(3), 239–250. doi:10.1177/1350650111428238

Bhandari, S., Singh, H., Kumar, H., & Rastogi, V. (2012). Slurry Erosion Performance Study of Detonation Gun-Sprayed WC-10Co-4Cr Coatings on CF8M Steel Under Hydro-Accelerated Conditions. *Journal of Thermal Spray Technology*, *21*(5), 1054–1064. doi:10.1007/s11666-012-9799-1

Bhatia, T., Ozturk, A., Xie, L., Jordan, E. H., Cetegen, B. M., Gell, M., & Padture, N. P. et al. (2002). Mechanisms of ceramic coating deposition in solution-precursor plasma spray. *Journal of Materials Research*, *17*(9), 2363–2372. doi:10.1557/JMR.2002.0346

Bianchi, L., Leger, A. C., Vardelle, M., Vardelle, A., & Fauchais, P. (1997). Splat formation and cooling of plasma-sprayed zirconia. *Thin Solid Films*, *305*(1-2), 35–47. doi:10.1016/S0040-6090(97)80005-3

Bielmann, M., Mahajan, U., Singh, R. K., Agarwal, P., Mischler, S., Rosset, E., & Landolt, D. (1999). Tribological experiments applied to tungstein chemical mechanical polishing. In *Material Research Symposium Proceedings*. Academic Press. doi:10.1557/PROC-566-97

Bitter, J. G. A. (1963). A study of erosion phenomena: Part I. *Wear*, *6*(1), 5–21. doi:10.1016/0043-1648(63)90003-6

Bitter, J. G. A. (1963). A study of erosion phenomena: Part II. *Wear*, *6*(3), 169–190. doi:10.1016/0043-1648(63)90073-5

Bjordal, M., Bardal, E., Rogne, T., & Eggen, T. G. (1995). Erosion and corrosion properties of WC coatings and duplex stainless steel in sand-containing synthetic sea-water. *Wear*, *186–187*, 508–514. doi:10.1016/0043-1648(95)07148-2

Bolelli, G., Bonferroni, B., Cannillo, V., Gadow, R., Killinger, A., Lusvarghi, L., & Stiegler, N. et al. (2010). Wear behaviour of high velocity suspension flame sprayed (HVSFS) Al_2O_3 coatings produced using micron- and nano-sized powder suspensions. *Surface and Coatings Technology*, *204*(16-17), 2657–2668. doi:10.1016/j.surfcoat.2010.02.018

Bolelli, G., Cannillo, V., Gadow, R., Killinger, A., Lusvarghi, L., & Rauch, J. (2009). Properties of high velocity suspension flame sprayed (HVSFS) TiO_2 coatings. *Surface and Coatings Technology*, *203*(12), 1722–1732. doi:10.1016/j.surfcoat.2009.01.006

Bolelli, G., Cannillo, V., Lusvarghi, L., & Manfredini, T. (2006). Wear behaviour of thermally sprayed ceramic oxide coatings. *Wear*, *261*(11-12), 1298–1315. doi:10.1016/j.wear.2006.03.023

Bolelli, G., Giovanardi, R., Lusvarghi, L., & Manfredini, T. (2006). Corrosion resistance of HVOF sprayed coatings for hard chrome replacment. *Corrosion Science*, *48*(11), 3375–3397. doi:10.1016/j.corsci.2006.03.001

Bolelli, G., Lusvarghi, L., Manfredini, T., Pighetti Mantini, F., Polini, R., Turunen, E., & Hannula, S.-P. et al. (2007a). Comparison between plasma- and HVOF-sprayed ceramic coatings. Part I: Microstructure and mechanical properties. *International Journal of Surface Science and Engineering*, *1*(1), 38–61. doi:10.1504/IJSURFSE.2007.013620

Bolelli, G., Lusvarghi, L., Manfredini, T., Pighetti Mantini, F., Turunen, E., Varis, T., & Hannula, S.-P. (2007b). Comparison between plasma- and HVOF-sprayed ceramic coatings. Part II: Tribological behaviour. *International Journal of Surface Science and Engineering*, *1*(1), 62–79. doi:10.1504/IJSURFSE.2007.013621

Bolelli, G., Rauch, J., Cannillo, V., Killinger, A., Lusvarghi, L. V., & Gadow, R. (2009). Microstructural and tribological investigation of high-velocity suspension flame sprayed (HVSFS) Al_2O_3 coatings. *Journal of Thermal Spray Technology*, *18*(1), 35–49. doi:10.1007/s11666-008-9279-9

Bolelli, G., Rauch, J., Cannillo, V., Killinger, A., Lusvarghi, L., & Gadow, R. (2008). Investigation of high-velocity suspension flame sprayed (HVSFS) glass coatings. *Materials Letters*, *62*(17-18), 2772–2775. doi:10.1016/j.matlet.2008.01.049

Bolognini, S., Feusier, G., Mari, D., Viatte, T., & Benoit, W. (2003). TiMoCN based cermets: High temperature deformation. *International Journal of Refractory Metals & Hard Materials*, *21*(1-2), 19–29. doi:10.1016/S0263-4368(02)00091-4

Börner, T., Berger, L.-M., Saaro, S., & Thiele, S. (2010). Systematische Untersuchung der Abrasionsbeständigkeit thermisch gespritzter Schichten aus dem Werkstoffsystem Al_2O_3-TiO_2-Cr_2O_3. In B. Wielage (Ed.), *Tagungsband zum 13. Werkstofftechnischen Kolloquium in Chemnitz, 30. September - 01. Oktober 2010, Werkstoffe und werkstofftechnische Anwendungen* (Vol. 37, pp. 178–187). Chemnitz: TU Chemnitz.

Bouyer, E., Gitzhofer, F., & Boulos, M. (1996). Parametric study of suspension plasma sprayed hydroxyapatite. In BerndtC.C. (Ed.), *Proceedings of International Thermal Spray Conference 1996* (pp. 683-691). Materials Park, OH: ASM International.

Bouyer, E., Gitzhofer, F., & Boulos, M. I. (1997). Powder processing by suspension thermal spraying. In *Thermal spray: A united forum for scientific and technological advances* (pp. 353–359). Materials Park, OH: ASM International.

Bozzini, B., De Gaudenzi, G. P., Serra, M., Fanigliulo, A., & Bogani, F. (2002). Corrosion behaviour of WC-Co based hardmetal in neutral chloride and acid sulphate media. *Werkstoffe und Korrosion*, *53*(5), 328–334. doi:10.1002/1521-4176(200205)53:5<328::AID-MACO328>3.0.CO;2-G

Branco, J. R. T., Gansert, R., Sampath, S., Berndt, C., & Herman, H. (2004). Solid Particle Erosion of Plasma Sprayed Ceramic Coatings. *Materials Research*, *7*(1), 147–153. doi:10.1590/S1516-14392004000100020

Bromark, M., Larsson, M., Hedenqvist, P., & Hogmark, S. (1997). Wear of PVD Ti/TiN multilayer coatings. *Surface and Coatings Technology*, *90*(3), 217–223. doi:10.1016/S0257-8972(96)03141-6

Bu Qian, W., & Luer, K. (1994). The erosion-oxidation behavior of HVOF Cr3C2-NiCr cermet coating. *Wear*, *174*(1-2), 177–185. doi:10.1016/0043-1648(94)90100-7

Buchmann, M., & Gadow, R. (2001). Mechanical and tribological characterization of APS and HVOF sprayed TiO2 coatings on light metals, In C.C. Berndt, K.A. Khor, & E.F. Lugscheider (Eds.), *Thermal Spray 2001: New Surfaces for A New Millennium, Proc. Int. Thermal Spray Conf.* Materials Park, OH: ASM International.

Budinski, K. G. (1988). *Surface engineering for wear resistance* (2nd ed.; Vol. 32). Englewood Cliffs, NJ: Prentice Hall.

Bull, S. J. (2005). Nanoindentation of coatings. *Journal of Physics. D, Applied Physics*, *38*(24), 393–413. doi:10.1088/0022-3727/38/24/R01

Burke, J. E. (1962). The science and technology of sintering. In C. Klingsberg (Ed.), *Physics and chemistry of ceramics* (pp. 167–178). New York: Gordon Beach.

Busso, E. P., Wright, L., Evans, H. E., McCartney, L. N., Saunders, S. R. J., Osgerby, S., & Nunn, J. (2007). A physics-based life prediction methodology for thermal barrier coating systems. *Acta Materialia*, *55*(5), 1491–1503. doi:10.1016/j.actamat.2006.10.023

Cantera, E. L., & Mellor, B. G. (1998). Fracture toughness and crack morphologies in eroded WC-Co-Cr thermally sprayed coatings. *Materials Letters*, *37*(4-5), 201–210. doi:10.1016/S0167-577X(98)00092-5

Celik, E., Demirkıran, A. S., & Avc, E. (1999). Effect of grit blasting of substrate on the corrosion behaviour of plasma-sprayed Al_2O_3 coatings. *Surface and Coatings Technology*, *116–119*, 1061–1064. doi:10.1016/S0257-8972(99)00238-8

Celik, E., Sengil, I. A., & Avc, E. (1997). Effects of some parameters on corrosion behaviour of plasma-sprayed coatings. *Surface and Coatings Technology*, *97*(1-3), 355–360. doi:10.1016/S0257-8972(97)00208-9

Chattopadhyay, R. (2004). *Advanced thermally assisted surface engineering processes, XII*. Dordrecht, The Netherlands: Kluwer Academic Publishers.

Chen, H., & Ding, C. X. (2002). Nanostructured zirconia coating prepared by atmospheric plasma spraying. *Surface and Coatings Technology*, *150*(1), 31–36. doi:10.1016/S0257-8972(01)01525-0

Chen, H., & Hutchings, I. M. (1998). Abrasive wear resistance of plasma-sprayed tungsten carbide cobalt coatings. *Surface and Coatings Technology*, *107*(2-3), 106–114. doi:10.1016/S0257-8972(98)00581-7

Chen, H., Xu, C., Qu, J., Hutchings, I. M., Shipway, P. H., & Lie, J. (2005). Sliding wear behaviour of laser clad coatings based upon a nickel-based self-fluxing alloy co-deposited with conventional and nanostructured tungsten carbide–cobalt hardmetals. *Wear*, *259*(7-12), 801–806. doi:10.1016/j.wear.2005.02.066

Chen, H., Xu, C., Zhou, Q., Hutchings, I. M., Shipway, P. H., & Liud, J. (2005). Micro-scale abrasive wear behaviour of HVOF sprayed and laser-remelted conventional and nanostructured WC-Co coatings. *Wear*, *258*(1-4), 333–338. doi:10.1016/j.wear.2004.09.044

Chen, H., Zeng, Y., & Ding, C. (2003). Microstructural characterization of plasma-sprayed nanostructured zirconia powders and coatings. *Journal of the European Ceramic Society*, *23*(3), 491–497. doi:10.1016/S0955-2219(02)00096-1

Chen, H., Zhang, Y., & Ding, C. (2002). Tribological properties of nanostructured zirconia coatings deposited by plasma spraying. *Wear*, *253*(7-8), 885–893. doi:10.1016/S0043-1648(02)00221-1

Chen, H., Zhou, X., & Ding, C. (2003). Investigation of the thermomechanical properties of a plasma-sprayed nanostructured zirconia coating. *Journal of the European Ceramic Society*, *23*(9), 1449–1455. doi:10.1016/S0955-2219(02)00345-X

Chen, J. F., & Bogaerts, W. F. (1996). Electrochemical emission spectroscopy for monitoring uniform and localized corrosion. *Corrosion*, *52*(10), 753–759. doi:10.5006/1.3292068

Chen, J., & Lu, G. (2012). Finite element modelling of nanoindentation based methods for mechanical properties of cells. *Journal of Biomechanics*, *45*(16), 2810–2816. doi:10.1016/j.jbiomech.2012.08.037 PMID:23017378

Cho, J. E., Hwang, S. Y., & Kim, K. Y. (2006). Corrosion behaviour of thermal sprayed WC cermet coatings having various metallic binders in strong alkaline environment. *Surface and Coatings Technology*, *200*(8), 2653–2662. doi:10.1016/j.surfcoat.2004.10.142

Chráska, P., Dubsky, J., Neufuss, K., & Písacka, J. (1997). Alumina-base plasma-sprayed materials, Part I: Phase stability of alumina and alumina-chromia. *Journal of Thermal Spray Technology*, *6*(3), 320–326. doi:10.1007/s11666-997-0066-9

Christman, T., & Shewmon, P. G. (1979). Erosion of strong Al alloy. *Wear*, *51*(1), 57–70. doi:10.1016/0043-1648(79)90196-0

Clark, H. M. (1992). The influence of the flow field in slurry erosion. *Wear*, *152*(2), 223–240. doi:10.1016/0043-1648(92)90122-O

Clark, H. M., Hawthorne, H. M., & Xie, Y. (1999). Wear rates and specific energies of some ceramic, cermet and metallic coatings determined in the Coriolis erosion tester. *Wear*, *233-235*, 319–327. doi:10.1016/S0043-1648(99)00213-6

Clyne, T. W., & Gill, S. C. (1996). Residual stresses in thermal spray coatings and their effect on interfacial adhesion: A review of recent work. *Journal of Thermal Spray Technology*, *5*(4), 401–418. doi:10.1007/BF02645271

Cockeram, B. V. (2006). The mechanical properties and fracture mechanisms of wrought low carbon arc cast (LCAC), molybdenum–0.5pct titanium–0.1pct zirconium (TZM), and oxide dispersion strengthened (ODS) molybdenum flat products. *Materials Science and Engineering A*, *418*(1-2), 120–136. doi:10.1016/j.msea.2005.11.030

Coyle, T. W., & Wang, Y. (2007). Solution precursor plasma spray (SPPS) of Ni-YSZ SOFC anode coatings. In B.R. Marple, M.M. Hyland, Y.-C. Lau, C.-J. Li, R.S. Lima, & G. Montavon (Eds.), Thermal spray 2007: Global coating solutions (pp. 699-704). Materials Park, OH: ASM International.

Crawmer, E. D. (2004). Introduction to coatings, equipment, and theory. In Handbook of thermal spray technology. Materials Park, OH: ASM International.

Davis, A. G., Boone, D. H., & Levy, A. V. (1986). Erosion of ceramic thermal barrier coatings. *Wear*, *110*(2), 101–116. doi:10.1016/0043-1648(86)90141-9

Dejang, N., Limpichaipanit, A., Watcharapasorn, A., Wirojanupatump, S., Niranatlumpong, P., & Jiansirisomboon, S. (2011). Fabrication and properties of plasma-sprayed Al_2O_3/ZrO_2composite coatings. *Journal of Thermal Spray Technology*, *20*(6), 1259–1268. doi:10.1007/s11666-011-9672-7

Della-Corte, C. (1996). The effect of counterface on the tribological performance of a high temperature solid lubricant composite from 25 to 650°C. *Surface and Coatings Technology*, *86–87*, 486–492. doi:10.1016/S0257-8972(96)02959-3

Deng, T., Chaudhry, A. R., Patel, M., Hutchings, I., & Bradley, M. S. A. (2005). Effect of particle concentration on erosion rate of mild steel bends in a pneumatic conveyor. *Wear*, *258*(1-4), 480–487. doi:10.1016/j.wear.2004.08.001

Dent, A. H., DePalo, S., & Sampath, J. (2002). Examination of the wear properties of HVOF sprayed nanostructured and conventional WC-Co cermets with different binder phase contents. *Journal of Thermal Spray Technology*, *11*(4), 551–558. doi:10.1361/105996302770348691

Diomidis, N., Mischler, S., More, N. S., & Roy, M. (2012). Tribo-Eelectrochemical characterisation of metallic biomaterial for total joint replacement. *Acta Biomaterialia*, *8*(2), 852–859. doi:10.1016/j.actbio.2011.09.034 PMID:22005332

Dom, R., Sivakumar, G., Hebalkar, N. Y., Joshi, S. V., & Borse, P. H. (2012). Deposition of nanostructured photocatalytic zinc ferrite films using solution precursor plasma spraying. *Materials Research Bulletin*, *47*(3), 562–570. doi:10.1016/j.materresbull.2011.12.044

Dubsky, J., Chraska, P., Kolman, B., Stahr, C. C., & Berger, L.-M. (2011). Phase formation control in plasma sprayed alumina-chromia coatings. *Ceramics-Silikáty*, *55*(3), 95–301.

Du, H., Sun, C., Hua, W. G., Wang, T. G., Gong, J., Jiang, X., & Lee, S. W. (2007). Structure, mechanical and sliding wear properties of WC–Co/MoS_2–Ni coatings by detonation gun spray. *Materials Science and Engineering A*, *445-446*, 122–134. doi:10.1016/j.msea.2006.09.011

Dunleavy, C. S., Golosnoy, I. O., Curran, J. A., & Clyne, T. W. (2009). Characterization of discharge events during plasma electrolytic oxidation. *Surface and Coatings Technology*, *203*(22), 3410–3419. doi:10.1016/j.surfcoat.2009.05.004

Dutta Majumdar, J., & Manna, I. (2012). Introduction to laser-assisted fabrication of materials. In Laser assisted fabrication-Present status and future scope in laser assisted fabrication of materials. Hiedelberg, Germany: Springer Verlag.

Economou, S., De Bonte, M., Celis, J. P., Smith, R. W., & Lugscheider, E. (2000). Tribological behaviour at room temperature and at 550°C of TiC-based plasma sprayed coatings in fretting gross slip conditions. *Wear*, *244*(1-2), 165–179. doi:10.1016/S0043-1648(00)00455-5

Eden, D. A., Hladky, K., John, D. G., & Dawson, J. L. (1986). Simultaneous monitoringof potential and current noise signals from corroding electrodes. In *Proceedings of the CORROSION'86* (Paper 274). National Association of Corrosion Engineers.

Erickson, L., Hawthorne, H., & Troczynski, T. (2001). Correlations between microstructural parameters, micromechanical properties and wear resistance of plasma sprayed ceramic coatings. *Wear*, *250*(1-12), 569–574. doi:10.1016/S0043-1648(01)00608-1

Eschnauer, H. (1980). Hard material powders and hard alloy powders for plasma surface coating. *Thin Solid Films*, *73*(1), 1–17. doi:10.1016/0040-6090(80)90323-5

Evans, A. G., Gulden, M. E., & Rosenblatt, M. (1978). Impact damage in brittle materials in the elastic-plastic response regime. *Proceedings of the Royal Society of London. Series A, Mathematical and Physical Sciences*, *361*(1706), 343–365. doi:10.1098/rspa.1978.0106

Evans, A. G., & Wilshaw, T. R. (1977). Dynamic solid particle damage in brittle materials: An appraisal. *Journal of Materials Science*, *12*(1), 97–116. doi:10.1007/BF00738475

Factor, M., & Roman, I. (2002). Microhardness as a simple means of estimating relative wear resistance of carbide thermal spray coatings part 1. Characterization of cemented carbide coatings. *Journal of Thermal Spray Technology*, *11*(4), 468–481. doi:10.1361/105996302770348600

Fasching, M., Prinz, F. B., & Weiss, L. E. (1995). Smart coating: A technical note. *Journal of Thermal Spray Technology*, *4*(2), 133–136. doi:10.1007/BF02646101

Fauchais, P. L., Heberlein, J., Boulos, V. R., & Maher, I. (2014). Thermal spray fundamentals. Springer Science + Business Media.

Fauchais, P. L., Herberlein, J. V. R., & Boulos, M. I. (2014). Thermal spray fundamentals: From powder to part. New York, NY: Springer. doi:10.1007/978-0-387-68991-3

Fauchais, P., Montavon, G., Lima, R. S., & Marple, B. R. (2011). Engineering of new class of thermal spray nano-based microstructures from agglomerated nanostructured particles, suspensions and solutions: An invited review. Journal of Physics D: Applied Physics, 44, 093001.

Fauchais, P. L., & Vardelle, A. (2000). Heat, mass and momentum transfer in coating formation by plasma spraying. *International Journal of Thermal Sciences*, *39*(9-11), 852–870. doi:10.1016/S1290-0729(00)01195-9

Fauchais, P., Coudert, J. F., Vardelle, M., Vardelle, A., & Denoirjean, A. (1992). Diagnostics of thermal spraying plasma jets. *Journal of Thermal Spray Technology*, *1*(2), 117–128. doi:10.1007/BF02659011

Fauchais, P., Etchart-Salas, R., Rat, V., Coudert, J. F., Caron, N., & Wittmann-Teneze, K. (2008). Parameters controlling liquid plasma spraying: Solutions, sols, or suspensions. *Journal of Thermal Spray Technology*, *17*(1), 31–59. doi:10.1007/s11666-007-9152-2

Fauchais, P., & Montavon, G. (2010). Latest developments in suspension and liquid precursor thermal spraying. *Journal of Thermal Spray Technology*, *19*(1-2), 226–239. doi:10.1007/s11666-009-9446-7

Fauchais, P., Montavon, G., Lima, R. S., & Marple, B. (2011). Engineering a new class of thermal spray nano-based microstructures from agglomerated nanostructured particles, suspensions and solutions: An invited review. *Journal of Physics. D, Applied Physics*, *44*(9), 1–53. doi:10.1088/0022-3727/44/9/093001

Fauchais, P., Rat, V., Delbos, C., Coudert, J. F., Chartier, T., & Bianchi, L. (2005). Understanding of suspension DC plasma spraying of finely structured coatings for SOFC. *IEEE Transactions on Plasma Science*, *33*(2), 920–930. doi:10.1109/TPS.2005.845094

Fauchais, P., Vardelle, A., & Dussoubs, B. (2001). Quo vadis thermal spraying? *Journal of Thermal Spray Technology*, *10*(1), 44–66. doi:10.1361/105996301770349510

Fauchais, P., & Verdelle, M. (2010). Sensors in spray processes. *Journal of Thermal Spray Technology*, *19*(4), 668–694. doi:10.1007/s11666-010-9485-0

Fedrizzi, L., Rossi, S., Cristel, R., & Bonora, P. L. (2004). Corrosion and wear behaviour of HVOF cermet coatings used to replace hard chromium. *Electrochimica Acta*, *49*(17-18), 2803–2814. doi:10.1016/j.electacta.2004.01.043

Fedrizzi, L., Valentinelli, L., Rossi, S., & Segna, S. (2007). Tribocorrosion behaviour of HVOF cermet coatings. *Corrosion Science*, *49*(7), 2781–2799. doi:10.1016/j.corsci.2007.02.003

Fervel, V., Normand, B., & Coddet, C. (1999). Tribological behavior of plasma sprayed Al_2O_3-based cermet coatings. *Wear*, *230*(1), 70–77. doi:10.1016/S0043-1648(99)00096-4

Fink, M. (1930). Wear oxidation, a new component of wear. *Transaction of American Society of Steel*, *18*, 1026–1034.

Finnie, I. (1979). Fundamental mechanism of the erosive wear of ductile metals by solid particles. In W. F. Adler (Ed.), Erosion, prevention, and useful applications, ASTM STP 664 (pp. 36-58). American Society for Testing and Materials. doi:10.1520/STP35794S

Finnie, I. (1995). Some reflections on the past and future of erosion. *Wear*, *186-187*, 1–10. doi:10.1016/0043-1648(95)07188-1

Fischer-Cripps, A. C. (2004). *Nanoindentation* (2nd ed.). Springer-Verlag.

Fontalvo, G. A., & Mitterrer, C. (2005). The effect of oxide-forming alloying elements on the high temperature wear of a hot work steel. *Wear*, *258*(10), 1491–1499. doi:10.1016/j.wear.2004.04.014

French, D., Hurst, H. J., & Marvig, P. (2001). Comments on the use of molybdenum components for slag viscosity measurements. *Fuel Processing Technology*, *72*(3), 215–225. doi:10.1016/S0378-3820(01)00188-6

Fukuda, Y., & Kumon, M. (Eds.). (1995). Thermal spraying current status and future trends. High Temperature Society of Japan.

Gadow, R., Killinger, A., & Rauch, J. (2008). New results in high velocity suspension flame spraying (HVSFS). *Surface and Coatings Technology*, *202*(18), 4329–4336. doi:10.1016/j.surfcoat.2008.04.005

Gadow, R., Killinger, A., Rempp, A., & Manzat, A. (2010). Advanced ceramic tribological layers by thermal spray routes. *Advances in Science and Technology*, *66*, 106–119. doi:10.4028/www.scientific.net/AST.66.106

Galliano, F., Galvanetto, E., Mischler, S., & Landolt, D. (2001). Tribocorrosion behavior of plasma nitrided Ti–6Al–4V alloy in neutral NaCl solution. *Surface and Coatings Technology, 145*(1-3), 121–131. doi:10.1016/S0257-8972(01)01309-3

Gang, J., Morniroli, J.-P., & Grosdidier, T. (2003). Nanostructures in thermal spray coatings. *Scripta Materialia, 48*(12), 1599–1604. doi:10.1016/S1359-6462(03)00142-8

Gant, A. J., & Gee, M. G. (2006). Abrasion of WC hardmetals using hard countefaces. *International Journal of Refractory Metals & Hard Materials, 24*(1-2), 189–198. doi:10.1016/j.ijrmhm.2005.05.007

Gaona, M., Lima, R. S., & Marple, B. (2008). Influence of particle temperature and velocity on the microstructure and mechanical behaviour of high velocity oxy-fuel (HVOF)-sprayed nanostructured titania coatings. *Journal of Materials Processing Technology, 198*(1-3), 426–435. doi:10.1016/j.jmatprotec.2007.07.024

Garcia, E., Zhang, Z. B., Coyle, T. W., Hao, S. E., & Mu, S. L. (2007). Liquid precursors plasma spraying of TiO2 and Ce- Doped Ba(Zr0.2Ti0.8)O3 coatings. In B.R. Marple, M.M. Hyland, Y.-C. Lau, C.-J. Li, R.S. Lima, & G. Montavon (Eds.), Thermal spray 2007: Global coating solutions (pp. 650-654). Materials Park, OH: ASM International.

Gardos, M. N. (1988). The effect of anion vacancies on the tribological properties of rutile (TiO_{2-x}). *Tribology Transactions, 31*(4), 427–436. doi:10.1080/10402008808981844

Gärtner, F., Kreye, H., & Richter, H. J. (2006). HVOF Spraying with powder and wire. In Proceedings 7th Kolloquium Hochgeschwindigkeitsflammspritzen (pp. 39-56). Unterschleißheim, Gemeinschaft Thermisches Spritzen.

Gee, M. G., Gant, A., Hutchings, I., Bethke, R., Schiffman, K., Van Acker, K., & von Stebut, J. et al. (2003). Progress towards standardization of ball cratering. *Wear, 255*(1-6), 1–13. doi:10.1016/S0043-1648(03)00091-7

Gell, M., Xie, L., Ma, X., Jordan, E.H., & Padture, N.P. (2004). Highly durable thermal barrier coatings made by the solution precursor plasma spray process. *Surface and Coating Technology, 177 –178*, 97–102.

Gell, M., Jordan, E. H., Sohn, Y. H., Goberman, D., Shaw, L., & Xiao, T. D. (2001). Development and implementation of plasma sprayed nanostructured ceramic coatings. *Surface and Coatings Technology, 146-147*, 48–54. doi:10.1016/S0257-8972(01)01470-0

Gell, M., Jordan, E. H., Teicholz, M., Cetegen, B. M., Padture, N. P., Xie, L., & Roth, J. et al. (2008). Thermal barrier coatings made by the solution precursor plasma spray process. *Journal of Thermal Spray Technology, 17*(1), 124–135. doi:10.1007/s11666-007-9141-5

GFE Schmalkalden E. V. (n.d.). *Principles of coating technology*. Retrieved on 27th February 2014, from http://www.fh-schmalkalden.de/schmalkaldenmedia/-p-16916

Ghabchi, A. (2013). Wear resistant carbide-based thermal sprayed coatings: Process, properties, mechanical degradation and wear. VTT Science, 29.

Godet, M., Berthier, Y., Lancaster, J., & Vincent, L. (1991). Wear modelling: Using fundamental understanding or practical experience. *Wear*, *149*(1-2), 325–345. doi:10.1016/0043-1648(91)90383-6

Gougeon, P., & Moreau, C. (1993). In-flight particle surface temperature measurement: Influence of the plasma light scattered by the particles. *Journal of Thermal Spray Technology*, *2*(3), 229–233. doi:10.1007/BF02650470

Grewal, H. S., Agrawal, A., & Singh, H. (2013a). Slurry erosion mechanism of hydroturbine steel: Effect of operating parameters. *Tribology Letters*, *52*(2), 287–303. doi:10.1007/s11249-013-0213-z

Grewal, H. S., Agrawal, A., & Singh, H. (2013b). Design and development of high-velocity slurry erosion test rig using CFD. *Journal of Materials Engineering and Performance*, *22*(1), 152–161. doi:10.1007/s11665-012-0219-y

Grewal, H. S., Agrawal, A., & Singh, H. (2013d). Slurry erosion performance of Ni–Al2O3 based composite coatings. *Tribology International*, *66*(0), 296–306. doi:10.1016/j.triboint.2013.06.010

Grewal, H. S., Agrawal, A., & Singh, H. (2013g). Identifying erosion mechanism: A novel approach. *Tribology Letters*, *51*(1), 1–7. doi:10.1007/s11249-013-0156-4

Grewal, H. S., Agrawal, A., Singh, H., & Shollock, B. A. (2013e). Slurry erosion performance of Ni-Al_2O_3 based thermal-sprayed coatings: Effect of angle of impingement. *Journal of Thermal Spray Technology*, *23*(3), 389–401. doi:10.1007/s11666-013-0013-x

Grewal, H. S., Bhandari, S., & Singh, H. (2012). Parametric study of slurry-erosion of hydroturbine steels with and without detonation gun spray coatings using taguchi technique. *Metallurgical and Materials Transactions. A, Physical Metallurgy and Materials Science*, *43*(9), 3387–3401. doi:10.1007/s11661-012-1148-y

Grewal, H. S., Singh, H., & Agrawal, A. (2013c). Microstructural and mechanical characterization of thermal sprayed nickel–alumina composite coatings. *Surface and Coatings Technology*, *216*, 78–92. doi:10.1016/j.surfcoat.2012.11.029

Grewal, H. S., Singh, H., & Agrawal, A. (2013f). Understanding liquid impingement erosion behaviour of nickel-alumina based thermal spray coatings. *Wear*, *301*(1-2), 424–433. doi:10.1016/j.wear.2013.01.063

Grujicic, M., Zhao, C. L., DeRosset, W. S., & Helfritch, D. (2004). Adiabatic shear instability based mechanism for particles/substrate bonding in the cold-gas dynamic-spray process. *Materials & Design*, *25*(8), 681–688. doi:10.1016/j.matdes.2004.03.008

Gruner, R. (1986), Vacuum plasma sprayed composite coatings: Advances in thermal spraying. In *Proceedings of the Eleventh International Thermal SprayingConference (ITSC86)*. Montreal, Canada: ASM International.

Guest, C. J. S. (1986). Plasma and detonation gun coatings. *Trans IMF*, *64*, 33–38.

Guilemany, J. M., Cinca, N., Ferna'ndez, J., & Sampath, S. (2008). Erosion, abrasive, and friction wear behavior of iron aluminide coatings sprayed by HVOF. *Journal of Thermal Spray Technology*, *17*(5-6), 762–773. doi:10.1007/s11666-008-9252-7

Guilemany, J. M., Dosta, S., & Miguel, J. R. (2006). The enhancement of the properties of WC-Co HVOF coatings through the use of nanostructured and microstructured feedstock powders. *Surface and Coatings Technology*, *201*(3-4), 1180–1190. doi:10.1016/j.surfcoat.2006.01.041

Guilemany, J. M., Fernandez, J., Delgado, J., Benedetti, A. V., & Climent, F. (2002). Effects of thickness coating on the electrochemical behaviour of thermal spray Cr_3C_2–NiCr coatings. *Surface and Coatings Technology*, *153*(2-3), 107–113. doi:10.1016/S0257-8972(01)01679-6

Guilemany, J. M., Miguel, J. M., Vizcaino, S., Lorenzana, C., Delgado, J., & Sanchez, J. (2002). Role of heat treatments in the improvement of the sliding wear properties of Cr_3C_2–NiCr coatings. *Surface and Coatings Technology*, *157*(2-3), 207–213. doi:10.1016/S0257-8972(02)00148-2

Guilemany, J. M., Paco, J. M., Nutting, J., & Miguel, J. R. (1999). Characterisation of W_2C phase formed during HVOF spraying of a WC-12Co powder. *Metallurgical and Materials Transactions. A, Physical Metallurgy and Materials Science*, 1913–1921. doi:10.1007/s11661-999-0002-3

Habib, K. A., Saura, J. J., Ferrer, C., Damra, M. S., Gimenez, E., & Cabedo, L. (2006). Comparison of flame sprayed Al_2O_3/TiO_2 coatings: Their microstructure, mechanical properties and tribology behavior. *Surface and Coatings Technology*, *201*(3-4), 1436–1443. doi:10.1016/j.surfcoat.2006.02.011

Hager, C. H. Jr, Sanders, J., Sharma, S., Voevodin, A., & Segall, A. (2009). The effect of temperature on gross slip fretting wear of cold-sprayed nickel coatings on Ti6Al4V interfaces. *Tribology International*, *42*(3), 491–502. doi:10.1016/j.triboint.2008.08.009

Halling, J., & Arnell, R. D. (1984). Ceramic coatings in the war on wear. *Wear*, *100*(1-3), 367–380. doi:10.1016/0043-1648(84)90022-X

Hammitt, F. G. (1976). Erosive wear testing. In R. G. Bayer (Ed.), Selection and use of wear tests for metals, ASTM STP 615 (pp. 45-67). American Society for Testing and Materials. doi:10.1520/STP38612S

Hansen, J. S. (1979). Relative erosion resistance of several materials. In W. F. Adler (Ed.), Erosion: Prevention and useful Applications ASTM STP 664 (pp. 148-162). American Society for Testing and Materials. doi:10.1520/STP35800S

Harju, M., Levänen, E., & Mäntylä, T. (2004). Surface characterization of plasma sprayed oxide materials: Estimation of surface acidity using mass titration. *Journal of Colloid and Interface Science*, *276*(2), 346–353. doi:10.1016/j.jcis.2004.03.058 PMID:15271562

Hartfield-Wunsch, S. E., & Tung, S. C. (1994). The effect of microstructure on the wear behaviour of thermal spray coatings. In C. C. Berndt & S. Sampath (Eds.), *Thermal spray industrial applications* (pp. 19–29). Materials Park, OH: ASM International.

Hawthorne, H. M., Arsenault, B., Immarigeon, J. P., Legoux, J. G., & Parameswaran, V. R. (1999). Comparison of slurry and dry erosion behaviour of some HVOF thermal sprayed coatings. *Wear*, *225-229*(Part 2), 825–834. doi:10.1016/S0043-1648(99)00034-4

Hawthrone, H. M., Arsenault, B., Immarigeon, J. P., Legoux, J. G., & Parameswaran, V. R. (1999). Comparison of slurry and dry erosion behaviour of some HVOF thermal sprayed coatings. *Wear*, *225-229*, 835–834.

Hay, J. L., & Pharr, G. M. (2000). Instrumented indentation testing. In ASM handbook, mechanical testing and evaluation. ASM International.

Hearley, A., Little, J. A., & Sturgeon, A. J. (1999). The erosion behaviour of NiAl intermetallic coatings produced by high velocity oxy-fuel thermal spraying. *Wear*, *328*, 233–235.

Heberlein, J., Rao, N. P., Neuman, A., Blum, J., Tymiak, N., McMurry, P. H., & Girshick, S. L. (1997). Thermal spraying of nanostructured coatings by hypersonic plasma particle deposition. In *Thermal spray: A united forum for scientific and technological advances* (pp. 393–397). Materials Park, OH: ASM International.

Herrmann, M., Toma, F.-L., Berger, L.-M., Kaiser, G., & Stahr, C. C. (2014). Comparative study of the corrosion resistance of thermally sprayed ceramic coatings and their bulk ceramic counterparts. *Journal of the European Ceramic Society*, *34*(2), 493–504. doi:10.1016/j.jeurceramsoc.2013.08.033

Hochstrasser, S. (2006). Mechanistic study of corrosion reactions on WC-Co hardmetal in aqueous solution- an investigation by electrochemical methods and elemental solution analysis. (Doctor of Science Thesis). ETH Zurich, Zurich, Switzerland.

Hochstrasser-Kurz, S., Mueller, Y., Latkoczy, C., Virtanen, S., & Schmutz, P. (2007). Analytical characterization of the corrosion mechanisms of WC–Co by electrochemical methods and inductively coupled plasma mass spectroscopy. *Corrosion Science*, *49*, 2002–2020. doi:10.1016/j.corsci.2006.08.022

Holmberg, K., & Matthews, A. (2009). Coatings tribology: Properties, mechanisms, techniques and applications in surface engineering (2nd ed.). Amsterdam: Elsevier.

Holmberg, K., & Matthews, A. (1994). *Coatings tribology: Contact mechanisms, deposition techniques and application*. Amsterdam: Elsevier.

Houdková, Š., Zahálka, F., Kašparová, M., & Berger, L.-M. (2011). Comparative study of thermally sprayed coatings under different types of wear conditions for hard chromium replacement. *Tribology Letters*, *43*(2), 139–154. doi:10.1007/s11249-011-9791-9

Hovis, S. K., Anand, K., Conrad, H., & Scattergood, R. O. (1985). A new method of velocity calibration for erosion testing. *Wear*, *101*(1), 69–76. doi:10.1016/0043-1648(85)90211-X

Hu, H. X., Jiang, S. L., Tao, Y. S., Xiong, T. Y., & Zheng, Y. G. (2011). Cavitation erosion and jet impingement erosion mechanism of cold sprayed Ni–Al2O3 coating. *Nuclear Engineering and Design*, *241*(12), 4929–4937. doi:10.1016/j.nucengdes.2011.09.038

Hurricks, P. L. (1972). Some aspects of the metallurgy and wear resistance of surface coatings. *Wear*, *22*(3), 291–320. doi:10.1016/0043-1648(72)90391-2

Hussain, T., McCartney, D. G., Shipway, P. H., & Zhang, D. (2009). Bonding mechanisms in cold spraying: The contributions of metallurgical and mechanical components. *Journal of Thermal Spray Technology*, *18*(3), 364–379. doi:10.1007/s11666-009-9298-1

Hussein, M. F., & Tabakoff, W. (1974). Computation and plotting of solid particle flow in rotating cascades. *Computers & Fluids*, *2*(1), 1–15. doi:10.1016/0045-7930(74)90002-4

Hutchings, I. M. (1979). Mechanism of the erosion of metals by solid particles. In W. F. Adler (Ed.), Erosion: Prevention and useful applications, ASTM STP 664 (pp. 59-76). American Society for Testing and Materials. doi:10.1520/STP35795S

Hutchings, I. M. (1997). Abrasive and erosive wear tests for thin coatings: A unified approach, new directions in tribology. Mechanical Engineering Publications.

Hutchings, I. M. (1977). Deformation of metal surfaces by the oblique impact of square plates. *International Journal of Mechanical Sciences*, *19*(1), 45–52. doi:10.1016/0020-7403(77)90015-7

Hutchings, I. M. (1981). A model for the erosion of metals by spherical particles at normal incidence. *Wear*, *70*(3), 269–281. doi:10.1016/0043-1648(81)90347-1

Hutchings, I. M., & Winter, R. E. (1974). Particle erosion of ductile metals: A mechanism of material removal. *Wear*, *27*(1), 121–128. doi:10.1016/0043-1648(74)90091-X

Hutchings, I. M., & Winter, R. E. (1975). The erosion of ductile metals by spherical particles, *Journal of Physics (D) Applied Physics (Berlin)*, *8*, 9–14.

Hwang, B., Lee, S., & Ahn, J. (2004). Correlation of microstructure and wear resistance of molybdenum blend coatings fabricated by atmospheric plasma spraying. *Materials Science and Engineering A*, *366*(1), 152–163. doi:10.1016/j.msea.2003.09.062

Ibrahim, A., & Berndt, C. C. (2007). Fatigue and deformation of HVOF sprayed WC–Co coatings and hard chrome plating. *Materials Science and Engineering A*, *456*(1-2), 114–119. doi:10.1016/j.msea.2006.12.030

Ilavsky, J., Berndt, C. C., Herman, H., Chraska, P., & Dubsky, J. (1997). Alumina-base plasma-sprayed materials, Part II: Phase transformations in aluminas. *Journal of Thermal Spray Technology*, *6*(4), 439–444. doi:10.1007/s11666-997-0028-2

Inman, I. A., Rose, S. R., & Dutta, P. K. (2006). Development of a simple 'temperature versus sliding speed' wear map for the sliding wear behaviour of dissimilar metallic interfaces. *Wear*, *260*(9-10), 919–932. doi:10.1016/j.wear.2005.06.008

Inman, I. A., Rose, S. R., & Dutta, P. K. (2006). Studies of high temperature sliding wear of metallic dissimilar interfaces II: Incoloy MA956 versus Stellite. *Tribology International*, *39*(11), 1361–1375. doi:10.1016/j.triboint.2005.12.001

Irons, G., Kratochvil, W. R., Bullock, W. R., & Berndt, R. A. (1994). Thermal spray industrial applications, investigation of erosion resistant chromium carbide coatings sprayed with the high pressure HVOF process. ASM International.

Iwasaki, K. (1997). Production of a functionally graded artificial tooth root by unique sequence of processes. *Materials Research Innovative*, *1*(3), 180–187. doi:10.1007/s100190050038

Jacobs, L., Hyland, M. M., & De Bonte, M. (1999). Study of the influence of microstructural properties on the sliding-wear behavior of HVOF and HVAF sprayed WC-cermet coatings. *Journal of Thermal Spray Technology*, *8*(1), 125–132. doi:10.1361/105996399770350656

Jia, K., & Fischer, T. E. (1996). Abrasion resistance of nanostructured and conventional cemented carbides. *Wear*, *200*(1-2), 206–214. doi:10.1016/S0043-1648(96)07277-8

Jia, K., & Fischer, T. E. (1997). Sliding wear resistance of conventional and nanostructured cemented carbides. *Wear*, *203*, 310–318. doi:10.1016/S0043-1648(96)07423-6

Jia, K., Fischer, T. E., & Gallois, B. (1998). Microstructure, hardness and toughness of nano-structured and conventional WC-Co composites. *Nanostructured Materials*, *10*(5), 875–891K. doi:10.1016/S0965-9773(98)00123-8

Jiang, J., & Stack, M. M. (2006). Modelling sliding wear: From dry to wet environments. *Wear*, *261*(9), 954–965. doi:10.1016/j.wear.2006.03.028

Jiang, J., Stott, F. H., & Stack, M. M. (1994). Some frictional features associated with the sliding wear of the nickel-base alloy N80A at temperatures to 250 °C. *Wear*, *176*(2), 185–194. doi:10.1016/0043-1648(94)90146-5

Jiang, J., Stott, F. H., & Stack, M. M. (1998). The role of triboparticulates in dry sliding wear. *Tribology International*, *31*(5), 245–256. doi:10.1016/S0301-679X(98)00027-9

Jiang, X., Liu, W., Dong, S. Y., & Xu, B. S. (2005, April). High temperature tribology behaviors of brush plated Ni–W–Co/SiC composite coating. *Surface and Coatings Technology*, *194*(1), 10–15. doi:10.1016/j.surfcoat.2004.04.095

JiangX.LiuW.DongS.YXuB.S., (2005);

Ji, G. C., Li, C. J., Wang, Y. Y., & Li, W. Y. (2006). Microstructural characterization and abrasive wear performance of HVOF sprayed Cr_3C_2–NiCr coating. *Surface and Coatings Technology*, *200*(24), 6749–6757. doi:10.1016/j.surfcoat.2005.10.005

Jordan, E. H., Gell, M., Sohn, Y. H., Goberman, D., Shaw, L., Jiang, S., & Strutt, P. et al. (2001). Fabrication and evaluation of plasma sprayed nanostructured alumina-titania Coatings with superior properties. *Materials Science and Engineering A*, *301*(1), 80–89. doi:10.1016/S0921-5093(00)01382-4

Jordan, E. H., Xie, L., Ma, X., Gell, M., Padture, N. P., Cetegen, B., & Bryant, P. E. C. et al. (2004). Superior thermal barrier coatings using solution precursor plasma spray. *Journal of Thermal Spray Technology*, *13*(1), 57–65. doi:10.1007/s11666-004-0050-6

K, W. R. J., & J, S. A. (2004). Erosion–corrosion of candidate HVOF aluminium-based marine coatings *Wear, 256*, 545–556.

Kagimato, Y. (1987). Experimental study on the relation between erosion wear rate and particle impact velocity measured by LDV. In *Proceedings of 7th International Conference on Erosion by Liquid and Solid Impact.* Cambridge, UK: Academic Press.

Kalish, H. S. (1987). *Corrosion of cemented carbides* (9th ed.; Vol. 13). ASM International.

Karimi, A., & Verdon, C. (1993). Hydroabrasive wear behaviour of high velocity oxyfuel thermally sprayed WC-M coatings. *Surface and Coatings Technology*, *62*(1-3), 493–498. doi:10.1016/0257-8972(93)90289-Z

Karimi, A., Verdon, C., Martin, J., & Schmid, R. (1995). Slurry erosion behaviour of thermally sprayed WC-M coatings. *Wear*, *186-187*, 480–486. doi:10.1016/0043-1648(95)07154-7

Karthikeyan, J., Berndt, C. C., Tikkanen, J., Reddy, S., & Herman, H. (1997). Plasma spray synthesis of nanomaterial powders and deposits. *Materials Science and Engineering A*, *238*(2), 275–286. doi:10.1016/S0921-5093(96)10568-2

Karthikeyan, J., Berndt, C. C., Tikkanen, J., Wang, J. Y., King, A. H., & Herman, H. (1997). Nanomaterial powders and deposits prepared by flame spray processing of liquid precursors. *Nanostructured Materials*, *8*(1), 61–74. doi:10.1016/S0965-9773(97)00066-4

Kassman, A., Jacobson, S., Erickson, L., Hedengvist, P., & Olson, M. (1991). A new test method for the intrinsic abrasion resistance of thin coatings. *Surface and Coatings Technology*, *50*(1), 75–89. doi:10.1016/0257-8972(91)90196-4

Kawasaki, A., & Watanabe, R. (1997). Concept and P/M fabrication of functionally gradient materials. *Ceramics International*, *23*(1), 73–83. doi:10.1016/0272-8842(95)00143-3

Kennedy, D. M., & Hashmi, M. S. (1998). Methods of wear testing for advanced surface coatings and bulk materials. *Journal of Materials Processing Technology*, *77*(1-3), 246–253. doi:10.1016/S0924-0136(97)00424-X

Kharlamov, Y. A. (1986). Detonation spraying of protective coatings. *Materials Science and Engineering*, *93*, 1–37. doi:10.1016/0025-5416(87)90409-5

Khor, K. A., & Gu, Y. W. (2000). Thermal properties of plasma-sprayed functionally graded thermal barrier coatings. *Thin Solid Films*, *372*(1-2), 104–113. doi:10.1016/S0040-6090(00)01024-5

Kieback, B., Neubrand, A., & Riedel, H. (2003). Processing techniques for functionally graded materials. *Materials Science and Engineering A*, *362*(1-2), 81–106. doi:10.1016/S0921-5093(03)00578-1

Killinger, A., Kuhn, M., & Gadow, R. (2006). High-velocity suspension flame spraying (HVSFS), a new approach for spraying nanoparticles with hypersonic speed. *Surface and Coatings Technology*, *201*(5), 1922–1929. doi:10.1016/j.surfcoat.2006.04.034

Kim, J. H., Kim, M. C., & Park, C. G. (2003). Evaluation of functionally graded thermal barrier coatings fabricated by detonation gun spray technique. *Surface and Coatings Technology*, *168*(2), 275–280. doi:10.1016/S0257-8972(03)00011-2

Kim, J. Y., & Lim, D. S. (1994). The effect of annealing temperature and environment on room temperature wear behaviour of plasma sprayed partially stabilised zirconia coating. *Journal of Korean Ceramic Society*, *31*(10), 1176–1180.

Kingswell, R., Rickerby, D. S., Bull, S. J., & Scot, K. T. (1991). Erosive wear of thermally sprayed tungsten coatings. *Thin Solid Films*, *198*(1-2), 139–148. doi:10.1016/0040-6090(91)90332-R

Kleis, I., & Kulu, P. (2008). *Solid Particle Erosion: Occurrence Predictio and Control*. Springer Verlag London.

Knuuttila, J., Ahmaniemi, S., & Mäntylä, T. (1999). Wet abrasion and slurry erosion resistance of thermally sprayed oxide coatings. *Wear*, *232*(2), 207–212. doi:10.1016/S0043-1648(99)00147-7

Knuuttila, J., Sorsa, P., Mäntylä, T., Knuuttila, J., & Sorsa, P. (1999). Sealing of thermal spray coatings by impregnation. *Journal of Thermal Spray Technology*, *8*(2), 249–257. doi:10.1007/s11666-999-0002-2

Kny, E., Bader, T., Hohenrainer, Ch., & Schmid, L. (1986). Korrosionsresistente, hochverschleißfeste Hartmetalle. *Werkstoffe und Korrosion*, *37*(5), 230–235. doi:10.1002/maco.19860370505

Koizumi, M., & Niino, M. (1995). Overview of FGM research in Japan. *MRS Bulletin*, *20*(1), 19–26.

Koon Aw, P., & Kuan Tan, A. (2008). Corrosion resistance of tungsten carbide based cermet coatings deposited by high velocity oxy-fuel spray process. *Thin Solid Films*, *516*(16), 5710–5715. doi:10.1016/j.tsf.2007.07.065

Kosel, T. H. (1992). Solid particle erosion, friction, lubrication and wear technology. In ASM handbook (Vol. 18). ASM International.

Kosel, T. H., & Anand, K. (1989). An opto-electronic erodent particle velocitimeter. In Proceedings of Corrosion and Particle Erosion at High Temperatures. The Minerals, Metals & Materials Society.

Kosel, T. H., Mao, Z. Y., & Prasad, S. V. (1984). Erosion debris particle observations and the micromachining mechanism of erosion. *ASLE Transactions*, *28*(2), 268–276. doi:10.1080/05698198508981621

Krishnamoorthy, P. R., Seetharamu, S., & Sampathkumaran, P. (1993). Influence of the mass flux and impact angle of the abrasive on the erosion resistance of materials used in pulverized fuel bends and other components in thermal power stations. *Wear*, *165*(2), 151–157. doi:10.1016/0043-1648(93)90330-O

Krömmer, W., Heinrich, P., & Kreye, H. (2009). High velocity oxy-fuel flame spraying: Past – present – future. In Proceedings 8th Kolloquium Hochgeschwindigkeitsflammspritzen (pp. 9–16). Unterschleißheim: Gemeinschaft Thermisches Spritzen.

Krumova, M., Klingshirn, C., Haupert, F., & Friedrich, K. (2001). Microhardness studies on functionally graded polymer composites. *Composites Science and Technology*, *61*(4), 557–563. doi:10.1016/S0266-3538(00)00228-1

Kulkarni, A., Vaidya, A., Goland, A., Sampath, S., & Herman, H. (2003). Processing effects on porosity-property correlations in plasma sprayed yttria-stabilized zirconia coatings. *Materials Science and Engineering A*, *359*(1-2), 100–111. doi:10.1016/S0921-5093(03)00342-3

Kulu, P., Hussainova, I., & Veinthal, R. (2005). Solid particle erosion of thermal sprayed coatings. *Wear*, *258*(1-4), 488–496. doi:10.1016/j.wear.2004.03.021

Kumar Goyal, D., Singh, H., Kumar, H., & Sahni, V. (2012). Slurry erosion behaviour of HVOF sprayed WC–10Co–4Cr and Al2O3+13TiO2 coatings on a turbine steel. *Wear*, *289*(0), 46–57. doi:10.1016/j.wear.2012.04.016

Kumar Roop, R., & Wang, M. (2002). Functionally graded bioactive coatings of hydroxyapatite/titanium oxide composite system. *Materials Letters*, *55*(3), 133–137. doi:10.1016/S0167-577X(01)00635-8

Kumar, A., Boy, J., Zatorski, R., & Stephenson, L. D. (2005). Thermal spray and weld repair alloys for the repair of cavitation damage in turbines and pumps: A technical note. *Journal of Thermal Spray Technology*, *14*(2), 177–182. doi:10.1361/10599630523737

Kuroda, S., & Clyne, T. W. (1991). The quenching stress in thermally sprayed coatings. *Thin Solid Films*, *200*(1), 49–66. doi:10.1016/0040-6090(91)90029-W

Kuroda, S., Watanabe, M., Kim, K., & Katanoda, H. (2011). Current status and future prospect of warm spray technology. *Journal of Thermal Spray Technology*, *20*(4), 653–673. doi:10.1007/s11666-011-9648-7

Lai, G. Y. (1978). Evaluation of sprayed chromium carbide coatings for gas cooled reactor application. *Thin Solid Films*, *53*(3), 343–351. doi:10.1016/0040-6090(78)90227-4

Lai, G. Y. (1979). Factors affecting the performances of sprayed chromium carbide coatings for gas cooled reactor heat exchangers. *Thin Solid Films*, *64*(2), 271–280. doi:10.1016/0040-6090(79)90520-0

Landa, J., Illaramendi, I., Kelling, N., Woydt, M., Skopp, A., & Hartelt, M. (2007). Potential of thermal sprayed Ti_nO_{2n-1}-coatings for substituting molybdenum-based ring coatings. *Industrial Lubrication and Tribology*, *59*(5), 217–229. doi:10.1108/00368790710776801

Landes, K. (2006). Diagonistics in plasma spraying techniques. *Surface and Coatings Technology*, *201*(5), 1948–1954. doi:10.1016/j.surfcoat.2006.04.036

Landolt, D. (2006). Electrochemical and materials aspects of tribo-corrosion systems. *Journal of Physics. D, Applied Physics*, *39*(15), 3121–3127. doi:10.1088/0022-3727/39/15/S01

Landolt, D., Mischler, S., Stemp, M., & Barril, S. (2004). Third body effects and material fluxes in tribocorrosion systems involving a sliding contact. *Wear*, *256*(5), 517–524. doi:10.1016/S0043-1648(03)00561-1

Landolt, D., Stemp, M., & Mischler, S. (2001). Electrochemical methods in tribo-corrosion: A critical appraisal. *Electrochimica Acta*, *46*(24-25), 3913–3929. doi:10.1016/S0013-4686(01)00679-X

Lathabai, S., Ottmuller, M., & Fernandez, I. (1998). Solid particle erosion behavior of thermal sprayed ceramic, metallic and polymer coatings. *Wear*, *221*(2), 93–108. doi:10.1016/S0043-1648(98)00267-1

Lawn, B., & Wilshaw, R. (1975). Review Indentation fracture: Principles and applications. *Journal of Materials Science*, *10*(6), 1049–1081. doi:10.1007/BF00823224

Lee, C. S., Ahn, S. H., DeJonghe, L. C., & Thomas, G. (2006). Effect of functionally graded material (FGM) layers on the residual stress of polytypoidally joined Si_3N_4–Al_2O_3. *Materials Science and Engineering A*, *434*(1-2), 160–165. doi:10.1016/j.msea.2006.06.139

Lee, M. K., Hong, S. M., Kim, G. H., Kim, K. H., & Kim, W. W. (2004). Structural properties in flame quenched Cu-9Al-4.5Ni-4.5Fe alloy. *Metals and Materials International*, *10*(4), 313–319. doi:10.1007/BF03185979

Lee, W. Y., Stinton, D. P., Berndt, C. C., Erdogan, F., Lee, Y. D., & Mutasim, Z. (1996). Concept of functionally graded materials for advanced thermal barrier coating applications. *Journal of the American Ceramic Society*, *79*(12), 3003–3012. doi:10.1111/j.1151-2916.1996.tb08070.x

Leibhard, M., & Levy, A. (1991). The effect of erodent particle characteristics on the erosion of steels. *Wear*, *151*(2), 381–390. doi:10.1016/0043-1648(91)90263-T

Leivo, E. M., Vippola, M. S., Sorsa, P. P. A., Vuoristo, P. M. J., & Mäntylä, T. A. (1997). Wear and corrosion properties of plasma sprayed Al_2O_3 and Cr_2O_3 coatings sealed by aluminum phosphates. *Journal of Thermal Spray Technology*, *6*(2), 205–210. doi:10.1007/s11666-997-0014-8

Leroy, C., Schiffmann, K. I., van Acker, K., & von Stebut, J. (2005). Ball cratering an efficient tool for 3 body microabrasion of coated samples. *Surface and Coatings Technology*, *200*(1-4), 153–156. doi:10.1016/j.surfcoat.2005.02.076

Levin, B. F., DuPont, J. N., & Marder, A. R. (2000). The effect of second phase volume fraction on the erosion resistance of metal-matrix composites. *Wear*, *238*(2), 160–167. doi:10.1016/S0043-1648(99)00363-4

Levy, A. (1986). The platelet mechanism of erosion of ductile metals. *Wear*, *108*(1), 1–21. doi:10.1016/0043-1648(86)90085-2

Levy, A. (1988). The erosion-corrosion behavior of protective coating. *Surface and Coatings Technology*, *36*(1-2), 387–406. doi:10.1016/0257-8972(88)90168-5

Levy, A. V. (1995). *Solid particle erosion and erosion-corrosion of materials*. ASM International.

Levy, A., & Wang, B. Q. (1988). Erosion of hard material coating systems. *Wear*, *121*(3), 325–346. doi:10.1016/0043-1648(88)90209-8

Li, C. C., & Lai, G. Y. (1979), Sliding wear studies of sprayed chromium carbide-nichrome coatings for gas-cooled reactor applications. In *Proc. Int. Conf. on Wear of Materials*. Dearborn, MI: ASME Newyork.

Li, Y., Huang, J., & Ding, C. (1998). Mechanical and tribological properties of plasma-sprayed Cr_3C_2-NiCr, WC-Co and Cr_2O_3 coatings. *Journal of Thermal Spray Technology*, *7*, 242-246.

Liao, H., Normand, B., & Coddet, C. (2000). Influence of coating microstructure on the abrasive wear resistance of WC/Co cermet coatings. *Surface and Coatings Technology*, *124*(2-3), 235–242. doi:10.1016/S0257-8972(99)00653-2

Li, C. C. (1980). Characterisation of thermally sprayed coatings for high temperature wear protection. *Thin Solid Films*, *73*(1), 59–77. doi:10.1016/0040-6090(80)90329-6

Li, C.-J., Yang, G.-J., & Ohmori, A. (2006). Relationship between particle erosion and lamellar microstructure for plasma-sprayed alumina coatings. *Wear*, *260*(2), 1166–1172. doi:10.1016/j.wear.2005.07.006

Li, J. F., & Watanabe, R. (1996). $BaFe_{12-2x}Co_xTi_xO_{19}$ (x=0, 0.6, 0.9) films synthesized by the spray-ICP technique. *Materials Transactions*, *37*, 1171–1181. doi:10.2320/matertrans1989.37.1171

Lima, M. M., Godoy, C., Avelar-Batista, J. C., & Modenesi, P. J. (2003). Toughness evaluation of HVOF WC-Co coatings using non-linear regression analysis. *Materials Science and Engineering A*, *357*(1-2), 337–345. doi:10.1016/S0921-5093(03)00204-1

Lima, R. S., & Bergmann, C. P. (1996). Phase transformations on flame sprayed alumina. In BerndtC.C. (Ed.), *Proceedings of the National Thermal Spray Conference*. Cincinnati, OH: ASM International.

Lima, R. S., Kucuk, A., & Berndt, C. C. (2001). Evaluation of microhardness and elastic modulus of thermally sprayed nanostructured zirconia coatings. *Surface and Coatings Technology*, *135*(2-3), 166–172. doi:10.1016/S0257-8972(00)00997-X

Lima, R. S., Kucuk, A., & Berndt, C. C. (2001). Integrity of nanostructured partially stabilized zirconia after plasma spray processing. *Materials Science and Engineering A*, *313*(1), 75–82. doi:10.1016/S0921-5093(01)01146-7

Lima, R. S., Kucuk, A., & Berndt, C. C. (2002). Bimodal distribution of mechanical properties on plasma sprayed nanostructured partially stabilized zirconia. *Materials Science and Engineering A*, *327*(2), 224–232. doi:10.1016/S0921-5093(01)01530-1

Lima, R. S., & Marple, B. (2007). Thermal spray coatings engineered from nanostructured ceramic agglomerated powders for structural, thermal barrier and biomedical applications. *Journal of Thermal Spray Technology*, *16*(1), 40–63. doi:10.1007/s11666-006-9010-7

Lima, R. S., & Marple, B. (2008). Nanostructured YSZ thermal barrier coatings engineered to counteract sintering effects. *Materials Science and Engineering A*, *485*(1-2), 182–193. doi:10.1016/j.msea.2007.07.082

Lima, R. S., & Marple, B. R. (2006). From APS to HVOF spraying of conventional and nanostructured titania feedstock powders: A study on the enhancement of the mechanical properties. *Surface and Coatings Technology*, *200*(11), 3428–3437. doi:10.1016/j.surfcoat.2004.10.137

Lima, R. S., Moreau, C., & Marple, B. R. (2007). HVOF-sprayed coatings engineered from mixtures of nanostructured and submicron Al_2O_3-TiO_2 powders: An enhanced wear performance. *Journal of Thermal Spray Technology*, *16*(5-6), 866–872. doi:10.1007/s11666-007-9092-x

Lim, S. C., & Ashby, M. F. (1987). Wear-mechanism maps. *Acta Materialia*, *30*(1), 1–24. doi:10.1016/0001-6160(87)90209-4

Liu, G. H., Robbevalloire, F., Gras, R., & Blouet, J. (1993). Improvement in tribological properties of chromium oxide coating at high temperature by solid lubricants. *Wear*, *160*(1), 181–189. doi:10.1016/0043-1648(93)90419-M

Liu, H., Tao, J., Xu, J., Chen, Z., & Gao, Q. (2009). Corrosion and tribological behaviors of chromium oxide coatings prepared by the glow-discharge plasma technique. *Surface and Coatings Technology*, *204*(1-2), 28–36. doi:10.1016/j.surfcoat.2009.06.020

Liu, Z., Cabrero, J., Niang, S., & Al-Taha, Z. Y. (2007). Improving corrosion and wear performance of HVOF-sprayed Inconel 625 and WC-Inconel 625 coatings by high power diode laser treatments. *Surface and Coatings Technology*, *201*(16-17), 7149–7158. doi:10.1016/j.surfcoat.2007.01.032

Li, W. S., Wang, Z. P., Lu, Y., Jin, Y. H., Yuan, L. H., & Wang, F. (2006). Mechanical and tribological properties of a novel aluminum bronze material for drawing dies. *Wear*, *261*(2), 155–163. doi:10.1016/j.wear.2005.09.032

Li, W.-Y., Zhang, C., Guo, X. P., Zhang, G., Liao, H. L., Li, C.-J., & Coddet, C. (2008). Effect of standoff distance on coating deposition characteristics in cold spraying. *Materials & Design*, *29*(2), 297–304. doi:10.1016/j.matdes.2007.02.005

Lofti, B., Shipway, P. H., McCartney, D. G., & Eldris, H. (2003). Abrasive wear behaviour of NiCr-TiB2 coatings deposited by HVOF spraying of SHS-derived cermet powders. *Wear*, *254*(3-4), 340–349. doi:10.1016/S0043-1648(03)00014-0

Longo, F. N., & Davis, J. R. (Eds.). (2005). *Handbook of thermal spray technology*. Materials Park, OH: ASM International.

Lopez, E., Beltzung, F., & Zambelli, G. (1989). Measurement of cohesion and adhesion strengths in alumina coatings produced by plasma spraying. *Journal of Materials Science Letters*, *8*(3), 346–348. doi:10.1007/BF00725519

Lu, F., Li, H., Ji, Q., Zeng, R., Wang, S., Chi, J., & Sola, A. et al. (2007). Characterization of glass-alumina functionally graded coatings obtained by plasma spraying. *Journal of the European Ceramic Society*, *27*(4), 1935–1943. doi:10.1016/j.jeurceramsoc.2006.05.105

Luo, H., Goberman, D., Shaw, L., & Gell, M. (2003). Indentation fracture behavior of plasma-sprayed nanostructured Al_2O_3-13wt.%TiO_2 coatings. *Materials Science and Engineering A*, *346*(1-2), 237–245. doi:10.1016/S0921-5093(02)00523-3

Machio, C. N., Akdogan, G., Witcomb, M. J., & Luyckx, S. (2005). Performance of WC-VC-Co thermal spray coatings in abrasion and slurry erosion tests. *Wear*, *258*(1-4), 434–442. doi:10.1016/j.wear.2004.09.033

Manish, R. (2000). Wear testing: An assessment of test methodology. *Transactions of the Indian Institute of Metals*, *62*, 623–638.

Mann, B. S. (2000). High-energy particle impact wear resistance of hard coatings and their application in hydroturbines. *Wear*, *237*(1), 140–146. doi:10.1016/S0043-1648(99)00310-5

Mann, B. S., & Arya, V. (2001). Abrasive and erosive wear characteristics of plasma nitriding and HVOF coatings: Their application in hydro turbines. *Wear*, *249*(5-6), 354–360. doi:10.1016/S0043-1648(01)00537-3

Mann, B. S., & Arya, V. (2002). An experimental study to corelate water jet impingement erosion resistance and properties of metallic materials and coatings. *Wear*, *253*(5-6), 650–661. doi:10.1016/S0043-1648(02)00118-7

Mari, D., Bolognini, S., Feusier, G., Cutard, T., Verdon, C., Viatte, T., & Benoit, W. (2003). TiMoCN based cermets part I: Morphology and phase composition. *International Journal of Refractory Metals & Hard Materials*, *21*(1-2), 37–46. doi:10.1016/S0263-4368(03)00010-6

Marple, B. R., Voyer, J., & Béchard, P. (2001). Sol infiltration and heat treatment of alumina-chromia plasma-sprayed coatings. *Journal of the European Ceramic Society*, *21*(7), 861–868. doi:10.1016/S0955-2219(00)00276-4

Martın, A., Rodr'ıguez, J., Fernández, J. E., & Vijande, R. (2001). Sliding wear behaviour of plasma sprayed WC-NiCrBSi coatings at different temperatures. *Wear*, *251*(1-12), 1017–1022. doi:10.1016/S0043-1648(01)00703-7

Matejicek, J., & Sampath, S. (2003). In situ measurement of residual stresses and elastic moduli in thermal sprayed coatings: Part 1: Apparatus and analysis. *Acta Materialia*, *51*(3), 863–872. doi:10.1016/S1359-6454(02)00478-0

Mateos, J., Cuetos, J. M., Vijande, R., & Fernandej, E. (2001). Tribological properties of plasma sprayed and laser remelted 75/25 Cr_3C_2/NiCr coatings. *Tribology International*, *34*(5), 345–351. doi:10.1016/S0301-679X(01)00023-8

Mathesius, S., & Krömmer, W. (2009). *Praxis des thermischen Spritzens: Anleitung für das Fachpersonal*. Düsseldorf: DVS-Media.

Mathews, S. J., James, B. J., & Hyland, M. M. (2007). Microstructural influence on erosion behavior of thermal spray coatings. *Materials Characterization*, *58*(1), 59–64. doi:10.1016/j.matchar.2006.03.014

Matthews, S., James, B., & Hyland, M. (2009). The role of microstructure in the mechanism of high velocity erosion of Cr3C2–NiCrthermal spray coatings: Part 2 - Heat treated coatings. *Surface and Coatings Technology*, *203*(8), 1094–1100. doi:10.1016/j.surfcoat.2008.10.013

Mattox, D. M. (1964). Film deposition using accelerated ions. *Electrochemical Technology, 2*(9-10), 295–298.

Ma, X. Q., Ge, S. H., Zhang, Y. D., Roth, J., & Xiao, T. D. (2004). Solution plasma spray synthesis of niznfe2o4 magnetic nanocomposite thick films. In *Proceedings of International Thermal Spray Conference 2004*. Materials Park, OH: ASM International.

Ma, X. Q., Roth, J., Gandy, D. W., & Frederick, G. J. (2006). A new high-velocity oxygen fuel process for making finely structured and highly bonded inconel alloy layers from liquid feedstock. *Journal of Thermal Spray Technology, 15*(4), 670–675. doi:10.1361/105996306X147090

McCandlish, L. E., Kear, B. H., & Kim, B. K. (1992). Processing and properties of nanostructured WC-Co. *Nanostructured Materials, 1*(2), 119–123. doi:10.1016/0965-9773(92)90063-4

McPherson, R. (1973). Formation of metastable phases in flame- and plasma-prepared alumina. *Journal of Materials Science, 8*(6), 851–858. doi:10.1007/BF02397914

McPherson, R. (1980). On the formation of thermally sprayed alumina coatings. *Journal of Materials Science, 15*(12), 3141–3149. doi:10.1007/BF00550387

Melendez, N. M., Narulkar, V. V., Fisher, G. A., & McDonald, V. (2013). Effect of reinforcing particles on the wear rate of low-pressure cold-sprayed WC-based MMC coatings. *Wear, 306*(1-2), 185–195. doi:10.1016/j.wear.2013.08.006

Mellor, B. G. (Ed.). (2006). Surface coatings for protection against wear. Cambridge, UK: Woodhead Publishing Limited. doi:10.1201/9781439823422

Menčík, J., & Swain, M. V. (1995). Errors associated with depth-sensing microindentation Tests. *Journal of Materials Research, 10*(6), 1491–1501. doi:10.1557/JMR.1995.1491

Miguel, J. M., Guilemany, J. M., Mellor, B. G., & Xu, Y. M. (2003). Acoustic emission study on WC-Co thermal sprayed coatings. *Materials Science and Engineering A, 352*(1-2), 55–63. doi:10.1016/S0921-5093(02)00546-4

Mischler, S. (2008). Triboelectrochemical techniques and interpretation methods in tribocorrosion: A comparative evaluation. *Tribology International, 41*(7), 573–583. doi:10.1016/j.triboint.2007.11.003

Mischler, S., Debaud, S., & Landolt, D. (1998). Wear-accelerated corrosion of passive metals in tribocorrosion systems. *Journal of the Electrochemical Society, 145*(3), 750–758. doi:10.1149/1.1838341

Mischler, S., & Ponthiaux, P. (2001). A round robin on combined electrochemical and friction tests on alumina/stainless steel contacts in sulphuric acid. *Wear, 248*(1-2), 211–225. doi:10.1016/S0043-1648(00)00559-7

Mischler, S., Spiegel, A., & Landolt, D. (1999). The role of passive oxide films on the degradation of steel in tribocorrosion systems. *Wear, 225–229*, 1078–1087. doi:10.1016/S0043-1648(99)00056-3

Mischler, S., Spiegel, A., Stemp, M., & Landolt, D. (2001). Influence of passivity on the tribocorrosion of carbon steel in aqueous solutions. *Wear*, *251*(1-12), 1295–1307. doi:10.1016/S0043-1648(01)00754-2

Mishra, S. C., Sahu, A., Das, R., Satapathy, A., Sen, S., Ananthapadmanabhan, P. V., & Sreekumar, K. P. (2008). Microstructure, adhesion, and erosion wear of plasma sprayed alumina-titania composite coatings. *Journal of Reinforced Plastics and Composites*. doi:10.1177/0731684407087758

Misra, A., & Finnie, I. (1980). A classification of three-body abrasive wear and design of a new tester. *Wear*, *60*(1), 111–118. doi:10.1016/0043-1648(80)90252-5

Miyamoto, Y., Kaysser, W., Rabin, B., Kawasaki, A., & Ford, R. G. (1999). Functionally graded materials: Design, processing and applications. Dordrecht, The Netherlands: Kluwer Academic Publishers.

Mohanty, M., Smith, R. W., De Bonte, M., Celis, J. P., & Lugscheider, E. (1996). Sliding wear behavior of thermally sprayed 75/25 Cr_3C_2/NiCr wear resistant coatings. *Wear*, *198*(1-2), 251–266. doi:10.1016/0043-1648(96)06983-9

Monticelli, C., Balbo, A., & Zucchi, F. (2010). Corrosion and tribocorrosion behaviour of cermet and cermet/nanoscale multilayer CrN/NbN coatings. *Surface and Coatings Technology*, *204*(9-10), 1452–1460. doi:10.1016/j.surfcoat.2009.09.046

Monticelli, C., Balbo, A., & Zucchi, F. (2011). Corrosion and tribocorrosion behaviour of thermally sprayed ceramic coatings on steel. *Surface and Coatings Technology*, *205*(12), 3683–3691. doi:10.1016/j.surfcoat.2011.01.023

Monticelli, C., Frignani, A., & Zucchi, F. (2004). Investigation of corrosion process of carbon steel coated by HVOFWC-Co cermet in neutral solution. *Corrosion Science*, *46*(5), 1225–1237. doi:10.1016/j.corsci.2003.09.013

More, N., Diomidis, N., Paul, S. N., Roy, M., & Mischler, S. (2011). TriboCorrosion behaviour of β titanium alloys in physiological solution containing joint lubricants. *Materials Science and Engineering C*, *31*(2), 400–408. doi:10.1016/j.msec.2010.10.021

Moro, A., Kuroda, Y., & Kusaka, K. (2002). Development status of the reusable high-performance engines with functionally graded materials. *Acta Astronautica*, *50*(7), 427–432. doi:10.1016/S0094-5765(01)00174-6

Mortensen, A., & Suresh, S. (1995). Functionally graded metals and metal-ceramic composites: Part 1. *International Materials Reviews*, *40*(6), 239–265. doi:10.1179/imr.1995.40.6.239

Moya, J. S., Sanchezherencia, A. J., Requena, J., & Moreno, R. (1992). Functionally gradient ceramics by sequential slip casting. *Materials Letters*, *14*(5-6), 333–335. doi:10.1016/0167-577X(92)90048-O

Mrdak, M., Vencl, A., & Ćosić, M. (2009). Microstructure and mechanical properties of the Mo-NiCrBSi coating deposited by atmospheric plasma spraying. *FME Transactions*, *37*, 27–32.

Müller, J.-H., & Kreye, H. (2001). Mikrostruktur und Eigenschaften von thermisch gespritzten Aluminiumoxidschichten. *Schweissen und Schneiden*, *53*(6), 336–345.

Murthy, J. K. N., Rao, D. S., & Venkataraman, B. (2001). Effect of grinding on the erosion behaviour of a WC-Co-Cr coating deposited by HVOF and detonation gun spray processes. *Wear*, *249*(7), 592–600. doi:10.1016/S0043-1648(01)00682-2

Murthy, J. K. N., & Venkataraman, B. (2006). Abrasive wear behaviour of WC–CoCr and Cr3C2–20(NiCr) deposited by HVOF and detonation spray processes. *Surface and Coatings Technology*, *200*(8), 2642–2652. doi:10.1016/j.surfcoat.2004.10.136

Nadeau, A., Pouliot, L., Nadeau, F., Blain, J., Berube, S. A., Moreau, C., & Lamontagne, M. (2006). A new approach to online thickness measurement of thermal spray coatings. *Journal of Thermal Spray Technology*, *15*(4), 744–749. doi:10.1361/105996306X147054

Naerheim, Y., Coddet, C., & Droit, P. (1995). Effect of thermal spray process selection on tribological performances WC–Co and Al_2O_3–TiO_2 coatings. *Surface Engineering*, *11*(1), 66–70. doi:10.1179/sur.1995.11.1.66

Nakamura, T., Qian, G., & Berndt, C. C. (2000). Effects of pores on mechanical properties of plasma-sprayed ceramic coating. *Journal of the American Ceramic Society*, *83*(3), 578–584. doi:10.1111/j.1151-2916.2000.tb01236.x

Nath, S., Manna, I., & Dutta Majumdar, J. (2013). Compositionally graded thermal barrier coating by hybrid thermal spraying route and its non-isothermal oxidation behavior. *Journal of Thermal Spray Technology*, *22*(6), 901–917. doi:10.1007/s11666-013-9937-4

Neilson, J. H., & Gilchrist, A. (1968). Erosion by a stream of solid particles. *Wear*, *11*(2), 111–122. doi:10.1016/0043-1648(68)90591-7

Nerz, J., Kushner, B., Kaufold, R., & Rotolico, A. (1991). Thermal spray coatings: Properties, processes and applications. In Reduction of solid particle erosion by using HVOF and HEP coating deposition methods. Pittsburgh, PA: ASM International.

Niemi, K. (2009). *Abrasion wear characteristics of thermally sprayed alumina based coatings*. Doctoral Thesis, Publication 820. Tampere: Tampere University of Technology.

Niemi, K., Vuoristo, P., & Mäntylä, T. (1991). Chromium oxide coatings deposited by plasma spraying and detonation gun spraying. In S. Blum-Sandmeier, H. Eschnauer, P. Huber, & A.R. Nicoll (Eds.), Proceedings 2nd Plasma-Technik-Symposium (pp. 311-322). Wohlen, Switzerland: Plasma-Technik AG.

Niemi, K., Vuoristo, P., & Mäntylä, T. (1994). Properties of alumina-based coatings deposited by plasma spray and detonation gun spray processes. *Journal of Thermal Spray Technology, 3*(2), 199–203.

Niemi, K., Vuoristo, P., Mäntylä, T., Lugscheider, E., Knuuttila, J., & Jungklaus, H. (1995). Wear characteristics of oxide coatings deposited by plasma spraying, high power plasma spraying and detonation gun spraying. In C.C. Berndt & S. Sampath (Eds.), *Advances of Thermal Spray Science and Technology,Proceedings of the 8th National Thermal Spray Conference* (pp. 645-650). Materials Park, OH: ASM International.

Niihara, K. (1983). A fracture mechanics analysis of indentation induced Palmqvist cracks in ceramics. *Journal of Materials Science Letters*, *2*(5), 221–223. doi:10.1007/BF00725625

Ninham, A. J., & Hutchings, I. M. (1983). A computer model for particle velocity calculation in erosion testing. In *Proceedings of 6th International Conference On Erosion by Liquid and Solid Impact.* Cavendish Laboratory.

Ninham, A. J., & Levy, A. V. (1988). The erosion of carbide metal matrix composites. *Wear*, *121*(3), 347–361. doi:10.1016/0043-1648(88)90210-4

Nohava, J., Bonferroni, B., Bolelli, G., & Lusvarghi, L. (2010). Interesting aspects of indentation and scratch methods for characterization of thermally sprayed coatings. *Surface and Coatings Technology*, *205*(4), 1127–1131. doi:10.1016/j.surfcoat.2010.08.086

Nordin, M., Larsson, M., & Hogmark, S. (1998). Mechanical and tribological properties of multilayered PVD TiN/CrN, TiN/MoN, TiN/NbN, TiN/TaN coatings on cemented carbide. *Surface and Coatings Technology*, *106*(2-3), 234–241. doi:10.1016/S0257-8972(98)00544-1

Normand, B., Fervel., V., Coddet, C., Nikitine, V. (2000). Tribological properties of plasma sprayed alumina-titania coatings: Role and control of the microstructure. *Surface and Coatings Technology, 123*(2-3), 278-287.

Normand, B., Fervel, V., Coddet, C., & Nikitine, V. (2000). Tribological properties of plasma sprayed alumina–titania coatings: Role and control of the microstructure. *Surface and Coatings Technology*, *123*(2-3), 278–287. doi:10.1016/S0257-8972(99)00532-0

Nusair Khan, A., & Lu, J. (2009). Manipulation of air plasma spraying parameters for the production of ceramic coatings. *Journal of Materials Processing Technology*, *209*(5), 2508–2514. doi:10.1016/j.jmatprotec.2008.05.045

Oliver, W. C., & Pharr, G. M. (1992). An improved technique for determining hardness and elastic modulus using load and depth sensing indentation experiments. *Journal of Materials Research*, *7*(6), 1564–1583. doi:10.1557/JMR.1992.1564

Ouyang, J. H., & Sasaki, S. (2002). Microstructure and tribological characteristics of ZrO_2–Y_2O_3 ceramic coatings deposited by laser-assisted plasma hybrid spraying. *Tribology International*, *35*(4), 255–264. doi:10.1016/S0301-679X(02)00005-1

Ouyang, J. H., Sasaki, S., & Umeda, K. (2001). Low-pressure plasma-sprayed ZrO_2–CaF_2 composite coating for high temperature tribological applications. *Surface and Coatings Technology*, *137*(1), 21–30. doi:10.1016/S0257-8972(00)00918-X

Ouyang, J. H., Sasaki, S., & Umeda, K. (2002). The friction and wear characteristics of low-pressure plasma-sprayed ZrO_2-$BaCrO_4$ composite coating at elevated temperatures. *Surface and Coatings Technology*, *154*(2-3), 131–139. doi:10.1016/S0257-8972(02)00024-5

Padhy, M. K., & Saini, R. P. (2009). Effect of size and concentration of silt particles on erosion of Pelton turbine buckets. *Energy*, *34*(10), 1477–1483. doi:10.1016/j.energy.2009.06.015

Padture, N. P., Gell, M., & Jordan, E. H. (2002). Thermal barrier coatings for gas-turbine engine applications. *Science*, *296*(5566), 280–284. doi:10.1126/science.1068609 PMID:11951028

Padture, N. P., Schlichting, K. W., Bhatia, T., Ozturk, A., Cetegen, B., Jordan, E. H., & Osendi, M. I. et al. (2001). Towards durable thermal barrier coatings with novel microstructures deposited by solution precursor plasma spray. *Acta Materialia*, *49*(12), 2251–2257. doi:10.1016/S1359-6454(01)00130-6

Papageorgiou, N., & Mischler, S. (2012). Electrochemical Simulation of the Current and Potential Response in Sliding Tribocorrosion. *Tribology Letters*, *48*(3), 271–283. doi:10.1007/s11249-012-0022-9

Papini, M., Ciampini, D., Krajac, T., & Spelt, J. K. (2003). Computer modelling of interference effects in erosion testing: Effect of plume shape. *Wear*, *255*(1-6), 85–97. doi:10.1016/S0043-1648(03)00125-X

Papyrin, A. (2001). Cold spray technology. *Advanced Materials and Processes*, *159*(9), 49–51.

Park, J. J. (1999). Creep strength of a tungsten–rhenium–hafnium carbide alloy from 2200 to 2400 K. *Materials Science and Engineering A*, *265*(1-2), 174–178. doi:10.1016/S0921-5093(98)01134-4

Pauschitz, A., Roy, M., & Franek, F. (2003). Comparism of high temperature wear behaviour of chromia forming Fe-based and Ni-based alloy. *Tribol Schmierungstech*, *50*, 40–49.

Pauschitz, A., Roy, M., & Franek, F. (2003). On the chemical composition of the layers formed during sliding of metallic alloys at high temperature. *Tribologia*, *188*, 127–143.

Pauschitz, A., Roy, M., & Franek, F. (2004). Identification of the mechanisms of wear during sliding of metallic materials at elevated temperature using an optical interferometer. In D. Dowson (Ed.), *Transient phenomena in tribological processes* (pp. 721–730). Elsevier.

Pauschitz, A., Roy, M., & Franek, F. (2008). Mechanisms of sliding wear of metals and alloys at elevated temperatures. *Tribology International*, *41*(7), 584–602. doi:10.1016/j.triboint.2007.10.003

Pawlowski, L. (1995). *The science and engineering of thermal spray coatings*. London: Wiley.

Pawlowski, L. (1996). Technology of thermally sprayed Anilox rolls: State of the art, problems, and perspectives. *Journal of Thermal Spray Technology*, *5*(3), 317–334. doi:10.1007/BF02645884

Pejryd, L., Wigren, J., Greving, D. J., Shadley, J. R., & Rybicki, E. F. (1995). Residual stresses as a factor in the selection of tungsten carbide coatings for a jet engine application. *Journal of Thermal Spray Technology*, *4*(3), 268–274. doi:10.1007/BF02646970

Pennefather, R.C. (1988). Recent observations of the erosion of hard materials. *Materials Science and Engineering, A*(105/106).

Perry, J. M., Neville, A., Wilson, V. A., & Hodgkiess, T. (2001). Assessment of the corrosion rates and mechanisms of a WC–Co–Cr HVOF coating in static and liquid–solid impingement saline environments. *Surface and Coatings Technology*, *137*(1), 43–51. doi:10.1016/S0257-8972(00)01062-8

Pina, J., Diasb, A., & Lebrun, J. (2003). Study by x-ray diffraction and mechanical analysis of the residual stress generation during thermal spraying. *Materials Science and Engineering A, 347*(1-2), 21–31. doi:10.1016/S0921-5093(02)00580-4

Ponthiaux, P., Wenger, F., Drees, D., & Celis, J.-P. (2004). Electrochemical techniques for studying tribocorrosion processes. *Wear, 256*(5), 459–468. doi:10.1016/S0043-1648(03)00556-8

Pourbaix, M. (1974). *Atlas of electrochemical equilibria in aqueous solutions*. Houston, TX: NACE-Cebelcor.

Prasad, B., Jha, A., Modi, O., & Yegneswaran, A. H. (2004). Effect of sand concentration in the medium and travel distance and speed on the slurry wear response of a zinc-based alloy alumina particle composite. *Tribology Letters, 17*(2), 301–309. doi:10.1023/B:TRIL.0000032468.27468.2c

Praxair. (n.d.). *Powders*. Retrieved from http://www.praxairsurfacetechnologies.com/na/us/pst/pst.nsf/0/085C6CF7641FE446852576A50058300A?OpenDocument

Prchlik, L., Sampath, S., Gutleber, J., Bancke, G., & Ruff, A. W. (2001). Friction and wear properties of WC-Co and Mo-Mo_2C based functionally graded materials. *Wear, 249*(12), 1103–1115. doi:10.1016/S0043-1648(01)00839-0

Qadir, T., & Shewmon, P. (1981). Solid particle erosion mechanism in copper and two copper alloys. *Metallurgical Transactions, 12A*(7), 1163–1176. doi:10.1007/BF02642330

Qiao, Y., Fischer, T. E., & Dent, A. (2003). The effects of fuel chemistry and feedstock powder structure on the mechanical and tribological properties of HVOF thermal-sprayed WC-Co coatings with very fines structures. *Journal of Surface Coatings and Technology, 172*(1), 24–41. doi:10.1016/S0257-8972(03)00242-1

Qiao, Y., Liu, Y. R., & Fischer, T. E. (2001). Sliding and abrasive wear resistance of thermal-sprayed WC/Co coatings. *Journal of Thermal-Sprayed Materials, 10*(1), 118–125. doi:10.1361/105996301770349583

Qi, X., Eigen, N., Aust, E., Gartner, F., Klassen, T., & Bormann, R. (2006). Two-body abrasive wear of nano- and microcrystalline TiC–Ni based thermal spray coatings. *Surface and Coatings Technology, 200*(16-17), 5037–5047. doi:10.1016/j.surfcoat.2005.05.007

Quinn, T. F. J., Sullivan, J. L., & Rowson, D. M. (1984). Origins and development of oxidational wear at low ambient temperatures. *Wear, 94*(2), 175–191. doi:10.1016/0043-1648(84)90053-X

Rabiei, A., Mumm, D. R., Hutchinson, J. W., Schweinfest, R., Ruhle, M., & Evans, A. G. (1999). Microstructure, deformation and cracking characteristics of thermal spray ferrous coatings. *Materials Science and Engineering A, 269*(1-2), 152–165. doi:10.1016/S0921-5093(99)00132-X

Radu, I., & Li, D. Y. (2007). The wear performance of yttrium-modified Stellite 712 at elevated temperatures. *Tribology International, 40*(2), 259–265. doi:10.1016/j.triboint.2005.09.027

Rainforth, W. M. (2004). The wear behaviour of oxide ceramics - A review. *Journal of Materials Science*, *32*(22), 6705–6721. doi:10.1023/B:JMSC.0000045601.49480.79

Rajasekaran, B., Mauer, G., & Vassen, R. (2012). Enhanced characteristics of HVOF sprayed MCrAlY bond coats for TBC applications. *Journal of Thermal Spray Technology*, *20*(6), 1209–1216. doi:10.1007/s11666-011-9668-3

Ramnathan, V., & Jayaraman, N. (1989). Characterisation and wear performance of plasma sprayed WC-Co coatings. *Materials Science and Technology*, *5*(4), 382–388. doi:10.1179/mst.1989.5.4.382

Rayon, E., Bonache, V., Salvador, M. D., Roa, J. J., & Sanchez, E. (2011). Hardness and Young's modulus distributions in atmospheric plasma sprayed WC-Co coatings using nanoindentation. *Surface and Coatings Technology*, *205*(17-18), 4192–4197. doi:10.1016/j.surfcoat.2011.03.012

Richard, C. S., Lu, J., Beranger, G., & Decomps, F. (1995). Study of Cr_2O_3 coatings: Part I: Microstructure and modulus. *Journal of Thermal Spraying Technology (Elmsford, N.Y.)*, *4*(4), 342–346.

Rico, A., Poza, P., & Rodríguez, J. (2013). High temperature tribological behavior of nanostructured and conventional plasma sprayed alumina-titania coatings. *Vacuum*, *88*, 149–154. doi:10.1016/j.vacuum.2012.01.008

Rico, A., Rodriguez, J., Otero, E., Zeng, P., & Rainforth, W. M. (2009). Wear behaviour of nanostructured alumina-titania coatings deposited by atmospheric plasma spray. *Wear*, *267*(5-8), 1191–1197. doi:10.1016/j.wear.2009.01.022

Riedel, R. (Ed.). (2000). *Handbook of ceramic hard materials 1*. Weinheim: WILEY-VCH Verlag GmbH. doi:10.1002/9783527618217

Rogne, T., Bjordal, M., Solem, T., & Bardal, E. (1996). *The importance of corrosion on the erosion-corrosion performance of thermal spray ceramic-metallic coatings*. ASM International.

Roy, M. (2014). Solid particle erosion behaviour of WC coating obtained by electro spark technique and detonation spraying. *Communicated to Tribology Transactions.*

Roy, M., Pauschitz, A., & Franek, F. (2002), Elevated temperature wear of 253 MA alloy. In *Proceedings of the Third Aimeta International Tribology Conference*. Vietri sul Mare, Italy: Academic Press.

Roy, M. (2006). Elevated temperature erosive wear of metallic material. *Journal of Physics. D, Applied Physics*, *39*(6), R101–R124. doi:10.1088/0022-3727/39/6/R01

Roy, M. (2006). Mode of deformation during tribological degradation. In *Proceedings of the International Conference on Advances in Materials and Materials Processing* (pp. 202-209). Kharagpur, India: Academic Press.

Roy, M., Narkhede, B. E., & Paul, S. N. (Eds.). (1999). *Proceedings of 2nd international conference on industrial tribology*. Hyderabad, India: Academic Press.

Roy, M., Pauschitz, A., & Franek, F. (2004). The influence of temperature on the wear of Cr_3C_2-25(Ni20Cr) coating – comparism between nanocrystalline grains and conventional grains. *Wear*, *257*(7-8), 799–811. doi:10.1016/j.wear.2004.05.001

Roy, M., Pauschitz, A., Polak, R., & Franek, F. (2008), High temperature wear of thermal sprayed Cr3C2-25(Ni-20Cr) coating against Fe based and Ni based mating surfaces. In *Proc. 16th Intl. Colloquium Tribology*. Stuttgart, Germany: Academic Press.

Roy, M., Pauschitz, A., Wernisch, J., & Franek, F. (2004). Influence of mating surface on elevated temperature wear of 253 MA alloy. *Materials & Corrosion*, *55*(4), 259–273. doi:10.1002/maco.200303715

Roy, M., Rao, C. V. S., Rao, S., & Sundararajan, G. (1999). Abrasive wear behaviour of detonation sprayed WC-Co coatings on mild steel. *Surface Engineering*, *15*(2), 129–136. doi:10.1179/026708499101516470

Roy, M., Subramaniyam, M., & Sundararajan, G. (1992). Room temperature erosion behavior of a precipitation hardened stainless steel. *Tribology International*, *25*(4), 271–280. doi:10.1016/0301-679X(92)90064-T

Roy, M., Tirupataiah, Y., & Sundararajan, G. (1993). Effect of particle shape on the erosion of Cu and its alloys. *Materials Science and Engineering A*, *165*(1), 51–63. doi:10.1016/0921-5093(93)90626-P

Roy, M., Tirupataiah, Y., & Sundararajan, G. (1995). Influence of solid solution and dispersion strengthening mechanism on room temperature erosion behavior of nickel. *Materials Science and Technology*, *11*(8), 791–797. doi:10.1179/mst.1995.11.8.791

Roy, M., Vishwanathan, B., & Sundararajan, G. (1994). The solid particle erosion of polymer matrix composites. *Wear*, *171*(1-2), 149–161. doi:10.1016/0043-1648(94)90358-1

Ruff, A. W., & Ives, L. K. (1975). Measurement of solid particle velocity in erosive wear. *WEA*, *35*(1), 195–199. doi:10.1016/0043-1648(75)90154-4

Ruff, A. W., & Wiederhorn, S. M. (1979). Erosion by solid particle impact. In C. M. Preece (Ed.), *Treatise on material science and technology* (Vol. 16, pp. 69–123). Academic Press.

Russo, L., & Dorfmann, M. (1995). Structure evaluation of HVOF sprayed NiCr-Cr_3C_2 coatings. In A. Ohmori (Ed.), *Thermal spraying: Current status and future trends* (pp. 681–686). Osaka, Japan: High Temperature Society of Japan.

Rybicki, E. F., Shadley, J. R., Xiong, Y., & Greving, D. J. (1995). A cantilever beam method of evaluating Young's modulus and Poisson's ratio of thermal spray coatings. *Journal of Thermal Spray Technology*, *4*(4), 377–384. doi:10.1007/BF02648639

Saeed, S. S., & Gross, K. A. (2009). Nanoindentation reveals mechanical properties within thermally sprayed hydroxyapatite coatings. *Surface and Coatings Technology*, *203*(12), 1660–1664. doi:10.1016/j.surfcoat.2008.12.025

Sahraoui, T., Fenineche, N. E., Montavon, G., & Coddet, C. (2003). Structure and wear behaviour of HVOF sprayed Cr_3C_2–NiCr and WC–Co coatings. *Materials & Design*, *24*(5), 309–313. doi:10.1016/S0261-3069(03)00059-1

Sampath, S. (Ed.). (1988). *Proceedings of the national thermal spray conference*. Cincinnati, OH: ASM International.

Sánchez, E., Moreno, A., Vicent, M., Salvador, M. D., Bonache, V., Klyatskina, E., & Moreno, R. et al. (2010). Preparation and spray drying of Al_2O_3-TiO_2 nanoparticle suspensions to obtain nanostructured coatings by APS. *Surface and Coatings Technology*, *205*(4), 987–992. doi:10.1016/j.surfcoat.2010.06.002

Sanikommu, N., Sivakumar, G., Joshi, A. S., Aruna, N. S., Rao, D., & Sundararajan, G. (2013). A computer-based approach for developing functionally graded and layered coatings with detonation spray coating process. *Journal of Scientific &Industrial Research*, *72*, 477-480. Retrieved from http://www.oerlikon.com/metco/en/products-services/coating-materials/coating-materials-thermal-spray/

Santa, J. F., Espitia, L. A., Blanco, J. A., Romo, S. A., & Toro, A. (2009). Slurry and cavitation erosion resistance of thermal spray coatings. *Wear*, *267*(1-4), 160–167. doi:10.1016/j.wear.2009.01.018

Sapate, S. G., & Rama Rao, A. V. (2004). Effect of carbide volume fraction on erosive wear behaviour of hardfacing cast irons. *Wear*, *256*(7-8), 774–786. doi:10.1016/S0043-1648(03)00527-1

Sapate, S. G., & RamaRao, A. V. (2006). Erosive wear behaviour of weld hardfacing high chromium cast irons: Effect of erodent particles. *Tribology International*, *39*(3), 206–212. doi:10.1016/j.triboint.2004.10.013

Saravanan, P., Selvarajan, V., Srinivasa Rao, D., Joshi, S. V., & Sundararajan, G. (2000). Influence of process variables on the quality of detonation gun sprayed alumina coatings. *Surface and Coatings Technology*, *123*(1), 44–54. doi:10.1016/S0257-8972(99)00252-2

Sarikaya, O. (2005). Effect of some parameters on microstructure and hardness of alumina coatings prepared by the air plasma spraying process. *Surface and Coatings Technology*, *190*(2-3), 388–393. doi:10.1016/j.surfcoat.2004.02.007

Sarkar, S., Sekharan, V., Mitra, R., & Roy, M. (2009). The solid particle erosion behavior of carbon/carbon and carbon/phenolic composite used in re-entry vehicles. *Tribology Transactions*, *52*(6), 777–787. doi:10.1080/10402000903097411

Savarimuthu, A. C., Taber, H. F., Megat, I., Shadley, J. R., Rybicki, E. F., Cornell, W. C., & Nuse, J. D. et al. (2001). Sliding wear behaviour of tungstein carbide thermal spray coatings for replacement of chromium elctroplate in aircraft applications. *Journal of Thermal Spray Technology*, *10*(3), 502–510. doi:10.1361/105996301770349286

Schiffmann, K. I., Bethke, R., & Kristen, N. (2005). Analysis of perforating and non-perforating micro-scale abrasion tests on coated substrates. *Surface and Coatings Technology*, *200*(7), 2348–2357. doi:10.1016/j.surfcoat.2005.01.015

Schimdt, G., & Steinhauser, S. (1996). Characterzation of wear protective coatings. *Tribology International*, *29*(3), 207–214. doi:10.1016/0301-679X(95)00090-Q

Schmidt, T., Gärtner, F., Assadi, H., & Kreye, H. (2006). Development of a generalized parameter window for cold spray deposition. *Acta Materialia*, *54*(3), 729–742. doi:10.1016/j.actamat.2005.10.005

Schneider, K. E., Belashchenko, V., Dratwinski, M., Siegmann, S., & Zagorski, A. (2006). *Thermal spraying for power generation components*. Weinheim: WILEY-VCH. doi:10.1002/3527609342

Schoop, M. U. (1909). *Verfahrenzur Herstellung vondichten metallischen Schichten.* CH-Patent No. 49278.

Schoop, M. U., Guenther, H. & Das. (1917). *Schoopsche, Metallspritz-Verfahren* [The Schoop Metal Spraying Technique]. Stuttgart, Germany: Franckh Publishers.

Schulz, U., Peters, M., Bach, F.-W., & Tegeder, G. (2003). Graded coatings for thermal, wear and corrosion barriers. *Materials Science and Engineering A, 362*(1-2), 61–80. doi:10.1016/S0921-5093(03)00579-3

Serre, I., Celati, N., & Pradeilles-Duval, R. M. (2002). Tribological and corrosion wear of graphite ring against Ti_6Al_4V disk in artificial sea water. *Wear, 252*(9-10), 711–718. doi:10.1016/S0043-1648(02)00030-3

Serre, I., Pradeilles-Duval, R. M., & Celati, N. (2006). Tribological and corrosion experiments of graphite ring against Ti–6Al–4V disk: Influence of electrochemical and mechanical parameters. *Wear, 260*(9-10), 1129–1135. doi:10.1016/j.wear.2005.07.007

Sheldon, G. L., & Finnie, I. (1966). The mechanism of material removal in the erosive cutting of brittle materials. *Journal of Engineering for Industry, 88*(4), 393–401. doi:10.1115/1.3672667

Shenhar, A., Gotmana, I., Radinb, S., Ducheyne, P., & Gutmanasa, E. Y. (2000). Titanium nitride coatings on surgical titanium alloys produced by a powder immersion reaction assisted coating method: Residual stresses and fretting behavior. *Surface and Coatings Technology, 126*(2-3), 210–218. doi:10.1016/S0257-8972(00)00524-7

Shen, M., & Bever, M. B. (1972). Gradients in polymeric materials. *Journal of Materials Science, 7*(7), 741–746. doi:10.1007/BF00549902

Shetty, D. K., Wright, I. G., Mincer, P. N., & Clauer, A. H. (1985). Indentation fracture of WC-Co cermets. *Journal of Materials Science, 20*(5), 1873–1882. doi:10.1007/BF00555296

Shin, L. D. S., Lim, D.-S., & Ahn, H.-S. (2000). Effect of annealing and Fe O addition on the high temperature tribological behavior of the plasma sprayed yttria-stabilized zirconia coating. *Surface and Coatings Technology, 133-134*, 403–410. doi:10.1016/S0257-8972(00)00965-8

Shipway, P. H., & Howell, L. (2005a). Microscale abrasion-corrosion behaviour of WC-Co hardmetals and HVOF sprayed coatings. *Wear, 258*(1-4), 303–312. doi:10.1016/j.wear.2004.04.003

Shipway, P. H., McCartney, D. G., & Sudaprasert, T. (2005). Sliding wear behaviour of conventional and nanostructured HVOF sprayed WC-Co coatings. *Wear, 259*(7-12), 820–827. doi:10.1016/j.wear.2005.02.059

Shipway, P. H., & Wirojanupatump, S. (2002). The role of lubrication and corrosion in abrasion of materials in aqueous environments. *Tribology International*, *35*(10), 661–667. doi:10.1016/S0301-679X(02)00057-9

Shivamurthy, R. C., Kamaraj, M., Nagarajan, R., Shariff, S. M., & Padmanabham, G. (2009). Influence of microstructure on slurry erosive wear characteristics of laser surface alloyed 13Cr–4Ni steel. *Wear*, *267*(1-4), 204–212. doi:10.1016/j.wear.2008.12.027

Shrestha, S., Hodgkiess, T., & Neville, A. (2001). The effect of post-treatment of a high-velocity oxy-fuel Ni-Cr-Mo-Si-B coating part 2: Erosion-corrosion behavior. *Journal of Thermal Spray Technology*, *10*(4), 656–665. doi:10.1361/105996301770349187

Shrestha, S., Hodgkiess, T., & Neville, A. (2005). Erosion–corrosion behaviour of high-velocity oxy-fuel Ni–Cr–Mo–Si–B coatings under high-velocity seawater jet impingement. *Wear*, *259*(1–6), 208–218. doi:10.1016/j.wear.2005.01.038

Shukla, S., Seal, S., Vij, R., & Bandyopadhyay, S. (2003). Reduced activation energy for grain growth in nanocrystalline Yttria-Stabilized Zirconia. *Nano Letters*, *3*(3), 397–401. doi:10.1021/nl0259380

Singh, H., Sidhu, T. S., & Kalsi, S. B. S. (2012). Cold spray technology: Future of coating deposition processes. *Integrità Strutturale Frattura*, *22*, 69–84.

Singh, V. P., Sil, A., & Jayaganthan, R. (2011). Tribological behavior of plasma sprayed Cr_2O_3–3%TiO_2 coatings. *Wear*, *272*(1), 149–158. doi:10.1016/j.wear.2011.08.004

Sivakumar, G., & Joshi, S. V. (2013). *An improved hybrid methodology for producing composite, multi-layered and graded coatings by plasma spraying utilizing powder and solution precursor feedstock*. Indian Patent No. 2965/DEL/2011 dt. 17th Oct, 2011 and United Patent Application No. US2013/0095340A1 dt. 18th Apr, 2013.

Sivakumar, G., Ramakrishna, L., Jain, V., Srinivasa Rao, D., & Sundararajan, G. (2001) The influence of the process parameters on the properties of detonation sprayed WC-12Co coatings. In *Proceedings of International Thermal Spray Conference*. ASM Materials.

Sivakumar, G., Dusane, R. O., & Joshi, S. V. (2011). *In situ* particle generation and splat formation during solution precursor plasma spraying of yttria-stabilized zirconia coatings. *Journal of the American Ceramic Society*, *94*(12), 4191–4199. doi:10.1111/j.1551-2916.2011.04773.x

Skandan, G., Jain, M., Fischer, T. E., Kear, B. H., Rigney, W. R., Shropshire, R., & Brunhouse, S. (2000). *Thermal spray: Surface engineering via applied research*. Materials Park, OH: ASM International.

Skopp, A., Kelling, N., Woydt, M., & Berger, L.-M. (2007). Thermally sprayed titanium suboxide coatings for piston ring/cylinder liners under mixed lubrication and dry-running conditions. *Wear*, *262*(9-10), 1061–1070. doi:10.1016/j.wear.2006.11.012

Sliney, H. E. (1982). Solid lubricant materials for high temperatures—A review. *Tribology International*, *15*(5), 303–315. doi:10.1016/0301-679X(82)90089-5

Soh, M. T. K., Fischer-Cripps, A. C., Musca, C. A., & Faraone, L. (2006). Nanoindentation of plasma-deposited nitrogen-rich silicon nitride thin films. *Journal of Applied Physics*, *100*(2), 1–8. doi:10.1063/1.2217105

Sōmiya, S., Hirano, S., & Kamiya, S. (1978). Phase relations of the Cr_2O_3-TiO_2 system. *Journal of Solid State Chemistry*, *25*(3), 273–284. doi:10.1016/0022-4596(78)90112-3

Souza, V. A. D., & Neville, A. (2003). Linking electrochemical corrosion behaviour and corrosion mechanisms of thermal spray cermet coatings (WC–CrNi and WC/CrC–CoCr). *Materials Science and Engineering A*, *352*(1-2), 202–211. doi:10.1016/S0921-5093(02)00888-2

Souza, V. A. D., & Neville, A. (2003a). Corrosion and erosion damage mechanisms during erosion-corrosion of WC-Co-Cr cermet coatings. *Wear*, *255*(1-6), 146–156. doi:10.1016/S0043-1648(03)00210-2

Souza, V. A. D., & Neville, A. (2005). Corrosion and synergy in a WC-Co-Cr HVOF thermal spray coating-understanding their role in erosion-corrosion degradation. *Wear*, *259*(1-6), 171–180. doi:10.1016/j.wear.2004.12.003

Souza, V. A. D., & Neville, A. (2007). Aspects of microstructure on the synergy and overall material loss of thermal spray coatings in erosion–corrosion environments. *Wear*, *263*(1–6), 339–346. doi:10.1016/j.wear.2007.01.071

Spear, K. E. (1989). Diamond—ceramic coating of the future. *Journal of the American Ceramic Society*, *72*(2), 171–191. doi:10.1111/j.1151-2916.1989.tb06099.x

Stachowiak, G. W., & Batchelor, A. W. (2005). *Engineering tribology*. Butterworth-Heinemann.

Stack, M. M., & Abd El-Badia, T. M. (2008). Some comments on mapping the combined effects of slurry concentration, impact velocity and electrochemical potential on the erosion-corrosion of WC/Co-Cr coatings. *Wear*, *264*(9-10), 826–837. doi:10.1016/j.wear.2007.02.025

Stack, M. M., & Badia, T. M. A. E. (2006). On the construction of erosion-corrosion maps for WC/Co-Cr based coatings in aqueous conditions. *Wear*, *261*(11-12), 1181–1190. doi:10.1016/j.wear.2006.03.038

Stack, M. M., & Chi, K. (2003). Mapping sliding wear of steels in aqueous conditions. *Wear*, *255*(1-6), 456–465. doi:10.1016/S0043-1648(03)00203-5

Stahr, C. C., Saaro, S., Berger, L.-M., Dubsky, J., Neufuss, K., & Herrmann, M. (2007). Dependance of the stabilization of α-alumina on the spray process. *Journal of Thermal Spray Technology*, *16*(5-6), 822–830. doi:10.1007/s11666-007-9107-7

Standard Test Method for Conducting Erosion Tests by Solid Particles Impingement Using Gas Jets, G 76-95. (1995). *Annual book of ASTM standards* (vol. 3.02). Wear and Erosion, Metal Corrosion ASTM.

Stark, H. C. (n.d.) *AMPERIT® thermal spray powders*. Retrieved from http://www.hcstarck.com/en/products/amperitreg_thermal_spray_powders.html

Stein, K. J., Schorr, B. S., & Marder, A. (1999). Erosion of thermal spray MCr–Cr_3C_2 cermet coatings. *Wear*, *224*(1), 153–159. doi:10.1016/S0043-1648(98)00298-1

Stevenson, A. N. J., & Hutchings, I. M. (1995). Scaling laws for particle velocity in the gas blast erosion test. *Wear*, *181-183*, 56–62. doi:10.1016/0043-1648(95)90008-X

Stewart, D. A., Shipway, P. H., & McCartney, D. G. (1999). Abrasive wear behaviour of conventional and nanocomposite HVOF-sprayed WC–Co coatings. *Wear*, *225–229*, 789–798. doi:10.1016/S0043-1648(99)00032-0

Stewart, D. A., Shipway, P. H., & McCartney, D. G. (2000). Microstructural evolution in thermally sprayed WC-Co coatings: Comparison between nanocomposite and conventional starting powders. *Acta Materialia*, *48*(7), 1593–1604. doi:10.1016/S1359-6454(99)00440-1

Stewart, D., Shipway, P., & McCartney, D. G. (2000). Microstructural evolution in thermally sprayed WC-Co coatings: Comparison between nanocomposite and conventional starting powder. *Acta Materialia*, *40*, 1596–1604.

Stokes, J., & Looney, L. (2004). Residual stress in HVOF thermally sprayed thick deposits. *Surface and Coatings Technology*, *177-178*, 18–23. doi:10.1016/j.surfcoat.2003.06.003

Storz, O., Gasthuber, H., & Woydt, M. (2001). Tribological properties of thermal-sprayed Magnéli-type coatings with different stoichiometries (Ti_nO_{2n-1}). *Surface and Coatings Technology*, *140*(2), 76–81. doi:10.1016/S0257-8972(01)01024-6

Stott, F. H. (2002). High-temperature sliding wear of metals. *Tribology International*, *35*(8), 489–495. doi:10.1016/S0301-679X(02)00041-5

Suarez, M., Bellayer, S., Traisnel, M., Gonzalez, W., Chicot, D., Lesage, J., & Staia, M. H. et al. (2008). Corrosion behavior of Cr_3C_2–NiCr vacuum plasma sprayed coatings. *Surface and Coatings Technology*, *202*(18), 4566–4571. doi:10.1016/j.surfcoat.2008.04.043

Sudaprasert, T., Shipway, P. H., & McCartney, D. G. (2003). Sliding wear behaviour of HVOF sprayed WC-Co coatings deposited with both gas-fuelled and liquid fuelled systems. *Wear*, *255*(7-12), 943–949. doi:10.1016/S0043-1648(03)00293-X

Sue, J. A., & Tucker, R. C. Jr. (1987). High temperature erosion behavior tungsten- and chromium-carbide-based coatings. *Surface and Coatings Technology*, *32*(1-4), 237–248. doi:10.1016/0257-8972(87)90110-1

Sun, D., Wharton, J. A., & Wood, R. J. K. (2009). Abrasive size and concentration effects on the tribo-corrosion of cast CoCrMo alloy in simulated body fluids. *Tribology International*, *42*(11-12), 1595–1604. doi:10.1016/j.triboint.2009.03.018

Sundararajan, G., Sivakumar, G., & Srinivasa Rao, D. (2001), The interrelationship between particle velocity and temperature, splat formation and deposition efficiency in detonation sprayed alumina coatings. In *Proceedings of International Thermal Spray Conference*. Academic Press.

Sundararajan, G., & Roy, M. (1997). Solid particle erosion behavior of metallic materials at room and elevated temperatures. *Tribology International*, *30*(5), 339–359. doi:10.1016/S0301-679X(96)00064-3

Sundararajan, G., Roy, M., & Venkataraman, B. (1990). Erosion efficiency-A new parameter to characterize the dominant erosion micromechanism. *Wear, 140*(2), 369–381. doi:10.1016/0043-1648(90)90096-S

Sundararajan, G., Sen, D., & Sivakumar, G. (2005). The tribological behaviour of detonation sprayed coatings: The importance of coating process properties. *Wear, 258*(1-4), 377–391. doi:10.1016/j.wear.2004.03.022

Sundararajan, G., & Shewmon, P. G. (1983). A new model for the erosion of metals a normal incidence. *Wear, 84*(2), 237–258. doi:10.1016/0043-1648(83)90266-1

Sundararajan, G., Sivakumar, G., Kavitha, R., & Phanikrishna, A. R. (2006), Erosion behaviour of composite cermet coatings. In *Proc. ACUN-5 "Developments in Composites: Advanced, Infrastructural, Natural, and Nanocomposites*. UNSW.

Sundararajan, G., Sivakumar, G., Sen, D., Srinivasa Rao, D., & Ravichandra, G. (2010). The tribological behaviour of detonation sprayed TiMo(CN) based cermet coatings. *International Journal of Refractory Metals & Hard Materials, 28*(1), 71–81. doi:10.1016/j.ijrmhm.2009.07.007

Sundararajan, G., Srinivasa Rao, D., Sen, D., & Somaraju, K. R. C. (1998). *Tribological behavior of thermal sprayed coatings*. London: Institute of Materials.

Sundararajan, G., Srinivasa Rao, D., Sivakumar, G., & Joshi, S. V. (1998). A comparative study of tribological behavior of plasma and D-gun sprayed coatings under different wear modes. *Journal of Materials Engineering and Performance, 7*(3), 343–351. doi:10.1361/105994998770347783

Sundararajan, G., Srinivasa Rao, D., Sivakumar, G., & Joshi, S. V. (2013). Detonation spray coatings. In *Encyclopedia of tribology*. Springer.

Sundararajan, G., & Suresh Babu, P. (2009). Detonation sprayed WC-Co coatings: Unique aspects of their structure and mechanical behavior. *Transactions of the Indian Institute of Metals, 62*(2), 1–9. doi:10.1007/s12666-009-0013-1

Suresh Babu, P., Srinivasa Rao, D., Rao, G.V.N., & Sundararajan, G. (2007). Effect of feedstock size and its distribution on the properties of detonation sprayed coatings. *Journal of Thermal Spray Technology, 16*(2), 281-290.

Suresh Babu, P., Basu, B., & Sundararajan, G. (2008). Processing–structure–property correlation and decarburization phenomenon in detonation sprayed WC–12Co coatings. *Acta Materialia, 56*(18), 5012–5026. doi:10.1016/j.actamat.2008.06.023

Taguchi, G., & Konishi, S. (1987). *Taguchi methods: Orthogonal arrays and linear graphs*. Dearborn, MI: ASI Press.

Takeda, M., Morihiro, N., Ebara, E., Harada, Y., Wang, R., & Kido, M. (2002). Corrosion behavior of thermally sprayed WC coating in Na_2SO_4 aqueous solution. *Materials Transactions, 33*(11), 2860–2865. doi:10.2320/matertrans.43.2860

Taktak S., (2006), Tribological behaviour of borided bearing steels at elevated temperatures. *Surface & Coatings Technology, 201*, 2230-2239.

Tampieri, A., Celotti, G., Sprio, S., Delcogliano, A., & Franzese, S. (2001). Porosity-graded hydroxyapatite ceramics to replace natural bone. *Biomaterials*, *22*(11), 1365–1370. doi:10.1016/S0142-9612(00)00290-8 PMID:11336309

Tang, C. H., Cheng, F. T., & Man, T. H. (2006). Laser surface alloying of a marine propeller bronze using aluminium powder part I: Microstructural analysis and cavitation erosion study. *Surface and Coatings Technology*, *200*(8), 2602–2609. doi:10.1016/j.surfcoat.2004.12.021

Tan, K. S., Wharton, J. A., & Wood, R. J. K. (2005). Solid particle erosion-corrosion behaviour of a novel HVOF nickel aluminium bronze coating for marine applications-correlation between mass loss and electrochemical measurements. *Wear*, *258*(1-4), 629–640. doi:10.1016/j.wear.2004.02.019

Tan, K., Wood, R. J. K., & Stokes, K. R. (2003). The slurry erosion behaviour of high velocity oxy-fuel (HVOF) sprayed aluminium bronze coatings. *Wear*, *255*(1-6), 195–205. doi:10.1016/S0043-1648(03)00088-7

Taylor, T. A., Overs, M. P., Quets, J. M., & Tucker, R. C. Jr. (1983). Development of several new nickel alluminide and chromium carbide coatings for use in high temperature nuclear reactors. *Thin Solid Films*, *107*(4), 427–435. doi:10.1016/0040-6090(83)90304-8

Telfer, C. G., Stack, M. M., & Jana, B. D. (2012). Particle concentration and size effects on the erosion-corrosion of pure metals in aqueous slurries. *Tribology International*, *53*(0), 35–44. doi:10.1016/j.triboint.2012.04.010

Thakare, M. R., Wharton, J. A., Wood, R. J. K., & Menger, C. (2007). Exposure effects of alkaline drilling fluid on the microscale abrasion-corrosion of WC-based hardmetals. *Wear*, *263*(1-6), 125–136. doi:10.1016/j.wear.2006.12.047

Thakare, M. R., Wharton, J. A., Wood, R. J. K., & Menger, C. (2009). Investigation of micro-scale abrasion-corrosion of WC-based sintered hardmetals and sprayed coatings using *in situ* electrochemical current-noise measurements. *Wear*, *267*(11), 1967–1977. doi:10.1016/j.wear.2009.06.006

Thakare, M. R., Wharton, J. A., Wood, R. J. K., & Menger, C. (2012). Effect of abrasive particle size and the influence of microstructure on the wear mechanisms in wear-resistant materials. *Wear*, *276-277*, 16–28. doi:10.1016/j.wear.2011.11.008

Thakur, L., & Arora, N. (2013). A comparative study on slurry and dry erosion behaviour of HVOF sprayed WC–CoCr coatings. *Wear*, *303*(1–2), 405–411. doi:10.1016/j.wear.2013.03.028

Thakur, L., Arora, N., Jayaganthan, R., & Sood, R. (2011). R. An investigation on erosion behavior of HVOF sprayed WC–CoCr coatings. *Applied Surface Science*, *258*(3), 1225–1234. doi:10.1016/j.apsusc.2011.09.079

Thomas, H. A. J., & Stevens, R. (1989). Aluminium titanate – A literature review part 2: Engineering properties and thermal stability. *British Ceramic. Transactions and Journal*, *88*(5), 184–190.

Tian, B. R., & Cheng, Y. F. (2007). Electrolytic deposition of Ni–Co–Al2O3 composite coating on pipe steel for corrosion/erosion resistance in oil sand slurry. *Electrochimica Acta, 53*(2), 511–517. doi:10.1016/j.electacta.2007.07.013

Tilly, G. P. (1969). Erosion caused by airborne particles. *Wear, 14*(1), 63–79. doi:10.1016/0043-1648(69)90035-0

Tilly, G. P. (1973). A two stage mechanism of ductile erosion. *Wear, 23*(1), 87–96. doi:10.1016/0043-1648(73)90044-6

Toma, F.-L., Berger, L.-M., Stahr, C.C., Naumann, T., & Langner, S. (n.d.). Thermisch gespritzte Al_2O_3-Schichten mit einem hohen Korundgehalt ohne eigenschaftsmindernde Zusätze und Verfahren zu ihrer Herstellung, DE 10 2008 026 101 B4, filed: 30.5.2008, granted: 18.2.2010; EP 2300630 A1; WO 2009/146832 A1; US 8,318,261, B2; CA 2726434 A1; JP 2011-522115 A.

Toma, D., Brandl, W., & Marginean, G. (2001). Wear and corrosion behaviour of thermally sprayed cermet coatings. *Surface and Coatings Technology, 138*(2-3), 149–158. doi:10.1016/S0257-8972(00)01141-5

Toma, F.-L., Berger, L.-M., Langner, S., & Naumann, T. (2010). Suspension spraying – The potential of a new spray technology. *Thermal Spray Bulletin, 3*(1), 24–29.

Toma, F.-L., Berger, L.-M., Stahr, C. C., Naumann, T., & Langner, S. (2010). Microstructures and functional properties of suspension-sprayed Al_2O_3 and TiO_2 coatings: An overview. *Journal of Thermal Spray Technology, 19*(1-2), 262–274. doi:10.1007/s11666-009-9417-z

Tomlinson, W. J., & Linzell, C. R. (1988). Anodic polarisation and corrosion of cemented carbides with cobalt and nickel binders. *Journal of Materials Science, 23*(3), 914–918. doi:10.1007/BF01153988

Trache, R., Berger, L.-M., & Leyens, C. (n.d.). Comparison of alumina-titania coatings deposited by plasma and HVOF spray techniques. In *Proceedings of the International Thermal Spray Conference* (pp. 368-372). Düsseldorf: DVS Media.

Tsai, P. C., Lee, J. H., & Chang, C. L. (2007). Improving erosion resistance of plasma sprayed zirconia thermal barrier coatings by laser glazing. *Surface and Coatings Technology, 201*(4-7), 719–724. doi:10.1016/j.surfcoat.2007.07.005

Tucker, R. C. (1994). Thermal sprays coatings. In ASM handbook, surface engineering (vol. 5, pp. 497-509). Materials Park, OH: ASM International.

Tucker, R. C., Jr. (Ed.). (2013). ASM handbook: Thermal spray technology (vol. 5A). Materials Park, OH: ASM International.

Tucker, R. C. (1999). Thermal spray coatings. In *ASM handbook for surface engineering* (Vol. 5). ASM International.

Tu, D., Chang, S., Chao, C., & Lin, C. (1985). Tungsten carbide phase transformation during the plasma spray process. *Journal of Vacuum Science & Technology. A, Vacuum, Surfaces, and Films, 3*(6), 2479–2482. doi:10.1116/1.572862

Tu, J. P., Jie, X. H., Mao, Z. Y., & Matsumara, M. (1998). The effect of temperature on the unlubricated sliding wear of 5 CrNiMo steel against 40 MnB steel in the range 400–600°C. *Tribology International*, *31*(7), 347–353. doi:10.1016/S0301-679X(98)00039-5

Tummala, R., Guduru, R. K., & Mohanty, P. S. (2012). Solution precursor plasma deposition of nanostructured CdS thin films. *Materials Research Bulletin*, *47*(3), 700–707. doi:10.1016/j.materresbull.2011.12.018

Turenne, S., Fiset, M., & Masounave, J. (1989). The effect of sand concentration on the erosion of materials by a slurry jet. *Wear*, *133*(1), 95–106. doi:10.1016/0043-1648(89)90116-6

Turner, T. H., & Budgen, N. E. (1926). *The origin, development and applications of the metal spray process of metallization*. London: Charles Griffin & Co.

Turunen, E., Varis, T., Gustafsson, T. E., Keskinen, J., Fält, T., & Hannula, S.-P. (2006). Parameter optimization of HVOF sprayed nanostructured alumina and alumina–nickel composite coatings. *Surface and Coatings Technology*, *200*(16–17), 4987–4994. doi:10.1016/j.surfcoat.2005.05.018

Turunen, E., Varis, T., Hannula, S.-P., Vaidya, A., Kulkarni, A., Gutleber, J., & Herman, H. et al. (2006). On the role of particle state and deposition procedure on mechanical, tribological and dielectric response of high velocity oxy-fuel sprayed alumina coatings. *Materials Science and Engineering A*, *415*(1-2), 1–11. doi:10.1016/j.msea.2005.08.226

Uhlig, H. (1954). Mechanism of Fretting corrosion. *Journal of Applied Mechanics*, *21*(4), 401–407.

Usmani, S., Sampath, S., Houck, D. L., & Lee, D. (1997). Effect of carbide grain size on the sliding and abrasive wear behavior of thermally sprayed WC-Co coatings. *Tribology Transactions*, *40*(3), 470–478. doi:10.1080/10402009708983682

Uuemo, H., & Kleis, I. (1975). A critical analysis of erosion problems which have been little studied. *Wear*, *31*(2), 359–371. doi:10.1016/0043-1648(75)90169-6

Vaidya, A., Streibl, T., Li, L., Sampath, S., Kovarik, O., & Greenlaw, R. (2005). An integrated study of thermal spray process-structure-property correlations: A case study for plasma sprayed molybdenum coatings. *Materials Science Engineering A. Structural Materials: Properties, Microstructure, and Processing*, *403*(1-2), 191–204.

Valarezo, A., Giovanni, B., Wanhuk, B. C., Sanjay, S., Valeria, C., Luca, L., & Roberto, R. (2010). Damage tolerant functionally graded WC–Co/stainless steel HVOF coatings. *Surface and Coatings Technology*, *205*(7), 2197–2208. doi:10.1016/j.surfcoat.2010.08.148

Valentinelli, L., Valente, T., Casadei, F., & Fedrizzi, L. (2004). Mechanical and tribocorrosion properties of HVOF sprayed WC-Co coatings. *Corrosion Engineering, Science, and Technology*, *39*, 301–307.

Vargas, F., Ageorges, H., Fournier, P., Fauchais, P., & López, M. E. (2010). Mechanical and tribological performance of Al_2O_3-TiO_2 coatings elaborated by flame and plasma spraying. *Surface and Coatings Technology*, *205*(4), 1132–1136. doi:10.1016/j.surfcoat.2010.07.061

Vassen, R., Kassner, H., Stuke, A., Mack, D. E., Maria, J. O., & Stöver, D. (2010). Functionally graded thermal barrier coatings with improved reflectivity and high-temperature capability. *Materials Science Forum, 631/632*, 73–78. doi:10.4028/www.scientific.net/MSF.631-632.73

Venugopal Reddy, A., & Sundarajan, G. (1986). Erosion behavior of ductile materials with a spherical non-friable erodent. *Wear, 111*(3), 313–323. doi:10.1016/0043-1648(86)90190-0

Venugopal Reddy, A., & Sundararajan, G. (1986). Selection of steels for erosion resistant applications. *Tool and Alloy Steels, 20*(12), 379–388.

Verdon, C., Karimi, A., & Martin, J.-L. (1997). Microstructural and analytical study of thermally sprayed WC-Co coatings in connection with their wear resistance. *Materials Science and Engineering A, 234-236*, 731–734. doi:10.1016/S0921-5093(97)00377-8

Verdon, C., Karimi, A., & Martin, J.-L. (1998). A study of high velocity oxy-fuel thermally sprayed tungsten carbide based coatings: Part 1: Microstructures. *Materials Science and Engineering A, 246*(1-2), 11–24. doi:10.1016/S0921-5093(97)00759-4

Vieira, A. C., Rocha, L. A., Papageorgiou, N., & Mischler, S. (2012). Mechanical and electrochemical deterioration mechanisms in the tribocorrosion of Al alloys in NaCl and in NaNO3 solutions. *Corrosion Science, 54*, 26–35. doi:10.1016/j.corsci.2011.08.041

Vieria, A. C., Ribeiro, A. R., Rocha, L. A., & Celis, J. P. (2006). Influence of pH and corrosion inhibitors on the tribocorrosion of titanium in artificial saliva. *Wear, 261*(9), 994–1004. doi:10.1016/j.wear.2006.03.031

Vinayo, M. E., Kassabji, F., Guyonnet, J., & Fauchais, P. (1985). Plasma sprayed WC-Co coating: Influence of spray condition (atmospheric and low pressure plasma spraying) on the crystal structure, porosity and hardness. *Journal of Vacuum Science & Technology. A, Vacuum, Surfaces, and Films, 3*(6), 2483–2489. doi:10.1116/1.572863

Vos, F., Delaey, L., De Bonte, M., & Froyen, L. (1998), Plasma-sprayed self-lubricatinng Cr2O3–CaF2 coatings: Friction and wear properties. In *Proceedings of the 15th International Thermal Spray Conference*. Nice, France: Academic Press.

Voyer, J., & Marple, B. R. (1999). Sliding wear behavior of high velocity oxy-fuel and high power plasma spray-processed tungsten carbide-based cermet coatings. *Wear, 225–229*, 135–145. doi:10.1016/S0043-1648(99)00007-1

Vuoristo, P., Määttä, A., Mäntylä, T., Berger, L.-M., & Thiele, S. (2002). Properties of ceramic coatings prepared by HVOF and plasma spraying of titanium suboxide powders. In E. Lugscheider (Ed.), *Proceedings of the International Thermal Spray Conference 2002* (pp. 488-493). Düsseldorf: DVS-Verlag.

Vuoristo, P. (2014). Thermal spray coating processes. In D. Cameron (Ed.), *Comprehensive materials processing* (Vol. 4, pp. 229–276). Elsevier. doi:10.1016/B978-0-08-096532-1.00407-6

Vuoristo, P. M. J., Niemi, K. J., & Mäntylä, T. A. (1992). On the properties of detonation gun sprayed and plasma sprayed ceramic coatings. In C. C. Berndt (Ed.), *Thermal spray: International advances in coatings technology* (pp. 171–175). Materials Park, OH: ASM International.

Wada, S., & Ritter, J. E. (Eds.). (1992). *Erosion of ceramic materials*. Zurich: Trans Tech Publications.

Wagner, C. D., Naumkin, A. V., Kraut-Vass, A., Allison, J. W., & Powell, C. J., Jr. (2006). NIST x-ray photoelectron spectroscopy database. Measurement Services Division of the National Institute of Standards and Technology (NIST) Technology Services.

Walsh, P., & Tabakoff, W. (1990). Comparative erosion resistance of coatings intended for steam turbine components. In *Proceedings of International Joint Power Generation Conference*. Boston: Academic Press.

Walsh, P. (1992). Erosion resistant coatings at steam turbine temperatures. *PWR-Steam Turbine Generator Developments for the Power Generation Industry*, *18*, 123–128.

Wang, B. (1996). The dependence of erosion-corrosion wastage on carbide/metal binder proportion for HVOF carbide-metal cerment coatings. *Wear*, *196*(1-2), 141–146. doi:10.1016/0043-1648(95)06883-X

Wang, B. Q., Geng, G. Q., & Levy, A. V. (1990). Article. *Erosion-Corrosion of Thermal Spray Coatings Surface and Coating Technology*, *43-44*, 859–874.

Wang, B. Q., & Lee, S. W. (1997). Elevated temperature erosion of several thermal-sprayed coatings under the simulated erosion conditions of in-bed tubes in a fluidized bed combustor. *Wear*, *203–204*, 580–587. doi:10.1016/S0043-1648(96)07381-4

Wang, B., & Lee, S. W. (2000). Erosion-corrosion behaviour of HVOF NiAl-Al2O3 intermetallic-ceramic coating. *Wear*, *239*(1), 83–90. doi:10.1016/S0043-1648(00)00309-4

Wang, L., & Li, D. Y. (2003). Effects of yttrium on microstructure, mechanical properties and high-temperature wear behavior of cast Stellite 6 alloy. *Wear*, *255*(1-6), 535–544. doi:10.1016/S0043-1648(03)00057-7

Wang, Q., Hu, R., & Shao, Y. (2012). The combined effect of surface roughness and internal stresses on nanoindentation tests of polysilicon thin films. *Journal of Applied Physics*, *112*(4), 1–8. doi:10.1063/1.4748176

Wang, X., Hwang, K. S., Koopman, M., Fang, Z., & Zhang, L. (2013). Mechanical properties and wear resistance of functionally graded WC-Co. *International Journal of Refractory Metals & Hard Materials*, *36*, 46–55. doi:10.1016/j.ijrmhm.2012.04.011

Wang, Y. (1993). Friction and wear performances of detonation-gun and plasma-sprayed ceramic and cermet hard coatings under dry friction. *Wear*, *161*(1-2), 69–78. doi:10.1016/0043-1648(93)90454-T

Wang, Y., Hsu, S. M., & Jones, P. (1998). Evaluation of thermally-sprayed ceramic coatings using a novel ball-on-inclined plane scratch method. *Wear*, *218*(1), 96–102. doi:10.1016/S0043-1648(98)00182-3

Wang, Y., Jiang, S., Wang, M., Wang, S., Xiao, T. D., & Strutt, P. R. (2000). Abrasive wear characteristics of plasma sprayed nanostructured alumina/titania coatings. *Wear*, *237*(2), 176–185. doi:10.1016/S0043-1648(99)00323-3

Wang, Y., Jin, Y., & Wen, S. (1988). The analysis of the friction and wear mechanisms of plasma-sprayed ceramic coatings at 450 °C. *Wear*, *128*(3), 265–276. doi:10.1016/0043-1648(88)90063-4

Watari, F., Yokoyama, A., Saso, F., Uo, M., & Kawasaki, T. (1995). Functionally gradient dental implant composed of titanium and hydroxyapatite. In *Proc. 3rd Int. Symp. Structural & Functional Gradient Materials* (pp. 703–8). Lausanne, Switzerland: Polytech. Univ. Romand.

Watari, F., Yokoyama, A., Saso, F., Uo, M., & Kawasaki, T. (1997). Fabrication and properties of functionally graded dental implant. *Composites. Part B, Engineering*, *28B*(1-2), 5–11. doi:10.1016/S1359-8368(96)00021-2

Waterhouse, R. B. (1972). *Fretting corrosion*. Oxford, UK: Pergamon Press.

Watreme, M., Bricout, J. P., Marguet, B., & Qudin, J. (1996). Friction, temperature, and wear analysis for ceramic coated brake disks. *Journal of Tribology*, *118*(3), 457–465. doi:10.1115/1.2831558

Watson, S. W., Friedersdorf, F. J., Madsen, B. W., & Cramer, S. D. (1995). Methods of measuring wear-corrosion synergism. *Wear*, *181–183*, 476–484. doi:10.1016/0043-1648(94)07108-X

Wayne, S. F., Baldoni, J. G., & Buljan, S. T. (1990). Abrasion and erosion of WC-Co with controlled microstructures. *Tribology Transactions*, *33*(4), 611–617. doi:10.1080/10402009008981996

Wayne, S. F., & Sampath, S. (1992). Structure/property relationships in sintered and thermally sprayed WC-Co. *Journal of Thermal Spray Technology*, *1*(4), 307–315. doi:10.1007/BF02647158

Weiss, L. E., Prinz, F. B., Adams, D., & Siewiorek, D. P. (1992). Thermal spray shape deposition. *Journal of Thermal Spray Technology*, *1*(3), 231–237. doi:10.1007/BF02646778

Wenger, F., & Galland, J. (1990). Analysis of local corrosion of large metallic structures or reinforced concrete structures by electrochemical impedance spectroscopy (EIS). *Electrochimica Acta*, *35*(10), 1573–1578. doi:10.1016/0013-4686(90)80012-D

Werner, H.-D. (1974). Phasenbeziehungen im system TiO_2-Cr_2O_3. *Neues Jahrbuch für Mineralogie Monatshefte*, 218–234.

Westergård, R., Axén, N., Wiklund, U., & Hogmark, S. (1998). The erosion and abrasion characteristics of alumina coatings plasma sprayed under different spraying conditions. *Tribology International*, *31*(5), 271–279. doi:10.1016/S0301-679X(98)00033-4

Westergard, R., Axen, N., Wiklund, U., & Hogmark, S. (2000). An evaluation of plasma sprayed ceramic coatings by erosion, abrasion and bend testing. *Wear*, *246*(1-2), 12–19. doi:10.1016/S0043-1648(00)00506-8

Westergard, R., & Hogmark, S. (2004). Sealing to improve the wear properties of plasma sprayed alumina by electro-deposited Ni. *Wear*, *256*(11-12), 1153–1162. doi:10.1016/j.wear.2003.07.005

Wheeler, D. W., & Wood, R. J. K. (2005). Erosion of hard surface coating for use in offshore gate valves. *Wear*, *258*(1-4), 526–536. doi:10.1016/j.wear.2004.03.035

Winter, R. E., & Hutchings, I. M. (1974). Solid particle erosion studies using single angular particles. *Wear*, *29*(2), 181–194. doi:10.1016/0043-1648(74)90069-6

Wlodek, S. T. (1985). Development and testing plasma sprayed coatings for solid particle erosion resistance. In *Proceedings of SPE of Utility Steam Turbines*. EPRI CS4683.

Wong, K. K., & Clark, H. M. (1993). A model of particle velocities and trajectories in a slurry pot erosion tester. *Wear*, *160*(1), 95–104. doi:10.1016/0043-1648(93)90409-F

Wood, R. J. K. (2007). Tribo-corrosion of coatings: A review. *Journal of Physics. D, Applied Physics*, *40*(18), 5502–5521. doi:10.1088/0022-3727/40/18/S10

Wood, R. J. K. (2010). Tribology of thermal sprayed WC–Co coatings. *International Journal of Refractory Metals & Hard Materials*, *28*(1), 82–94. doi:10.1016/j.ijrmhm.2009.07.011

Wood, R. J. K., Mellor, B. G., & Binfield, M. L. (1997). Sand erosion performance of detonation gun applied tungsten carbide/cobalt-chromium coatings. *Wear*, *211*(1), 70–83. doi:10.1016/S0043-1648(97)00071-9

Wood, R. J. K., Wharton, J. A., Speyer, A. J., & Tan, K. S. (2002). Investigation of erosion–corrosion processes using electrochemical noise measurements. *Tribology International*, *35*(10), 631–641. doi:10.1016/S0301-679X(02)00054-3

Wood, R. K. (2006). Erosion–corrosion interactions and their effect on marine and offshore materials. *Wear*, *261*(9), 1012–1023. doi:10.1016/j.wear.2006.03.033

Woydt, M. (2005). Triboaktive Werkstoffe für hohe Gleitgeschwindigkeiten bis 800°C. *Tribologie und Schmierungstechnik*, *52*(4), 5–12.

Woydt, M., & Wäsche, R. (2010). The history of the Stribeck curve and ball bearing steels: The role of Adolf Martens. *Wear*, *268*(11-12), 1542–1546. doi:10.1016/j.wear.2010.02.015

Wu, G., Li, N., Zhou, D., & Mitsuo, K. (2004). Electrodeposited Co–Ni–Al2O3 composite coatings. *Surface and Coatings Technology*, *176*(2), 157–164. doi:10.1016/S0257-8972(03)00739-4

Wu, P.-Q., & Celis, J.-P. (2004). Electrochemical noise measurements on stainless steel during corrosion–wear in sliding contacts. *Wear*, *256*(5), 480–490. doi:10.1016/S0043-1648(03)00558-1

Xia, Z., Zhang, X., & Song, J. (1999). Erosion resistance of plasma sprayed coatings. *Journal of Materials Engineering and Performance*, *8*(6), 716–718. doi:10.1361/105994999770346495

Xie, Y., & Hawthorne, H. M. (1999). The damage mechanisms of several-sprayed ceramic coatings in controlled scratching. *Wear*, *233-235*, 293–305. doi:10.1016/S0043-1648(99)00211-2

Yang, K., Feng, J. W., Zhou, X. M., & Tao, S. Y. (2012). Microstructural characterization and strengthening-toughening mechanism of plasma-sprayed Al_2O_3-Cr_2O_3 composite coatings. *Journal of Thermal Spray Technology*, *21*(5), 1011–1024. doi:10.1007/s11666-012-9796-4

Yang, Q., Senda, T., & Hirose, A. (2004). Sliding wear behavior and tribofilm formation of ceramics at high temperatures. *Surface and Coatings Technology*, *184*(2-3), 270–277. doi:10.1016/j.surfcoat.2003.10.157

Yang, Q., Senda, T., & Hirose, A. (2006). Sliding wear behavior of WC–12% Co coatings at elevated temperatures. *Surface and Coatings Technology*, *200*(14-15), 4208–4212. doi:10.1016/j.surfcoat.2004.12.032

Yang, Q., Senda, T., & Ohmori, A. (2003). Effect of carbide grain size on microstructure and sliding wear behavior of HVOF-sprayed WC–12% Co coatings. *Wear*, *254*(1-2), 23–34. doi:10.1016/S0043-1648(02)00294-6

Yan, Y., Neville, A., Dowson, D., & Williams, S. (2006). Tribocorrosion in implants—Assessing high carbon and low carbon Co–Cr–Mo alloys by in situ electrochemical measurements. *Tribology International*, *39*(12), 1509–1517. doi:10.1016/j.triboint.2006.01.016

Yılmaz, R., Kurt, A. O., Demir, A., & Tatlı, Z. (2007). Effects of TiO_2 on the mechanical properties of the Al_2O_3–TiO_2 plasma sprayed coating. *Journal of the European Ceramic Society*, *27*(2-3), 1319–1323. doi:10.1016/j.jeurceramsoc.2006.04.099

Yu, S.H., & Wallar, H. (n.d.). Chromia spray powders and a process for making the same, US 6,774,076, B2, filed: 3.09.2002, granted: 10.08.2004, WO 03/086974 A1.

Yuanyuan, L., Tungwai, L. N., & Wei, X. (1996). Mechanical, friction and wear behaviors of a novel high-strength wear-resisting aluminum bronze. *Wear*, *197*(1-2), 130–136. doi:10.1016/0043-1648(95)06890-2

Yu, Q., Rauf, A., Wang, N., & Zhou, C. (2011). Thermal properties of plasma-sprayed thermal barrier coating with bimodal structure. *Ceramics International*, *37*(3), 1093–1099. doi:10.1016/j.ceramint.2010.11.033

Zeng, Y., Lee, S. W., Gao, L., & Ding, C. X. (2002). Atmospheric plasma sprayed coatings of nanostructured zirconia. *Journal of the European Ceramic Society*, *22*(3), 347–351. doi:10.1016/S0955-2219(01)00291-6

Zhang, J., & Kobayashi, A. (2008). Corrosion resistance of the Al_2O_3+ZrO_2 thermal barrier coatingson stainless steel substrate. *Vacuum*, *83*(1), 92–97. doi:10.1016/j.vacuum.2008.03.090

Zhao, H. X., Goto, H., Matsumura, M., Takahashi, T., & Yamamoto, M. (1999). Slurry erosion of plasma-sprayed ceramic coatings. *Surface and Coatings Technology*, *115*(2-3), 123–131. doi:10.1016/S0257-8972(99)00164-4

Zhijian, Y., Shunyan, T., & Xiaming, Z. (2010). Effect of the thickness on properties of Al_2O_3 coatings deposited by plasma spraying. *Materials Characterization*, *62*, 90–93.

Zhou, Y. H., Hermelin, M., & Bigot, J. (1989). Sintering behaviour of ultra-fine Fe, Ni and Fe-25wt%Ni powders. *Scripta Metalia*, *23*(8), 1391–1396. doi:10.1016/0036-9748(89)90065-3

Zhu, Y., Huang, M., Huang, J., & Ding, C. (1999). Vacuum-plasma sprayed nanostructured titanium oxide films. *Journal of Thermal Spray Technology*, *8*(2), 219–222. doi:10.1361/105996399770350430

Znamirowski, Z., Pawlowski, L., Cichy, T., & Czarczynski, W. (2004). Low macroscopic field electron emission from surface of plasma sprayed and laser engraved TiO_2, Al_2O_3+13TiO_2 and Al_2O_3+40TiO_2 coatings. *Surface and Coatings Technology*, *187*(1), 37–46. doi:10.1016/j.surfcoat.2004.01.001

Zou, Y., Goldbaum, D., Szpunar, J., & Yue, S. (2010). Microstructure and nanohardness of cold-sprayed coatings: Electron backscattered diffraction and nanoindentation studies. *Scripta Materialia*, *62*(6), 395–398. doi:10.1016/j.scriptamat.2009.11.034

Zum Gahr, H. (1987). Microstructure and wear of materials. Elsevier.

About the Contributors

Manish Roy graduated from IIT, Kharagpur. After a short stay with an integrated steel plant as Junior Manager and his post graduation, he joined Defence Metallurgical Research Laboratory as scientist. His areas of research interest are materials and surface engineering for tribological application and nanotribology of self-lubricating films. He is particularly interested in various interactions between oxidation/corrosion and tribology. Meanwhile, he has visited various European universities and research laboratories under prestigious programmes of Royal Society (London), Austrian Science Fund (FWF), and European Commission. Dr. Roy has around 100 papers in various reviewed national and international journals. He edited a book on *Surface Engineering for Enhanced Performance against Wear*. He has been guest editor of several special issues of several international journals. He is also on the editorial board of several international journals. Dr. Roy has organised several international symposium in India and abroad. He is also distinguished visiting professorof Indian National Academy of Engineering at Indian Institute of Technology, Bombay.

J. Paulo Davim received his PhD degree in Mechanical Engineering from the University of Porto in 1997 and the Aggregate title from the University of Coimbra in 2005. Currently, he is Professor at the Department of Mechanical Engineering of the University of Aveiro. He has more 27 years of teaching and research experience in manufacturing, materials, and mechanical engineering with special emphasis in Machining and Tribology. He also has interest in Industrial Engineering and Sustainable Manufacturing. He is the editor in chief of six international journals, guest editor of journals, books editor, book series editor and scientific advisory for many international journals and conferences. Presently, he is an editorial Board member of 20 international journals and acts as reviewer for than 70 prestigious ISI Web of Science journals. In addition, he has also published as author and co-author more than 40 book chapters and 350 articles in journals and conferences (more 180 articles in ISI Web of Science, h-index 30+).

* * *

Salim Barbhuiya is a Lecturer in the Department of Civil Engineering at Curtin University of Technology, Australia. He completed his Bachelor degree in Civil Engineering from National Institute of Technology Silchar, India (1999), Master's degree in Structural Engineering from Asian Institute of Technology, Thailand (2005) and obtained his PhD in Structural Materials from Queen's University Belfast (2009). One of his research areas at Curtin is nanotriblogical behaviour of anti-corrosion and anticarbonation coatings.

Lutz-Michael Berger obtained his Masters of Engineering degree in ceramic engineering and his PhD in the field of environmental protection from the Moscow "D. I. Mendeleev" Institute of Chemical Technology (today: Russian University "D. I. Mendeleev" of Chemical Technology). Since 1992 he has worked for several institutes of the Fraunhofer Society of Applied Research in Dresden. He is currently group manager at the Fraunhofer Institute for Ceramic Technologies and Systems. His research interests include thermal spray coating solutions, feedstock powders for surface modification technologies, solid materials syntheses, ceramics, hardmetals and characterization methods. He has published 68 papers in refereed journals and books, 150 papers in conference proceedings and holds 15 patents.

Ikbal Choudhury has recently graduated from National Institute of Technology Silchar, India in Mechanical Engineering. His research interests include nanotribological behaviour of various coatings.

Traugott Fischer is Professor Emeritus at Stevens Institute of Technology in Hoboken, USA. He studied physics at the Federal Institute of Technology (ETH) in Zurich, Switzerland where he obtained his Doctorate in Natural Sciences with a thesis in Field Emission from Silicon. He then moved to the Bell Laboratories in Murray Hill, USA, where he conducted research in the electronic structure of semiconductor bulk and surfaces. He continued this line of research as Associate Professor at Yale University in New Haven. Later he joined the Corporate Research Laboratories of Exxon where he lead a group performing research in surface science; adsorption; the fundamentals of catalysis; the surface aspects of metallurgy; and tribology. He finally became Professor at the Stevens Institute of Technology, in Hoboken, where his research group investigated the mechanical properties and tribological performance of novel materials, including ceramics and coatings. Dr. Fischer is a fellow of the American Physical Society, of the Society of Tribologists and Lubrication Engineers and a past member of the Materials Research Society.

Harpreet Singh Grewal is a research scholar in the School of Mechanical, Materials and Energy Engineering of Indian Institute of Technology (IIT) Ropar, India. His research interests includes Tribology and Surface modification of materials. He is a life member of Tribological Society of India and has published over 23 research articles in peer-reviewed international journals. He was awarded Young Tribologist award by Taiho Kogyo Tribology Research Foundation in Nov. 2013 and prestigious Swiss Government Excellence Postdoctoral fellowship by Swiss Federal Commission. He did his Bachelor's degree in Mechanical engineering and Master's degree in Machine Design from Punjab Technical University, India. He is doing his PhD in Mechanical engineering at IIT Ropar. Currently he is working as a research associate in the Department of BioMicrosystems of Korea Institute of Science and Technology, Seoul.

Shrikant Joshi has been at the International Advanced Research Centre for Powder Metallurgy and New Materials (ARCI), Hyderabad since 2001 and is currently an Additional Director. Following his M.S. and Ph.D. degrees in Chemical Engineering from the Rensselaer Polytechnic Institute (1984) and the University of Idaho (1987), Dr. Joshi had a decade-long stint at the Defence Metallurgical Research Laboratory before assuming a challenging role as head of the Centre for Laser Processing of Materials (CLPM) and then moving to ARCI. Apart from heading the Centre for Solar Energy Materials at ARCI, Dr.Joshi's areas of research have spanned the fields of Surface Engineering and Laser Materials Processing. Presently, he is also responsible for the technical programmes pursued at the Centres for Nanomaterials, Non-Oxide Ceramics and Structural & Mechanical Characterization at ARCI. Dr. Joshi has nearly 100 research publications to his credit and 10 patent applications.

Lucy Liu, a senior material scientist and process engineer at Chromalloy Gas Turbine Corporation, Orangeburg, NY, obtained a Ph.D. at material science and engineering, Tsinghua University, China, in 1998. She worked on tribology properties of thermal sprayed metallic/ceramic coatings as a Post Doctor at material science and engineering, Stevens Institute of Technology, USA from 1998 to 2001. Since 2001, she has been working with Chromally New York on Electron Beam Physical Vapor Deposition (EBPVD), Air Plasma Spray (APS), and Low Pressure Plasma Spray (LPPS) ceramic and metallic coatings for aircraft/ land base gas turbine engine. She invented new thermal barrier coating materials for gas turbine engine components and published 6 patents in USA and international.

Dutta Majumdar completed her PhD in Engineering in May, 1999 from Indian Institute of Technology Kharagpur and subsequently, Dr.-Ing. in the faculty of Mechanical, Mining and Metallurgical Engineering from Technical University of Clausthal in April 2000 (defended thesis in December 1999). Currently, she holds the position of Professor at Indian Institute of Technology Kharagpur. Her areas of speciallization include corrosion and environmental degradation, laser materials processing, surface engineering, bio-materials and advanced processing of materials. She is the recipient of the Friedrich Wilhelm Bessel Award (awarded by the Alexander von Humboldt Foundation), MRSI Medal (awarded by the Materials Research Society of India in 2013), Metallurgist of the year (awarded by the Ministry of Steel, India in 2012), Young Scientist Award (awarded by the Indian Science Congress Association in 2000), Young Metallurgists Award (awarded by the Ministry of Steel, India in 2000), Young Engineer's Award (awarded by the Indian National Academy of Engineering in 2003), FAST TRACK Grant for young researchers (awarded by the Department of Science and Technology in 2001), BOYSCAST Fellowship for carrying out research on laser materials processing at the University of Manchester, UK (awarded by the Department of Science and Technology in 2004) and DAAD fellowship (awarded by the German Academic Exchange Service, N. Delhi during 1996-1998 for carrying out research at Technical University of Clausthal, Germany). She has published 100 papers in international journal of repute, one book (edited), 5 book chapters (invited), 34 in the Conference Proceedings and delivered 32 invited lectures till date.

Indranil Manna obtained his B.E degree from Calcutta University (B.E. College) in 1983 and M Tech degree in 1984 from Indian Institute of Technology (IIT), Kanpur. After a brief stint at Mishra Dhatu Nigam, Hyderabad, he joined IIT, Kharagpur in 1985 as Lecturer and rose to the position of Full Professor in 2003. He was in this position till he joined CGCRI as Director. Presently he is director of Indian Institute of Technology, Kanpur. Prof Manna is a metallurgical engineer, a renowned educator and a prolific researcher. His significant contributions in the studies of amorphous/nanocrystalline Al-alloys, nano-fluid and laser/plasma assisted surface engineering greatly inspired the scientific community. His early contributions in moving boundary phase transformation are still widely cited. As a teacher, Prof Manna has developed several new courses and taught subjects related to phase transformation, materials engineering, surface engineering, thermodynamics and X-ray diffraction. His research interest spans from structure-property correlation in engineering materials including synthesis/application of nano-materials, surface coating/engineering, phase transition, fuel cells, sensors, bainitic steel and mathematical modeling. Prof Manna worked as guest scientist in several institutions of excellence and universities abroad such as Max Planck Institute at Stuttgart, Technical University of Clausthal, Nanyang Technological University, Liverpool University, University of Ulm, etc. Prof. Manna is a recipient of INSA Medal for Young Scientists, Young Metallurgist and Metallurgist of the Year Awards, MRSI Medal, AICTE career award, DAAD and the prestigious Humboldt Fellowships. He is also a recipient of the Acta Materialia Best Referee Award for years 1999 and 2003, respectively. The German Academic Exchange Service (DAAD), Bonn appointed Prof Manna as Honorary DAAD Advisor in India for three years, INAE selected him for Indian National Academy of Engineering-All India Council of Technical Education (INAE-AICTE) Distinguished Industry Professorship for 2007-08 to work with Tata Steel, and the Indian Institute of Metals (IIM) conferred upon him the GD Birla Gold Medal for 2008. Prof. Manna is a Fellow (FNAE) of the Indian National Academy of Engineering, New Delhi; a Fellow (FNASc) of The National Academy of Sciences of India, Allahabad; and a Fellow (FASc) of the Indian Academy of Sciences, Bangalore.

Stefano Mischler obtained the diploma in materials science in 1983 at the Swiss Federal Institute of Technology ETHZ in Zurich. He accomplished his PhD thesis in the field of surface analysis and corrosion at the Swiss Federal Institute of Technology EPFL in Lausanne in 1988. In the years 1989-1990, he held a postdoctoral position at the United Kingdom Atomic Energy Establishment within the Materials Development Division. In 1991, he joined the newly created Tribology group at the Laboratory for Metallurgical Chemistry of the EPFL where he developed research activities in the field of wear-corrosion interactions (tribocorrosion). He is currently head of the Tribology and Interface Chemistry group at EPFL where he develops research and training activities devoted to modern aspects of tribology and surface science.

Andreas Pauschitz graduated from the Technical University Graz with a degree in Mechanical and Economical Engineering in 1988. From 1989 – 1992, he served as a project manager in the car producing industry and from 1992 – 06/2003, he served as assistant at the Institute for Micro Technique and Precision Engineering (IMFT) at Vienna University of Technology. In 1996, he graduation to Dr.techn. at the Vienna University of Technology, Austria with a dissertation thesis titled Tribodesign of Journal Bearings. From 1998 – 06/2003, he served as Head of the Tribology Laboratory at IMFT and from 1999 – 06/2003, he served as Head of Tribology and Design Department at IMFT. In 2001, he participated in the Organization of the 2. World- Tribology-Congress, and in 2002, he participated in the Formation of the Austrian Center of Competence for Tribology (AC^{2}T research GmbH). Since 07/2003, he has worked for AC^{2}T research GmbH (till 09/2006: Key Researcher; since 10/2006: General Manager). In 2010, he established the project "COMET K2 Centre of Excellence for Tribology".

George (Yunfei) Qiao is Senior Scientist at Mogas Industries in Houston TX. He obtained a B.S. at the Beijing University for Aeronautics and Astronautics, and a M.S. at the Chinese University of Mining and Technology. He then worked as Instructor and Senior Engineer at the First Automobile Works in Changchung where he was responsible for the evaluation of materials for production and tribological performance in automobile parts. He obtained his PhD in Materials Science at the Stevens Institute of Technology in Hoboken, USA. The present chapter is based on his PhD thesis. He then joined Philips Electronics as a Senior Scientist and developed methods for the manufacture of superconducting materials. He finally joined Mogas Industries in Houston where he is developing materials for high-pressure, abrasive and corrosive environments.

D. Srinivasa Rao, right from inception of International Advanced Research Centre for Powder Metallurgy and New Materials (ARCI), Hyderabad, has been at ARCI since 1991 after his M.S. in mechanical engineering from the 'National Technical Institute (Kiev)'. As a team leader of 'Centre for Engineered Coatings', he played a pivotal role in establishment of state of art surface engineering facilities, carrying out R&D studies, prototype development of coatings on wide range of engineering components, indigenous development of coating technologies and its transfer & implementation at industries in India. D. Srinivasa Rao has nearly 50 research publications in journal/proceedings/book chapter articles to his credit.

S. G. Sapate is currently working as Professor at Department of Metallurgical and Materials Engineering m Visvesvaraya National Institute of Technology, Nagpu, India. Dr S G Sapate obtained his Ph.D. degree in the year 2001 from Nagpur University. His research work was in the area of solid particle erosion of high strength steels and hard coatings used in industry. He was awarded 'Technical Co-operation Training (TCT)' fellowship under UK- India REC Exchange program by British Council, India. He visited Sheffield Hallam University, Sheffield (UK). He has more than 40 publications in international and national journals and conferences to his credit. He has extensively worked in the area of solid particle erosion of weld hardfacing alloys used in the industry. His current research interests are erosion and abrasion of metallic materials, weld hardfacing alloys and coatings used in the industry, material selection, etc.

Debajyoti Sen is presently working as a technical officer at the 'Centre for Engineered Coatings', International Advanced Research Centre for Powder Metallurgy & New materials, Hyderabad, India. He received University "Gold medal" for securing first position in order of merit in "Diploma in computer integrated manufacturing" from Indira Gandhi National Open University, New Delhi in 2009. His research interests include detonation spray, Electron Beam Physical Vapour Deposition (EBPVD) and tribology. He has published over 16 journal/proceedings to his credit.

Harpreet Singh is currently working as an Associate Professor in Mechanical Engineering at Indian Institute of Technology Ropar, Punjab. After doing B.E. and M.E. in Mechanical Engineering, he did his Ph. D from Indian Institute of Technology Roorkee (India) in the year 2005. His areas of interest include Surface Engineering-Degradation of Materials, High Temperature Corrosion and its Prevention, Slurry Erosion of Hydraulic Turbines and its Control, Biomedical Coatings; and Friction-Stir Processing/Welding. He has authored one monograph and two book chapters, and contributed more than one-hundred fifty research papers in various international/national journals and conferences. He has produced six Ph.Ds. He is a recipient of many Indian research-related awards and has been involved in several sponsored research projects. He has visited several countries including USA, UK, Netherlands, Canada, South Korea and Germany and delivered several invited talks.

G. Sivakumar is presently a Scientist at the Centre for Engineered Coatings, International Advanced Research Centre for Powder Metallurgy & New Materials, Hyderabad, India. He received his PhD degree from Indian Institute of Technology, Mumbai in 2014. His research interests include solution precursor plasma spray, thermal barrier coatings, tribology and detonation spray. He has published over 25 journal/proceedings/book chapter articles and has 1 patent to his credit.

Mandar Thakare is a Senior Applications Engineer with Element Six Ltd developing structure-performance relationships for polycrystalline diamond based drill-bit cutters. Dr Thakare has ten years of R&D experience in developing an understanding of tribological performance of coatings, ceramics and alloy-steels in a wide range of engineering applications. He has published several journal papers, two book chapters and has been invited to present at international conferences. Since completing his PhD at University of Southampton in 2008, Dr Thakare has worked in Schlumberger plc and University of Oxford. Dr Thakare also holds a college lectureship at St Catherine's College, University of Oxford and is a reviewer for *Tribology International, ASME Journal of Tribology,* and *Wear.*

Robert Wood is Professor of Surface Engineering and Tribology within Engineering Sciences of the Faculty of Engineering and the Environment at the University of Southampton and has 25 years research experience in the field of tribology and surface engineering. He has spent several years at BP Research researching into erosion and corrosion resistant coatings but returned to Southampton in 1993 to re-establish surface engineering/tribology research. His group was awarded a EPSRC S&I award in 2008 (£10M) to create the National Centre for Advanced Tribology at Southampton (nCATS) and was awarded a further £3M for research into Green Tribology under a platform Grant from EPSRC. The Centre has 9 academics, 12 postdoctoral research fellows and 20 research students. Professor Robert Wood has research interests in rain, cavitation and solid particle erosion, erosion-corrosion interactions and modelling; tribological and multifunctional coating design and performance, biomimetic coatings for anti-fouling; electrochemical control of interfacial friction; particle modelling in pipe bends by swirl dispersion of particles. He has been involved in working on erosion of helicopter blades, nuclear slurry handling systems at Sellafield, reverse thrust actuators on aero engines, polymer coatings for potable water systems as well as offshore choke valves. He has teaching experience in Surface Engineering, Tribological Engineering and Fluid Mechanics at Undergraduate and MSc level.

Index

N

O

P

R

www.ingramcontent.com/pod-product-compliance
Lightning Source LLC
Chambersburg PA
CBHW080501051126
54379CB00005B/134
* 9 7 8 1 4 6 6 6 7 4 8 9 9 *